Intelligent Industrial Alarm Systems

Jiandong Wang · Wenkai Hu · Tongwen Chen

Intelligent Industrial Alarm Systems

Advanced Analysis and Design Methods

Springer

Jiandong Wang
College of Electrical Engineering and Automation
Shandong University of Science and Technology
Qingdao, Shandong, China

Wenkai Hu
School of Automation
China University of Geosciences
Wuhan, Hubei, China

Tongwen Chen
Department of Electrical and Computer Engineering
University of Alberta
Edmonton, AB, Canada

ISBN 978-981-97-6515-7 ISBN 978-981-97-6516-4 (eBook)
https://doi.org/10.1007/978-981-97-6516-4

This Springer imprint is published by the registered company Springer Nature Singapore Pte Ltd.
The registered company address is: 152 Beach Road, #21-01/04 Gateway East, Singapore 189721, Singapore

Preface

In modern industrial plants, hundreds and thousands of process variables are monitored in real time every second or faster by modern computerized monitoring systems such as distributed control systems or supervisory control and data acquisition systems. Safe and efficient conditions of industrial plants can be revealed by looking at these process variables. For instance, if the level of a vessel is too high, then an abnormality may be present and corrective actions should be taken to control the abnormality so that the level could return back to normal ranges; if the alarm is ignored, then the vessel may run into unsafe or inefficient conditions. However, due to the large number of process variables being monitored, it is infeasible to look at these variables all the time (24 hours, 7 days) by industrial plant operators in a manual manner. As an alternative, it is indispensable to conduct monitoring tasks in an automatic manner, which leads to industrial alarm systems.

A basic alarm generator is to announce an alarm event when the value of a process variable overpasses an alarm threshold. Once an alarm event is announced automatically, industrial plant operators will pay attention to corresponding process variables and investigate them in detail. Hence, industrial alarm systems are integrated parts of modern industrial plants, and play an important role for the operation safety and efficiency. However, contemporary industrial alarm systems often suffer from an unsatisfactory performance on one hand, e.g., with too many nuisance alarms being announced, and desire for advanced techniques to assist in designing and analyzing alarms on the other hand, e.g., root-cause analysis for occurring alarms. Hence, the subject of industrial alarm systems receives increasing attention from academic researchers and industrial engineers.

There are three major reasons motivating us to come up with this monograph. First, an academic-style book on industrial alarm systems is not available in the literature, to the best of our knowledge. There are two engineering-style books: *Alarm Management: A Comprehensive Guide* written by B. Hollifield and E. Habibi, and *Alarm Management for Process Control* written by D. Rothenberg. For a long term, industrial alarm systems have been regarded as a subject full of technical know-hows and ad-hoc skills, without in-depth scientific research methodologies and theories. This is not true, as proven by recently published academic papers. For instance, Markov chain theory and Bayesian estimation rules have been exploited to design optimal values of alarm thresholds. We believe that this monograph will be a necessary complement from an academic viewpoint to two existing engineering-style books on industrial alarm systems.

Second, students and researchers need a monograph on industrial alarm systems as a quick starting point. We started the research work on the subject of industrial alarm systems in 2008, stumbled through the stage of many journals taking this subject as out of scope and refusing reviews for submitted papers, and witnessed the amount of journal and conference papers on this subject going up from a dozen to more than 300 at the beginning of 2024. The amount of these papers is not too large. Reading through all of them is a demanding but feasible way to get acquainted with this subject. As a result, we hesitated for a while to work on this monograph. However, a few casual talks with some graduate students and researchers alleviate our hesitation, because reading more than 300 papers is definitely not a good choice for a quick start. In fact, graduate students and researchers are looking forward to a monograph based on a small amount of key papers being screened from a large pool and being collated in a systematic way.

Third, engineers need a monograph with systematic methodologies and practical techniques to improve the performance of industrial alarm systems. Most of industrial softwares on alarm systems are confined to very basic tools such as counting the number of alarms in a certain time period, and do not have advanced tool-kits such as the design of optimal parameters in alarm generators. This monograph provides numerical and industrial examples to illustrate all methodologies, techniques, and theories being proposed. By doing so, we hope that engineers could apply these theoretical results in practice by following the steps in these illustrative examples, despite that some theoretical results look complicated with mathematical equations. Visual analytic tools for industrial alarm systems are also provided in the last chapter; it is expected to see these analytic tools to be implemented in some industrial softwares in the near future.

The monograph is composed of six chapters. The first chapter provides an overview of industrial alarm systems, by presenting basic concepts and background information, investigating alarm overloading and its main causes, and summarizing current research status in the literature. Chapters 2–5 cover the generation of alarm variables, the analysis of root causes leading to a single occurring alarm, and the analysis of multiple occurring ones in alarm floods. In particular, the second and third chapters are on the optimal design of alarm systems to generate alarm variables from single or multiple process variables. The fourth chapter is to analyze the root causes of an occurring alarm based on alarm and/or process variables. The fifth chapter is to detect the presence of alarm floods, find similar alarm foods, and predict upcoming alarm floods, where alarm floods are referred to as sequences of multiple occurring alarms in a short time period (e.g., 10 min). The last chapter complements the analysis and design methods in previous four chapters with alarm visual analytics, in order to evaluate the performance of industrial alarm systems and extract patterns in alarms from process changes, alarm messages, and operator responses.

The subject of industrial alarm systems origins from practice. It is sensible that the proposed methods, techniques, and theories in this monograph and published papers could be implemented in practice to improve the safety and efficiency operation of industrial plants, rather than solely staying on papers. By deploying and improving the proposed results in practice, an ultimate goal is to formulate an intelligent industrial alarm system. Such a new generation of industrial alarm systems is intelligent enough to extract and exploit useful but hidden information from historical data and process knowledge, for achieving satisfactory performance in the design of alarm generators, the analysis of occurring alarms, and the prediction of upcoming alarm floods. We hope that this monograph attracts more people to work on industrial alarm systems, so that the ultimate goal could be reached sooner.

We would like to sincerely thank our collaborators, Profs. Sirish Shah (University of Alberta), Fan Yang (Tsinghua University), and Iman Izadi (Isfahan University of Technology), industrial partners, graduate students, and postdoctoral fellows for their contributions. In particular, the subject of industrial alarm systems is indeed problem-oriented. The problems to be tackled should come from needs in practice, and their solutions need to be validated in practice. Hence, we are indebted to our industrial collaborators for sharing extensive data and providing validation opportunities at industrial plants. The subject of industrial alarm systems, being practical without well-formulated problems, is challenging for graduate students and postdoctoral fellows to keep a good balance between empirical and theoretical results. Many innovative ideas come from insightful and inspiring discussions with our academic colleagues and visitors. We are very grateful to all of them whose contributions indeed make the monograph possible.

Last but not the least, we are grateful for the financial support received from funding agencies and industrial partners. In particular, the work of Jiandong Wang and Wenkai Hu was mainly supported by the National Natural Science Foundation of China; the work of Tongwen Chen was mainly supported by the Natural Sciences and Engineering Research Council of Canada.

Qingdao, China — Jiandong Wang
Wuhan, China — Wenkai Hu
Edmonton, Canada — Tongwen Chen
June 2024

Contents

Acronyms

A&E	Alarm & Event
AAD	Averaged Alarm Delay
ALM	Alarm occurrence
ATG	Adaptive Time Gradient
BLAST	Basic Local Alignment Search Tool
BPCS	Basic Process Control System
CCF	Cross-Correlation Function
CDF	Cumulative Density Function
CHARM	Closed Association Rule Mining
CSTR	Continuous Stirred Tank Reactor
DBSCAN	Density-Based Spatial Clustering of Applications with Noise
DCS	Distributed Control System
DTE	Direct Transfer Entropy
DTW	Dynamic Time Warping
EEMUA	Engineering Equipment and Materials Users' Association
FAR	False Alarm Rate/Ratio
GIS	Generalized Iterative Scaling
HDAP	High Density Alarm Plot
HMI	Human-Machine Interface
HSP	High Scoring segment Pair
ICA	Inverse Coefficient of Variation
IGDTE	Information Granulation based Direct Transfer Entropy
IGTE	Information Granulation based Transfer Entropy
IID	Independent and Identically Distributed
ISA	International Society of Automation
KPI	Key Performance Indicator
MAR	Missed Alarm Rate/Ratio
MLR	Multiple Linear Regression
NDTE	Normalized Direct Transfer Entropy
NOZ	Normal Operating Zone
NTE	Normalized Transfer Entropy

OLS	Ordinary Least Squares
OPTICS	Ordering Points To Identify the Clustering Structure
PCA	Principle Component Analysis
PDF	Probability Density Function
PLR	Piece-wise Linear Representation
QTA	Qualitative Trend Analysis
ROC	Receiver Operating Characteristic
RTN	Returns-To-Normal
SCADA	Supervisory Control And Data Acquisition
SIS	Safety Instrumented System
SW	Smith-Waterman
TE	Transfer Entropy
TEP	Tennessee Eastman Process
TID	Transaction IDentifier
VIF	Variance Inflation Factor

Chapter 1
Overview of Industrial Alarm Systems

Abstract This chapter provides an overview of industrial alarm systems on main causes for alarm overloading, research status, and open problems [179]. First, basic concepts and important roles of alarm systems are presented. Second, alarm overloading is investigated as the most observable phenomenon for poor performance suffered by many existing industrial alarm systems, and four main causes are identified as the culprits for alarm overloading. Third, the current research status for industrial alarm systems is summarized, by focusing on the existing studies related to these main causes. Finally, eight fundamental research problems to be solved are formulated for the complete lifecycle of alarm variables composed of three stages including alarm configuration, alarm design, and alarm removal.

1.1 Basic Concepts and Background Information

According to the industrial standard ANSI/ISA-18.2 (page 16) [92], an alarm system is the collection of hardware and software that detects an alarm state, communicates the indication of that state to operators, and records changes in the alarm state. Alarm systems are integrated parts of modern computerized monitoring systems such as distributed control systems (DCS) and supervisory control and data acquisition (SCADA) systems. A basic way in detecting an alarm state is to compare the value of a process variable x to a constant high (low) alarm threshold x_{th},

$$x_a(t) = \begin{cases} 1, \text{ if } x(t) \geq (\leq)\, x_{th}, \\ 0, \text{ if } x(t) < (>)\, x_{th}. \end{cases} \tag{1.1}$$

Alarm states are mathematically represented by a discrete-valued alarm variable x_a. That is, the high (low) alarm variable x_a takes the value of 1 when x is higher (lower) than x_{th}, and the value of 0 otherwise.

As an example, x represents the drum level (in the unit of mm) of a large-scale thermal power plant. Figure 1.1a presents one-hour samples of x with sampling period 1 sec. For the safe drum operation, the drum level is configured with the high alarm threshold 100 mm and the low alarm threshold -100 mm. Alarms are announced if

J. Wang et al., *Intelligent Industrial Alarm Systems*,
https://doi.org/10.1007/978-981-97-6516-4_1

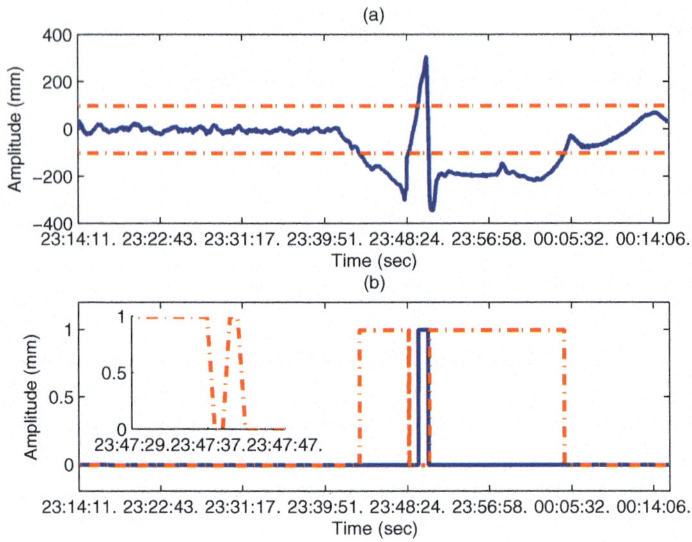

Fig. 1.1 **a** Data samples of a process variable $x(t)$ (solid) with alarm thresholds (dot-dash), **b** the alarm variables $x_a(t)$ with high (solid) and low (dot-dash) alarm thresholds, reprinted from Ref. [179], copyright 2016, with permission from IEEE

the drum level goes beyond the high and low alarm thresholds. That is, the high (low) alarm variable $x_a(t)$ takes the value of 1 when $x(t)$ is higher (lower) than the value 100 (-100) mm, and the value of 0 otherwise. Figure 1.1b shows the samples of two alarm variables associated with $x(t)$ in Fig. 1.1a. For the drum level, the basic alarm generator in (1.1) is implemented via the High Low Alarm (HLALM) function in Fig. 1.2 that is an embedded function in the DCS. The high and low alarm variables

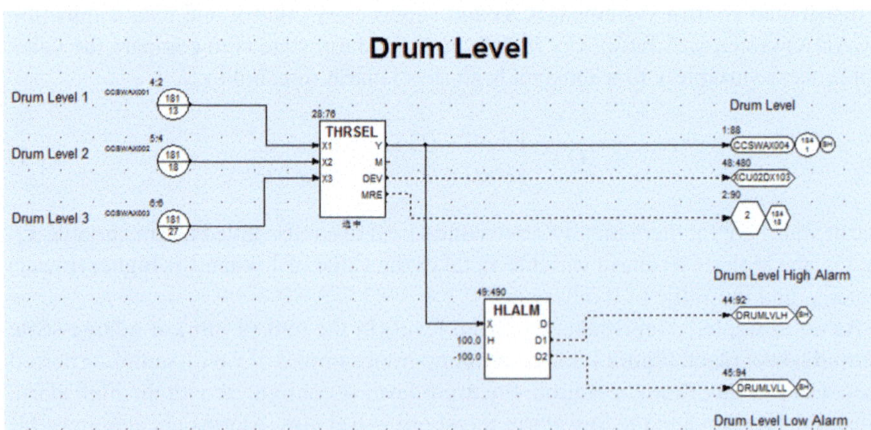

Fig. 1.2 The alarm generation function in the DCS for drum levels

Table 1.1 Events of alarm occurrences and clearances for high and low alarm variables of the drum level in Fig. 1.1, reprinted from Ref. [179], copyright 2016, with permission from IEEE

Date	Time	Status	Tagname	Description
2014-01-24	23:42:42	Alarm on	DRUMLVLL	Drum level low
2014-01-24	23:47:38	Alarm off	DRUMLVLL	Drum level low
2014-01-24	23:47:40	Alarm on	DRUMLVLL	Drum level low
2014-01-24	23:47:42	Alarm off	DRUMLVLL	Drum level low
2014-01-24	23:48:38	Alarm on	DRUMLVLH	Drum level high
2014-01-24	23:49:39	Alarm off	DRUMLVLH	Drum level high
2014-01-24	23:49:45	Alarm on	DRUMLVLL	Drum level low
2014-01-25	00:03:27	Alarm off	DRUMLVLL	Drum level low

are with the tagnames ‘DRUMLVLH’ and ‘DRUMLVLL’, respectively. Changes of alarm variables from 0 to 1 and from 1 to 0 are respectively referred to as alarm occurrences and alarm clearances. They appear as events in the alarm list with the corresponding time stamps and descriptions, as shown in Table 1.1. Note that the low alarm variable experiences two quick changes between 23:47:38 and 23:47:42, which are visible by the enlarged plot in Fig. 1.1b.

Alarm systems are critically important for safe and efficient operations of modern industrial plants such as oil refineries, petrochemical facilities, and power plants [23, 139], where hundreds and thousands of process variables are configured with alarms. First, alarm systems are the tools to detect the near misses that are defined as departures from and subsequent returns to normal operating ranges for process variables [134]. For instance, the variations of two alarm variables in Fig. 1.1 indicate that the drum level departs from and returns to the normal operating range $[-100, 100]$ mm. The safety pyramid in Fig. 1.3 says that every accident is associated with a number of near misses as precursors. Alarm systems promptly indicate the occurrences of

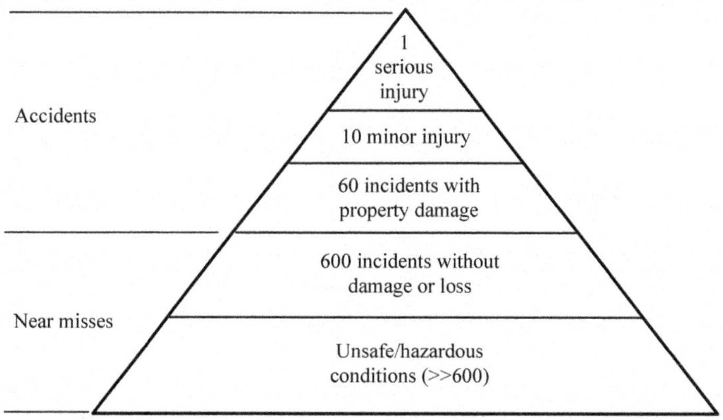

Fig. 1.3 Safety pyramid with typical historical statistics [134]

near misses, so that operators can take corrective actions to drive processes back to normal operating ranges. Therefore, alarm systems are indeed the safeguards to prevent the deterioration of near misses to accidents.

Second, retrospective investigation on a large number of accidents support the important role played by alarm systems. For instance, the Buncefield accident in 2005 was caused by a series of explosions and fire of oil storages at the Buncefield oil depot, one of the most severe industrial accidents in Europe. The final report of the Buncefield accident [79] provided 25 recommendations, among which the 8th recommendation was to develop high-high level alarms for overfill prevention, and the 23rd recommendation was to collect accident data to find alarm system defects. However, similar accidents repeatedly occur worldwide, despite of omnipresent recommendations on improving alarm systems in accident reports.

Finally, alarm systems also play a prominent role in maintaining efficiency of plant operation. It is a well-known fact that the deviation of process variables from normal/optimal operating zones usually imply negative consequences such as off-specification products as well as excessive consumption of raw materials and energy. Alarm systems with satisfactory performance are able to assist operators to reduce the probability and time duration of deviations of key process variables from their normal/optimal operating zones. For instance, Figs. 1.4 and 1.5 show time sequences of the active power E, main steam pressure P_T, steam control valve μ_T, and coal flow rate B in a thermal power plant on two days. The increments of E are similar from 184 MW to 248 MW on May 13, and from 188 MW to 250 MW on July 7. However, the operational efficiencies are very different. The difference is difficult to

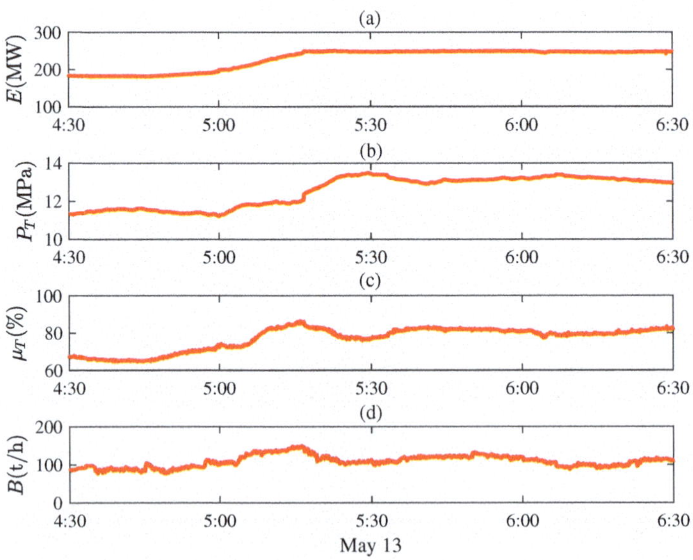

Fig. 1.4 Time sequences of the active power E **a**, main steam pressure P_T **b**, steam control valve μ_T **c**, and coal flow rate B on May 13

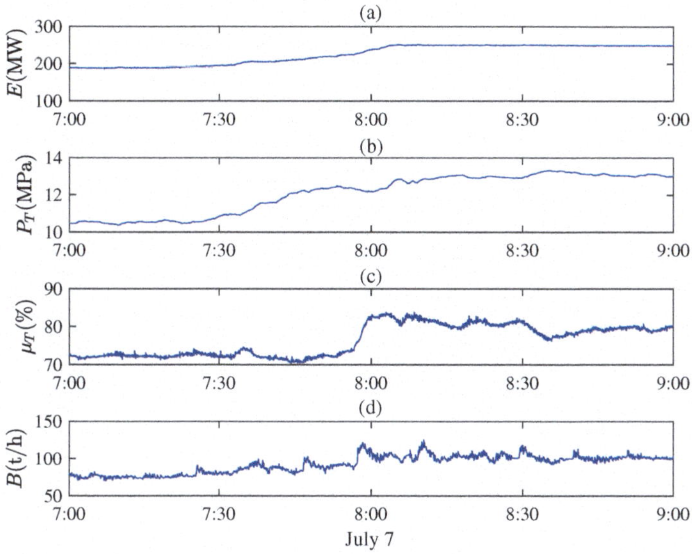

Fig. 1.5 Time sequences of the active power E **a**, main steam pressure P_T **b**, steam control valve μ_T **c**, and coal flow rate B on July 7

reveal from Figs. 1.4 and 1.5, but is transparent from an operating zone formulated by E and B in Fig. 1.6. In the operating zone, black boundary lines encompass a safety region of E and B, and lighter colors indicate the higher operational efficiency. The red path on May 13 is clearly less efficient than the blue path on July 7, because a similar power increment is achieved by more coal consumption. A multivariate alarm system can be designed to monitor the efficiency variations, so that the red path will be detected and adjusted promptly to a higher efficiency path like the blue one.

On one hand, alarm systems play significant roles for operational safety and efficiency of industrial plants [64, 120, 150], as shown in Fig. 1.7; on the other hand, most existing industrial alarm systems suffer from poor performance, most noticeably with alarm overloading (to be clarified later in Sect. 1.2). Driven by the big gap between these facts, industrial standards and guidelines have been proposed for alarm systems by industrial societies and professional organizations. The Nuclear Regulatory Commission from United States published the document NUREG/CR-61056684 to give guidance and technical basis for advanced alarm systems [133]. The Engineering Equipment and Materials Users' Association presented the guideline EEMUA-191 for the design, management and procurement of alarm systems [54]. The Standardization Association for Measurement and Control in Chemical Industries issued the standard NAMUR-NA-102 for alarm management [128]. The USA based Electric Power Research Institute produced the document EPRI-1010076 as the requirements and implementation guidance for advanced alarm systems [55]. The Abnormal Situations Management Consortium proposed a set of guidelines for

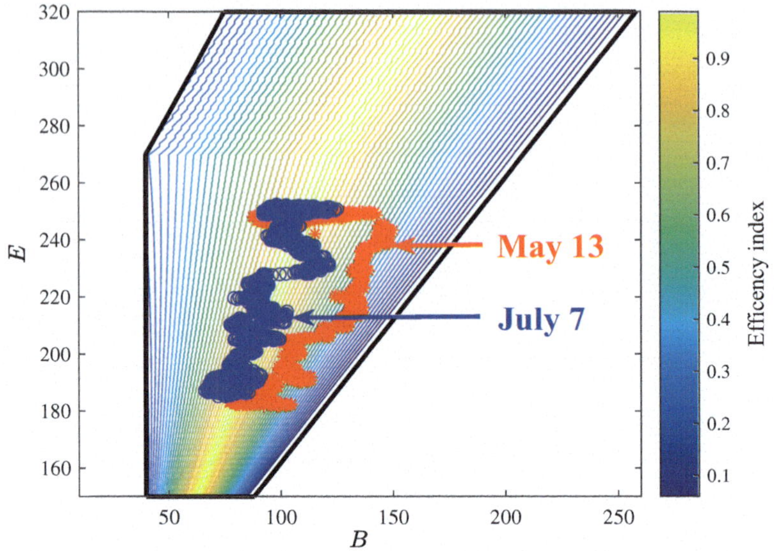

Fig. 1.6 An operating zone formulated by the active power E and the coal flow rate B with one increment of E from 184 MW to 248 MW along the red path on May 13, and another increment from 188 MW to 250 MW along the blue path on July 7

Fig. 1.7 Layers of protection and their impact [150]

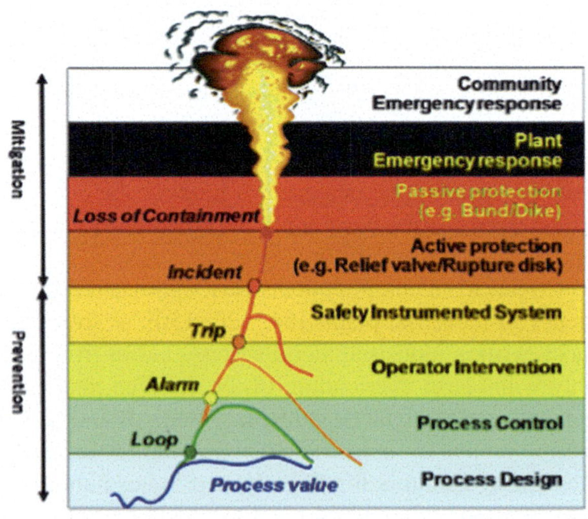

effective alarm management practice [15]. The International Society of Automation and the International Electrotechnical Commission respectively issued the standards ANSI/ISA-18.2 [92] and IEC-68682, for management of alarm systems in process industries. The American Petroleum Institute published the standard API-1167 for pipeline SCADA alarm management [11]. The China International Association for

Promotion of Science and Technology released a national group standard T/CI-020-2021 for guidelines of designing alarm delay timers, alarm deadbands and state-based alarm generators to detect and reduce nuisance alarms [185]. The China Machinery Industry Federation released the national standard GB/T 41261-2022 [118], similar to the ANSI/ISA-18.2 standard [92]. Most of these standards and guidelines impose specific requirements on the performance of alarm systems, for example, "the average number of alarms per day should be no more than 144 [54]", but do not provide methodologies and technical details on how to achieve these requirements. Hence, there is a big gap to be fulfilled between industrial requirements in practice and available techniques in literature.

1.2 Alarm Overloading and Its Main Causes

This section investigates alarm overloading as the most observable phenomenon in practice, and finds the main causes leading to alarm overloading with industrial examples as evidences.

Many existing industrial alarm systems are associated with poor performance, where the most common phenomenon is that there are far too many alarms to be handled by industrial plant operators, referred to as alarm overloading in the sequel. This phenomenon is clearly revealed from Table 1.2 [139], which lists statistics of several basic performance metrics of alarms systems, based on a study of 39 industrial plants ranging from oil and gas, petrochemical, power and other industries. The corresponding benchmarks in the guideline EEMUA-191 [54] are also provided in Table 1.2 for comparison. Obviously, the statistics of performance metrics from various industries are much greater than the EEMUA benchmarks. Another industrial survey was provided by Bransby & Jenkinson [23] for 15 plants including oil refineries, chemical plants, pharmaceutical plants, gas terminal, and power stations; the average alarm rate per 10 minutes under normal operation ranged from 2 to 33, and the peak alarm rate per 10 minutes in plant upsets varied from 72 to 625. The number of operators was also recorded for the industrial survey by Bransby & Jenkinson [23]; for most cases, however, the statistics of alarm systems were not normalized by the number of operators. Kirschen & Wollenberg [102] presented the estimates of peak numbers of alarms triggered by some abnormal events at regional control centers of Hydro Quebec Inc, Canada, which were up to 20 alarms per second during a thunderstorm. Brown [26] provided similar results for BP Oil plants: the average rate of alarms per 10 minutes was in the range of [17, 60], while the maximum rate of alarms per 10 minutes in upset conditions was in the range of [150, 560]. Liu et al. [116] and Srinivasan et al. [149] stated that the daily alarm number in abnormal operation was about 11000 (76 alarms per 10 minutes) at a major Singapore refinery. Noda et al. [132] reported that 15 out of 29 Japanese chemical plants had the monthly average alarm rates per operator larger than the EEMUA benchmark, with the maximum value of 7.5 alarms per operator for 10 minutes, where the number of

Table 1.2 Cross-industry study [139]

	EEMUA	Oil-Gas	PetroChem	Power
Average alarms/day	144	1200	1500	2000
Peak alarms/10 min	10	220	180	350
Average alarms/10 min	1	6	9	8

operators was taken into consideration. Soares et al. [146] gave the monthly alarm occurrences as 102917 in one year at a natural gas processing plant.

To have an example of industrial alarm systems with more specifics, we investigated the alarm system of an industrial power plant [183]. The DCS of the power plant measures real-time values of 24079 process variables every 0.2 sec. Among these process variables, there are 8145 analog variables, of which 208 variables are configured with alarms, and 15934 digital variables among which 2018 variables are configured with alarms. The basic way in (1.1) is used to generate all the 2226 alarm variables in the DCS. In words, an alarm occurs when an analog variable configured with alarm is exceeding the corresponding high (high-high) or low (low-low) alarm threshold, or when a digital variable configured with alarm changes the value from 0 to 1 (or from 1 to 0). Note that some digital variables configured with alarms are essentially generated by comparing the measurements of analog variables with alarm thresholds. Figure 1.8 presents the numbers of alarm occurrences during non-overlapping consecutive 10-minute periods for one month. The maximum, average and minimum numbers of the alarm occurrence rate per 10 minutes are 342, 119 and 19, respectively. Figure 1.9 provides the daily distributions (in the form of boxplots) for the number of alarm occurrences for each month in one year. In particular, the average number of alarms per day is up to 18280. Clearly, the alarm occurrence rates are much larger than the EEMUA benchmarks in Table 1.2. In the industrial power plant, there are 25 operators, 5 operators per shift and 5 shifts in total, and the number of alarms is too enormous to be promptly handled by these operators.

The occurred alarms can be classified into two groups, namely, nuisance alarms and true alarms. A nuisance alarm is one that does not require a specific action or response from operators [54, 139]. An alternative definition of a nuisance alarm is an alarm that annunciates excessively, unnecessarily, or does not return to normal after the correct response is taken [92] (page 18). Hence, nuisance alarms in a process are defined as the occurred alarms that do not affect the process, even if these alarms are ignored by operators. The industrial standard ANSI/ISA-18.2 [92] (page 14) defines an alarm as "an audible and/or visible means of indicating to the operator an equipment malfunction, process deviation, or abnormal condition requiring a response". Thus, the key point to distinguish such an alarm from nuisance alarms is on the requirement of operator response. Rothenberg [139] gives a brief definition of an alarm as "Alarm activation = Operator action". Therefore, in contrast to the nuisance alarms, a true alarm is defined as the one that requires operators to pay attention or to take action in a prompt manner; otherwise, abnormal situations associated with true

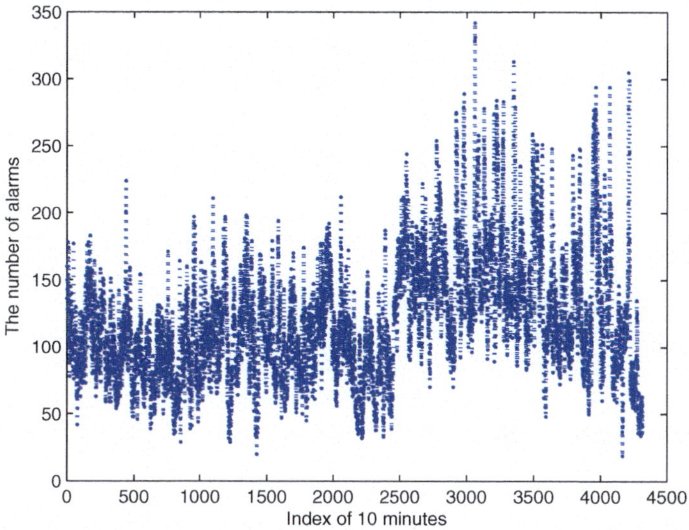

Fig. 1.8 The number of alarms per 10 minutes in a power plant for one month, reprinted from Ref. [179], copyright 2016, with permission from IEEE

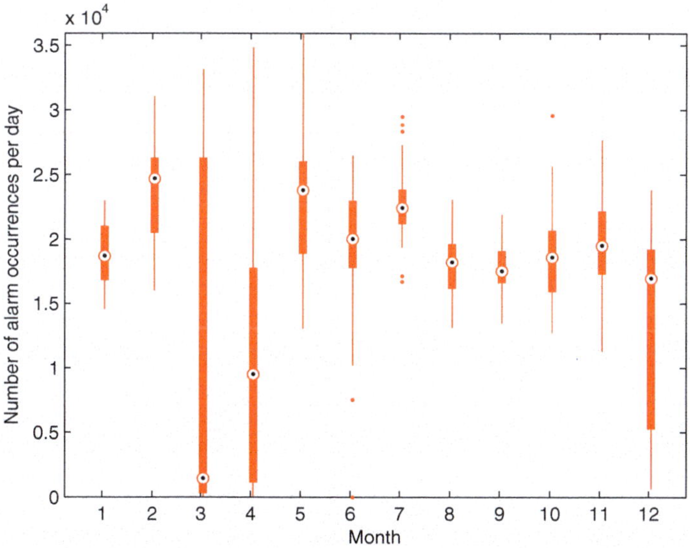

Fig. 1.9 The daily distributions (in the form of boxplots) for the number of alarm occurrences for one year, reprinted from Ref. [179], copyright 2016, with permission from IEEE

alarms would have negative effects on operation safety and/or efficiency. Nuisance alarms are the major culprits for the phenomenon of alarm overloading. On the other hand, there are also the scenarios of having too many true alarms, referred to as alarm floods (to be clarified later at Cause #4 in this section).

The consequences of alarm overloading are extremely detrimental to the important role played by industrial alarm systems. Firstly, a large number of alarms belong to the nuisance alarm group; they provide no useful information and only serve as distractions to industrial plant operators. Due to the "cry wolf" effect, nuisance alarms lead to confidence crisis of alarm systems. As a result, a true alarm may be buried among such nuisance alarms and may consequently be overlooked by operators. Secondly, even if all occurred alarms are true alarms, e.g., those in alarm floods, the alarm rate may be too high to be manageable for operators. When the alarm rate is too high, operators have no choice but to ignore many of the occurred alarms. In this case, the designed functionality of alarm systems is completely discredited. As an example, two operators received 275 different alarms during the 10.7 minutes before the explosion accident occurred at the Texaco Refinery in Milford Haven [78].

In order to alleviate the phenomenon of alarm overloading, the very first step is to find the main causes leading to such a phenomenon. Here, four main causes are identified, with industrial examples as supportive evidences. Note that the four causes could be mixed or present simultaneously in practice. For instance, the alarm floods given later in Fig. 1.15 also involved the main cause #2, because Tags # 25-27 (superheater temperature high alarm variables) are similar to those in Fig. 1.11, and should be integrated to configure one alarm variable, instead of three individual alarm variables.

Cause #1: Chattering alarms due to noises/disturbances

Chattering alarms are the mostly-encountered nuisance alarms and may account for 10–60% of alarm occurrences [139] (page 123). An analysis based on 75 alarm systems showed that on average over 70% of the alarm occurrences came from chattering alarms [77] (page 83). The industrial standard ANSI/ISA-18.2 defines a chattering alarm as one that repeatedly transitions between the alarm state and the normal state in a short period of time [92] (page 16). As a result, there is no time or necessity for operators to analyze such alarms and take actions. Two closely related nuisance alarms are the fleeting and repeating alarms. Fleeting alarms also have short-time alarm duration, but do not immediately repeat [92] (page 74). Repeating alarms are alarms rising and clearing repeatedly over a period of time [54] (page 95). Chattering alarms are also named as cycling alarms [139] (page 444). These types of alarms are typically generated due to random noises and/or disturbances on process variables configured with alarms, especially when the process variables are operating close to their alarm thresholds [54] (page 95). In addition, chattering alarms can be induced by repeated on-off actions of control loops or regular oscillatory disturbances in process variables [176]; in this case, chattering alarms repeatedly make transitions between alarm and non-alarm states with regular (possibly large) time periods. To have a unifying terminology, all of these alarms are referred to as chattering alarms in this context, with a refined definition as: A chattering alarm is the one that transitions between the alarm state and the normal state very quickly or with a constant time period.

Bransby and Jenkinson [23] (Appendix-10 therein) discussed some industrial examples of chattering alarms caused by noises in a process variable that was operating close to the alarm threshold, and by the oscillations from repeated on-off control actions having a regular oscillation period of 43 min. Hollifield & Habibi [77] (page 84) listed top 10 chattering alarms based on 150 days of data. Ahnlund, Bergquist & Spaanenburg [7] showed chattering alarms associated with periodic signals or outliers in signals. Wang & Chen [176, 177] presented industrial examples of chattering alarms due to noises and oscillations at petro-chemical and thermal power plants.

One example of chattering alarms from a thermal power plant is presented here for illustration. A process variable x is the difference between the maximum and minimum values of measurements from 54 temperature sensors installed at stator outlet pipes at a thermal power plant. x is configured with a high alarm variable x_a with an alarm threshold 8.0 °C. That is, x_a takes the value 1 if x is greater than 8.0, and the value 0 otherwise. Due to the measurement noise aggregation in 54 sensors, high-frequency noises contaminate x and lead to a large number of alarm occurrences, as shown in Fig. 1.10a, b. Among the samples of x collected in one hour with sampling period 1 sec, there are 105 alarm occurrences in Fig. 1.10b. The

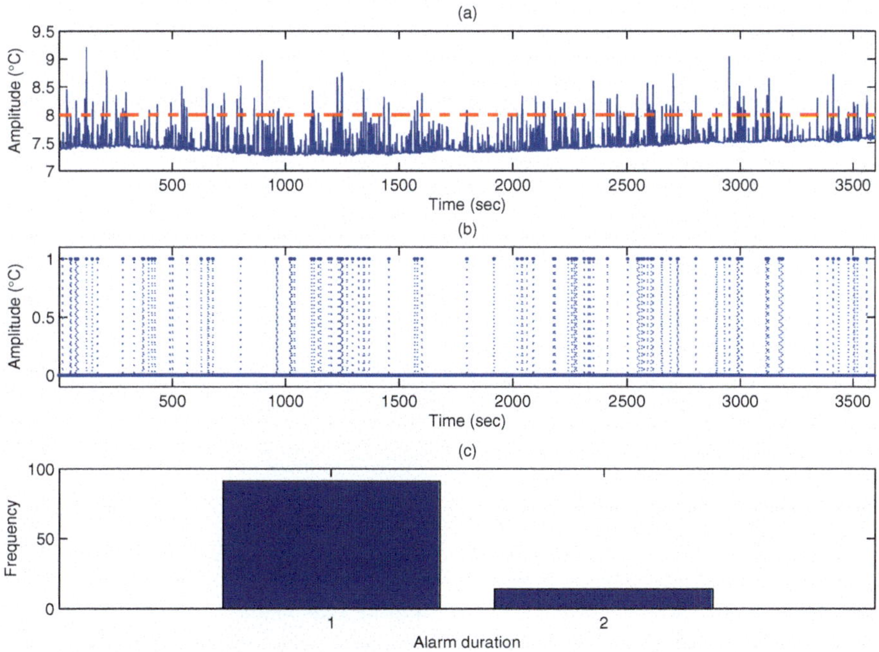

Fig. 1.10 An example of chattering alarms due to measurement noises: **a** process variable (solid) and alarm threshold (dash), **b** alarm variable, **c** histogram of alarm durations, reprinted from Ref. [179], copyright 2016, with permission from IEEE

alarm durations of these 105 alarms were all no larger than 2 samples, as shown in the histogram of alarm durations in Fig. 1.10c. Here the alarm duration is calculated as the number of consecutive samples taking the value 1 between each alarm occurrence and its corresponding clearance. Therefore, these occurred alarms are clearly nuisance alarms.

Cause #2: Incorrectly configured alarm variables

Before the appearance of modern computerized monitoring systems such as the DCS and SCADA systems, each alarm variable was realized by hardware devices with high investment costs. As a result, each variable to be configured with alarm was carefully selected and thoroughly justified, and the total number of alarm variables was very limited, e.g., about 30 to 50 per process unit. As a comparison, alarm variables in modern computerized monitoring systems can be easily realized in a technical sense by clicking a mouse and entering alarm threshold values at a computer, so that alarm configuration is regarded as "free" without any cost. In addition, configuring more alarm variables is often believed to be beneficial in improving operation safety. In particular, once accidents occur, the accident reports often recommend the configuration of more alarm variables. Hence, the number of alarm variables increases dramatically. For instance, Nimmo [130] reported that the number of alarm variables in one plant was increased from 150 for the hardware-based alarm variables to 14000 for the computerized alarm variables. Hollifield & Habibi [77] (page 13) indicated that the configured alarms per operator had increased exponentially from less than 100 alarm variables in 1960 to about 4000 alarm variables in 2000. Many variables are configured with alarms without a careful study on the necessity of configuring alarms and on the alarm priorities. As a result, there are a large number of variables that should not be configured with alarms or are configured with alarms in an incorrect manner. For instance, Timms [165] stated that some industrial facilities were with more than 15000 configured alarm variables, and over 50% of the configured alarm variables had been removed after alarm rationalization.

An example from a thermal power plant is presented here to illustrate incorrectly configured alarm variables. As a critical process variable, the main steam temperature is monitored by three independent sensors. The measurement of each sensor is configured with high-alarm threshold 546 °C and low-alarm threshold 530 °C. Thus, there are three high (low) alarm variables associated with the main steam temperature. Figure 1.11 presents the measurements of the three sensors and the corresponding six alarm variables in 24 hours. The high (low) alarm variables are very similar for most of times; sometimes only one alarm variable runs into the alarm status for a short period of time due to noise effects, which certainly should not lead to operator's action such as changing the attemperator water flow in order to adjust the main steam temperature. Clearly, it is not necessary to configure an individual high (low) alarm variable for each sensor; a more reasonable way is to design two alarm variables (one for the high alarm and another for the low alarm) by integrating the measurements from three sensors.

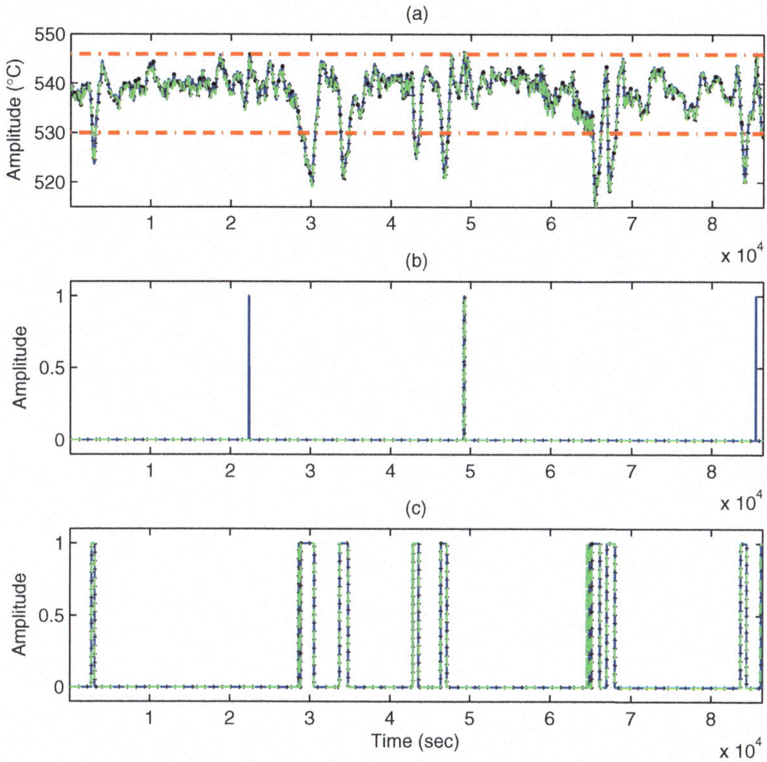

Fig. 1.11 **a** Measurements of the steam temperature from 3 sensors, **b** samples of three high alarm variables, **c** samples of three low alarm variables, reprinted from Ref. [179], copyright 2016, with permission from IEEE

This is another typical example of incorrectly configured alarm variables. There are more than 100 control loops at an industrial power plant, playing critical roles in the plant operation. Each control loop is associated with two digital variables, namely, the manual mode variable and the switch-to-manual variable. Figure 1.12 shows historical data samples of the two variables for a feed air control loop at a thermal power plant. The manual mode variable indicates whether the control loop is in the manual mode. For each control loop, the manual mode variable is configured with an alarm. Thus, when a control loop is switched from auto mode to manual mode, an alarm occurs. However, the switching from auto to manual mode is often done by operators. When an alarm occurs under such a circumstance, operators certainly take no actions to address the alarm. Therefore, there is no need to configure such a mode change with an alarm variable. By contrast, the switch-to-manual variable is the one forcing a control loop switching from the auto mode to manual, when some conditions are satisfied. It should be configured with an alarm under the condition that the control loop is in the auto mode. That is, when an alarm for the switch-to-manual

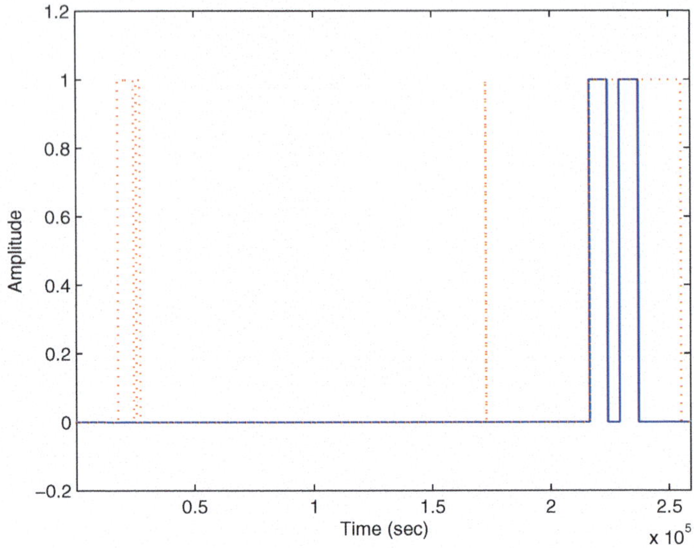

Fig. 1.12 Data samples of the manual mode variable (dash) and the switch-to-manual variable (solid) for a feed air control loop, reprinted from Ref. [179], copyright 2016, with permission from IEEE

variable occurs, operators need to check the corresponding abnormal conditions and take corrective actions in order to return the control loop back to the auto mode.

Cause #3: Alarm design isolated from related variables

The main task in alarm design is to determine the mechanisms to generate alarm variables, and choose proper parameters of the mechanisms. For instance, the most common alarm variables are generated by comparing the continuous-valued measurements of analog variables with the high or low alarm thresholds; thus, alarm thresholds are the design parameters to be chosen. However, in contemporary alarm systems, alarm design is usually isolated from other related variables, e.g., alarm thresholds are constant values and do not vary with other variables. It is a well-known fact that many process variables are related to each other via physical principles such as mass and energy conservation laws. Hence, a proper design of alarm generation mechanisms should be dependent on the related variables; otherwise, some types of nuisance alarms, including false alarms and missed alarms, may occur. As an illustration, Fig. 1.13 presents a schematic diagram of a normal operating zone of two correlated variables configured with alarms. If their alarm thresholds are designed in an isolated manner, then a rectangular area is formulated, being inconsistent with the normal operating zone. As a result, false alarms (missed alarms) are possibly present, shown as the star (circle) point in Fig. 1.13.

This is an example to illustrate an alarm design isolated from related variables. The inlet flow of a feedwater pump at a thermal power plant is configured with a high-alarm threshold 560 t/h and a low-alarm threshold 50 t/h. In one abnormal situation,

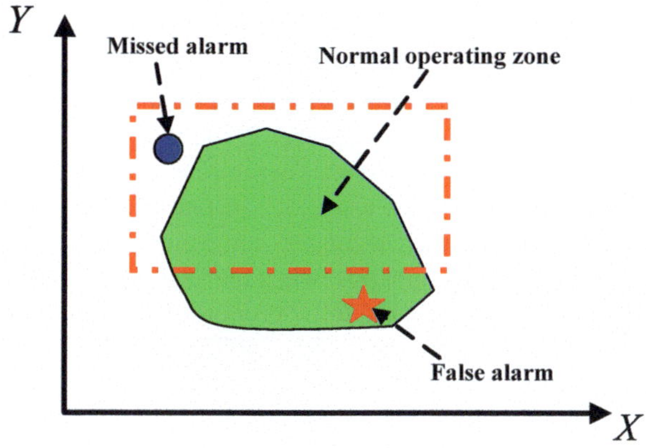

Fig. 1.13 A normal operating zone with isolated alarm thresholds, reprinted from Ref. [179], copyright 2016, with permission from IEEE

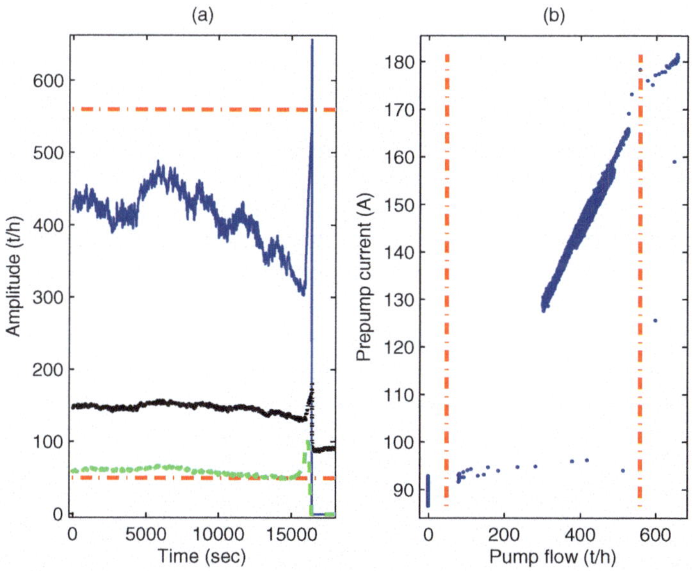

Fig. 1.14 **a** Time trends of inlet flow (blue solid), pre-pump current (black dotted), operator demand (green dashed) and high/low-alarm thresholds (red dash-dotted), **b** the scatter plot (dotted) of inlet flow and pre-pump current with high/low-alarm thresholds (red dash-dotted), reprinted from Ref. [179], copyright 2016, with permission from IEEE

the drum water level was decreasing abruptly so that the operating demand for the feedwater pump was increasing quickly, shown as the dashed line in Fig. 1.14a. As a result, the inlet flow (the solid line in Fig. 1.14a) exceeded the high-alarm threshold 560, raising a high alarm. However, the increment of inlet flow was induced by operator's demand. That is, the feedwater pump performed normally as requested, instead of being at an abnormal condition as implied by the occurred high alarm. The pre-pump current (the dash-dotted line in Fig. 1.14a) was closely related to the inlet flow. As a matter of fact, the same proportional relation between the pre-pump current and the inlet flow was unchanged, even if the inlet flow exceeded the high-alarm threshold 560, as shown in Fig. 1.14b. Therefore, the occurred high alarm of the inlet flow was a false alarm due to the usage of constant alarm threshold, which was isolated to the related variables such as the pre-pump current. A remedy is to design dynamic alarm thresholds for the inlet flow when the feedwater pump is running. The dynamic alarm thresholds are expected to be proportional to the pre-pump current, and form a normal operating band that could tolerate a certain level of uncertainties for such a proportional relation due to process variations.

After the high-alarm occurrence, another low alarm occurred as shown in Fig. 1.14a. This was owing to the shutdown of the feedwater pump, since the operating demand in Fig. 1.14a took the value of zero at the end. Such a low alarm did not say that the feedwater pump was in an abnormal condition, and no operator actions were needed. Therefore, the occurred low alarm was also a false alarm, which could be removed by incorporating the states of related variables such as the on/off state of the feedwater pump into the design of alarm variables, as described by the state-based alarming in the industrial standard ANSI/ISA-18.2 [92].

Cause #4: Abnormality propagation via physical connections

A large-scale industrial process is usually composed of upstream and downstream devices, which are physically connected. An abnormal condition in one process unit is very likely to propagate to the downstream devices or the upstream devices owing to recycling connections. As a result, a large number of true alarms may arise in a short period of time for the process variables associated with these devices configured with alarms; these alarms appear to "flood" operators. Hence, this phenomenon is referred to as an alarm flood. In other words, an alarm flood is defined as the situation when the number of alarm activations exceeds the operator's ability to process them [139] (page 440). There is no unique quantitative definition for alarm floods. A widely-accepted definition is that an alarm flood begins when 10 or more alarms occur within a 10-minute time period until the alarm rate drops below 5 alarm occurrences in 10 minutes [77, 92, 139]. Rothenberg [139] (page 120) also defines serious alarm floods as the situations having no less than 100 alarms within 10 minutes, or having 10 consecutive time periods, where each time period has no less than 10 alarms within 10 minutes. Generally, alarm floods include a large amount of nuisance alarms. However, the true reason that "loss incidents frequently involved the operator being overloaded with alarm floods" [54] (page 139) is that too many true alarms arise in a short time period. Taking the explosion accident at the Texaco refinery as an example, there were 275 alarms in the 10.7 minutes before the explosion [78]. Large numbers of true alarms are usually due to the abnormality propagation. That is, a primary abnormal event results in consequential abnormal events; these events raise the related alarms [165].

Let us look at a specific example for a concrete view of alarm floods. One alarm flood occurred at an industrial power plant. The alarm flood started with a high alarm of the main steam pressure (alarm tag #1 in Table 1.3) that was raised at the time instant 23:36:07. According to related process knowledge, the high value of the main steam pressure could be decreased either by reducing the feed coal flow or enlarging the opening of main steam control valves. However, industrial plant operators mistakenly adjusted the feed coal flow in a wrong direction, so that the main steam pressure kept increasing. The boiler water level decreased to an even lower position -298 mm, which almost reached a low-low alarm threshold -300 mm. Then, the operator reduced the feed coal flow after realizing such a mistake by shutting down two coal grinding mills. The main steam pressure decreased dramatically; as a result, the boiler drum level increased too quickly to arrive at a high-high alarm threshold at 300, which automatically triggered an emergency shutdown (alarm tag #80 in Table 1.3) of the entire power generation unit at the time instant 23:49:13. During the 13 minutes from 23:36:07 to 23:49:13, 80 alarm variables ran into the alarm status, as shown in Fig. 1.15. The descriptions of the 80 alarm variables are given in Table 1.3.

A retrospective investigation revealed that this alarm flood involved the propagation of several abnormalities. The time trends of several major process variables are presented in Fig. 1.16. The abnormalities of the steam pressure (tagname THRPRS in Fig. 1.16) and the boiler drum level (tagname CCSWAX004 in Fig. 1.16) were the two most critical ones. Because the physical connection of the steam pipeline and the boiler drum, the boiler drum pressure (tagname 4U09PT06 in Fig. 1.16) was directly affected so that another abnormality appeared in the boiler drum due to the unbalanced inlet water flow (tagname CCSWAX005 in Fig. 1.16) and the main steam flow (tagname MSFLOW in Fig. 1.16); such an abnormality caused a large variation of the boiler drum level, leading to the emergency shutdown of the power plant. As some less critical abnormalities, the control loops of two steam-driven feedwater pumps were switched automatically into the manual mode due to the abnormal boiler drum level; in order to reduce the main steam pressure via manual operations, operators switched several control loops into the manual mode such as the feed air flow control loop, and adjusted the feed coal flow by switching off two coal grinding mills. The 80 alarms associated with all abnormalities were raised in 13 minutes, resulting in an alarm flood. Apparently, the operators failed in diagnosing and handling the abnormalities, so that the abnormalities led to the incident of emergency shutdown of the process unit. It is worthy to point out that the alarm occurred first in time does not always indicate the origin of abnormalities, because the order of occurrences depends on several factors such as the configuration of alarm thresholds and the variation speeds of process variables. Clearly, it is very demanding and challenging to deal with such an alarm flood and abnormalities therein.

Table 1.3 Alarm variables involved in an alarm flood at an industrial power plant, reprinted from Ref. [179], copyright 2016, with permission from IEEE

#	Description	#	Description
1	Main steam pressure high	41	Mill B1 coal speed low
2	Mill C1 separator temperature #1 high	42	Mill B2 coal speed low
3	Coal system abnormal #1	43	Mill B2 #2 coal valve close
4	Coal system abnormal #2	44	Mill B1 #4 coal valve close
5	Mill C1 separator temperature #2 high	45	Mill B2 #1 coal valve close
6	MW control in manual	46	Mill B1 #2 coal valve close
7	MW control switch to manual	47	Mill B2 #3 coal valve close
8	Feed water A control error high	48	Mill B1 #3 coal valve close
9	Feedwater B control error high	49	Mill B2 seal/primary air pressure low
10	Feed air A in manual	50	Mill B1 seal/primary air pressure low
11	Feed air B in manual	51	Mill B1 separator temperature #2 high
12	O2 control in manual	52	Fire B1-2 off
13	Primary frequency regulation on	53	Fire B1-3 off
14	Boiler main control in manual	54	Fire B2-2 off
15	CCS mode off	55	Fire B2-3 off
16	Drum level #2 low	56	Mill B1 separator temperature #1 high
17	Turbine main control in manual	57	Mill B2 separator temperature #1 high
18	ATC invalid	58	Reheater temperature high
19	Drum level abnormal	59	Reheated steam temperature high
20	Drum level #3 low	60	Mill B2 separator temperature #2 high
21	Drum level low	61	Fire B1-4 off
22	Feedwater A control in manual	62	Fire B2-1 off
23	Feedwater B control in manual	63	Condensor flow high
24	Drum level #1 low	64	Drum level #1 high
25	Superheater temperature #1 high	65	Drum level #3 high
26	Superheater temperature #2 high	66	Drum level high
27	Superheater temperature #3 high	67	Drum level #2 high
28	Main steam temperature high	68	Feedwater A ETS trip
29	Mill C MTR	69	Feedwater A remote control off
30	Mill C1 #4 coal valve close	70	Feedwater A valve close
31	Mill C1 #3 coal valve close	71	Feedwater A emergency valve close
32	Mill C1 #1 coal valve close	72	Feedwater A/B speed error high
33	Mill C1 #2 coal valve close	73	Feedwater B inlet flow high
34	Fire C1-4 off	74	Feedwater A MEH trip
35	Fire C1-2 off	75	Feedwater A trip
36	Fire C1-3 off	76	Feedwater A inlet flow high
37	Fire C1-1 off	77	Drum level #2 high high
38	Mill C1 seal/primary air pressure low	78	Drum level #1 high high
39	Mill C2 seal/primary air pressure low	79	Drum level #3 high high
40	Mill B MTR	80	MFT

Fig. 1.15 Time trends of alarm variables in an alarm flood at an industrial power plant (the lower level for each alarm variable represents the value 0, and the higher level represents the value 1), reprinted from Ref. [179], copyright 2016, with permission from IEEE

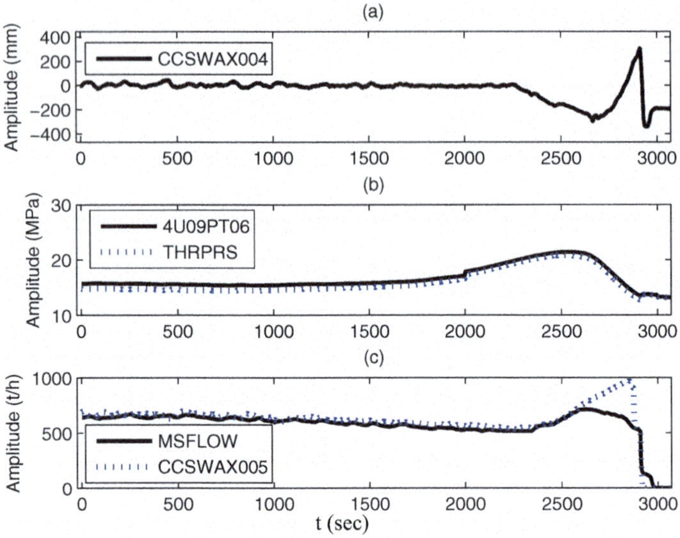

Fig. 1.16 Time trends of main process variables during an alarm flood at an industrial power plant: **a** boiler drum level (tagname CCSWAX004) as the process variable associated with the alarm variables with Tags #16,19-21,24,64-67 and 77-79 in Table 1.3, **b** steam pressure (tagname THRPRS) as the process variable associated with the alarm variable Tag #1, boiler drum pressure (tagname 4U09PT06), **c** inlet water flow (tagname CCSWAX005) as the process variable associated with the alarm variables with Tags # 73 and 76, main steam flow (tagname MSFLOW), reprinted from Ref. [179], copyright 2016, with permission from IEEE

1.3 Current Research Status in Literature

This section summarizes the current research status for industrial alarm systems. The summary focuses on the state of the art of methodologies related to the four main causes given in Sect. 1.2 that lead to the phenomenon of alarm overloading. Hence, the cited references are not going to represent a complete list of publications on industrial alarm systems.

State of the art in dealing with chattering alarms

The first step in dealing with chattering alarms is to detect the presence of chattering alarms. Yuki [219] detected chattering or unnecessary alarms by focusing on the balance between alarm occurrences and operator actions. Noda et al. [132] applied event correlation analysis to detect statistical similarities between alarm events. Naghoosi et al. [127] estimated a chattering index based on statistical properties of process variables. Kondaveeti et al. [105] quantified the degree of chattering alarms based on the alarm run lengths. Wang & Chen [176] revised the chattering index by taking the number of data samples into consideration. Instead of using chattering indices, Wang & Chen [177] formulated two hard rules to detect chattering alarms based on time durations in the alarm state. Wang & Wang [190] evaluated performance indices for false and missed alarms based on normal and abnormal data segments being classified based on qualitative trends.

The second step in handling chattering alarms is to design alarm systems to reduce the number of chattering alarms in the future. There are quite a few works on the design of alarm delay timers, alarm deadbands and alarm thresholds. Alarm delay timers are widely used in practice to eliminate chattering alarms. Xu et al. [198] designed optimal parameters of alarm delay timers based on analytical expressions of performance indices including the false alarm rate (FAR) and missed alarm rate (MAR). Wang & Chen [176, 177] proposed online methods to remove chattering and repeating alarms by adjusting alarm thresholds or using alarm delay timers. Afzal & Chen [4] captured the configuration of alarm delay timers in multiple modes of a process using a hidden Markov model, which effectively reduced false and nuisance alarms. Zeng et al. [221] evaluated the performance of a generalized alarm delay timer by designing generalized formulas for false and missed alarm rates. Tulsyan et al. [168] expressed the design of alarm delay timers based on a stochastic nonlinear time series model for a nonlinear chemical process. Su et al. [151] increased the number of univariate alarm setpoints to design alarm delay timers to optimize the false alarm rate, missed alarm rate, and average alarm delay. Taheri-Kalani et al. [154] and Kaced et al. [97] designed and improved generalized alarm delay timers based on Markov chains to improve the alarm system performance. Sompura et al. [147] provided a simple alarm delay timer strategy to handle chattering alarms. Wang et al. [186] and Gyasi & Wang [70] analyzed statistics of alarm durations to design alarm delay timers by exploiting Bayesian estimation theory.

Alarm deadbands are also an effective tool to minimize disturbing alarms, but there has been limited research on alarm deadbands compared to alarm delay timers. Hugo [89] designed adaptive alarm deadbands to reduce the number of chattering alarms. Naghoosi et al. [127] designed optimal alarm limits and deadbands by minimizing the chattering index. Afzal et al. [5] configured time-deadbands using a mathematical model based on Markov processes. Wang et al. [189] proposed a deadband width design method to achieve an optimal weighted balance between the duration of nuisance alarms and the delay in detecting abnormalities. Tulsyan & Gopaluni [169] relaxed the assumptions of data uncorrelation and stationarity to design alarm deadbands to minimize the false and missed alarm rates. Wang et al. [187] designed an alarm deadband to remove false alarms based on the maximum magnitude deviation, for which Gyasi et al. [72] extended to adaptive alarm deadbands for non-stationary process variables.

For some process variables, alarm thresholds are subject to design in a certain range. Yu et al. [214] selected optimal alarm thresholds based on the alarm probability graph to reduce false alarms. Tian et al. [163] used principal component analysis weights and Johnson transform to optimize the alarm thresholds in chemical processes. Toshkova et al. [167] applied generalized extreme value theory to the baseline of different operating conditions to determine alarm thresholds and alarms to avoid triggering false alarms. Aslansefat et al. [14] evaluated the performance of a variable threshold alarm system using prioritization with gates and a semi-Markov process.

Zhang et al. [222] optimized alarm thresholds by using a sensitivity clustering algorithm and ROC curves based on the weighted FAR and MAR.

In addition to alarm delay timers and alarm deadbands, other alarm generation mechanisms have been used to reduce the number of chattering alarms. Henningsen & Kemmerer [75] logically combined alarms with the current state of the process to filter out duplicate alarms. Burnell & Dicken [28] introduced an auto-shelving mechnasium to deal with repeating alarms. Bransby & Jenkinson [23] (Appendix 10) recommended alarm deadbands, alarm delay timers, and shelving mechanisms to handle repeating alarms. Ahnlund, et al. [7] separated process variables into 14 classes and selected proper filters according to the classes to deal with chattering alarms. Srinivasan et al. [149] temporarily changed alarm thresholds or put alarm variables in a shelving condition based on statistical process control techniques. Hwang et al. [90] applied control charts to design a pre-alarm system to reduce the alarm frequency. Xu et al. [200] dynamically obtain the optimal parameters of the designed alarm system for updating and optimizing the online alarm evidence. Tan et al. [157] applied rank-order filters to the design of alarm systems. Mannani et al. [121] considered the difference between point-based data and interval-based data to eliminate chattering alarms. Roohi et al. [138] designed a new state feedback controller to fulfill the performance of an alarm system. Xu et al. [201] and Weng et al. [196] dealt with uncertainty in monitoring process variables based on alarm evidence updating in the Dempster-Shafer theory of evidence to reduce false and interfering alarms. Lee et al. [113] proposed an innovative AI-assisted false alarm detection and diagnosis system to troubleshoot false alarms. Dorgo et al. [48] [49] measured the alarm system performance in reducing nuisance alarms considering the informativeness and operability of the alarms. Valipoori et al. [170] extracted two probability density functions of historical data under normal and abnormal conditions to improve the performance of alarm systems. Li et al. [115] built a data-driven model of normal drilling behavior and qualitative trend analysis to reduce false alarms during drilling. Raei et al. [136] introduced uplink/downlink counters for univariate alarm systems to derive parsing expressions for performance metrics. Gyasi & Wang [69, 71] established series and parallel connections of alarm delay timers and alarm deadbands in order to deal with different types of false and missed alarms.

State of the art in handling incorrectly-configured alarm variables

In terms of configuring process variables with alarms, there are mainly two types of alarm configuration methods. The first type is to establish relations between process variables and abnormal situations based on *process knowledge* in order to configure alarm variables accordingly. Yan et al. [205] drew up an abnormality propagation diagram and selected suitable alarm variables for malfunctions. Yang et al. [206] allocated sensors for alarm variables to optimize fault detection reliability based on process topology represented by signed directed graphs. Takeda et al. [155, 156] selected process variables to be configured with alarms based on cause-effect models from process topology. Dalapatu et al. [44] assigned alarm variables to a group of process variables based on process knowledge to identify abnormal events.

Another type of alarm configuration methods is to analyze correlated or consequential alarm variables based on *historical data*, to remove redundant alarm variables or redesigning alarm variables. Geng et al. [62] and Zhu & Geng [230] grouped alarm variables into different clusters and ranked the priorities of alarm variables in each cluster via a fuzzy clustering-ranking algorithm. Kordic et al. [106] found correlated sets of alarm variables by using time intervals between alarm occurrence and disappearance time instants. Kondaveeti et al. [104] built an alarm similarity map based on the Jaccard similarity index for binary-valued alarm variables. Yang et al. [208] clustered correlated alarm variables based on correlation coefficients of pseudo continuous time series generated from binary alarm data samples. Yang et al. [211] exploited the Sorgenfrei similarity coefficient for binary-valued alarm variables and the distribution of correlation delays to detect correlated alarm variables. Wang et al. [178] found consequential alarms based on correlation delays and Granger causality measures. Rao et al. [137] yielded alarm correlations and their statistical features being visualized as network graphs to indicate process interactions and identify alarms for prioritized analysis.

In terms of assigning different priorities to alarm variables, qualitative approaches are the common practice. Timms [164] discussed alarm prioritization based on classifying the consequences on personnel safety, financial loss and environmental consequences into several categories. Very few quantitative approaches appeared recently. Chang et al. [33] proposed a quantitative risk-based approach to prioritize alarm variables by integrating the process safety time together with the probability and impact of potential hazards. Arifin & Choudhury [12] quantified the system failure probability to sort the importance of alarm variables. Venkidasalapathy et al. [172] obtained quantitative information from mixed integer linear programming and linear multi-objective optimization formulations to consider potential fault prioritization order.

State of the art in alarm design by considering related variables

As stated in Sect. 1.2, the design of alarm generation mechanisms often needs to take related variables into consideration. Since multiple process and alarm variables are involved, the alarm generation mechanisms are diverse, without a unifying framework. Yamanaka & Nishiya [204] used qualitative descriptions to develop an alarm monitoring system for process variable trends. Bristol [24] suggested automatic adaptation of alarm thresholds to varying process situations. Brooks et al. [25] proposed a geometric process control method to obtain dynamic alarm thresholds from multivariate best operating zones. Charbonnier et al. [34] extracted the trends of several related process variables to formulate different episodes for generating alarm variables. Izadi et al. [94] and Kondaveeti, Shah & Izadi [103] applied multivariate statistics to generate alarms more efficiently. ANSI/ISA-18.2 [92] and Hollifield & Habibi [77] described state-based alarming to avoid long-standing alarms by changing alarm settings according to process states. Yang et al. [207] performed correlation analysis for alarm signals to optimize alarm thresholds. Charbonnier & Poret [35] adaptively calculated qualitative trends of relevant process variables based on estimated noise variances. Nihlwing & Kaarstad [129] developed a state-based alarm system that dynamically presented alarms based on 19 different process states for a nuclear power plant simulator. Jang et al. [95] developed different rules to filter or suppress alarms for a nuclear power plant reactor. Gupta et al. [68] exploited wavelet analysis and principal component analysis to alleviate noise effects and detect faults in pharmaceutical manufacturing processes. Zhu et al. [227] obtained dynamic alarm thresholds depending upon multiple steady states and transitions between these states. Zang & Li [220] optimized alarm thresholds by minimizing FAR and MAR based on a joint probability density of multiple process variables. Alrowaie et al. [10] proposed a model-based alarm design method based on particle filtering for multivariate nonlinear stochastic systems. Gao et al. [59] obtained multivariate optimal alarm thresholds based on correlation consistency of process data and alarm data. Wang & Chen [180] designed a dynamic state-based alarm system that effectively reduces long-term alarms. Yu et al. [215] proposed a method to detect abnormal data segments from historical data samples of multivariate systems. Chen & Wang [37] [38] dealt with variational directions of multiple process variables to design multivariate alarm systems. Kim et al. [101] described multivariate control charts based on a new optimal false alarm control support vector technique. Xiong et al. [197] observed a new feature on variation rates of model parameters to distinguish between normal and abnormal conditions in order to design a multivariate alarm system for time-varying processes. Tian et al. [162] clustered multiple variables using the objective functions of FAR, MAR and average alarm delay to obtain optimal alarm thresholds. Yu et al. [216] and Yu & Wang [217] designed alarm systems with dynamic alarm thresholds for multivariate processes based on historical normal data points to establish normal operating zones. Gao et al. [61] constructed a multivariate alarm trippoints design framework based on causal modeling to select alarm trippoints for key variables. Kaced et al. [98] statistically synthesized information from multiple variables based on principal component analysis to manage disturbance alarms. Wang et al. [188] established an adaptive alarm threshold setting method based on model matching by considering industrial processes under multiple operating conditions. Zhao et al. [223] captured the variation of variables within the normal operating region using a dynamic multivariate alarm threshold optimization method based on time series clustering. Sudarshan et al. [152] evaluated the quality of multivariate alarm systems to focus on quality alarms for rare abnormal events. Wang et al. [192] provided a multivariate alarm system for non-convex operating zones being described by search cones.

State of the art in dealing with alarm flooding

Abnormality propagation is a major reason for multiple alarms and even alarm floods. Some techniques have been developed to deal with this case in alarm floods. Varga et al. [171] detected the development of unsafe situations and suggested the operators with necessary safety actions based on the prediction of physical models. Tchamova & Dezert [158] estimated the degree of danger for alarms based on Dezert-Smarandache theory to fuse conflicting evidences. Basu et al. [18] ordered alarms on the power grid according to different severity measures. Blaauwgeers et al. [22] and Zhu et al. [228] used the BowTie diagram and Bayesian network respectively, to perform real-time risk analysis for operators to prioritize alarm handling. Xu et al. [199] predicted the impending alarms based on hybrid models utilizing both first principles and data, so that operators could have more time to handle alarms. Laberge et al. [107] addressed alarm floods through a new alarm summary display design by showing alarms in a time series represented by icons together with short alarm descriptions. Satuf et al. [140] validated the utility of an advanced intelligent alerting system capable of dynamically assigning real-time priorities to alarms. Zhu et al. [229] predicted the probability of critical alarm occurrence using a dynamic alarm prediction algorithm. Hu et al. [85] present work on designing visual graphs of alarm and event data in response to the great need for effective visual forms for a large number of alarms. Goel et al. [65] developed a methodology to benchmark alarm systems by calculating key performance indicators to manage alarm flooding. Zhou et al. [224] determined dynamic alarm limits for online monitoring based on an offline design framework and data related to normal start-up operations. Cai et al. [29, 30] utilized the information contained in the alarm log itself and applied word embedding techniques to capture relationships between alarms so that operators could make critical decisions in alarm flooding situations. Niyazmand & Izadi [131] and Lai et al. [110] mined similar patterns in alarm floods using an improved PrefixSpan algorithm and an accelerated multiple sequence comparison algorithm. Schirru et al. [141] used quantum evolutionary algorithms to create decision support systems that provide critical information in the presence of alarm flooding. Xu et al. [202] and Xu & Wang [203] predicted alarms for upcoming alarm floods based on historical alarm flood sequences. Shang & Chen [143] and Alinezhad et al. [8] investigated online early classification of alarm floods by employing a data-driven analysis. Parvez et al. [135] conducted a similarity analysis of online alarm floods to achieve early prediction of alarm floods. Wang et al. [191] described the operation of a large number of alarm variables in an industrial alarm system through two analytic diagrams. Alinezhad et al. [9] considered key features of alarms including temporal information to develop a vectorized model of alarm flooding. Bantay et al. [16] combined sequence compression and sequence alignment to predict the most probable alarm progression. Hu et al. [87, 88] proposed a pattern matching method based on word embedding and dynamic time warping to reveal the sequence similarity between alarm floods.

To reduce the number of alarms due to abnormality propagation, suppression of consequential alarms or redesign of alarm variables is recommended during alarm floods. Hollender & Beuthel [76] suggested hiding consequential alarms based on causal relations between process variables. The guideline in EEMUA 191 [54] suggested reviewing consequential alarms and using alarm grouping to reduce the number of alarm activations during alarm floods. Beebe et al. [21] suggested state-based alarm rationalization to control alarm floods whose occurrences are typically due to a change of process states. Han et al. [73] optimized alarm thresholds to control alarm flooding based on FAP, MAP and correlation analysis. Hu & Yi [80] proposed a two-level intelligent alarm management framework for alarm filtering of alarm floods. Lai & Chen [108] obtained a common pattern of multiple alarm flood sequences based on a pattern mining approach across multiple alarm flood sequences. Lai et al. [109] and Guo et al. [66] analyzed the similarity of alarm flooding based on a pattern matching approach to suppress alarm floods. Wang et al. [182] formed a new criterion to detect the presence of an occurring alarm flood based on the number of new alarm variables in the alarm state. Hu et al. [83] [84] utilized a data-driven approach to detect alarm patterns in alarm floods to reduce nuisance alarms. Luo et al. [119] constructed a simplified multilayer Bayesian network based on active transfer entropy to adjust thresholds in real time to improve the accuracy of alarm propagation analysis.

State of the art in finding root cause alarms

A solution to consequential alarms owing to abnormality propagation is to find one or more abnormalities as the root cause(s) of a set of occurred alarms [51, 210]. One type of methods to find the root causes are based on the process knowledge and/or learning algorithms. Young et al. [213], McDonald et al. [124], Cheon et al. [41], Miao et al. [126], Dahlgren et al. [43] exploited process knowledge of power systems to formulate rules or expert systems to analyze faults from alarms. Wen & Chang [194] and Wen et al. [195] respectively used a Tabu-search method and a refined genetic algorithm to find the abnormal events to explain a given set of reported alarms from knowledge-based tables describing the relations between abnormal events and alarm variables for power systems. Cauvin et al. [31] used causal graphs and models to interpret the root causes of alarms. Dashlstrand [45], Souza et al. [148] and Larsson et al. [112] introduced multilevel flow models or fuzzy neural networks to analyze root causes of alarms. Kezunovic & Guan [99] used fuzzy reasoning Petri-net techniques to diagnose the root cause of alarms. Liu et al. [117] introduced an operator model as a virtual subject to evaluate the performance in diagnosing root causes of alarms. Dubois et al. [53] and Lee et al. [114] adopted logic diagrams to perform real-time cause-effect analysis for alarms. Guo et al. [67] and Wei et al. [193] determined fault/disturbance causes based on rule networks or temporal constraint networks between cause hypothesis and alarms for digital power substations. Tolga et al. [166] dealt with causal relations by utilizing functional modeling techniques. Simeu-Abazi et al. [145] exploited dynamic fault trees to locate faults from alarms with application to avionic systems. Abele et al. [1] exploited Bayesian networks to analyze the root cause of alarms in an online manner. Gao et al. [60] introduced a spatial interpretive structure model to represent the hierarchical organization of spatial unit blocks and to show causal relationships at different levels of granularity. Wang et al. [175] and Wang et al. [184] described the relationship between alarm variables through Bayesian networks to determine the root cause of alarms occurring. Meng et al. [125] quantified the degree of interaction between variables to form a new Bayesian network learning score function. Zhu et al. [231] performed online monitoring and root cause analysis based on a distributed alarm analysis model with multiple correlated chunked partial least squares. He et al. [74] analyzed the root cause through a novel regular variable analysis of multicorrelated blocks and enhanced contribution graphs. Zhu et al. [232] proposed a new multimodular Bayesian network with a recurrent structure to find the root cause of alarms for resolving the alarm flooding problem. Sudarshan et al. [152] developed multivariate alarm systems to recognize unpostulated abnormal events and suggest response actions using process modeling and path-sampling. The above methods need rather complete and accurate process knowledge, which is sometimes difficult to obtain in practice, especially when the process is large and complex.

Alternatively, historical alarm series can be used to extract time patterns of alarms. For this purpose, sequential pattern matching has been introduced for examining alarm series. Folmer et al. [56], Folmer & Vogel-Heuser [57] and Folmer et al. [58] used this idea to cluster frequent occurring sub-sequences in alarm logs and identify alarms with causal relations to redesign alarm systems for reducing the number of alarm variables. Similar historical alarm floods can be exploited to extract representative information. Ahmed et al. [6] located similar alarm floods based on the consecutive alarm frequencies and used dynamic time warping to obtain optimal matching between two alarm floods. Cheng et al. [40] proposed a modified Smith-Waterman algorithm for local alignment of two alarm flood sequences so that common alarm sequence segments could be extracted. Charbonnier et al. [36] extracted the fault sequence template from alignments of alarm sequences from the same fault, and compared a new alarm sequence to the template for fault isolation. Wang et al. [181] combined fuzzy sets and extracted potential alarm association rules from a quantitative database of temporal alarm sequences to find out the root alarm. Zhu et al. [233] determined patterns between similar sequences by extracting alarm flood sequences from alarm logs. Zhou et al. [225, 226] matched or extracted alarm sequence patterns from historical flood alarm sequences to compare alarm floods of different processes. Venkidasalapathy & Kravaris [173, 174] utilized a hidden Markov model based alarm sequence classification method to identify the root cause of alarms. Shirshahi & Aliyari-Shoorehdeli [144] diagnosed root causes of faults based on alarm flood classification using transfer entropy and multi-sensor fusion approaches. This type of methods rely on sufficient historical alarm data, including all possible patterns. However, all the abnormalities may not be encountered in the past, leading to failures for new patterns in online practice.

Another type of methods is to capture the plant topology in advance to describe the intrinsic structure of the process. When the process enters an abnormality, a backtrack or hypothesis test can be employed based on the current symptom to find the root causes. Bauer et al. [20] used time-delayed cross correlation to identify the propagation paths and then build a causal map. Similarly, a series of causality identification methods have been proposed, such as Granger causality [218], partial directed coherence [63], transfer entropy [19], direct transfer entropy [50], and transfer zero-entropy [52]. Hu et al. [81] proposed a new local comparison algorithm based on the basic local comparison search tool to reduce the computational complexity. Hu et al. [82] analyzed the causal relationship between alarm variables using transfer entropy considering the mutual independence between random delays and alarm occurrence. Abreu et al. [46] detected causal relationships between industrial alarm variables based on the transfer entropy theory and Bayesian network K2 algorithm. Nonlinearity can also be used as an indicator in causality analysis, e.g., Thornhill [161] used a nonlinearity index to find the root cause because nonlinearity is strongest at its source. Cecilio et al. [32] used the nearest neighbors method to identify abnormality in each series and showed the abnormality propagation order as a color plot. By the total contribution plot, one can observe how a fault is spreading across the process [17]. The limitations of these data-based methods are assumptions of model linearity or data stationarity, and the high computational burden. In addition, these methods are usually used for obtaining a short list of possible root causes that cannot be confirmed without resorting to process knowledge. Therefore, in addition to process data, another important resource for building plant topology is plant connectivity information that describes physical, mass or information linkage between process units. Adjacency matrices [96], signed directed graphs [122], XML description [212], and semantic description [210] are efficient representations of plant connectivity, which provide a physical foundation of plant topology with extra and redundant information. These connectivity-based methods should be integrated with data-based methods for validation. Thambirajah et al. [159] combined the cause-and-effect matrix derived from measurements and qualitative information about the process layout. Yang et al. [209] validated the data-based and connectivity-based results mutually. Alarm data can also be used, e.g., Schleburg et al. [142] combined plant connectivity and alarm logs. Landman et al. [111] used a dedicated search algorithm to validate the quantitative results of the data-driven causality using the qualitative information on plant connectivity.

The combination of different resources and methods is a proper choice in real applications according to availability of resources and application objectives. Chiang & Braatz [42] integrated the statistical analysis with the causal map. Thornhill et al. [160] enhanced data-based analysis process understanding. Maurya [123] combined signed directed graphs with qualitative trend analysis. Di Geronimo Gil et al. [47] merged first-principles structural models with plant topology derived from a process drawing. Hu et al. [86] analyzed time-varying cause-effect relations based on qualitative trends, change amplitudes and process knowledge.

1.4 Major Research Problems to Be Solved

This section first presents the lifecycle of alarm variables into three stages, namely, alarm configuration, alarm design, and alarm removal, and connects the stages with the four main causes in Sect. 1.2. Next, for each stage, some fundamental research problems to be solved are proposed on the basis of research status summarized in Sect. 1.3.

The industrial standard ANSI/ISA-18.2 [92] and the guideline in EEMUA-191 [54] defined desired performance benchmarks for industrial alarm systems, e.g., the 2nd column in Table 1.2. In order to achieve the benchmarks, the industrial standard ANSI/ISA-18.2 presented 10 stages for an alarm management lifecycle, namely, alarm philosophy, identification, rationalization, detailed design, implementation, operation, maintenance, monitoring and assessment, management of change, and audit [92] (page 22). Hollifield & Habibi [77] listed 7 steps to achieve a highly effective alarm system:

step 1: develop, adopt and maintain alarm philosophy,
step 2: benchmark alarm system,
step 3: find bad actor alarms,
step 4: perform alarm documentation and rationalization,
step 5: implement alarm audit and enforcement technology,
step 6: implement real-time alarm management,
step 7: control and maintain improved alarm system.

These stages and steps have been proven to be effective pathways to follow in some practical applications, but specific techniques need to be developed to support these stages and steps. The techniques, as the expected outcomes of the fundamental research problems to be proposed here, are better to be presented in a different way with respect to alarm variables.

The lifecycle of alarm variables may be formulated as three stages, namely, alarm configuration, alarm design, and alarm removal, as shown in Fig. 1.17. That is, some process variables in the set Y are selected in the stage of alarm configuration to formulate the set X as the process variables to be configured with alarms, X_a is the set of alarm variables based on the alarm design for X, and operator action needs to be decided in the stage of alarm removal once X_a is in the alarm state. In particular, each stage has its own alarm performance monitoring techniques in order to have a systematic solution to the problems therein and to maintain a satisfactory performance of each stage in a long run. The three stages are closely related to the four main causes discussed in Sect. 1.2. The 2nd cause "alarm variables are incorrectly configured" is about alarm configuration; the 1st and 3rd causes are for nuisance alarms, which are the main concerns in the alarm design; the 4th cause "abnormality propagates owing to physical connections" is related to the alarm removal, since one of key steps in alarm removal is to find the root causes of alarms, especially during an alarm flood. The three stages in Fig. 1.17 are also connected with the 10 stages in [92] and the 7 steps in [77], e.g., the alarm configuration are in line with steps 1 and 4, the alarm design are related to steps 2, 3, 5 and 7, and the alarm removal encompasses step 6.

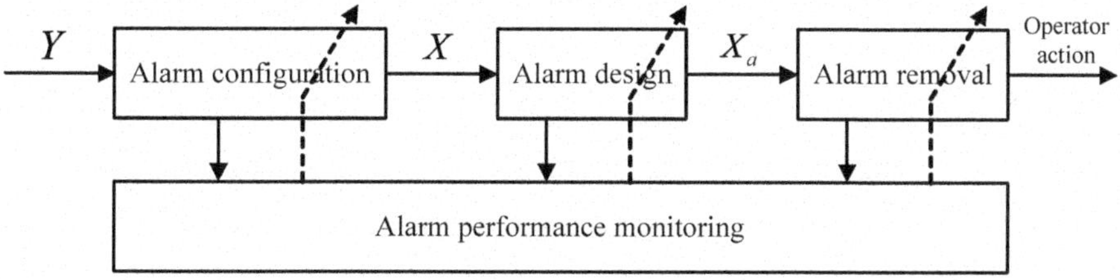

Fig. 1.17 The lifecycle of alarm variables, reprinted from Ref. [179], copyright 2016, with permission from IEEE

Research problems in alarm configuration

Alarm configuration mainly involves the selection of process variables to be configured with alarms, and the determination of alarm priorities. There are three fundamental research problems to be solved:

Problem 1: Should an alarm variable be configured for a given process variable?
Problem 2: Which priority should an alarm variable be assigned?
Problem 3: Are there any incorrectly-configured alarm variables in an existing alarm system?

In terms of Problem 1, the main concern is the determination of relations between abnormal events and process variables so that proper process variables are selected to identify abnormal events. Mathematically, the relation can be described as

$$X = g(Y, A), \tag{1.2}$$

where A stands for the set of abnormal events, Y is the set of candidate process variables, and X is the set of selected process variables to be configured with alarms. If the relation between abnormal events and process variables is established, then the process variables, whose variations reveal the presence of abnormalities, are selected to be configured with alarms. However, the existing approaches summarized in Sect. 1.3 to establish such a relation are based on process knowledge or obtained from historical operating data. Either way has its own limitations, e.g., the process knowledge is often incomplete and qualitative, while the data-based approaches are confined by the occurred phenomena in historical data sets and contaminated by noises/disturbances. Hence, the main challenge is to exploit process knowledge and historical data *simultaneously* in establishing relations and selecting proper process variables to be configured with alarms.

In terms of Problem 2, current approaches in industrial practice are mostly qualitative, based on the understanding of severity in missing alarms and a rough estimation of safety time in handling alarms. Quantitative approaches are the ones to be developed as alternatives or complements to the qualitative approaches. For instance, a more accurate estimation of the safety time could be obtained from historical data sets. Here the safety time is the time period allowed to handle an alarm so that the associated negative consequences do not occur. The risk-based approaches are promising quantitative approaches to assign priorities for alarm variables. Alarm priorities may not be static; instead, dynamic alarm priorities are perhaps more reasonable for different process states and operational scenarios [18, 22, 158, 228].

In terms of Problem 3, the incorrectly-configured alarm variables may present themselves in different forms. The redundant (duplicated) alarm variables are the ones that always run into the alarm status simultaneously or in a constant time delay. The redundant alarm variables yield no extra useful information and should not be configured with alarms. The correlated and consequential alarm variables may also be resulted from incorrect alarm configurations, e.g., the highly correlated alarm variables from three sensors for the main steam temperature in Fig. 1.11.

The correlated alarm analysis methods listed in Sect. 1.3 are able to effectively detect the redundant, correlated and consequential alarm variables. If operator actions cannot be clearly defined for the occurred alarms, then the corresponding alarm variables may be incorrectly configured, as shown by the alarms from the switch-to-manual variable in Fig. 1.12. However, the challenges for doing so is that many operator actions or responses are not recorded in the historical database, and even for the recorded ones, they are difficult to be associated with occurred alarms in an automatic manner.

Research problems in alarm design

The first objective in the stage of alarm design is to design an alarm generation mechanism that transforms process variables X to be configured with alarms into alarm variables X_a as shown in Fig. 1.17,

$$X_a = h(X, X_r; \theta).$$

Here X_r stands for the set of process variables related to X, and θ is the vector of design parameters in the alarm generation mechanism $h(\cdot; \theta)$. The alarm generation mechanism $h(\cdot; \theta)$ includes the simplest one in (1.1), alarm deadbands, and alarm delay timers, which have been commonly adopted in practice, as well as more complex ones such as logic/model-based alarming methods [54, 77, 92, 139]. Thus, the corresponding fundamental research problem to be solved is

Problem 4: How to design an alarm generation mechanism with good performance?

One main objective in the design of $h(\cdot; \theta)$ is to reduce the number of nuisance alarms caused by noises and/or disturbances (main cause #1 in Sect. 1.2); another main objective is to take the related variables into the generation of alarm variables (main cause #3). In particular, if related variables have to be considered, the alarm generation mechanisms have many different forms, as shown by the references cited in Sect. 1.3. However, there are two main challenges, which have not been well addressed in the existing studies, in order to solve Problem 4 in a systematic manner.

First, the normal and/or abnormal operating zones for process variables to be configured with alarms are difficult to obtain. The normal and/or abnormal operating zones are the pre-requisites for the design of $h(\cdot; \theta)$. Some physical principles and laws may be exploited to build up mathematical models for process variables in the normal or abnormal conditions. However, the uncertainties of model parameters have to be estimated from historical normal/abnormal data sets to formulate the operating zones, to tolerate the variations of normal operations and the effects from noises/disturbances. The operating zones could also be established directly from the normal and abnormal historical data sets. However, these data sets are often not available at hand, and have to be separated from historical data samples. One argument is to do the classification manually via visualization and consultation to plant operators, which is possible only for small sized data sets. Few existing studies have addressed this research challenge. Xu et al. [198] and Wang and Wang [190] detected the data sets with different sample means and to compare the sample means to alarm thresholds via hypothesis tests, in order to isolate normal and abnormal data sets of one single process variable. If multiple process variables are involved, the static alarm thresholds cannot truly reflect the normal operating zone, as shown in Fig. 1.13. As implied by the industrial example shown in Fig. 1.14 and several references [34, 35, 204] cited in Sect. 1.3, abnormalities may be detected by monitoring the consistency of changing directions of X and X_r. By incorporating this process knowledge, change detection methods and data clustering techniques need to be developed in the context of industrial alarm systems.

Second, the relation between $h(\cdot;\theta)$ and a performance index vector η is difficult to establish. The false alarm rate (FAR) and missed alarm rate (MAR) are the most commonly-adopted performance indices. It is a well-known fact that there usually is a tradeoff between the FAR and MAR. Thus, a loss function $l(\eta)$ can be formulated to balance the conflicting indices, e.g.,

$$l(\eta) = w\text{FAR} + (1 - w)\text{MAR},$$

where the real number $w \in [0, 1]$ is a weighting factor. An optimal design of $h(\cdot;\theta)$ is obtained by minimizing $l(\eta)$, i.e.,

$$h_{opt}\left(\cdot, \theta_{opt}\right) = \arg\min_{h(\cdot,\theta)} l\left(\eta\right).$$

One critical knowledge required in the optimization is the quantitative relation between $l(\eta)$ and $h(\cdot;\theta)$. For certain univariate alarm variables where the one-dimensional process variable X is independent and identically distributed (IID), this relation has been established for alarm deadbands, alarm delay timers and alarm filters [2, 3, 39, 198]. However, the relation is rather difficult to establish under more practical assumptions, e.g., $x(t)$ is not IID. As an attempt, Tulsyan et al. [168] [169] adopted particle filtering to estimate the FAR and MAR of alarm deadbands and delay timers for non-IID process variables based on nonlinear stochastic models; Wang et al. [186, 187] exploited alarm durations and deviations for designing alarm deadbands and delay timers for non-IID process variables. If multiple process variables are involved, the relation is much harder to be obtained. As a result, many existing studies in Sect. 1.3 were limited to the proposition of $h(\cdot;\theta)$ without theoretical analysis on the performance, but only with some examples showing the empirical effectiveness.

The second objective in the stage of alarm design is to detect the presence of nuisance alarms for industrial alarm systems being in service, in order to initiate a redesign of $h(\cdot;\theta)$ as described in Problem 4 to reduce the number of nuisance alarms in the future. Thus, the corresponding fundamental research problem is:

> *Problem 5:* Are there too many nuisance alarms, so that alarm generation mechanisms need to be redesigned?

Clearly, a solution of isolating historical normal and abnormal data sets partially solves Problem 5, because the availability of normal (abnormal) data segments directly classifies the occurred (disappeared) alarms in the normal (abnormal) data segments as false (missed) alarms. However, such a solution may be difficult or costly to be obtained, it would be desirable to attack Problem 5 via different approaches.

One approach to solve Problem 5 is based on some special characteristics of nuisance alarms. As an example, the rationale of the methods in [176, 177] to detect chattering and repeating alarms is to look at the statistical regularity of alarm durations or intervals in historical alarm data samples. The main challenge in doing so is from the diverse types of nuisance alarms. Each type of nuisance alarms has its own characteristics to be exploited for detection. Some characteristics are easier to be captured. For instance, chattering and stale alarms are the alarms whose time durations are very short (e.g., less than 20 sec) [177] or exceptionally long (e.g., more than 24 hours) [54, 92]. By contrast, the characteristics of some nuisance alarms are hard to be described, e.g., those due to the 3rd main cause in Sect. 1.2.

Another approach is to associate operator actions with occurred alarms [132, 219]. According to the definition of true alarms in Sect. 1.2, if an operator action is associated with the occurrence of an alarm, then it is a true alarm; otherwise, it belong to a nuisance alarm. However, such an approach may not be feasible in practice, as commented earlier for Problem 3.

A more feasible approach to solve Problem 5 is to connect the occurred alarms with their consequences. If no harmful consequences have been detected, then the occurred alarms are certainly nuisance. The relation between alarm variables and their consequences is the one to be established in the stage of alarm configuration. Owing to the presence of noises or disturbances, the detection of consequences may rely upon some hypothesis tests to make a statistical classification.

After the detection of nuisance alarms, the severity of nuisance alarms needs to be evaluated to determine whether a redesign of alarm systems is necessary or which alarm variables are addressed with high priorities. For instance, the chattering indices in [105, 127, 176] are used for this purpose.

The third objective in the stage of alarm design is to generate a predictive alarm to indicate upcoming critical abnormalities, so that operators could have more time to analyze the upcoming alarms and take proactive actions. Thus, the research problem is

> *Problem 6:* How to design mechanisms to generate predictive alarms in order to forecast upcoming critical abnormal events?

A standard approach is to generate predictive alarms based on time series modeling and prediction techniques, e.g.,

$$\hat{X}(t+i) = p\left(X(t), X(t-1), \cdots\right),$$

where $\hat{X}(t+i)$ with $i > 1$ is the prediction of a process variable $X(t)$ configured with alarms, from a predictor $p(\cdot)$ based on the current and past data sample $X(t)$, $X(t-$

1), ⋯. Such an approach may not work well due to complexity of process variables in practice.

A special attention has to be paid to alarm floods, which should be avoided as much as possible due to the equalities "floods = incidents = loss" [21]. Hence, it is important to predict the upcoming of alarm floods and take preventive actions to avoid the occurrence of alarm floods. One approach for the prediction of alarm floods is based on the physical or hybrid models, which predict the evolution of process variables [171, 199]. However, developing physical models are technically challenging and time consuming. Hence, such an approach may be feasible and worthwhile for some critical devices or equipments, not applicable to general alarm variables.

Since alarm floods are usually composed by alarm variables having physical connections as shown in Sect. 1.2, historical alarm floods may have certain regularities to be exploited. For instance, the switching-off events of coal grinding mills in Fig. 1.15 always lead to the occurrences of the fire-off alarms and the mill sear/primary air pressure low alarms. Thus, another approach to handle alarm floods is to detect similar historical alarm floods, extract regular patterns of these similar alarm floods, and predict an upcoming alarm flood by comparing the currently occurring alarms to the regular patterns. The related methods in [6, 36, 40] cited in Sect. 1.3 detected similar alarm floods and their regular patterns solely based on historical data, whose validation is rather difficult. A more convincing conclusion may be obtained by complementing these methods with the physical connections of alarm variables. Hence, the main challenge is to obtain the process knowledge related to similar alarm floods, and transform it into a form that could be incorporated together with the statistical regularity from historical data. Another challenge is about the computational speed of algorithms in the predication of alarm floods. The modified Smith-Waterman algorithm [40] is formulated for alignment of similar alarm floods; however, the algorithm is only suitable for offline usage, due to the slow computation speed, especially for alarm floods with long sequences. Hence, the computation has to be improved greatly, in order to be fast enough for matching an occurring alarm flood to its similar ones in the historical database in an online manner.

Research problems in alarm removal

Alarm removal is mainly concerned with analyzing root causes leading to the occurrence of alarms, and advising operators to take some proper actions to avoid the deterioration of negative consequences associated with alarms, and drive the process variables back to their normal operating zones so that the occurred alarms are eventually removed. There are two fundamental research problems to be solved in this stage:

Problem 7: What are the root causes of the occurring alarms?
Problem 8: What actions should operators take to address the occurring alarms?

The very first step in solving Problem 7 is to tell whether the occurring alarms have meaningful root causes, or in other words, they belong to nuisance alarms or true ones. Nuisance alarms require no operator action or response, and their removal is one of the main objectives for the stage of alarm design. If an alarm variable always produces nuisance alarms, then the removal of these nuisance alarms is done by removing the configuration of the alarm variable. A hard online classification of nuisance and true alarms is rather difficult, but a classification using statistical inference is possible. For instance, a hypothesis test can be formulated with the following null and alternative hypotheses,

$$H_0 : X_a(t) \in \text{nuisance alarms}; \;\; H_1 : X_a(t) \notin \text{nuisance alarms}.$$

Such a hypothesis test can be based on the statistical characters of alarm and/or process variables revealed in historical data sets, e.g., the probability mass functions of time durations of alarm variables.

For true alarms, the objective of Problem 7 is to find out the occurring abnormalities as the root causes of alarms. As pointed out in Fig. 1.15, the alarm occurred first in time does not always indicate the origin of abnormalities, which certainly complicates the root cause analysis. If the relation between the abnormalities and alarm variables can be established in some manner, then the root causes could be found backwards. That is, the root causes of occurring alarms are located based on the inverse model of the relation $g(\cdot)$ in (1.2), i.e.,

$$\hat{A} = g^{-1}(X).$$

Here $\hat{A}$ is the estimate of root causes for the alarms in X. There are a large variety of representations of $g(\cdot)$ and the approaches to yield $\hat{A}$; see the references cited in Sect. 1.3. In this sense, Problems 1 and 7 share a common objective to establish the relation $g(\cdot)$ between process variables and abnormalities. Therefore, as commented for Problem 1, the limitations of process knowledge and historical operating data are the main challenges in solving Problem 7, too.

Due to the large scale and complexity of industrial processes, the above mapping in (1.2) is more related to plant topology than statistical models. It is more fundamental to locate the root causes for an event or an alarm flood. As discussed in Sect. 1.3, a local metric of nonlinearity can be used to find the most possible root causes. To obtain the abnormality propagation, causality capturing methods are necessary based on process measurements as well as alarm series. All the nonlinearity- and causality-based methods need sufficient historical data. However, for a new process without historical data, or a process that do not have so much trouble in the past, it is impractical to obtain sufficient statistical data in all abnormal situations. As a result, plant connectivity should be taken into account to describe the interior relationship between process units and process variables. In practice, the information in the historical data and the plant connectivity should be integrated to improve their efficiency and accuracy.

The objective of Problem 8 is to provide operators quantitative or qualitative advices in taking proper actions to eventually remove the occurring alarms. Brooks et al. [25] and Yu et al. [216] suggested that corrective changes of manipulated variables could be advised from the geometric process control method based on best operating zones in a multivariate framework, and these advices would be valuable to help operators to take proper actions; however, technical details on the parallel coordinate techniques and the projective geometry theory are quite complicated and difficult to understand and generalize. These advices are indispensable, especially when many process and alarm variables are involved. The multiple occurring alarms have to be ordered on the basis of their severities of consequences, in order not to deviate further from the normal operating zones and lead to aggravation or even incidents [18, 22, 171, 228]. This is closely related to Problem 2, where the main challenge is to develop quantitative approaches along with the ones solely based on the process knowledge. In terms of alarm floods, the best way perhaps is to take preventative actions in order to avoid the occurrence of alarm floods, on the basis of the alarm flood prediction that has been discussed for Problem 6.

References

1. Abele, L., Anic, M., Gutmann, T., Folmer, J., Kleinsteuber, M., & Vogel-Heuser, B. (2013). Combining knowledge modeling and machine learning for alarm root cause analysis. *The 7th IFAC Conference on Manuafacturing Modeling, Management, and Control*, June 19-21, Saint Petersburg, Russia, 1843-1848.
2. Adnan, N. A., Izadi, I., & Chen, T. (2011). On expected detection delays for alarm systems with deadbands and delay-timers. *Journal of Process Control, 21*, 1318–1331.
3. Adnan, N. A., Cheng, Y., Izadi, I., & Chen, T. (2013). Study of generalized delay-timers in alarm configuration. *Journal of Process Control, 23*, 382–395.
4. Afzal, M. S., & Chen, T. (2017). Analysis and design of multimode delay-timers. *Chemical Engineering Research and Design, 120*, 179–193.
5. Afzal, M. S., Chen, T., Bandehkhoda, A., & Izadi, I. (2018). Analysis and design of time-deadbands for univariate alarm systems. *Control Engineering Practice, 71*, 96–107.
6. Ahmed, K., Izadi, I., Chen, T., Joe, D., & Burton, T. (2013). Similarity analysis of industrial alarm flood data. *IEEE Transactions on Automation Science and Engineering, 10*, 452–457.
7. Ahnlund, J., Bergquist, T., & Spaanenburg, L. (2003). Rule-based reduction of alarm signals in industrial control. *Journal of Intelligent & Fuzzy Systems, 14*, 73–84.
8. Alinezhad, H. S., Shang, J., & Chen, T. (2022). Early classification of industrial alarm floods based on semisupervised learning. *IEEE Transactions on Industrial Informatics, 18*, 1845–1853.
9. Alinezhad, H. S., Shang, J., & Chen, T. (2023). Open set online classification of industrial alarm floods with alarm ranking. *IEEE Transactions on Instrumentation and Measurement, 72*, 1–11.
10. Alrowaie, F., Gopaluni, R.B., & Kwok, K.E. (2014). Alarm design for nonlinear stochastic systems. *The 11th World Congress on Intelligent Control and Automation*, June 29-July 4, 2014, Shenyang, China, 472-479.
11. API (2010). *API-1167: Pipeline SCADA Alarm Management*. American Petroleum Institute.
12. Arifin, B. M. S., & Choudhury, M. A. A. S. (2011). An alternative approach of risk analysis for multivariable alarm system. *Journal of Chemical Engineering, 26*, 75–79.
13. Arjomandi, R.K., & Salahshoor, K. (2011). Development of an efficient alarm management package for an industrial process plant. *2011 Chinese Control & Decision Conference*, May 23-25, 2011, Mianyang, China, 1875-1880.
14. Aslansefat, K., Gogani, M. B., Kabir, S., Shoorehdeli, M. A., & Yari, M. (2020). Performance evaluation and design for variable threshold alarm systems through semi-Markov process. *ISA Transactions, 97*, 282–295.

15. ASM (2009). *ASM Consortium Guideline: Effective Alarm Management Practice*. Abnormal Situations Management Consortium.
16. Bantay, L., Sas, N., Dorgo, G., & Abonyi, J. (2023). Sequence compression and alignment-based process alarm prediction. *Industrial & Engineering Chemistry Research, 62*, 10577–10586.
17. Barocio, E., Pal, B.C., Fabozzi, D., & Thornhill, N.F. (2013). Detection and visualization of power system disturbances using principal component analysis. *2013 IREP Symposium-Bulk Power System Dynamics and Control—IX (IREP)*, Rethymnon, Greece, Aug. 25-30, 2013, 1-10.
18. Basu, C., Das, K., Hazra, J., & Seetharam, D.P. (2013). Enhancing wide-area monitoring and control with intelligent alarm handling. *The 4th IEEE PES European Innovative Smart Grid Technologies*, Oct. 6-9, 2013, Copenhagen, 1-5.
19. Bauer, M., Cox, J. W., Caveness, M. H., Downs, J. J., & Thornhill, N. F. (2007). Finding the direction of disturbance propagation in a chemical process using transfer entropy. *IEEE Transactions on Control Systems Technology, 15*, 12–21.
20. Bauer, M., & Thornhill, N. F. (2008). A practical method for identifying the propagation path of plant-wide disturbances. *Journal of Process Control, 18*, 707–719.
21. Beebe, D., Ferrer, S., & Logerot, D. (2013). The connection of peak alarm rates to plant incidents and what you can do to minimize. *Process Safety Progress, 32*, 72–77.
22. Blaauwgeers, E., Dubois, L., & Ryckaert, L. (2013). Real-time risk estimation for better situational awareness. *The 12th IFAC Symposium on Analysis, Design, and Evaluation of Human-Machine Systems*, LA, USA, Aug. 11-15, 2013, 232-239.
23. Bransby, M.L., & Jenkinson, J. (1998). *The Management of Alarm Systems*. Health and Safety Executive.
24. Bristol, E. H. (2001). Improved process control alarm operation. *ISA Transactions, 40*, 191–205.
25. Brooks, R., Thorpe, R., & Wilson, J. (2004). A new method for defining and managing process alarms and for correcting process operation when an alarm occurs. *Journal of Hazardous Materials, 115*, 169–174.
26. Brown, D.C.C. (1998). Alarm management experience in BP oil. *IEE Colloquium on Best Practices in Alarm Management*, March 25, 1998.
27. Brown, N. (2003). Alarm management/The EEMUA guidelines in practice. *Measurement and Control, 36*, 114–119.
28. Burnell, E., & Dicken, C. R. (1997). *Handling of repeating alarms*. London: IEE Colloquium on Stemming the Alarm Flood.
29. Cai, S., Palazoglu, A., Zhang, L., & Hu, J. (2019). Process alarm prediction using deep learning and word embedding methods. *ISA Transactions, 85*, 274–283.
30. Cai, S., Zhang, L., Palazoglu, A., & Hu, J. (2019). Clustering analysis of process alarms using word embedding. *Journal of Process Control, 83*, 11–19.
31. Cauvin, S., Cordier, M. O., Dousson, C., Laborie, P., Levy, F., Montmain, J., Porcheron, M., Servet, I., & Trave-Massuyes, L. (1998). Monitoring and alarm interpretation in industrial environments. *AI Communications, 11*, 139–173.
32. Cecilioa, I. M., Ottewillb, J. R., Pretlovec, J., & Thornhill, N. F. (2014). Nearest neighbors method for detecting transient disturbances in process and electromechanical systems. *Journal of Process Control, 24*, 1382–1393.
33. Chang, Y., Khana, F., & Ahmed, S. (2011). A risk-based approach to design warning system for processing facilities. *Process Safety and Environmental Protection, 89*, 310–316.
34. Charbonnier, S., Garcia-Beltan, C., Cadet, Catherine, & Gentil, S. (2005). Trends extraction and analysis for complex system monitoring and decision support. *Engineering Applications of Artificial Intelligence, 18*, 21–36.
35. Charbonnier, S., & Portet, F. (2012). A self-tuning adaptive trend extraction method for process monitoring and diagnosis. *Journal of Process Control, 22*, 1127–1138.
36. Charbonnier, S., Bouchair, N., & Gayet, P. (2014). Fault isolation by comparing alarm lists using a symbolic sequence matching algorithm. *The 19th IFAC World Congress*, Cape Town, South Africa, Aug. 24-29, 2014, 7085-7090.

37. Chen, K., & Wang, J. (2017). Normal and abnormal data segmentation based on variational directions and clustering algorithms. *Industrial & Engineering Chemistry Research, 56*, 7799–7813.
38. Chen, K., & Wang, J. (2017). Design of multivariate alarm systems based on online calculation of variational directions. *Chemical Engineering Research and Design, 122*, 11–21.
39. Cheng, Y., Izadi, I., & Chen, T. (2013). Optimal alarm signal processing: Filter design and performance analysis. *IEEE Transactions on Automation Science and Engineering, 10*, 446–451.
40. Cheng, Y., Izadi, I., & Chen, T. (2013). Pattern matching of alarm flood sequences by a modified Smith-Waterman algorithm. *Chemical Engineering Research and Design, 91*, 1085–1094.
41. Cheon, S. W., Chang, S. H., & Chung, H. Y. (1993). Development strategies of an expert system for multiple alarm processing and diagnosis in nuclear power plants. *IEEE Transactions on Nuclear Science, 40*, 21–30.
42. Chiang, L. H., & Braatz, R. D. (2003). Process monitoring using causal map and multivariate statistics: fault detection and identification. *Chemometrics and Intelligent Laboratory Systems, 65*, 159–178.
43. Dahlgren, R., Rosenwald, G., Liu, C. C., Muchlinski, S., Eide, A., & Sobajic, D. (1998). Model-based synthesis and suppression of transformer alarms in a control center environment. *IEEE Transactions on Power Delivery, 13*, 843–848.
44. Dalapatu, P., Ahmed, S., & Khan, F. (2013). Alarm allocation for event-based process alarm systems. *The 10th IFAC International Symposium on Dynamics and Control of Process Systems*, December 18-20, 2013, Mumbai, India, 815-820.
45. Dashlstrand, F. (2002). Consequence analysis theory for alarm analysis. *Knowledge-Based Systems, 15*, 27–36.
46. de Abreu, R., Nunes, Y., Guedes, L., & Silva, I. (2021). A method for detecting causal relationships between industrial alarm variables using Transfer Entropy and K2 algorithm. *Journal of Process Control, 106*, 142–154.
47. Di Geronimo Gil, G.J., Alabi, D.B., Iyun, O.E., & Thornhill, N.F. (2011). Merging process models and plant topology. *Advanced Control of Industrial Processes (ADCONIP 2011)*, May 23-26, 2011, Hangzhou, China,
48. Dorgo, G., Palazoglu, A., & Abonyi, J. (2021). Decision trees for informative process alarm definition and alarm-based fault classification. *Process Safety and Environmental Protection, 149*, 312–324.
49. Dorgo, G., Tandari, F., Szabó, T., Palazoglu, A., & Abonyi, J. (2021). Quality versus quantity of alarm messages-How to measure the performance of an alarm system. *Chemical Engineering Research and Design, 173*, 63–80.
50. Duan, P., Yang, F., Chen, T., & Shah, S. L. (2013). Direct causality detection via the transfer entropy approach. *IEEE Transactions on Control Systems Technology, 21*, 2052–2066.
51. Duan, P., Chen, T., Shah, S. L., & Yang, F. (2014). Methods for root cause diagnosis of plant-wide oscillations. *AIChE Journal, 60*, 2019–2034.
52. Duan, P., Chen, T., Shah, S. L., & Yang, F. (2015). Transfer zero-entropy and its application for capturing cause and effect relationship between variables. *IEEE Transactions on Control Systems Technology, 23*, 855–867.
53. Dubois,L., Forêt, J.M., Mack, P., & Ryckaert, L. (2010). Advanced logic for alarm and event processing: Methods to reduce cognitive load for control room operators. *The 11th IFAC Symposium on Analysis, Design, and Evaluation of Human-Machine Systems*, Aug. 31-Sep. 3, 2010, Valenciennes, France.
54. EEMUA. (2013). *EEMUA-191: Alarm Systems-A Guide to Design*. Management and Procurement: Engineering Equipment and Materials Users' Association.
55. EPRI (2005). *EPRI-1010076: Advanced Control Room Alarm System: Requirements and Implementation Guidance*. The Electric Power Research Institute.
56. Folmer, J., Pantforder, D., & Vogel-Heuser, B. (2011). An analytical alarm flood reduction to reduce operator's workload. Human-Computer Interactions, Part IV, HCII 2011. *LNCS, 6764*, 297–306.

57. Folmer, J., & Vogel-Heuser, B. (2012). Computing dependent industrial alarms for alarm flood reduction. *The 9th International Multi-Conference on Systems, Signals and Devices*, March 20-23, 2012, 1-6.
58. Folmer, J., Schuricht, F., & Vogel-Heuser, B. (2014). Detection of temporal dependencies in alarm time series of industrial plants. *The 19th IFAC World Congress*, Aug. 24-29, 2014, Cape Town, South Africa, 1802-1807.
59. Gao, H., Liu, F., & Zhu, Q. (2016). A correlation consistency based multivariate alarm thresholds optimization approach. *ISA Transactions, 65*, 37–43.
60. Gao, H., Xu, Y., & Zhu, Q. (2016). Spatial interpretive structural model identification and AHP-based multimodule fusion for alarm root-cause diagnosis in chemical processes. *Industrial & Engineering Chemistry Research, 55*, 3641–3658.
61. Gao, H., Wei, C., Huang, W., & Gao, X. (2021). Design of multivariate alarm trippoints for industrial processes based on causal model. *Industrial & Engineering Chemistry Research, 60*, 9128–9140.
62. Geng, Z., Zhu, Q., & Gu, X. (2005). A fuzzy clustering-ranking algorithm and its application for alarm operating optimization in chemical processing. *Process Safety Progress, 24*, 66–75.
63. Gigi, S., & Tangirala, A. K. (2010). Quantitative analysis of directional strengths in jointly stationary linear multivariate processes. *Biological Cybernetics, 103*, 119–133.
64. Goel, P., Datta, A., & Mannan, M. S. (2017). Industrial alarm systems: Challenges and opportunities. *Journal of Loss Prevention in the Process Industries, 50*, 23–36.
65. Goel, P., Pistikopoulos, E. N., Mannan, M. S., & Datta, A. (2019). A data-driven alarm and event management framework. *Journal of Loss Prevention in the Process Industries, 62*, 103959.
66. Guo, C., Hu, W., Lai, S., Yang, F., & Chen, T. (2017). An accelerated alignment method for analyzing time sequences of industrial alarm floods. *Journal of Process Control, 57*, 102–115.
67. Guo, W., Wen, F., Liao, Z., Wei, L., & Xin, J. (2010). An analytic model-based approach for power system alarm processing employing temporal constraint network. *IEEE Transactions on Power Delivery, 25*, 2435–2447.
68. Gupta, A., Giridhar, A., Venkatasubramanian, V., & Reklaitis, G. V. (2013). Intelligent alarm management applied o continuous pharmaceutical tablet manufacturing: an integrated approach. *Industrial & Engineering Chemistry Research, 52*, 12357–12368.
69. Gyasi, P., & Wang, J. (2022). Design of serial alarm systems based on deadbands and delay timers for removing false alarms. *Process Safety and Environmental Protection, 162*, 1033–1041.
70. Gyasi, P., & Wang, J. (2023). Design of alarm thresholds and delay timers for non-IID process variables based on alarm durations. *Process Safety and Environmental Protection, 170*, 1173–1187.
71. Gyasi, P., & Wang, J. (2023). Parallel-connected timers and deadbands for reducing the number of false and missed alarms. *Control Engineering Practice, 134*, 105478.
72. Gyasi, P., Wang, J., Yang, F., & Izadi, I. (2023). An adaptive method to update alarm deadbands for non-stationary process variables. *Process Safety and Environmental Protection, 179*, 493–502.
73. Han, L., Gao, H., Xu, Y., & Zhu, Q. (2016). Combining FAP, MAP and correlation analysis for multivariate alarm thresholds optimization in industrial process. *Journal of Loss Prevention in the Process Industries, 40*, 471–478.
74. He, Y., Zhao, Y., Zhu, Q., & Xu, Y. (2020). Online distributed process monitoring and alarm analysis using novel canonical variate analysis with multicorrelation blocks and enhanced contribution plot. *Industrial & Engineering Chemistry Research, 59*, 20045–20057.
75. Henningsen, A., & Kemmerer, J. P. (1995). Intelligent alarm handling in cement plants. *IEEE Industry Applications Magazine, 1*, 9–15.
76. Hollender, M., & Beuthel, C. (2007). Intelligent alarming. *ABB Review, Feburary, 2007*, 20–23.
77. Hollifield, B., & Habibi, E. (2010). *The Alarm Management Handbook* (2nd ed.). PAS: Houston.

78. HSE (1997). *The explosion and fires at the Texaco Refinery, Milford Haven, 24 July 1994: A report of the investigation by the Health and Safety Executive into the explosion and fires on the Pembroke Cracking Company Plant at the Texaco Refinery, Milford Haven on 24 July 1994*. Health and Safety Executive.
79. HSE (2008). *The final report of the Major Incident Investigation Board*. Health and Safety Executive.
80. Hu, J., & Yi, Y. (2016). A two-level intelligent alarm management framework for process safety. *Safety Science, 82*, 432–444.
81. Hu, W., Wang, J., & Chen, T. (2016). A local alignment approach to similarity analysis of industrial alarm flood sequences. *Control Engineering Practice, 55*, 13–25.
82. Hu, W., Wang, J., Chen, T., & Shah, S. L. (2017). Cause-effect analysis of industrial alarm variables using transfer entropies. *Control Engineering Practice, 64*, 205–214.
83. Hu, W., Chen, T., & Shah, S. L. (2018). Discovering association rules of mode-dependent alarms from alarm and event logs. *IEEE Transactions on Control Systems Technology, 26*, 971–983.
84. Hu, W., Chen, T., & Shah, S. L. (2018). Detection of frequent alarm patterns in industrial alarm floods using itemset mining methods. *IEEE Transactions on Industrial Electronics, 65*, 7290–7300.
85. Hu, W., Al-Dabbagh, A. W., Chen, T., & Shah, S. L. (2018). Design of visualization plots of industrial alarm and event data for enhanced alarm management. *Control Engineering Practice, 79*, 50–64.
86. Hu, W., Wang, J., Yang, F., Han, B., & Wang, Z. (2022). Analysis of time-varying cause-effect relations based on qualitative trends and change amplitudes. *Computers & Chemical Engineering, 162*, 107813.
87. Hu, W., Zhang, X., Wang, J., Yang, G., & Cai, Y. (2023). Pattern matching of industrial alarm floods using word embedding and dynamic time warping. *IEEE/CAA Journal of Automatica Sinica, 10*, 1096–1098.
88. Hu, W., Wang, Z., & Wang, J. (2023). A priority-aware sequential pattern mining method for detection of compact patterns from alarm floods. *Journal of Process Control, 129*, 103041.
89. Hugo, A.J. (2009). Estimation of alarm deadbands. *The 7th IFAC Symposium on Fault Detection, Supervision and Safety of Technical Processes*, June 30-July 3, 2009, Barcelona, Spain, 663-667.
90. Hwang, S. L., Lin, J. T., Liang, G. F., Yau, Y. J., Yenn, T. C., & Hsu, C. C. (2008). Application control chart concepts of designing a pre-alarm system in the nuclear power plant control room. *Nuclear Engineering and Design, 238*, 3522–3527.
91. IEC (2014). *IEC-68682: Management of Alarm Systems for the Process Industries*. International Electrotechnical Commission.
92. ISA (2009). *ANSI/ISA-18.2: Management of Alarm Systems for the Process Industries*. International Society of Automation.
93. Isermann (2011). *Fault-Diagnosis Applications: Model-bassed Condition Monitoring*. Springer.
94. Izadi, I., Shah, S.L., Shook, D., & Chen, T. (2009) An introduction to alarm analysis and design, *The 7th IFAC Symposium on Fault Detection, Supervision and Safty of Technical Processes*, June 30-July 3, 2009, Barcelona, Spain, 645-650.
95. Jang, G., Suh, S., Kim, S., Suh, Y., & Park, J. (2013). A proactive alarm reduction method and its human factors validation test for a main control room for SMART. *Annals of Nuclear Energy, 51*, 125–134.
96. Jiang, H., Patwardhan, R., & Shah, S. L. (2009). Root cause diagnosis of plant-wide oscillations using the concept of adjacency matrix. *Journal of Process Control, 19*, 1347–1354.
97. Kaced, R., Kouadri, A., & Baiche, K. (2019). Designing alarm system using modified generalized delay-timer. *Journal of Loss Prevention in the Process Industries, 61*, 40–48.
98. Kaced, R., Kouadri, A., Baiche, K., & Bensmail, A. (2021). Multivariate nuisance alarm management in chemical processes. *Journal of Loss Prevention in the Process Industries, 72*, 104548.

99. Kezunovic, M., & Guan, Y. (2009). Intelligent alarm processing: from data intensive to information rich. *The 42nd Hawaii International Conference on System Sciences*, Jan. 5-8, 2009, HI, USA, 1-8.
100. Kim, I. S. (1994). Computerized systems for online management of failures: A state-of-the-art discussion of alarm systems and diagnostic systems applied in the nuclear industry. *Reliability Engineering & System Safety, 44*, 279–295.
101. Kim, Y., & Kim, S. B. (2017). Optimal false alarm controlled support vector data description for multivariate process monitoring. *Journal of Process Control, 32*, 72–77.
102. Kirschen, D. S., & Wollenberg, B. F. (1992). Intelligent alarm processing in power systems. *Proceedings of the IEEE, 80*, 663–672.
103. Kondaveeti, S.R., Shah, S.L., & Izadi, I. (2009). Application of multivariate statistics for efficient alarm generation. *The 7th IFAC Symposium on Fault Detection, Supervision and Safety of Technical Processes*, June 30-July 3, 2009, Barcelona, Spain, 657-662.
104. Kondaveeti, S. R., Izadi, I., Shah, S. L., Black, T., & Chen, T. (2012). Graphical tools for routine assessment of industrial alarm systems. *Computers and Chemical Engineering, 46*, 39–47.
105. Kondaveeti, S. R., Izadi, I., Shah, S. L., Shook, D., Kadali, R., & Chen, T. (2013). Quantification of alarm chatter based on run length distributions. *Chemical Engineering Research and Design, 91*, 2550–2558.
106. Kordic, S., Lam, C. P., Xiao, J., & Li, H. (2009). Using event-intervals to analyze alarm sequences in a chemical plant. *Journal of Frontiers of Computer Science and Technology, 3*, 267–281.
107. Laberge, J. C., Bullermer, P., Tolsma, M., & Reising, D. V. C. (2014). Addressing alarm flood situations in the process industries through alarm summary display design and alarm response strategy. *International Journal of Industrial Ergonomics, 44*, 395–406.
108. Lai, S., & Chen, T. (2017). A method for pattern mining in multiple alarm flood sequences. *Chemical Engineering Research and Design, 117*, 831–839.
109. Lai, S., Yang, F., & Chen, T. (2017). Online pattern matching and prediction of incoming alarm floods. *Journal of Process Control, 56*, 69–78.
110. Lai, S., Yang, F., Chen, T., & Cao, L. (2019). Accelerated multiple alarm flood sequence alignment for abnormality pattern mining. *Journal of Process Control, 82*, 44–57.
111. Landman, R., Kortela, J., Sun, Q., & J ämsä-Jounela, S.-L. (2014). Fault propagation analysis of oscillations in control loops using data-driven causality and plant connectivity. *Computers and Chemical Engineering, 71*, 446–456.
112. Larsson, J.E., Ohman, B., Calzada, A., Nihlwing, C., Jokstad, H. Kristianssen, L.I., Kvalem, J., & Lind, M. (2006). A revival of the alarm system: making the alarm list useful during incidents. *The 5th International Topical Meeting on Nuclear Plant Instrumentation Controls, and Human Machine Interface Technology*, Nov. 12-16, 2006, Albuquerque, NM, 711-716.
113. Lee, D., Lai, C., Liao, K., & Chang, J. (2021). Artificial intelligence assisted false alarm detection and diagnosis system development for reducing maintenance cost of chillers at the data center. *Journal of Building Engineering, 36*, 102110.
114. Lee, J. W., Kim, J. T., Park, J. C., Hwang, I. K., & Lyu, S. P. (2010). Computer-based alarm processing and presentation methods in nuclear power plants. *World Academy of Science, Engineering and Technology, 41*, 594–598.
115. Li, Y., Cao, W., Gopaluni, R. B., Hu, W., Cao, L., & Wu, M. (2023). False alarm reduction in drilling process monitoring using virtual sample generation and qualitative trend analysis. *Control Engineering Practice, 133*, 105457.
116. Liu, J., Lim, K. W., Ho, W. K., Tan, K. C., Srinivasan, R., & Tay, A. (2003). The intelligent alarm management system. *IEEE Software, 20*, 66–71.
117. Liu, X., Noda, M., & Nishitani, H. (2010). Evaluation of plant alarm systems by behavior simulation using a virtual subject. *Computers and Chemical Engineering, 34*, 374–386.
118. Liu, Y., Wei, H., Zhu, M., Shi, X., Zhu, J., Wei, Z. (2022). *Management of Alarm Systems for the Process Industries*. China Machinery Industry Federation, GB/T 41261-2022, Mar. 9, 2022.

119. Luo, Y., Gopaluni, B., Cao, L., Wang, Y., & Cheng, J. (2023). Adaptive online optimization of alarm thresholds using multilayer Bayesian networks and active transfer entropy. *Control Engineering Practice, 137*, 105534.
120. Macdonald, D. (2004). *Practical Hazops*. Trips and Alarms: Elsevier.
121. Mannani, Z., Izadi, I., & Ghadiri, N. (2019). Preprocessing of alarm data for data mining. *Industrial & Engineering Chemistry Research, 58*, 11261–11274.
122. Maurya, M. R., Rengaswamy, R., & Venkatasubramanian, V. (2003). A systematic framework for the development and analysis of signed digraphs for chemical processes. 1. algorithms and analysis. *Industrial & Engineering Chemistry Research, 42*, 4789–4810.
123. Maurya, M. R., Rengaswamy, R., & Venkatasubramanian, V. (2007). A signed directed graph and qualitative trend analysis-based framework for incipient fault diagnosis. *Chemical Engineering Research & Design, 85*, 1407–1422.
124. McDonald, J. R., Burt, G. M., & Young, D. J. (1992). Alarm processing and fault diagnosis using knowledge based systems for transmission and distribution network control. *IEEE Transactions on Power Systems, 7*, 1292–1298.
125. Meng, Q., Zhu, Q., Gao, H., He, Y., & Xu, Y. (2019). A novel scoring function based on family transfer entropy for Bayesian networks learning and its application to industrial alarm systems. *Journal of Process Control, 76*, 122–132.
126. Miao, H., Sforna, M., & Liu, C. C. (1996). A new logic-based alarm analyzer for online operational environment. *IEEE Transactions on Power Systems, 11*, 1600–1606.
127. Naghoosi, E., Izadi, I., & Chen, T. (2011). Estimation of alarm chattering. *Journal of Process Control, 21*, 1243–1249.
128. NAMUR (2003). *NAMUR-NA-102: Alarm Management*. Standardization Association for Measurement and Control in Chemical Industries.
129. Nihlwing, C., & Kaarstad, M. (2012). The development and usability test of a state based alarm system for a nuclear power plant simulator. *The NPIC & HMIT 2012*, July 22-26, San Diego, USA.
130. Nimmo, I. (2005). Rescue your plant from alarm overload. *Chemical Processing*, 28-33.
131. Niyazmand, T., & Izadi, I. (2019). Pattern mining in alarm flood sequences using a modified PrefixSpan algorithm. *ISA Transactions, 90*, 287–293.
132. Noda, M., Higuchi, F., Takai, T., & Nishitani, H. (2011). Event correlation analysis for alarm system rationalization. *Asia-Pacific Journal of Chemical Engineering, 6*, 497–502.
133. NRC (2000). *NUREG/CR-61056684: Advanced Alarm Systems: Revision of Guidance and Its Technical Basis*. Nuclear Regulatory Commission.
134. Pariyani, A., Seider, W. D., Oktem, U. G., & Soroush, M. (2010). Incidents investigation and dynamic analysis of large alarm databases in chemical plants: A fluidized-catalytic-cracking unit case study. *Industrial & Engineering Chemistry Research, 49*, 8062–8079.
135. Parvez, M., Hu, W., & Chen, T. (2022). Real-time pattern matching and ranking for early prediction of industrial alarm floods. *Control Engineering Practice, 120*, 105004.
136. Raei, R., Izadi, I., & Kamali, M. (2023). Performance analysis of up/down counters in alarm design. *Process Safety and Environmental Protection, 170*, 877–885.
137. Rao, H. R. M., Zhou, B., Brown, K., Chen, T., & Shah, S. L. (2024). Alarm correlation analysis with applications to industrial alarm management. *Control Engineering Practice, 143*, 105812.
138. Roohi, M. H., Chen, T., & Izadi, I. (2020). Control and alarm interplay and robust state-feedback synthesis with an alarm performance constraint. *Industrial & Engineering Chemistry Research, 59*, 16708–16719.
139. Rothenberg, D. (2009). *Alarm Management for Process Control*, Momentum Press.
140. Satuf, E. N., Kaszkurewicz, E., Schirru, R., & de Campos, M. C. M. M. (2016). Situation awareness measurement of an ecological interface designed to operator support during alarm floods. *International Journal of Industrial Ergonomics, 53*, 179–192.
141. Schirru, R., Schirru, M., Boas, M. V. B., Nicolau, A. D., de Lima, A. M. M., & Pinheiro, V. H. C. (2020). Determination of alarm templates for decision support in nuclear power plants alarm floods using evolutionary computation. *Progress in Nuclear Energy, 123*, 103308.

142. Schleburg, M., Christiansen, L., Thornhill, N. F., & Fay, A. (2013). A combined analysis of plant connectivity and alarm logs to reduce the number of alerts in an automation system. *Journal of Process Control, 23*, 839–851.
143. Shang, J., & Chen, T. (2020). Early classification of alarm floods via exponentially attenuated component analysis. *IEEE Transactions on Industrial Electronics, 67*, 8702–8712.
144. Shirshahi, A., & Aliyari-Shoorehdeli, M. (2024). Diagnosing root causes of faults based on alarm flood classification using transfer entropy and multi-sensor fusion approaches. *Process Safety and Environmental Protection, 181*, 469–479.
145. Simeu-Abazi, Z., Lefebvre, A., & Derain, J. P. (2011). A methodology of alarm filtering using dynamic fault tree. *Reliability Engineering & System Safety, 96*, 257–266.
146. Soares, V. B., Pinto, J. C., & de Souza, M. B. (2016). Alarm management practices in natural gas processing plants. *Control Engineering Practice, 55*, 185–196.
147. Sompura, J., Joshi, A., Srinivasan, B., & Srinivasan, R. (2019). A practical approach to improve alarm system performance: Application to power plant. *Chinese Journal of Chemical Engineering, 27*, 1094–1102.
148. Souza, J. C. S., Meza, E. M., Schilling, M. T., & Filho, M. B. D. C. (2004). Alarm processing in electrical power systems through a neuro-fuzzy approach. *IEEE Transactions on Power Delivery, 19*, 537–544.
149. Srinivasan, R., Liu, J., Lim, K. W., Tan, K. C., & Ho, W. K. (2004). Intelligent alarm management in a petroleum refinery. *Hydrocarbon Processing, 11*, 47–52.
150. Stauffer, T., & Clarke, P. (2012). Using alarms as a layer of protection. *The 8th Global Congress on Process Safety*, April 1-4, 2012, Houston, USA.
151. Su, J., Guo, C., Zang, H., Yang, F., Huang, D., & Gao, X. (2018). A multi-setpoint delay-timer alarming strategy for industrial alarm monitoring. *Journal of Loss Prevention in the Process Industrie, 54*, 1–9.
152. Sudarshan, V., Seider, W. D., Patel, A. J., Oktem, U. G., & Arbogast, J. E. (2024). Multivariate alarm systems to recognize rare unpostulated abnormal events. *AIChE Journal, 70*, e18284.
153. Sudarshan, V., Seider, W. D., Patel, A. J., Oktem, U. G., & Arbogast, J. E. (2024). Alarm rationalization and dynamic risk analyses for rare abnormal events. *Computers & Chemical Engineering, 184*, 108633.
154. Taheri-Kalani, J., Latif-Shabgahi, G., & Shooredeli, M. A. (2018). On the use of penalty approach for design and analysis of univariate alarm systems. *Journal of Process Control, 69*, 103–113.
155. Takeda, K., Hamaguchi, T., Noda, M., Kimura, N., & Itoh, T. (2010). Use of two-layer cause-effect model to select source of signal in plant alarm system. *in R.Setchi et al. (Eds.): KES2010, Part II, LNAI 6277*, 381-388.
156. Takeda, K., Hamaguchi, T., Kimura, N., & Noda, M. (2013). A method of designing plant alarm system based on first alarm alternative signals for each assumed plant malfunction. *Proc. the 6th International Conference on Process System Engineering (PSE ASIA)*, June 25-27, 2013, Kuala Lumpur.
157. Tan, W., Sun, Y., Azad, I. I., & Chen, T. (2017). Design of univariate alarm systems via rank order filters. *Control Engineering Practice, 59*, 55–63.
158. Tchamova, A., & Dezert, J. (2012). Intelligent alarm classification based on DSmT. *The 6th IEEE International Conference on Intelligent Systems*, Sept. 6-8, 2012, Sofia, Bulgaria, 120-125.
159. Thambirajah, J., Benabbas, L., Bauer, M., & Thornhill, N. F. (2009). Cause-and-effect analysis in chemical processes utilizing XML, plant connectivity and quantitative process history. *Computers and Chemical Engineering, 33*, 503–512.
160. Thornhill, N. F., Cox, J. W., & Paulonis, M. A. (2003). Diagnosis of plant-wide oscillation through data-driven analysis and process understanding. *Control Engineering Practice, 11*, 1491–1490.
161. Thornhill, N. F. (2005). Finding the source of nonlinearity in a process with plant-wide oscillation. *IEEE Transactions on Control Systems Technology, 13*, 434–443.

162. Tian, W., Zhang, G., & Liang, H. (2018). Alarm clustering analysis and ACO based multivariable alarms thresholds optimization in chemical processes. *Process Safety and Environmental Protection, 113*, 132–140.
163. Tian, W., Zhang, G., Zhang, X., & Dong, Y. (2018). PCA weight and Johnson transformation based alarm threshold optimization in chemical processes. *Chinese Journal of Chemical Engineering, 26*, 1653–1661.
164. Timms, C. (1999). A methodology for alarm classification and prioritisation. *International Conference on Human Interfaces in Control Rooms, Cockpits and Command Centers*, June 21-23, 1999, Bath, U.K., 222-228.
165. Timms, C. (2009). Hazards equal trips or alarms or both. *Process Safety and Environment Protection, 87*, 3–13.
166. Tolga, U. S., Niels, J., Morten, L., & Bay, J. S. (2011). Fundamental principles of alarm design. *Nuclear Safety and Simulations, 2*, 44–51.
167. Toshkova, D., Asher, M., Hutchinson, P., & Lieven, N. (2020). Automatic alarm setup using extreme value theory. *Mechanical Systems and Signal Processing, 139*, 106417.
168. Tulsyan, A., Alrowaie, F., & Gopaluni, R. B. (2018). Design and assessment of delay timer alarm systems for nonlinear chemical processes. *AIChe Journal, 64*, 77–90.
169. Tulsyan, A., & Gopaluni, R. B. (2019). Univariate model-based deadband alarm design for nonlinear processes. *Industrial & Engineering Chemistry Research, 58*, 11295–11302.
170. Valipoori, A., Latif-Shabgahi, G., & Izadi, I. (2021). PDF-based technique for univariate alarm systems and its application to mixture distributions. *Industrial & Engineering Chemistry Research, 60*, 17626–17639.
171. Varga, T., Szeifert, F., & Abonyi, J. (2010). Detection of safe operating regions: a novel dynamic process simulator based predictive alarm management approach. *Industrial & Engineering Chemistry Research, 49*, 658–668.
172. Venkidasalapathy, J. A., Mannan, M. S., & Kravaris, C. (2018). A quantitative approach for optimal alarm identification. *Journal of Loss Prevention in the Process Industries, 55*, 213–222.
173. Venkidasalapathy, J. A., & Kravaris, C. (2021). Hidden Markov model based approach for diagnosing cause of alarm signals. *AIChE J, 67*, e17297.
174. Venkidasalapathy, J. A., & Kravaris, C. (2022). Hidden Markov model based fault diagnoser using binary alarm signals with an analysis on distinguishability. *Computers & Chemical Engineering, 160*, 107689.
175. Wang, H., Khan, F., & Abimbola, M. (2018). A new method to study the performance of safety alarm system in process operations. *Journal of Loss Prevention in the Process Industries, 56*, 104–118.
176. Wang, J., & Chen, T. (2013). An online method for detection and reduction of chattering alarms due to oscillation. *Computers and Chemical Engineering, 54*, 140–150.
177. Wang, J., & Chen, T. (2014). An online method to remove chattering and repeating alarms based on alarm durations and intervals. *Computers and Chemical Engineering, 67*, 43–52.
178. Wang, J., Li, H., Huang, J., & Su, C. (2015). A data similarity based analysis to consequential alarms of industrial processes. *Journal of Loss Prevention in the Process Industries, 35*, 29–34.
179. Wang, J., Yang, F., Chen, T., & Shah, S. L. (2016). An overview of industrial alarm systems: main causes for alarm overloading, research status, and open problems. *IEEE Transactions on Automation Science and Engineering, 13*, 1045–1061.
180. Wang, J., & Chen, T. (2016). Main causes of long-standing alarms and their removal by dynamic state-based alarm systems. *Journal of Loss Prevention in the Process Industries, 43*, 106–119.
181. Wang, J., Li, H., Huang, J., & Su, C. (2016). Association rules mining based analysis of consequential alarm sequences in chemical processes. *Journal of Loss Prevention in the Process Industries, 41*, 178–185.
182. Wang, J., Zhao, Y., & Bi, Z. (2018). Criteria and algorithms for online and offline detections of industrial alarm floods. *IEEE Transactions on Control Systems Technology, 26*, 1722–1731.

183. Wang, J., Yang, Z., Chen, K., & Zhou, D. (2017). Practices of detecting and removing nuisance alarms for alarm overloading in thermal power plants. *Control Engineering Practice, 67*, 21–30.
184. Wang, J., Yang, Z., Su, J., Zhao, Y., Gao, S., Pang, X., & Zhou, D. (2018). Root-cause analysis of occurring alarms in thermal power plants based on Bayesian networks. *International Journal of Electrical Power and Energy Systems, 103*, 67–74.
185. Wang, J., Wang, Z., Zhou, D., Yang, Z., Yang, F., et al. (2021). Guidelines for detection and clearance of nuisance alarms in industrial alarm systems for process industries. *China International Association for Promotion of Science and Technology*, T/CI020-2021, Dec. 10, 2021.
186. Wang, J., Wang, Z., Zhou, X., & Yang, F. (2022). Design of delay timers based on estimated probability mass functions of alarm durations. *Journal of Process Control, 110*, 154–165.
187. Wang, J., Sun, S., Wang, Z., Zhou, X., & Gyasi, P. (2023). Alarm deadband design based on maximum amplitude deviations and Bayesian estimation. *IEEE Transactions on Control Systems Technology, 31*, 1941–1948.
188. Wang, Y., Li, J., Yang, B., & Li, H. (2022). Stream-data-clustering based adaptive alarm threshold setting approaches for industrial processes with multiple operating conditions. *ISA Transactions, 129*, 594–608.
189. Wang, Z., Bai, X., Wang, J., & Yang, Z. (2019). Indexing and designing deadbands for industrial alarm signals. *IEEE Transactions on Industrial Electronics, 66*, 8093–8103.
190. Wang, Z., & Wang, J. (2020). Data classification and performance evaluation for the most commonly-used univariate alarm systems. *Journal of Loss Prevention in the Process Industries, 66*, 104208.
191. Wang, Z., & Wang, J. (2022). Analytical graphs to describe operating status of industrial alarm variables. *Control Engineering Practice, 118*, 104961.
192. Wang, Z., Wang, J., & Hou, J. (2024). Multivariate alarm monitoring for non-convex normal operating zones based on search cones. *IEEE Transactions on Automation Science and Engineering, 21*, 452–462.
193. Wei, L., Guo, W., Wen, F., Ledwich, G., Liao, Z., & Xin, J. (2011). An online intelligent alarm-processing system for digital substation. *IEEE Transactions on Power Delivery, 26*, 1615–1624.
194. Wen, F., & Chang, C. S. (1997). Tabu search approach to alarm processing in power systems. *IEE Proc-Gener. Transm. Distrib., 144*, 31–38.
195. Wen, F., Chang, C. S., & Fu, W. (1998). New approach to alarm processing in power systems based upon the set covering theory and a refined genetic algorithm. *Journal of Electric Machines & Power Systems, 26*, 53–67.
196. Weng, X., Xu, X., Bai, Y., Ma, F., Wang, G., & Dustdar, S. (2021). A data-driven industrial alarm decision method via evidence reasoning rule. *Journal of Process Control, 105*, 15–26.
197. Xiong, W., Wang, J., & Chen, K. (2018). Multivariate alarm systems for time-varying processes using Bayesian filters with applications to electrical pumps. *IEEE Transactions on Industrial Informatics, 14*, 504–513.
198. Xu, J., Wang, J., Izadi, I., & Chen, T. (2012). Performance assessment and design for univariate alarm systems based on FAR, MAR and AAD. *IEEE Transactions on Automation Science & Engineering, 9*, 296–307.
199. Xu, S., Adhitya, A., & Srinivasan, R. (2014). Hybrid mode-based framework for alarm anticipation. *Industrial & Engineering Chemistry Research, 53*, 5182–5193.
200. Xu, X., Li, S., Song, X., Wen, C., & Xu, D. (2016). The optimal design of industrial alarm systems based on evidence theory. *Control Engineering Practice, 46*, 142–156.
201. Xu, X., Weng, X., Xu, D., Xu, H., Hu, Y., & Li, J. (2020). Evidence updating with static and dynamical performance analyses for industrial alarm system design. *ISA Transactions, 99*, 110–122.
202. Xu, Y., Wang, J., & Yu, Y. (2020). Alarm event prediction from historical alarm flood sequences based on Bayesian estimators. *IEEE Transactions on Automation Science and Engineering, 17*, 1070–1075.

203. Xu, Y., & Wang, J. (2021). A maximum-entropy-based method for alarm flood prediction. *Journal of Process Control, 107*, 58–69.
204. Yamanaka, F., & Nishiya, T. (1997). Application of the intelligent alarm system for the plant operation. *Computers and Chemical Engineering, 21*, 625–630.
205. Yan, L., Liu, X., Noda, M., & Nishitani, H. (2007). Systematic design approach for plant alarm systems. *Journal of Chemical Engineering of Japan, 40*, 765–772.
206. Yang, F., Xiao, D., & Shah, S. L. (2009). Optimal sensor location design for reliable fault detection in presence of false alarms. *Sensor, 9*, 8579–8592.
207. Yang, F., Shah, S.L., & Xiao, D. (2010). Correlation analysis of alarm data and alarm limit design for industrial proceses. *2010 American Control Conference*, June 30-July 2, 2010, Baltimore, USA, 5850-5855.
208. Yang, F., Shah, S. L., Xiao, D., & Chen, T. (2012). Improved correlation analysis and visualization of industrial alarm data. *ISA Transactions, 51*(499–506), 2012.
209. Yang, F., Shah, S. L., & Xiao, D. (2012). Signed directed graph based modeling and its validation from process knowledge and process data. *International Journal of Applied Mathematis and Computer Science, 22*, 387–392.
210. Yang, F., Duan, P., Shah, S.L., & Chen, T. (2014). *Capturing Connectivity and Causality in Complex Industrial Processes*, Springer.
211. Yang, Z., Wang, J., & Chen, T. (2013). Detection of correlated alarms based on similarity coefficients of binary data. *IEEE Transactions on Automation Science and Engineering, 10*, 1014–1025.
212. Yim, S. Y., Ananthakumar, H. G., Benabbas, L., Horch, A., Drath, R., & Thornhill, N. F. (2006). Using process topology in plant-wide control loop performance assessment. *Computers and Chemical Engineering, 31*, 86–99.
213. Young, D. J., Lo, K. L., McDonald, J. R., Howard, R., & Rye, J. (1992). Development of a practical expert system for alarm processing. *IEE Proceedings-C, 139*, 437–447.
214. Yu, Y., Wang, J., & Yang, Z. (2017). Design of alarm trippoints for univariate analog process variables based on alarm probability plots. *IEEE Transactions on Industrial Electronics, 64*, 6496–6505.
215. Yu, Y., Zhu, D., Wang, J., & Zhao, Y. (2017). Abnormal data detection for multivariate alarm systems based on correlation directions. *Journal of Loss Prevention in the Process Industries, 45*, 43–55.
216. Yu, Y., Wang, J., & Ouyang, Z. (2020). Designing dynamic alarm limits and adjusting manipulated variables for multivariate systems. *IEEE Transactions on Industrial Electronics, 67*, 2314–2325.
217. Yu, Y., & Wang, J. (2020). Alarm monitoring for multivariate processes based on a convex-hull normal operating zone. *IEEE Transactions on Control Systems Technology, 28*, 2649–2656.
218. Yuan, T., & Qin, S. J. (2014). Root cause diagnosis of plant-wide oscillations using Granger causality. *Journal of Process Control, 24*, 450–459.
219. Yuki, Y. (2002). Alarm system optimization for increasing operations productivity. *ISA Transactions, 41*, 383–387.
220. Zang, H., & Li, H. (2014). Optimization of process alarm thresholds: a multidimensional kernel density estimation approach. *Process Safety Progress, 33*, 292–298.
221. Zeng, Z., Tan, W., & Zhou, R. (2017). Performance assessment for generalized delay-timers in alarm configuration. *Journal of Process Control, 57*, 80–101.
222. Zhang, G., Wang, Z., & Mei, H. (2020). Sensitivity clustering and ROC curve based alarm threshold optimization. *Process Safety and Environmental Protection, 141*, 83–94.
223. Zhao, Y., & Zhao, C. (2022). Dynamic multivariate threshold optimization and alarming for nonstationary processes subject to varying conditions. *Control Engineering Practice, 124*, 105180.
224. Zhou, B., Hu, W., & Chen, T. (2019). A new method for alarm monitoring of equipment start-up operations with applications to pumps. *Industrial & Engineering Chemistry Research, 58*, 11251–11260.

225. Zhou, B., Hu, W., Brown, K., & Chen, T. (2021). Generalized pattern matching of industrial alarm flood sequences via word processing and sequence alignment. *IEEE Transactions on Industrial Electronics, 68*, 10171–10179.
226. Zhou, B., Hu, W., & Chen, T. (2021). Pattern extraction from industrial alarm flood sequences by a modified clofast algorithm. *IEEE Transactions on Industrial Informatics, 18*, 288–296.
227. Zhu, J., Shu, Y., Zhao, J., & Yang, F. (2014). A dynamic alarm management strategy for chemical process transitions. *J. Loss Prevention in the Process Industries, 30*, 207–218.
228. Zhu, J., Zhao, J., & Yang, F. (2014). Dynamic risk analysis with alarm data to improve process safety using Bayesian network. *The 11th World Congress on Intelligent Control and Automation*, June 29-July 4, 2014, Shenyang, China, 461-466.
229. Zhu, J., Wang, C., Li, C., Gao, X., & Zhao, J. (2016). Dynamic alarm prediction for critical alarms using a probabilistic model. *Chinese Journal of Chemical Engineering, 24*, 881–885.
230. Zhu, Q., & Geng, Z. (2005). A new fuzzy clustering-ranking algorithm and its application in process alarm management. *Chinese Journal of Chemical Engineering, 13*, 477–483.
231. Zhu, Q., Luo, Y., & He, Y. (2019). Novel distributed alarm visual analysis using multi-correlation block-based PLS and its application to online root cause analysis. *Industrial & Engineering Chemistry Research, 58*, 20655–20666.
232. Zhu, Q., Ding, W., & He, Y. (2020). Novel multimodule Bayesian network with cyclic structures for root cause analysis: application to complex chemical processes. *Industrial & Engineering Chemistry Research, 59*, 12812–12821.
233. Zhu, Q., Jin, C., He, Y., & Xu, Y. (2021). Pattern mining of alarm flood sequences using an improved PrefixSpan algorithm with tolerance to short-term order ambiguity. *Industrial & Engineering Chemistry Research, 60*, 4375–4384.

Chapter 2
Optimal Design of Univariate Alarm Systems

Abstract This chapter is on the optimal design of three commonly-used univariate alarm systems: Basic alarm generators, alarm delay timers and alarm deadbands. Optimal values of alarm thresholds, delay factors and deadband widths are designed to achieve desired performance requirements on false alarms, missed alarms and detection delays. Different methods are proposed for dealing with process variables that are independent and identically distributed (IID) or non-IID. Numerical and industrial examples are provided to illustrate the proposed methods.

2.1 Alarm Delay Timers for IID Process Variables

This section is on the optimal design of alarm delay timers for IID process variables [33]. Transitions among alarm and non-alarm states of IID process variables are described by Markov chains. Analytical expressions of three performance indices are established for alarm delay timers to measure the levels of false alarms, missed alarms and averaged alarm delays. Optimal values of alarm thresholds and delay factors are designed by balancing trade-offs among performance indices.

2.1.1 Problem Description

For a basic alarm generator, an alarm variable x_a takes the value of '1'('0') if a process variable x is larger (less) than a high (low) alarm threshold x_{th} for the alarm (non-alarm) state, i.e.,

$$x_a(t) = \begin{cases} 1, \text{ if } x(t) \geq (\leq)\, x_{th}, \\ 0, \text{ if } x(t) < (>)\, x_{th}. \end{cases} \tag{2.1}$$

Here t is the sampling index being associated with a sampling period h (e.g., $h = 0.2$ sec). If high and low alarm thresholds coexist, (2.1) becomes

J. Wang et al., *Intelligent Industrial Alarm Systems*,
https://doi.org/10.1007/978-981-97-6516-4_2

$$x_a(t) = \begin{cases} 0, \text{ if } x_{th,l} < x(t) < x_{th,h}, \\ 1, \qquad \text{otherwise.} \end{cases} \tag{2.2}$$

Here $x_{th,h}$ and $x_{th,l}$ are respectively the high and low alarm thresholds. As an example, Fig. 1.1a presents a drum water level $x(t)$ configured with the high alarm threshold +100 mm and low alarm threshold –100 mm from a large-scale power generation plant. Figure 1.1b gives the alarm variables of $x_a(t)$ in (2.2).

Due to various reasons, including measurement noises of x and improper selection of x_{th}, two types of nuisance alarms may appear, namely, false and missed alarms. A false alarm is raised even though x is behaving normally; missed alarms occur when x is behaving abnormally but no alarm is raised. False alarms may lead to losing the trusty of alarm systems due to the "cry wolf" effect, while the missed alarms may severely degrade the designed functionality of alarm systems for operational safety and efficiency monitoring. Hence, the false alarm rate (FAR) and missed alarm rate (MAR) are two important indices to assess the performance of alarm systems.

Besides the FAR and MAR, the averaged alarm delay (AAD) is another important performance index for alarm systems. Suppose that the process variable x is experiencing a variation from the normal condition to abnormal one at the time instant t_0. Let t_a be the time instant when the first sample point of $x(t)$ overpasses the alarm threshold so that an alarm is raised. The time difference between t_0 and t_a is named as the alarm delay, i.e.,

$$T_d = t_a - t_0.$$

Because of sampling, T_d is a discrete random variable with the sample space $\{0, 1, 2, \cdots\}$. Note that $t_a \geq t_0$ so that T_d is always nonnegative since false alarms ahead of t_0 should not be considered in order to have a physically meaningful interpretation of the alarm delay. The AAD is defined as the expected value of T_d,

$$\overline{T}_d = E(T_d). \tag{2.3}$$

The AAD measures the alarm system promptness: if $\overline{T}_d$ is smaller, then the alarm is raised more promptly after $x(t)$ runs into the abnormal condition, which leaves operators more time to respond to the alarm.

Alarm delay timers are widely used in practice to remove false alarms and missed alarms [6]. The n-sample alarm delay timer raises (clears) an alarm if and only if n consecutive samples of x_a are '1's ('0's), i.e.,

$$x_a(t; n) = \begin{cases} 1, & \text{if } \sum_{i=t-n+1}^{t} x_a(i) = n \;\&\; x_a(t-1; n) = 0, \\ 0, & \text{if } \sum_{i=t-n+1}^{t} x_a(i) = 0 \;\&\; x_a(t-1; n) = 1, \\ x_a(t-1; n), & \text{otherwise.} \end{cases} \tag{2.4}$$

Here n is referred to as the delay factor. For the alarm delay timer, its input and output are $x_a(t)$ and $x_a(t; n)$, respectively; as shown in Fig. 2.1, the HLALM function and TIMER function together implement the alarm delay timer in distributed control systems. Clearly, an alarm delay timer with a larger value of n may lead to smaller

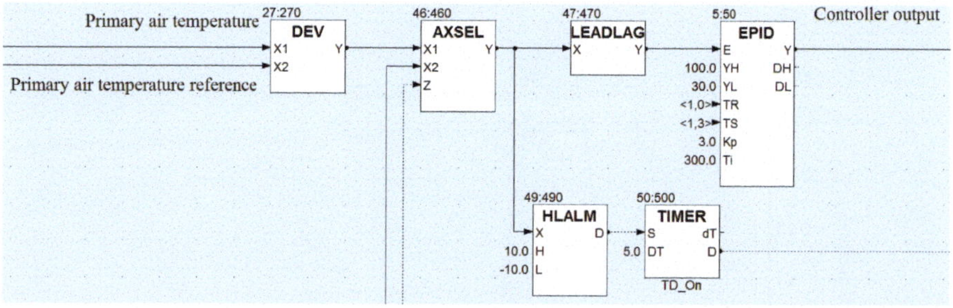

Fig. 2.1 An embedded alarm delay timer composed of the High Low Alarm (HLALM) function and TIMER function in distributed control systems

values of the FAR and MAR, but a larger value of the AAD. The problem to be solved is to design optimal values of n and x_{th} in order to satisfy certain requirements on the three performance indices of FAR, MAR and AAD. The process variable x is assumed to be IID with invariant statistical properties, and historical data of x in the normal and abnormal conditions are available.

2.1.2 *Expressions of Three Performance Indices*

It is ready to estimate probability density functions (PDFs) of the process variable x in the normal and abnormal conditions from historical data samples. Figure 2.2 illustrates the PDFs together with the high alarm threshold x_{th}; only high alarm thresholds are considered in the sequel to simplify presentations and upcoming results are equally applicable to low alarm thresholds with minor modifications. The PDF for the normal condition is denoted as $q(x)$ (solid line in Fig. 2.2). For $x_a(t)$ in (2.1), the FAR as the probability of false alarms is the area under the distribution curve $q(x)$ for the values of x greater than x_{th}, i.e.,

$$F(x_a(t)) = \int_{x_{th}}^{+\infty} q(x)\mathrm{d}x. \tag{2.5}$$

In Fig. 2.2, the FAR is denoted by q_1. Similarly, the MAR as the probability of missed alarms for $x_a(t)$ in (2.1) is

$$M(x_a(t)) = \int_{-\infty}^{x_{th}} p(x)\mathrm{d}x, \tag{2.6}$$

where $p(x)$ is the PDF of x in the abnormal condition (dashed line in Fig. 2.2). In Fig. 2.2, the MAR is denoted by p_2.

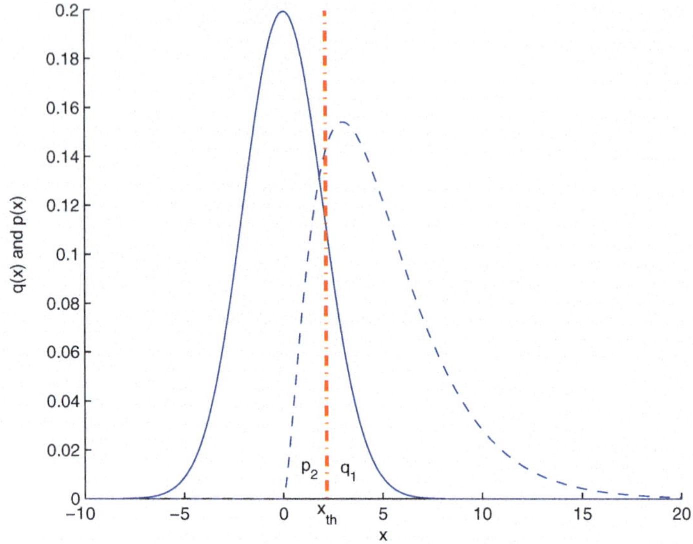

Fig. 2.2 The PDFs $q(x)$ and $p(x)$ of x under normal and abnormal conditions, together with a high alarm threshold x_{th}, reprinted from Ref. [33], copyright 2012, with permission from IEEE

In general, the computation of the AAD $\overline{T}_d$ requires the knowledge of multi-dimensional joint PDFs of $x(t_0)$, $x(t_0+1)$, $\cdots$. In this section, $x(t)$ is assumed to be IID. For $t \geq t_0$, x is under the abnormal condition, so that the PDF of x is $p(x)$ (the dashed line in Fig. 2.2), and as defined in (2.6), p_2 is the probability of x with the PDF $p(x)$ less than x_{th}. For ease of notations, $p_1 := 1 - p_2$. As $x(t)$ is IID, the probability mass function of $T_d = i$ is

$$\begin{aligned} P(T_d = i) = P(x(t_0) < x_{th}, \cdots, x(t_0+i-1) < x_{th}, \\ x(t_0+i) > x_{th}) = p_2^i p_1. \end{aligned}$$

Thus, the time delay T_d follows a geometric distribution; using the mean expression of a geometric random variable [28], the AAD is

$$\overline{T}_d = E(T_d) = \sum_{i=0}^{\infty} i p_2^i p_1 = \frac{p_2}{p_1}. \tag{2.7}$$

The next objective is to derive the expressions of FAR, MAR and AAD for alarm delay timers. The n-sample alarm delay timer in (2.4) involves n no-alarm states and n alarm states, and its working mechanism can be described by Markov chains. The transition among these states for x in the normal condition is represented by the Markov chain in Fig. 2.3. For x in the normal condition, the PDF of x is $q(x)$ (the solid line in Fig. 2.2); as defined earlier in (2.5), q_1 is the probability of x with the

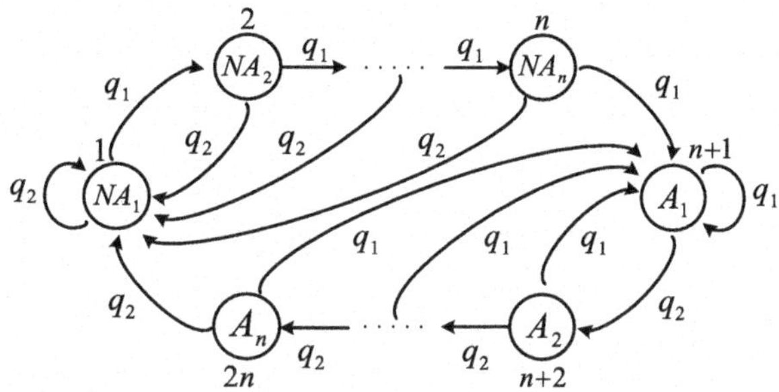

Fig. 2.3 Markov chain for the n-sample alarm delay timer in the case that x is in the normal condition, reprinted from Ref. [33], copyright 2012, with permission from IEEE

PDF $q(x)$ greater than x_{th}, and for ease of notations, $q_2 := 1 - q_1$. If the alarm delay timer for a certain sample $x(t_1)$ is at the i-th no-alarm state NA_i for $i = 1, 2, \cdots, n$, and the next sample $x(t_1 + 1)$ exceeds x_{th}, then the current state NA_i goes to the $(i+1)$th no-alarm state NA_{i+1} for $i < n$, or to the alarm state A_1 for $i = n$. If $x(t_1 + 1)$ is less than x_{th}, then the current state NA_i directly goes back to the first no-alarm state NA_1. Similarly, if the alarm delay timer for a certain sample $x(t_2)$ is at the i-th alarm state A_i for $i = 1, 2, \cdots, n$, and the next sample $x(t_2 + 1)$ is less than x_{th}, then the state A_i goes to the $(i+1)$th alarm state A_{i+1} for $i < n$, or to the no-alarm state NA_1 for $i = n$. If $x(t_2 + 1)$ is larger than x_{th}, then the state A_i goes back to the first alarm state A_1.

Proposition 2.1 *For the n-sample alarm delay timer, the FAR is*

$$F(x_a(t; n)) = \frac{q_1^n \sum_{i=0}^{n-1} q_2^i}{q_1^n \sum_{i=0}^{n-1} q_2^i + q_2^n \sum_{j=0}^{n-1} q_1^j}, \tag{2.8}$$

Proof of Proposition 2.1 To simplify the notation, the states $NA_1, NA_2, \cdots, NA_n$, $A_1, \cdots, A_n$ are defined as the states $1, 2, \cdots, n-1, n, \cdots, 2n$, respectively. Let $T_{i,k}$ be the number of steps taken for the transmission from state i to another state k, and $P_{i,k}^{(l)}$ as the probability of $T_{i,k} = l$, i.e.,

$$P_{i,k}^{(l)} := P\left(T_{i,k} = l\right).$$

When $l = 1$, $P_{i,k}^{(1)}$ denotes the one step transition probability. For the Markov chain in Fig. 2.3, the matrix $Q \in \mathbb{R}^{2n \times 2n}$ of one step transition probability is

$$Q = \begin{bmatrix} q_2 & q_1 & 0 & \cdots & 0 & 0 & 0 & \cdots & 0 \\ q_2 & 0 & q_1 & \cdots & 0 & 0 & 0 & \cdots & 0 \\ \vdots & \vdots & \vdots & \ddots & \vdots & \vdots & \vdots & \vdots & \vdots \\ q_2 & 0 & \cdots & 0 & q_1 & 0 & 0 & \cdots & 0 \\ 0 & 0 & \cdots & 0 & q_1 & q_2 & 0 & \cdots & 0 \\ 0 & 0 & 0 & \cdots & q_1 & 0 & q_2 & \cdots & 0 \\ \vdots & \vdots & \vdots & \vdots & \vdots & \vdots & \vdots & \ddots & \vdots \\ 0 & 0 & 0 & \cdots & q_1 & 0 & 0 & \cdots & q_2 \\ q_2 & 0 & 0 & \cdots & q_1 & 0 & 0 & \cdots & 0 \end{bmatrix}.$$

Here the element locating at the i-th row and the j-th column of Q is the one step transition probability of the state from i to j. For an irreducible ergodic Markov chain [21] (pages 727–729), the limiting probabilities

$$\pi_k = \lim_{l \to \infty} P_{ik}^{(l)} > 0, \forall k = 1, 2, \cdots, 2n$$

exist, being independent of the initial state, and satisfy the equality

$$\sum_{k=1}^{2n} \pi_k = 1. \tag{2.9}$$

Because the Markov chain has only a finite number of states, these limiting probabilities satisfy the equality

$$\Pi = \Pi Q, \tag{2.10}$$

where

$$\Pi = [\pi_1 \; \pi_2 \; \cdots \; \pi_{2n}].$$

Equation (2.10) can be rewritten as

$$\begin{cases} q_2(\pi_1 + \cdots + \pi_n) + q_2\pi_{2n} = \pi_1, \\ q_1\pi_1 = \pi_2, \\ q_1\pi_2 = \pi_3, \\ \vdots \\ q_1\pi_{n-1} = \pi_n, \\ q_1(\pi_n + \cdots + \pi_{2n}) = \pi_{n+1}, \\ q_2\pi_{n+1} = \pi_{n+2}, \\ \vdots \\ q_2\pi_{2n-1} = \pi_{2n}. \end{cases} \tag{2.11}$$

From (2.11), it is ready to obtain

$$\begin{cases} \pi_2 = q_1 \pi_1, \\ \vdots \\ \pi_n = q_1^{n-1} \pi_1, \end{cases} \tag{2.12}$$

and

$$\begin{cases} \pi_{n+2} = q_2 \pi_{n+1}, \\ \vdots \\ \pi_{2n} = q_2^{n-1} \pi_{n+1}. \end{cases} \tag{2.13}$$

Using (2.12) and (2.13), (2.9) and the first equality in (2.11) can be respectively written as

$$\begin{cases} \pi_1(1 + q_1 + \cdots + q_1^{n-1}) + \pi_{n+1}(1 + q_2 + \cdots + q_2^{n-1}) = 1 \\ q_2 \pi_1 (1 + q_1 + \cdots + q_1^{n-1}) + q_2^n \pi_{n+1} = \pi_1 \end{cases}$$

from which π_{n+1} is obtained as

$$\pi_{n+1} = \frac{q_1^n}{q_1^n(1 + q_2 + \cdots + q_2^{n-1}) + q_2^n(1 + q_1 + \cdots + q_1^{n-1})}. \tag{2.14}$$

From the Markov chain in Fig. 2.3, the FAR is the sum of probabilities of all the alarm states,

$$\begin{aligned} F(x_a(t; n)) &= P(A_1) + P(A_2) + \cdots + P(A_n) \\ &= \pi_{n+1} + \cdots + \pi_{2n} \\ &= \pi_{n+1}(1 + q_2 + \cdots + q_2^{n-1}), \end{aligned}$$

where the last equality is from (2.13). This result, together with (2.14), proves the FAR in (2.8). □

The state transition for $x(t)$ in the abnormal condition is presented by the Markov chain in Fig. 2.4, where p_2 is the probability of x with the PDF $p(x)$ (the dashed line in Fig. 2.2) less than the alarm threshold x_{th}, and $p_1 := 1 - p_2$. Thus, the MAR for $x_{a,n}$ in (2.4) is the sum of probabilities of all the no alarm states, i.e.,

$$M(x_a(t; n)) = P(NA_1) + P(NA_2) + \cdots + P(NA_n).$$

The MAR is associated with p_1, p_2, and n as given in the next proposition.

Proposition 2.2 *For the n-sample alarm delay timer, the MAR is*

$$M(x_a(t; n)) = \frac{p_2^n \sum_{i=0}^{n-1} p_1^i}{p_2^n \sum_{i=0}^{n-1} p_1^i + p_1^n \sum_{j=0}^{n-1} p_2^j}. \tag{2.15}$$

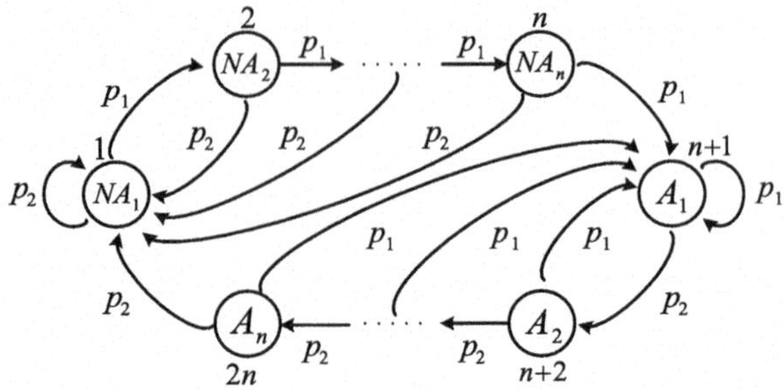

Fig. 2.4 Markov chain for the n-sample alarm delay timer in the case that x is in the abnormal condition, reprinted from Ref. [33], copyright 2012, with permission from IEEE

Proof of Proposition 2.2 For the Markov chain in Fig. 2.4, the matrix $P \in \mathbb{R}^{2n\times 2n}$ of one step transition probability is

$$P = \begin{bmatrix} p_2 & p_1 & 0 & \cdots & 0 & 0 & 0 & \cdots & 0 \\ p_2 & 0 & p_1 & \cdots & 0 & 0 & 0 & \cdots & 0 \\ \vdots & \vdots & \vdots & \ddots & \vdots & \vdots & \vdots & \vdots & \vdots \\ p_2 & 0 & \cdots & 0 & p_1 & 0 & 0 & \cdots & 0 \\ 0 & 0 & \cdots & 0 & p_1 & p_2 & 0 & \cdots & 0 \\ 0 & 0 & 0 & \cdots & p_1 & 0 & p_2 & \cdots & 0 \\ \vdots & \vdots & \vdots & \vdots & \vdots & \vdots & \vdots & \ddots & \vdots \\ 0 & 0 & 0 & \cdots & p_1 & 0 & 0 & \cdots & p_2 \\ p_2 & 0 & 0 & \cdots & p_1 & 0 & 0 & \cdots & 0 \end{bmatrix}.$$

Then, (2.15) can be proved by taking a procedure similar to that in the proof of Proposition 2.1. □

With respect to the AAD, the definition is the same as (2.3), except that t_a refers to the time instant when the first alarm is raised by using the n-sample alarm delay timer. It is reasonable to assume that the alarm is inactive at the time instant $(t_0 - 1)$ and do not consider the process of clearing the alarm once the alarm is raised. Hence, denote the state at the time instant $(t_0 - 1)$ as the first no-alarm state NA_1 and the process is carried forward to the alarm state A_1 with the same procedure in Fig. 2.4. Because our concern is the first alarm (when the state reaches to the alarm state A_1 for the first time), only the top half of the Markov chain given in Fig. 2.4 is considered. In particular, when the state reaches to the alarm state A_1, it will stay in the state A_1 with probability 1 for the computation of AAD. This operating mechanism is illustrated by the Markov chain in Fig. 2.5.

It is worthy to note that the state at the time instant $(t_0 - 1)$ does not have to be NA_1, and could be any one of the no-alarm states, NA_i for $i = 1, \cdots, n$; however, only the state NA_1 is considered for two reasons. (i) It is ready to obtain from the proof of Proposition 2.1 that the probability of being in the state NA_i is

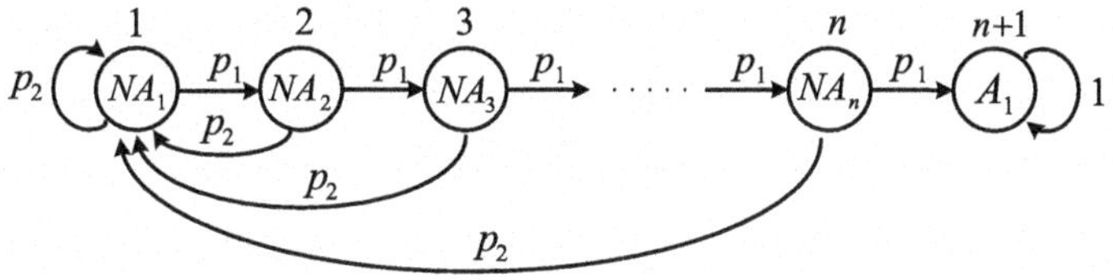

Fig. 2.5 Markov diagram for the n-sample alarm delay timer to calculate the AAD, reprinted from Ref. [33], copyright 2012, with permission from IEEE

$$P(NA_i) = \frac{q_1^{i-1} q_2^n}{q_1^n(1 + \cdots + q_2^{n-1}) + q_2^n(1 + \cdots + q_1^{n-1})},$$

where q_1 is the probability of x in the normal condition with the PDF $q(x)$ taking values greater than x_{th}. Thus, the probability of starting with NA_1 is much larger than the probability of the rest states combined, since q_1 is usually quite small. (ii) In practice, the normal and abnormal PDFs of x are usually not available and have to be estimated. If the state at the time instant $(t_0 - 1)$ is NA_i for $i > 1$, i.e., all the samples $x(t_0 - i), \cdots, x(t_0 - 1)$ are greater than x_{th}, and $x(t_0 - i - 1)$ is smaller than x_{th}, then it is very likely to classify $x(t_0 - i), \cdots, x(t_0 - 1)$ into the abnormal data section. Thus, the sample right before the detected change point from the normal to abnormal is less than x_{th}, and the corresponding state is NA_1. Owing to these two reasons, it is reasonable to ignore the case that the time instant $(t_0 - 1)$ is associated with the other no-alarm state NA_i for $i > 1$.

Proposition 2.3 *If the state at the time instant* $(t_0 - 1)$ *is* NA_1, *for the n-sample alarm delay timer, the AAD is*

$$\overline{T_d}(x_a(t; n)) = h\frac{(1 - p_1^n - p_2 p_1^n)}{p_2 p_1^n}. \tag{2.16}$$

Proof of Proposition 2.3 Define the states $NA_1, NA_2, \cdots, A_1$ as the states $1, 2, \cdots, n + 1$, respectively. According to the definition in (2.3), the AAD for the n-sample alarm delay timer is

$$\overline{T_d}(x_a(t; n)) = E(T_d(x_a(t; n))) = E(T_{1,n+1}) - 1, \tag{2.17}$$

where $T_{1,n+1}$ is the number of transmission steps taken from state 1 to the alarm state $n + 1$. The subtraction of 1 in (2.17) is owing to the fact that the state 1 occurs at the time instant $(t_0 - 1)$.

The computation of $E(T_{1,n+1})$ can be obtained as follows. First, the moment generating function of the discrete random variable $T_{i,n+1}$, for $i \in \{1, 2, \cdots, n + 1\}$, is

$$\Gamma_{i,n+1}(z) = \sum_{l=0}^{\infty} P_{i,n+1}^{(l)} z^l, \tag{2.18}$$

where $P_{i,n+1}^{(l)} := P(T_{i,n+1} = l)$ and $z := e^t$ [21]. Differentiating $\Gamma_{i,n+1}(z)$ with respect to z yields

$$\frac{d}{dz}\Gamma_{i,n+1}(z) = \sum_{l=0}^{\infty} l P_{i,n+1}^{(l)} z^{l-1},$$

which implies that the mean transmission step from state i to state $n+1$ is

$$E(T_{i,n+1}) = \sum_{l=0}^{\infty} l P_{i,n+1}^{(l)} = \left.\frac{d}{dz}\Gamma_{i,n+1}(z)\right|_{z=1}.$$

Thus, $E(T_{1,n+1})$ in (2.17) can be calculated as

$$E(T_{1,n+1}) = \left.\frac{d}{dz}\Gamma_{1,n+1}(z)\right|_{z=1}. \tag{2.19}$$

Second, using Chapman-Kolmogorov Equation ([21], page 705), namely,

$$P_{i,k}^{(l)} = \sum_{j\in I} P_{i,j}^{(1)} P_{j,k}^{(l-1)},$$

and the definitions in Markov chain theory

$$\begin{cases} P_{i,k}^{(0)} = 0, & \text{for } i \neq k, \\ P_{i,k}^{(0)} = 1, & \text{for } i = k, \end{cases}$$

(2.18) for $i \neq n+1$ becomes

$$\begin{aligned} \Gamma_{i,n+1}(z) &= \sum_{l=0}^{\infty} P_{i,n+1}^{(l)} z^l \\ &= P_{i,n+1}^{(0)} z^0 + \sum_{l=1}^{\infty}\sum_{j\in I} P_{i,j}^{(1)} P_{j,n+1}^{(l-1)} z^l \\ &= \left(\sum_{j\in I} z P_{i,j}^{(1)}\right)\left(\sum_{l=1}^{\infty} P_{j,n+1}^{(l-1)} z^{l-1}\right) \\ &= \sum_{j\in I} z P_{i,j}^{(1)} \Gamma_{j,n+1}(z), \end{aligned} \tag{2.20}$$

where I is the whole state space $\{1, 2, \cdots, n+1\}$. Since the state $n+1$ is a recurrent state, (2.18) for $i = n+1$ reduces to

$$\Gamma_{n+1,n+1}(z) = \sum_{l=0}^{\infty} P_{n+1,n+1}^{(l)} z^l = 1,$$

which is obtained based on two definitions in Markov chain theory for a recurrent state k,

$$\begin{cases} P_{k,k}^{(l)} = 0, & \text{for } l > 0, \\ P_{k,k}^{(l)} = 1, & \text{for } l = 0. \end{cases}$$

From (2.20) and the Markov chain in Fig. 2.5, it is ready to have

$$\begin{cases} \Gamma_{1,n+1}(z) = & zp_2\Gamma_{1,n+1}(z) + zp_1\Gamma_{2,n+1}(z), \\ \Gamma_{2,n+1}(z) = & zp_2\Gamma_{1,n+1}(z) + zp_1\Gamma_{3,n+1}(z), \\ & \vdots \\ \Gamma_{n,n+1}(z) = & zp_2\Gamma_{1,n+1}(z) + zp_1\Gamma_{n+1,n+1}(z). \end{cases} \tag{2.21}$$

As $\Gamma_{n+1,n+1}(z) = 1$, solving (2.21) for $\Gamma_{1,n+1}(z)$ gives

$$\Gamma_{1,n+1}(z) = \frac{z^n p_1^n}{1 - zp_2 - z^2 p_2 p_1 - \cdots - z^n p_2 p_1^{n-1}}.$$

Therefore, (2.19) becomes

$$E(T_{1,n+1}) = \frac{d}{dz}\Gamma_{1,n+1}(z)\bigg|_{z=1} = \frac{(1 - p_1^n)}{p_2 p_1^n},$$

and the AAD in (2.17) is

$$\overline{T_d}(x_a(t; n)) = E(T_{1,n+1}) - 1 = \frac{(1 - p_1^n - p_2 p_1^n)}{p_2 p_1^n}.$$

□

With the above results for the FAR, MAR and AAD, it is ready to design alarm systems in a systematic manner based on the three performance indices.

2.1.3 Design of Delay Factors and Alarm Thresholds

This subsection first estimates the PDFs of the process variable x in normal and abnormal conditions, based on which a design problem is solved afterwards to determine delay factors and alarm thresholds.

Given the collected data $\{x(t)\}_{t=1}^{T}$ of x, how to estimate the PDFs of x in normal and abnormal conditions? The critical step to solve this problem is to separate $\{x(t)\}_{t=1}^{T}$ into segments in the normal and abnormal conditions. Doing so is possible under two assumptions. Assumption A1: two PDFs of x in normal and abnormal conditions have different mean values. Assumption A2: The process knowledge that the mean values belong to either the normal condition or the abnormal one is known *a priori*. Assumption A1 says that x experiences some mean changes; hence, mean change detection techniques can be exploited to locate the mean change points in $\{x(t)\}_{t=1}^{T}$. For Assumption A2, the current value of the alarm threshold x_{th} usually is a good threshold for the mean values to be compared with. For instance, if x_{th} stands for a high-value alarm threshold, then a data segment having the mean value statistically larger than x_{th} belongs to the abnormal condition.

There are quite a few change detection techniques in literature, e.g., Shewhart chart, moving average charts, cumulative sum procedures, generalized likelihood ratio tests, Bayesian and information criterion approaches [4, 7]. Most of these techniques are parametric in assuming the initial distribution or the signal model structure to be known a priori, e.g., x is a Gaussian random process or takes the autoregressive model structure. However, usually it is very difficult to know this kind of information in practice. In addition, the collected data $\{x(t)\}_{t=1}^{T}$ may contain multiple mean changes instead of a single one.

The proposed method of estimating the PDFs is based on the non-parametric approach for one mean-change point detection proposed by Pettitt [22]. This non-parametric approach does not suffer from the above-mentioned problems for parametric change detection techniques, and is very effective in detecting change of mean values. However, it cannot be used directly to find multiple change points. Here it is revised by adopting the idea of bisection method. The proposed method consists of the following steps:

Step 1: For $\{x(t)\}_{t=1}^{T}$, find one mean change point as follows [22]:

- Calculate the test statistic $U_{1,T} = V_{1,T}$ and $U_{t,T} = U_{t-1,T} + V_{t,T}$ for $t = 2, 3, \cdots, T$, where

$$V_{t,T} = \sum_{j=1}^{T} sgn(x(t) - x(j)).$$

  Here $sgn(\cdot)$ takes the sign value $1, 0$ or -1 for a positive, zero or negative operand.
- Find the time instant t_{max} maximizing $|U_{t,T}|$ and compute the corresponding P-value as

$$P = 2\exp\left(\frac{-6\max\limits_{1\le t<T}|U_{t,T}|^2}{T^2+T^3}\right).$$

- Choose the probability of type I error α, e.g., $\alpha = 0.01$, and define the null hypothesis, namely, $x(t_{max})$ is not a mean change point. If $P < \alpha$, reject the null hypothesis so that $x(t_{max})$ is the change point of $\{x(t)\}_{t=1}^{T}$. If $P > \alpha$, the null hypothesis cannot be rejected and no change point can be found.

Step 2: Divide $\{x(t)\}_{t=1}^{T}$ into two sub-segments $\{x_1(t)\}_{t=1}^{t_{max}}$ and $\{x_2(t)\}_{t=t_{max}+1}^{T}$ according to t_{max}. Go to Step 1 for each sub-segment to find their own one change point. Repeat Steps 1 and 2 until no further change points can be found.

Step 3: After finding all the change points, $\{x(t)\}_{t=1}^{T}$ has been isolated into several data segments. For each data segment, the sample mean is calculated and compared with the alarm threshold x_{th}. Here a standard T-test [28] is used for the comparison as follows. Let one of the data segments be denoted as $\{x(t)\}_{t=t_0}^{t_1}$. The Student's t-distributed statistic is computed

$$t_x = \frac{\bar{x} - x_{tp}}{s/\sqrt{t_1 - t_0}}.$$

Here $\bar{x}$ and s are respectively the sample mean and standard deviation of x,

$$\bar{x} = \frac{\sum_{t=t_0}^{t_1} x(t)}{t_1 - t_0 + 1}, \quad s = \sqrt{\frac{\sum_{t=t_0}^{t_1} (x(t) - \bar{x})^2}{t_1 - t_0}}.$$

If t_x is larger (smaller) than the critical value t_{β,t_1-t_0} $(-t_{\beta,t_1-t_0})$, then the sample mean $\bar{x}$ is statistically larger (smaller) than x_{th}, and this data segment is regarded as the one in the abnormal (normal) condition; otherwise, $\bar{x}$ is statistically equal to x_{th}, and this data segment is discarded and not included in the subsequent PDF estimation. Here the probability of Type-I error is denoted as β, e.g., $\beta = 0.05$. All the data segments regarded in the abnormal (normal) conditions are put together into one single group, referred to as the abnormal (normal) data.

Step 4. The PDFs of x in the normal and abnormal conditions are now ready to be estimated based on the normal and abnormal data obtained in Step 3. Here the kernel-based method is used to estimate the PDFs using the Gaussian kernel function,

$$K(x) = \frac{1}{\sqrt{2\pi}} \exp\left(-x^2/2\right).$$

The design problem for delay factors and alarm thresholds is formulated as the following: Given the PDFs of $x(t)$ in the normal and abnormal conditions, how to choose the delay factor n and/or the alarm threshold x_{th} to satisfy certain requirements on the FAR, MAR and AAD? The design problem can be broken into three cases:

- Case I: Design x_{th} for a fixed value of n;
- Case II: Design n for a fixed value of x_{th};
- Case III: Design both n and x_{th}.

For the first case of having a fixed value of n, the design of x_{th} is based on two trade-offs between the FAR and MAR/AAD. From (2.8), (2.15) and (2.16), typical relations between FAR/MAR/AAD and x_{th} for a fixed value of n are shown as the three plots in Fig. 2.6. Clearly, there are two trade-offs between the FAR and MAR/AAD in Fig. 2.6. That is, when x_{th} gets larger, FAR is decreasing but MAR

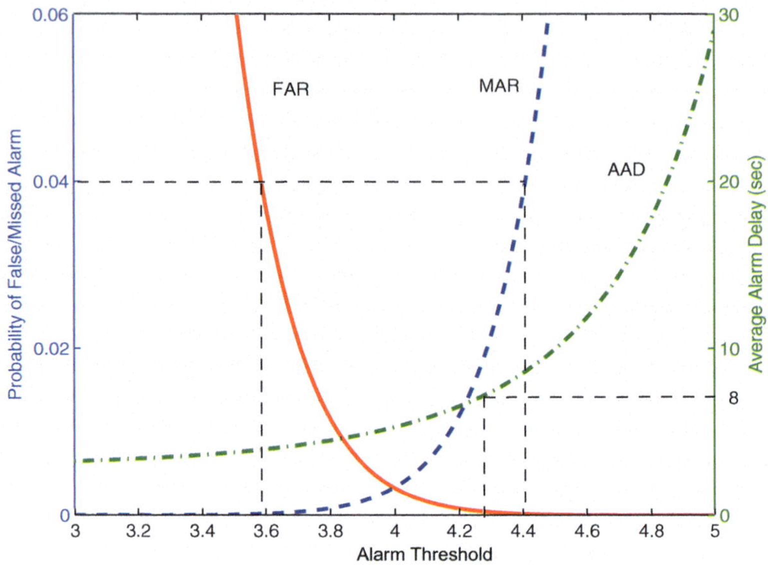

Fig. 2.6 The relation between FAR/MAR/AAD and the alarm threshold, reprinted from Ref. [33], copyright 2012, with permission from IEEE

and AAD are increasing, and vice versa. Hence, the requirements of FAR, MAR and AAD will impose their own valid ranges of x_{th} based on these trade-offs. The intersection of these ranges will be the final choice of x_{th} to meet all the requirements of FAR, MAR and AAD; if the intersection is empty, then there is no way to meet all the requirements by solely changing x_{th}.

For the second case of having a fixed value of x_{th}, the design of n is based on two trade-offs between the FAR/MAR and AAD. Similar to Fig. 2.6, the relations between FAR/MAR/AAD and n can be obtained from (2.8), (2.15) and (2.16) with a fixed value of x_{th}, shown as the three plots in Fig. 2.7. There are two trade-offs between the FAR/MAR and AAD. For a fixed value of x_{th}, a larger value of n leads to decrements of FAR and MAR, but to an increment of AAD. These trade-offs will confine the possible ranges of n, whose intersection is the final valid range of n.

For the third case where both n and x_{th} are free for design, the design procedure is to determine the valid values of n first and design x_{th} for each valid value of n. For the above three cases, valid ranges of n and x_{th} can be found. However, only one pair of n and x_{th} can be implemented. Thus, it is necessary to choose the optimal values of n and x_{th} in some manner. Depending on the preference on FAR, MAR and AAD in practice, different optimization criteria can be formulated to obtain optimal values of n and x_{th}. One possible choice is to minimize AAD subject to some upper bounds of FAR and MAR. Despite the variations of optimization criteria, the optimal values of n and x_{th} are obtained by following the same principle. Hence, a weighted-sum loss function is taken as

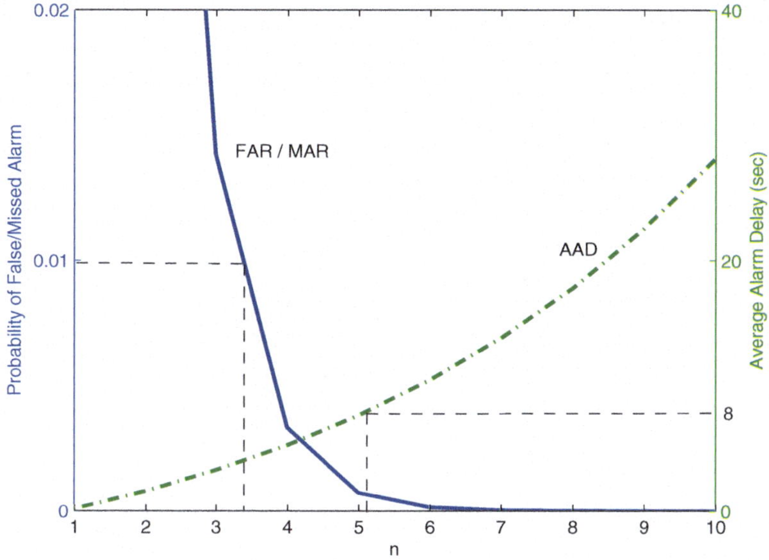

Fig. 2.7 The relation between FAR/MAR/AAD and the delay facotr n, reprinted from Ref. [33], copyright 2012, with permission from IEEE

$$J(x_{th}, n) = \omega_1 \frac{\text{FAR}}{\text{RFAR}} + \omega_2 \frac{\text{MAR}}{\text{RMAR}} + \omega_3 \frac{\text{AAD}}{\text{RAAD}} \tag{2.22}$$

to illustrate the principle. Here RFAR, RMAR and RAAD are the requirements of FAR, MAR and AAD, respectively; ω_1, ω_2 and ω_3 are the weights of FAR, MAR and AAD, respectively. Then, the optimal values of x_{th} and n are the ones minimizing the loss function in (2.22), i.e.,

$$(x_{th}, n)_{opt} = \underset{x_{th}, n}{\text{argmin}} J(x_{th}, n).$$

Since the FAR in (2.8), MAR in (2.15) and AAD in (2.16) are nonlinear functions of n and x_{th}, a two-dimensional grid search is implemented to find the optimal values of x_{th} and n. Here n is a positive integer, but x_{th} is continuous; by discretizing x_{th} in a certain resolution, it takes discrete values.

2.1.4 Numerical and Industrial Examples

This subsection presents two examples to validate theoretical expressions of performance indices and illustrate the optimal design of alarm delay timers in practice.

2.1.4.1 Numerical Example

This example is to validate the FAR in (2.8), MAR in (2.15) and AAD in (2.16) via simulation. The process variable x is generated as a white Gaussian random process with a mean change at t_0,

$$\begin{cases} x(t) \sim N(3, 1), \text{ if } t < t_0, \\ x(t) \sim N(5, 1), \text{ if } t \geq t_0. \end{cases}$$

The change time is $t_0 = 1000h$ with the sampling period $h = 1$ sec. The data length is 2000. The alarm variable is generated via an alarm delay timer with the factor $n = 3$ and the high alarm threshold $x_{th} = 4$. Eqs. (2.8), (2.15) and (2.16) give the theoretical values of the FAR, MAR and AAD, respectively,

$$F(x_a(t; n)) = 0.0142, \quad M(x_a(t; n)) = 0.0142, \quad \overline{T}_d(x_a(t; n)) = 3.2804. \quad (2.23)$$

To verify these theoretical values, 500 independent realizations of the sequence $\{x(t)\}_{t=1}^{2000}$ are generated; for each realization, a single estimate $\hat{F}(x_a(t; n))$ $(\hat{M}(x_a(t; n)))$ is provided as the ratio between the observed number of alarms (no-alarms) to the data length of $x(t)$ in the normal (abnormal) condition (equal to 1000 here). Based on the 500 independent realizations, the sample mean and standard deviation of $\hat{F}(x_a(t; n))$ $(\hat{M}(x_a(t; n)))$ can be obtained,

$$\mathrm{m}\left(\hat{F}(x_a(t; n))\right) = 0.0144, \quad \mathrm{s}\left(\hat{F}(x_a(t; n))\right) = 0.0082,$$
$$\mathrm{m}\left(\hat{M}(x_a(t; n))\right) = 0.0143, \quad \mathrm{s}\left(\hat{M}(x_a(t; n))\right) = 0.0085,$$

which are consistent with the theoretical values in (2.23).

Since the actual change time t_0 is available, the alarm delay for each realization is observed. The average of the alarm delays in the 500 independent realizations is regarded as one sample of the AAD, denoted as $\hat{\overline{T}}_d(x_a(t; n))$. To obtain a number of such samples, 500 sets of the above 500 independent realizations are generated, providing 500 samples of $\hat{\overline{T}}_d(x_a(t; n))$. The sample mean and standard deviation of these 500 samples are

$$\mathrm{m}\left(\hat{\overline{T}}_d(x_a(t; n))\right) = 3.2805, \quad \mathrm{s}\left(\hat{\overline{T}}_d(x_a(t; n))\right) = 0.0945,$$

which support the theoretical value $\overline{T}_d(x_a(t; n)) = 3.2804$ in (2.23).

2.1.4.2 Industrial Example

This industrial example is to design optimal values of n and x_{th} to meet with the requirements of FAR, MAR and AAD.

The process variable x is the pressure of the main steam driving the power turbine from a large-scale thermal power plant. Owing to the time-varying operating conditions, the steam pressure experiences a large scale of amplitude variations. The current alarm system for the steam pressure takes the basic alarm generator in (2.1) with a low alarm threshold $x_{th} = 23.6$ MPa. That is, when the steam pressure is less than 23.6 MPa, an alarm will be raised. From the DCS database, 6.8×10^4 data points shown in Fig. 2.8 are collected with the sampling period $h = 1$ sec, standing for the routine operation of the thermal power plant for 19 hours. Our objective is to design optimal values of n and x_{th} in order to satisfy the requirements: FAR$\leq 5\%$, MAR$\leq 5\%$, AAD≤ 5 sec.

First, data samples in Fig. 2.8 can be classified into the groups of normal and abnormal data by exploiting the proposed method in Sect. 2.1.3. The detected change points and the corresponding P-values are listed in Table 2.1; here the probability of type-I error is chosen to be $\alpha = 0.05$. By comparing with the alarm threshold $x_{th} = 23.6$ via the hypothesis tests (the probability of Type-I error is $\beta = 0.05$), the isolated data segments are classified into the groups of normal and abnormal data, as given in Table 2.2. The estimated PDFs based on the normal and abnormal data are shown in Fig. 2.9. By using the estimated PDFs, the current performance indices are respectively obtained by (2.5), (2.6) and (2.7) as FAR $= 14.86\%$, MAR $= 12.04\%$

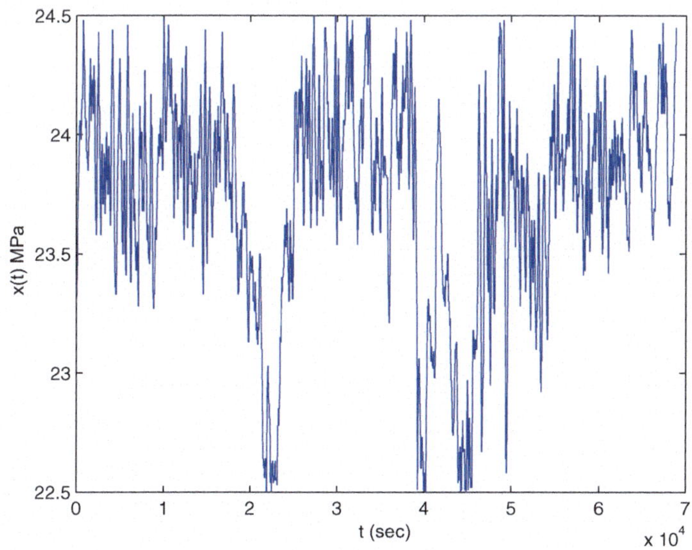

Fig. 2.8 Data samples of the industrial process variable, reprinted from Ref. [33], copyright 2012, with permission from IEEE

Table 2.1 The detected change points and the corresponding P-values, reprinted from Ref. [33], copyright 2012, with permission from IEEE

Change point	P-value
9395	1.10×10^{-144}
18354	0
25065	0
33886	6.80×10^{-72}
39063	0
46135	0
54410	0
63505	9.59×10^{-228}

Table 2.2 Classification of isolated data segments into normal and abnormal ones, reprinted from Ref. [33], copyright 2012, with permission from IEEE

Normal data	Abnormal data
$x(1{:}18353)$	$x(18354{:}25064)$
$x(25065{:}39062)$	$x(39063{:}46134)$
$x(46135{:}68873)$	

and AAD $= 0.1369$ sec. The FAR and MAR are much larger than their requirements, and the current alarm system needs to be redesigned.

Second, x_{th} is designed for the basic alarm generator in (2.1). This is a special case for the delay factor $n = 1$ in Case I at Sect. 2.1.3. According to the relation between FAR/MAR/AAD and x_{th}, the requirements of FAR$\leq 5\%$ and MAR$\leq 5\%$ impose

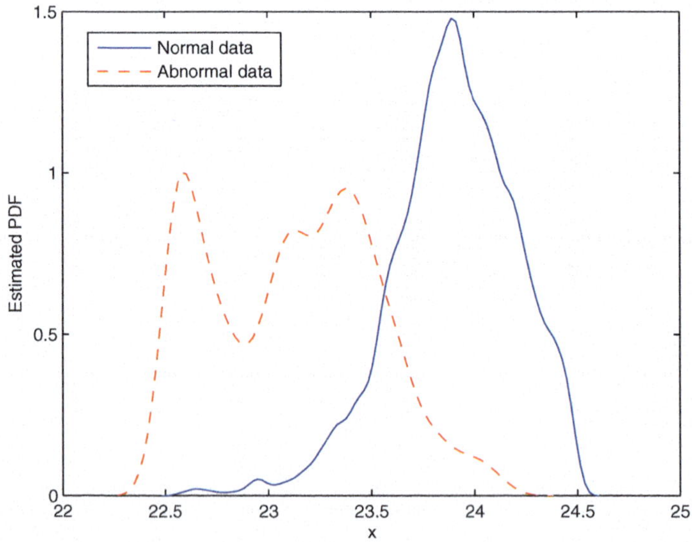

Fig. 2.9 Estimated PDFs from normal (blue solid) and abnormal data (red dash), reprinted from Ref. [33], copyright 2012, with permission from IEEE

Table 2.3 Performance indices for different values of n with the fixed alarm threshold $x_{th} = 23.6$, reprinted from Ref. [33], copyright 2012, with permission from IEEE

n	FAR	MAR	AAD
2	0.0468	0.0305	1.4294
3	0.0116	0.0060	2.8988
4	0.0025	0.0010	4.5694

Table 2.4 Valid ranges of the alarm threshold x_{th} for different values of n, reprinted from Ref. [33], copyright 2012, with permission from IEEE

n	x_{th} for FAR and MAR	x_{th} for AAD	x_{th}
2	[23.55,23.60]	[23.15,24]	[23.55,23.60]
3	[23.45,23.70]	[23.40,24]	[23.45,23.70]
4	[23.38,23.76]	[23.58,24]	[23.58,23.76]
5	[23.34,23.78]	[23.77,24]	[23.77,23.78]

the valid ranges of x_{th} to be $x_{th} \in [23, 23.37]$ and $x_{th} \in [23.79, 24]$, respectively. The two ranges have no intersection; as a result, it is impossible to satisfy all the three requirements by simply designing x_{th}.

Next, an alarm delay timer with the delay factor n is considered with the current alarm threshold $x_{th} = 23.6$ unchanged. Following the design procedure for the second case in Sect. 2.1.3, the three requirements FAR, MAR and AAD confine the valid ranges of n to be $n > 1$, $n > 1$ and $n < 5$ respectively, based on the relation between FAR/MAR/AAD and n. Hence, the valid range of n to meet with all the three requirements is $1 < n < 5$, and the performance indices after design are listed in Table 2.3.

The procedure for the third case in Sect. 2.1.3 is taken to design x_{th} and n, if both parameters are free to be changed. According to the relation between FAR and MAR as well as the relation between AAD and x_{th}, the valid ranges of x_{th} and n are given in Table 2.4.

Finally, the optimal values of n and x_{th} are determined. Here the weights in (2.22) are chosen as $\omega_1 = \omega_2 = \omega_3 = 1$. The two-dimensional grid search is implemented to find out the optimal values within the valid ranges given in Table 2.4. The minimal values of the loss function for different values of n and x_{th} are listed in Table 2.5, which says that the optimal parameters are $n = 4$ and $x_{th} = 23.59$. As a validation, the updated alarm system with $n = 4$ and $x_{th} = 23.59$ is applied to new data samples of x, some parts of which are in the normal condition as shown in Fig. 2.10. If the previous alarm system with $n = 1$ and $x_{th} = 23.59$ is used, the FAR is equal to 0.0849; by contrast, the FAR is reduced to 0.0463 for the updated alarm system.

Table 2.5 Minimal values of the loss function $J(x_{th}, n)$, reprinted from Ref. [33], copyright 2012, with permission from IEEE

n	x_{th}	Minimal value of $J(x_{th}, n)$
2	23.58	7.7036
3	23.58	2.2133
4	23.59	1.2548
5	23.77	3.8856

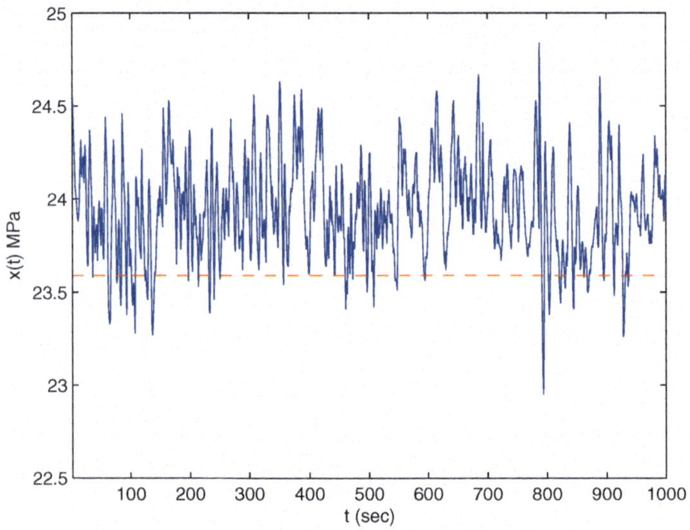

Fig. 2.10 Validation data samples of the industrial process variable

2.2 Alarm Delay Timers for Non-IID Process Variables

This section is on the optimal design of alarm delay timers for process variables that do not have to be IID [15]. A key observation is that all false alarms with alarm durations, or all missed alarms with alarm intervals, less than n samples are removed by a n-sample alarm delay timer. Thus, alarm durations and intervals provide the information for designing alarm delay timers to achieve required performance on false and missed alarms. Optimal values of the delay factor and alarm threshold are determined by minimizing distances between performance indices of false and missed alarms to their desired values. Bayesian estimation rule is exploited to estimate posterior probabilities of alarm durations and intervals, and to yield confidence intervals of the performance indices for false and missed alarms.

2.2.1 Problem Description

An alarm variable $x_{a,h}$ is usually generated by setting $x_{a,h}$ to '1' for the alarm state and '0' for the no-alarm state, when a process variable x crosses a high alarm threshold x_{th} [13, 25], i.e.,

$$x_{a,h}(t) = \begin{cases} 1, & \text{if } x(t) \geq x_{th}, \\ 0, & \text{otherwise.} \end{cases} \tag{2.24}$$

If x_{th} is a low alarm threshold, or both high and low alarm thresholds are involved, the counterpart of (2.24) can be obtained analogously. Without the loss of generality, a high-alarm threshold is considered in the sequel.

False alarms are the occurring alarms when x is in the normal condition. Missed alarms are the ones that do not occur when x is in the abnormal condition. Missed alarms are synonymous to quasi-false alarms if the high alarm threshold x_{th} is treated as a low alarm threshold for x in the abnormal condition. A quasi-alarm variable $x_{a,l}$ is generated as a counterpart of (2.24) to replace $x_{a,h}$ when x is in the abnormal condition, i.e.,

$$x_{a,l}(t) = \begin{cases} 1, & \text{if } x(t) \leq x_{th}, \\ 0, & \text{otherwise.} \end{cases} \tag{2.25}$$

Here x_{th} is taken as a low-alarm threshold. Consequently, by taking missed alarms as quasi-alarms, false and missed alarms can be treated in the same manner to greatly simplify mathematical notations.

False and missed alarms reduce the functionality, credibility and trustworthiness of an alarm system thus risking the operational safety of industrial processes. An optimally designed alarm threshold can reduce the numbers of false and missed alarms [6]. However, the alarm threshold alone sometimes may not be able to achieve a desired performance [33]. In this situation, the alarm threshold can jointly be applied with the n-sample alarm delay timer in (2.4). Taking $x_{a,h}$ in (2.24) as the input variable, the alarm delay timer yields the output variable,

$$x_{a,h}(t;n) = \begin{cases} 1, & \text{if } \sum_{i=t-n+1}^{t} x_{a,h}(i) = n \;\&\; x_{a,h}(t-1;n) = 0, \\ 0, & \text{if } \sum_{i=t-n+1}^{t} x_{a,h}(i) = 0 \;\&\; x_{a,h}(t-1;n) = 1, \\ x_{a,h}(t-1;n), & \text{otherwise.} \end{cases} \tag{2.26}$$

Analogously to (2.26), $x_{a,l}(t;n)$ can be generated by applying (2.4) to $x_{a,l}$.

The problem to be solved is to jointly design an alarm threshold x_{th} in (2.24) and the delay factor n of an alarm delay timer in (2.26) to achieve desired performance requirements on false and missed alarms. It is assumed that the process variable x is continuous-valued, being stationary with time-invariant statistical characteristics, and historical data of x in the normal and abnormal conditions are available. In particular, there is no need to make an assumption on whether the process variable x is IID or non-IID.

2.2.2 Observations on Alarm Durations and Intervals

This subsection presents key observations that a n-sample alarm delay timer removes all false alarms (missed alarms) with alarm durations (alarm intervals) less than n samples.

First, alarm durations and alarm intervals are defined. An alarm duration is the time duration from an alarm occurrence until its clearance [6, 29]. Here the alarm occurrence (alarm clearance) is the event that an alarm variable switches from '0' to '1' (from '1' to '0'). Mathematically, the k-th alarm duration for $x_{a,h}$ in (2.24) is

$$T_h(k) = t_0(k) - t_1(k), \tag{2.27}$$

where $t_1(k)$ and $t_0(k)$ are respectively the sampling indices of the k-th alarm occurrence and clearance,

$$x_a(t_1(k) - 1) = 0,\ x_a(t_0(k)) = 0,$$

$$\sum_{t=t_1(k)}^{t_0(k)} x_{a,h}(t) = t_0(k) - t_1(k),\ \text{for } t_0(k) > t_1(k).$$

As a counterpart of the alarm duration, an alarm interval is the time duration from an alarm clearance to the next alarm occurrence. Mathematically, the k-th alarm interval for $x_{a,h}$ in (2.24) is

$$T_h'(k) = t_1(k+1) - t_0(k). \tag{2.28}$$

Figures 2.11 and 2.12 illustrate the definitions of alarm durations and alarm intervals for a process variable x in the normal and abnormal conditions, respectively. Here x is the maximum difference among three sensor measurements for a drum level that is a very critical variable for safe operation of thermal power plants. If x is larger than 100 mm, then three sensors are not consistent, so that an alarm should be announced. In order to reduce the number of false (missed) alarms via alarm delay timers, alarm durations (alarm intervals) are the metrics to be exploited. Since missed alarms can be regarded as quasi-false alarms in $x_{a,l}$ in (2.25), the k-th alarm interval $T_h'(k)$ for $x_{a,h}$ become the k-th alarm duration $T_l(k)$ for $x_{a,l}$. Hence, only alarm durations are investigated in the sequel.

Second, there are two key observations: (i) the numbers of false and missed alarms depend on alarm thresholds; (ii) all false and missed alarms will be removed by a n-sample alarm delay timer if alarm durations T_h and T_l of these alarms are less than the delay factor n. As an illustration, Fig. 2.13a gives the time trend of a process variable x with a high-alarm threshold x_{th}. Figure 2.13b, c respectively show the false and missed alarms when a delay timer with $n = 1$ is applied. Figure 2.13d, e show the false and missed alarms when an alarm delay timer with $n = 3$ is applied to remove the false (missed) alarms whose durations (intervals) are less than 3 samples, as the red ones in Fig. 2.13b, c, respectively.

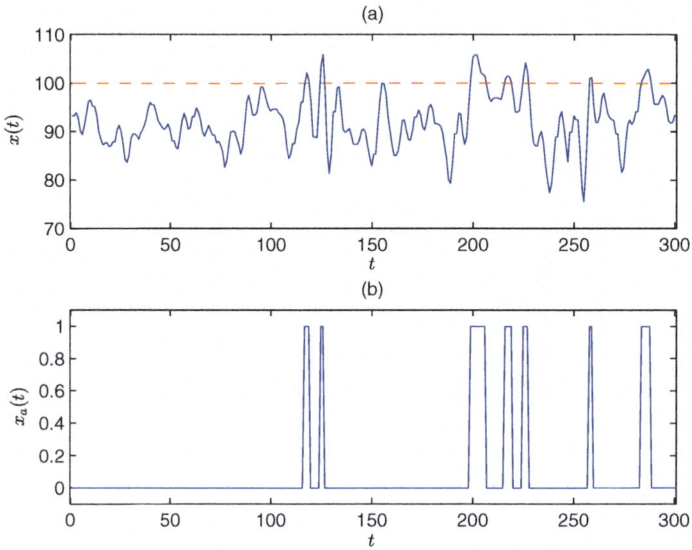

Fig. 2.11 Alarm durations and intervals for a process variable in the normal condition: **a** $x(t)$ (blue solid) with the high alarm threshold $x_{th,h}$ (red dash), **b** alarm variable $x_a(t)$

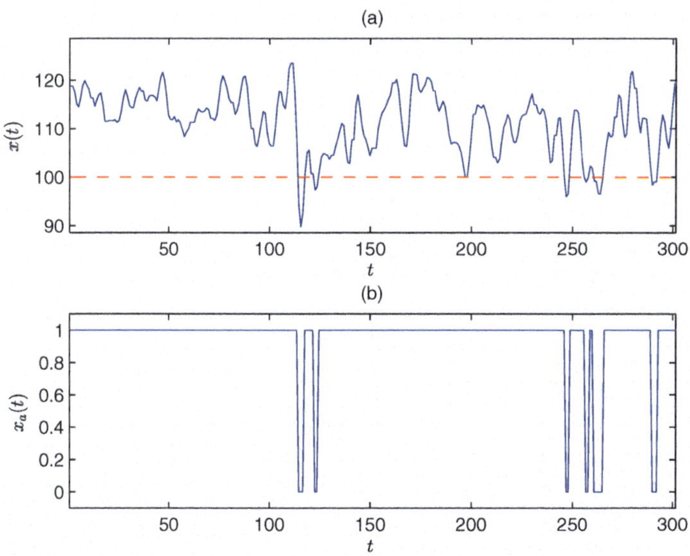

Fig. 2.12 Alarm durations and intervals for a process variable in the abnormal condition: **a** $x(t)$ (blue solid) with the high alarm threshold $x_{th,h}$ (red dash), **b** alarm variable $x_a(t)$

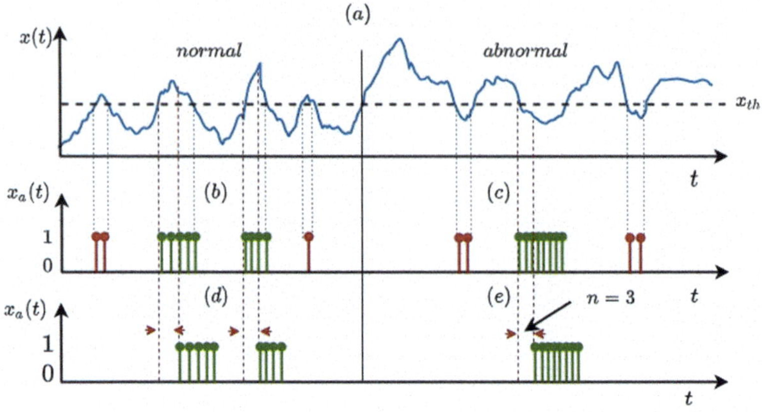

Fig. 2.13 **a** Process variable x in the normal and abnormal conditions with a high-alarm threshold x_{th}, **b** false alarms for $n = 1$, **c** missed alarms for $n = 1$, **d** false alarms for $n = 3$, **e** missed alarms for $n = 3$, reprinted from Ref. [15], copyright 2023, with permission from Elsevier

Finally, performance indices need to be defined. Let $\{T_h(k)\}_{k=1}^{K_h}$ and $\{T_l(k)\}_{k=1}^{K_l}$ be data samples of T_h and T_l, respectively. A false alarm ratio (FAR) is hereby defined as the ratio of false alarms to be left by the n-sample alarm delay timer,

$$F(x_{th}, n) = \frac{|\{T_h(k) \geq n\}_{k=1}^{K_h}|_c}{|\{T_h(k)\}_{k=1}^{K_h}|_c}. \tag{2.29}$$

Here $|\cdot|_c$ takes the cardinality of operand (the number of elements in a set). The numerator in (2.29) is the number of false alarms associated with $x_{a,h}$ in (2.24) before applying an alarm delay timer. Such a number depends on the alarm threshold x_{th}. The alarm delay timer removes all false alarms with durations less than n, so that the numerator in (2.29) is the number of false alarms in $\{T_h(k)\}_{k=1}^{K_h}$ to be left. Similarly, a missed alarm ratio (MAR) is defined for T_l as

$$M(x_{th}, n) = \frac{|\{T_l(k) \geq n\}_{k=1}^{K_l}|_c}{|\{T_l(k)\}_{k=1}^{K_l}|_c}. \tag{2.30}$$

The symbol $\{T_l(k) \geq n\}$ is the number of missed alarms in $\{T_l(k)\}_{k=1}^{K_l}$ whose T_l's are greater than the delay factor n, and $\{T_l(k)\}$ is the total number of data samples in $\{T_l(k)\}_{k=1}^{K_l}$.

At a certain pair of the alarm threshold x_{th} and the delay factor n, $F(x_{th}, n)$ in (2.29) and $M(x_{th}, n)$ in (2.30) may take values close to their desired ones. Figure 2.14a, b present scatter plots of T_h and T_l for different values of x_{th} by using an alarm delay timer with $n = 1$. The number of false alarms decreases as the value of x_{th} increases. Hence, $F(x_{th}, n)$ in (2.29) decreases with the increment in x_{th}. The opposite is, however, true for the missed alarms; thus, $M(x_{th}, n)$ in (2.30) increases

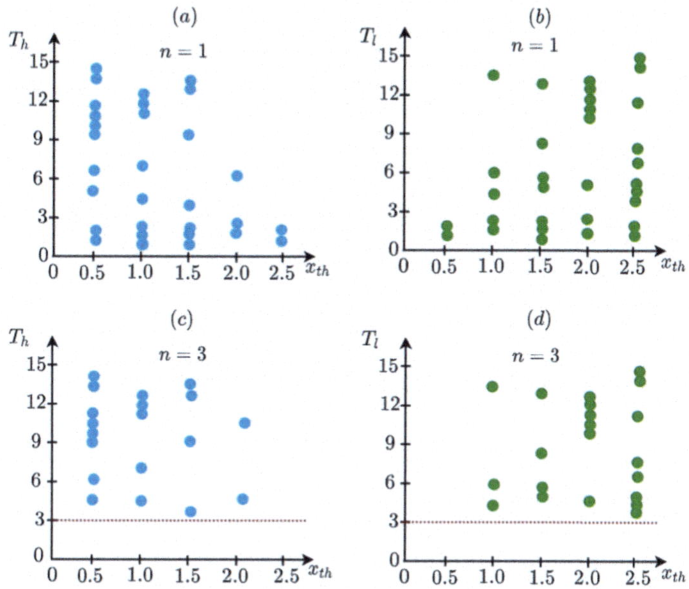

Fig. 2.14 **a** T_h (blue dots) for different values of the alarm threshold x_{th} by using an alarm delay timer with $n = 1$, **b** T_l (green dots) for different values of x_{th} with $n = 1$, **c** T_h for different values of x_{th} with $n = 3$, **d** T_l for different values of x_{th} with $n = 3$, reprinted from Ref. [15], copyright 2023, with permission from Elsevier

Table 2.6 FAR $F(x_{th}, n)$ and MAR $M(x_{th}, n)$ at different values of the alarm threshold x_{th} and the delay factor n, reprinted from Ref. [15], copyright 2023, with permission from Elsevier

$n = 1$			$n = 3$		
x_{th}	$F(x_{th}, n)$	$M(x_{th}, n)$	x_{th}	$F(x_{th}, n)$	$M(x_{th}, n)$
0.5	0.1	0.02	0.5	0.08	0
1	0.08	0.05	1	0.05	0.03
1.5	0.07	0.07	1.5	0.04	0.04
2	0.04	0.08	2	0.02	0.06
2.5	0.02	0.1	2.5	0	0.08

when the value of x_{th} increases. Table 2.6 lists the values of $F(x_{th}, n)$ and $M(x_{th}, n)$ for $n = 1$ at different values of x_{th}, shown in Fig. 2.14a, b. Let us take 0.05 as the desired values for both FAR and MAR. Table 2.6 says that there is no value of x_{th} at which both $F(x_{th}, n)$ and $M(x_{th}, n)$ for $n = 1$ are close to 0.05. This means that x_{th} alone cannot reduce the FAR and MAR to the desired value 0.05. As a result, an alarm delay timer should be used at the same time. Figure 2.14c, d show scatter plots of T_h (blue dots) for false alarms and T_l (green dots) for missed alarms at different values of x_{th} with $n = 3$. All false and missed alarms whose T_h and T_l are less than 3 are removed by an alarm delay timer with $n = 3$ (see Fig. 2.13), so that $F(x_{th}, n)$

and $M(x_{th}, n)$ decrease. The corresponding FAR and MAR are given at Table 2.6. The pair $x_{th} = 1.5$ and $n = 3$ gives $F(1.5, 3) = 0.04$ and $M(1.5, 3) = 0.04$, both of which are close to 0.05. Thus, $x_{th} = 1.5$ and $n = 3$ are selected as the final choices.

The definitions $F(x_{th}, n)$ in (2.29) and $M(x_{th}, n)$ in (2.30) are respectively equal to the cumulative probabilities of alarm durations T_h and T_l greater or equal to n. Hence, it is required to estimate the cumulative probabilities of T_h and T_l, whose estimation accuracy directly affects the FAR and MAR achieved by the designed values of x_{th} and n. All these issues are going to be solved in the next subsection by exploiting the Bayesian estimation rule.

2.2.3 Design of Delay Factors and Alarm Thresholds

It is ready to design the delay factor n and alarm threshold x_{th}. Given historical data of a process variable x in the normal and abnormal conditions, samples of alarm variables $x_{a,h}(t)$ and $x_{a,l}(t)$ are respectively obtained from (2.24) and (2.25) for different values of x_{th}. Their alarm durations are collected as $\{T_h(k)\}_{k=1}^{K_h}$ and $\{T_l(k)\}_{k=1}^{K_l}$. The optimal values of n and x_{th} are the solutions to a minimization problem,

$$(x_{th}, n)_{opt} = \underset{x_{th}, n}{\operatorname{argmin}}\ J(x_{th}, n), \tag{2.31}$$

where the loss function $J(x_{th}, n)$ is a weight sum of distances from the FAR and MAR to their desired values,

$$J(x_{th}, n) = \omega \cdot |F(x_{th}, n) - F_o| + (1 - \omega) \cdot |M(x_{th}, n) - M_o|. \tag{2.32}$$

Here F_o and M_o respectively are desired values of the FAR and MAR, and ω is a weighting factor in the range [0, 1].

Since false alarms occur when x is in the normal condition, they are equivalent to Type-I errors in statistics. Missed alarms occur when x is in the abnormal condition, so that they are the same as Type-II errors. Thus, $F(x_{th}, n)$ and $M(x_{th}, n)$ can be regarded as the probabilities of Type-I and Type-II errors, whose common choices are small positive values such as 0.05 [16].

The proposed method is composed of the following main steps:

1. For an alarm threshold x_{th} in (2.24) and a delay factor n of an alarm delay timer in (2.26), calculate samples of alarm durations T_h and T_l from normal and abnormal data to yield the alarm duration sets $\{T_h(k)\}_{k=1}^{K_h}$ and $\{T_l(k)\}_{k=1}^{K_l}$, respectively;
2. Obtain estimates of the FAR $F(x_{th}, n)$ in (2.29) and the MAR $M(x_{th}, n)$ in (2.30) respectively from the sets $\{T_h(k)\}_{k=1}^{K_h}$ and $\{T_l(k)\}_{k=1}^{K_l}$ via Bayesian estimation rule;
3. Repeat Steps 1 and 2 for different pairs of x_{th} and n;
4. Compute the solution of (2.31) as the optimal pair $(x_{th}, n)_{opt}$.

In these steps, different pairs of x_{th} and n are formulated as follows. Let T_{min} and T_{max} be the minimum and maximum values of alarm durations T_h's and T_l's. Since the parameter n takes positive integer values, candidate values of n are

$$T_{\min}, T_{\min}+1, T_{\min}+2, \ldots, T_{\max}. \tag{2.33}$$

The alarm threshold x_{th} is a continuous variable and has to be discretized. Let $[x_{th,min}, x_{th,max}]$ be the candidate range of x_{th} and Δx be the measurement resolution of x. Then, candidate values of x_{th} are

$$x_{th,\min}, x_{th,\min}+\Delta x, x_{th,\min}+2\Delta x, \ldots, x_{th,\max}. \tag{2.34}$$

The EEMUA guideline [13] recommends ranges of alarm deadband widths, which can be taken here as $x_{th,min} = (1-p)\cdot x_{th}$ and $x_{th,max} = (1+p)\cdot x_{th}$ with a ratio p. The value of p varies with the type of x, e.g., $p = 0.01$ for temperature variables and $p = 0.05$ for flow-rate variables.

Given the alarm duration sets, Step 2 estimates the FAR $F(x_{th}, n)$ and MAR $M(x_{th}, n)$ that are equivalent to cumulative density functions (CDFs). Various CDF estimators can be used here, such as moment-based estimators and maximum likelihood estimators. Here Bayesian estimators are exploited, because they yield probability distributions of the estimates, from which confidence intervals of the FAR and MAR can be readily obtained to measure estimation uncertainties.

The technical details are presented as follows to estimate $F(x_{th}, n)$ in (2.29) and $M(x_{th}, n)$ in (2.30). When the numbers of false and missed alarms are small, the uncertainties of $F(x_{th}, n)$ and $M(x_{th}, n)$ become large such that the designed alarm system may not achieve the desired FAR and MAR values for new alarm data. Notwithstanding, the frequency methods in (2.29) and (2.30) may not be able to measure the estimation uncertainties of $F(x_{th}, n)$ and $M(x_{th}, n)$. Therefore, the challenge is how to measure uncertainties in estimating $F(x_{th}, n)$ and $M(x_{th}, n)$. This technical challenge is solved here by exploiting Bayesian estimation rule.

For the presentation simplicity, the Bayesian estimate of $F(x_{th}, n)$ in (2.29) is given here, and the counterpart of $M(x_{th}, n)$ can be obtained analogously. Let us rewrite the definition of FAR in (2.29) as

$$F(x_{th}, n) = \sum_{i=1}^{n} \Theta(x_{th}, i). \tag{2.35}$$

where

$$\Theta(x_{th}, i) = \frac{|\{T_h(k) = i\}_{k=1}^{K_h}|_c}{|\{T_h(k)\}_{k=1}^{K_h}|_c}.$$

In Bayesian estimation rule, the parameter to be estimated is considered as a random variable. Thus, $\Theta(x_{th}, n)$ is not regarded as a constant value, but a random variable taking continuous values in the interval (0, 1). Because each alarm occurrence is usually independent, so is the alarm duration. As a result, the number of alarm durations being equal to n is defined as

$$Z_h(x_{th}, n) := |\{T_h(k) = n\}_{k=1}^{K_h}|_c, \tag{2.36}$$

and $Z_h(x_{th}, n)$ follows a binomial distribution with the probability of a success outcome as $\Theta(x_{th,n})$. Note that $Z_h(x_{th}, n)$ is explicitly dependent on n and implicitly dependent on x_{th}. If $z_h(x_{th}, n)$ and $\theta(x_{th}, n)$ are treated as random realizations of $Z_h(x_{th}, n)$ and $\Theta(x_{th}, n)$, respectively, then the conditional probability of $z_h(x_{th}, n)$ given $\theta(x_{th}, n)$ is,

$$\begin{aligned} &p_{Z_h(x_{th},n)|\Theta(x_{th},n)}\left(z_h(x_{th}, n) \mid \theta(x_{th}, n)\right) \\ &= \frac{K_h!(\theta(x_{th}, n))^{z_h(x_{th},n)}\,(1-\theta(x_{th}, n))^{K_h - z_h(x_{th},n)}}{z_h(x_{th}, n)!\,(K_h - z_h(x_{th}, n))!}. \end{aligned}$$

The joint probability distribution of $Z_h(x_{th}, n)$ and $\Theta(x_{th}, n)$ is

$$\begin{aligned} &p_{Z_h(x_{th},n),\Theta(x_{th},n)}\left(z_h(x_{th}, n), \theta(x_{th}, n)\right) \\ &= p_{Z_h(x_{th},n)|\Theta(x_{th},n)}\left(z_h(x_{th}, n) \mid \theta(x_{th}, n)\right) \cdot p_{\Theta(x_{th},n)}\left(\theta(x_{th}, n)\right), \end{aligned} \tag{2.37}$$

where $p_{\Theta(x_{th},n)}\left(\theta(x_{th}, n)\right)$ is a continuous uniform prior PDF,

$$p_{\Theta(x_{th},n)}\left(\theta(x_{th}, n)\right) = \begin{cases} 1, & \text{if } 0 < \theta(x_{th}, n) < 1, \\ 0, & \text{otherwise.} \end{cases} \tag{2.38}$$

Hence, the posterior probability distribution of $\Theta(x_{th}, n)$ given $z_h(x_{th}, n)$ is obtained by the Bayesian rule as

$$\begin{aligned} &p_{\Theta(x_{th},n)|Z_h(x_{th},n)}\left(\theta(x_{th}, n) \mid z_h(x_{th}, n)\right) \\ &= \frac{p_{Z_h(x_{th},n),\Theta(x_{th},n)}\left(z_h(x_{th}, n), \theta(x_{th}, n)\right)}{\int_0^1 p_{Z_h(x_{th},n),\Theta(x_{th},n)}\left(z_h(x_{th}, n), \theta(x_{th}, n)\right) \mathrm{d}\theta(x_{th}, n)}. \end{aligned} \tag{2.39}$$

Based on the posterior distribution in (2.39) and $\Theta_{(x_{th},n)}$, the Bayesian estimate of $\theta(x_{th}, n)$ usually takes the conditional mean as [23, 24]

$$\hat{\theta}(x_{th}, n) = \int_0^1 \theta(x_{th}, n) \cdot p_{\Theta(x_{th},n)|Z_h(x_{th},n)}\left(\theta(x_{th}, n) \mid z_h(x_{th}, n)\right) \mathrm{d}\theta(x_{th}, n). \tag{2.40}$$

The $(1-\alpha)100\%$ Bayesian interval bounding $\hat{\theta}(x_{th}, n)$ is obtained as the narrowest interval $\left[\hat{\theta}(x_{th}, n)_-, \hat{\theta}(x_{th}, n)_+\right]$ that satisfies an equality

$$\int_{\hat{\theta}(x_{th},n)_-}^{\hat{\theta}(x_{th},n)_+} p_{\Theta(x_{th},n)|z_h(x_{th},n)}\left(\theta(x_{th},n) \mid z_h(x_{th},n)\right) \mathrm{d}\theta(x_{th},n) = 1 - \alpha. \tag{2.41}$$

Here α is a certain significance level such as 0.05. Numerical methods are often used to solve (2.41) [12, 24], due to non-availability of analytical expressions of (2.37), (2.39) and (2.40). Here $\hat{\theta}(x_{th}, n)_+$ and $\hat{\theta}(x_{th}, n)_-$ respectively are the upper and lower bounds of FAR subject to a significance level α.

The smaller the deviation of $\hat{\theta}(x_{th}, n)_-$ or $\hat{\theta}(x_{th}, n)_+$ from $\hat{\theta}$, the more accurate $\hat{\theta}$ is [24]. To measure the credibility of $\hat{\theta}$, a credibility index $\eta_{(x_{th},n)}$ is formulated based on an inverse coefficient of variation (ICV). The ICV is defined as the ratio of an estimated mean to the standard deviation [14, 26]. The larger the value of ICV, the closer the estimate is to the true mean. Thus, $\eta(x_{th}, n)$ is formulated as

$$\eta(x_{th}, n) = \frac{\hat{\theta}(x_{th}, n)}{\max\left(\left|\hat{\theta}(x_{th}, n) - \hat{\theta}(x_{th}, n)_-\right), \left(\hat{\theta}(x_{th}, n)_+ - \hat{\theta}(x_{th}, n)\right|\right)}. \tag{2.42}$$

The estimate $\hat{\theta}(x_{th}, n)$ is credible if $\eta(x_{th}, n) \geq \mu$. A default value of μ is set to $\mu = 1$ to say that the uncertainty width is not larger than $\hat{\theta}(x_{th}, n)$.

In summary, the Bayesian estimate of the FAR $F(x_{th}, n)$ in (2.29) is obtained from $\hat{\theta}(x_{th}, n)$ in (2.40) as an estimate of the cumulative probability in (2.35),

$$\hat{F}(x_{th}, n) = \sum_{i=1}^{n} \hat{\theta}(x_{th}, i). \tag{2.43}$$

Counterparts of the MAR $M(x_{th}, n)$ in (2.30) can be estimated analogously from alarm intervals $\{T_l(k)\}_{k=1}^{K_l}$.

Finally, it is necessary to clarify the difference between the proposed method and existing methods for IID process variables (e.g., those in [2, 33]). Let us consider an IID process variable x being configured with a high-alarm threshold x_{th}. Equation (2.8) is an analytical relationship between the FAR and the parameter n of an alarm delay timer,

$$F_{iid}(x_{th}, n) = \frac{q_1^n \sum_{i=0}^{n-1} q_2^i}{q_1^n \sum_{i=0}^{n-1} q_2^i + q_2^n \sum_{j=0}^{n-1} q_1^j}, \tag{2.44}$$

where $q_1 = \int_{x_{th}}^{+\infty} q(x)\mathrm{d}x$, $q_2 = 1 - q_1$. Here $q(\cdot)$ is the probability distribution of x in the normal condition. Similarly, the MAR has an analytical expression in (2.15),

$$M_{iid}(x_{th}, n) = \frac{p_2^n \sum_{i=0}^{n-1} p_1^i}{p_2^n \sum_{i=0}^{n-1} p_1^i + p_1^n \sum_{j=0}^{n-1} p_2^j}, \tag{2.45}$$

where $p_2 = \int_{-\infty}^{x_{th}} p(x)\mathrm{d}x,\ p_1 = 1 - p_2$. Here $p(\cdot)$ is the probability distribution of x in the abnormal condition. The above expressions are developed based on the Markov chains in Figs. 2.3 and 2.4 that describe the transitions among no-alarm states and alarm states. Clearly, the Markov chain requires an assumption that x is IID. To be analogous to the proposed method in (2.31), the alarm threshold x_{th} and the delay factor n of an alarm delay timer can be designed by solving a minimization problem as

$$(x_{th}, n)_{opt} = \underset{x_{th}, n}{\operatorname{argmin}}\ J_{iid}(x_{th}, n), \tag{2.46}$$

where

$$J_{iid}(x_{th}, n) = \omega \cdot |F_{iid}(x_{th}, n) - F_o| + (1 - \omega) \cdot |M_{iid}(x_{th}, n) - M_o|. \tag{2.47}$$

Here ω, F_o and M_o are the same as those in (2.32). The existing method is based on probability distributions of process variables as shown by (2.44)–(2.47). By contrast, the proposed method is solely based on cumulative probabilities of alarm durations. In particular, none of (2.29)–(2.32) involve probability distributions of process variables. Hence, the proposed method is equally applicable to IID and non-IID process variables.

2.2.4 *Numerical and Industrial Examples*

This subsection provides two examples to support the Bayesian estimates and apply the proposed method to IID and non-IID industrial process variables.

2.2.4.1 Numerical Example

This example is used to illustrate the Bayesian estimation as Step 2 of the proposed method. Consider a discrete random variable $\mathbf{T_h}$ representing the alarm duration with a theoretical PMF as

$$p_{\mathbf{T}_h}(T_h) = \begin{cases} 0.1, & \text{if } T_h = 2, \\ 0.2, & \text{if } T_h = 3, \\ 0.4, & \text{if } T_h = 4, \\ 0.25, & \text{if } T_h = 5, \\ 0.05, & \text{if } T_h = 6. \end{cases} \tag{2.48}$$

The PMF is estimated by Step 2 of the proposed method. A set of sequences $\{T_h(k)\}_{k=1}^{K_h}$ for $K_h = 50$ is generated, and $Z_h(x_{th}, n)$ in (2.36) takes the values $8, 10, 20, 11, 1$ for $T_h = 2, 3, \cdots, 6$, respectively. Figure 2.15a presents the five estimated posterior PDFs in (2.39) for $T_h = 2, 3, \cdots, 6$. Figure 2.15b gives the Bayesian estimates of $\hat{\theta}(x_{th}, n)$ in (2.40) for $T_h = 2, 3, \cdots, 6$ together with the 95%

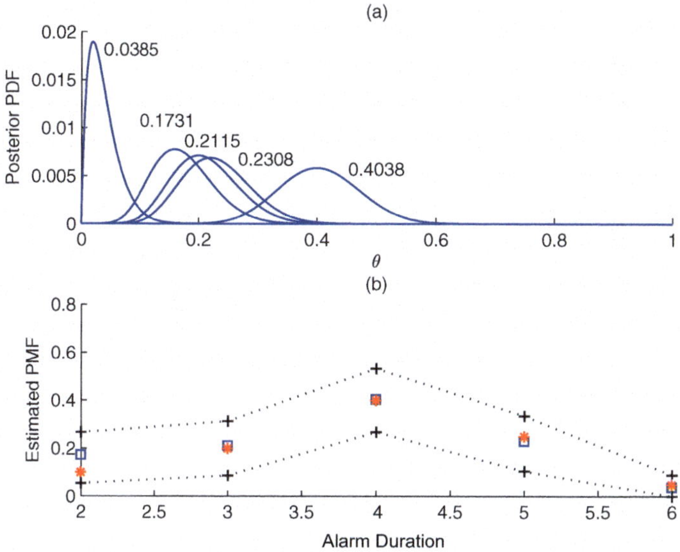

Fig. 2.15 **a** Estimated posterior PDFs, **b** Bayesian estimates (blue square) of $\hat{\theta}(x_{th}, n)$ with confidence intervals $\left[\hat{\theta}(x_{th}, n)_{-}, \hat{\theta}(x_{th}, n)_{+}\right]$ (black cross), and actual PMF values (red star) for $T_h = 2, 3, \cdots, 6$, reprinted from Ref. [30], copyright 2022, with permission from Elsevier

Table 2.7 Estimated PMFs and their confidence intervals, reprinted from Ref. [30], copyright 2022, with permission from Elsevier

T_h	$\hat{\theta}(x_{th}, n)$	$\left[\hat{\theta}(x_{th}, n)_{-}, \hat{\theta}(x_{th}, n)_{+}\right]$	$\eta(\hat{\theta}(x_{th}, n))$
2	0.1731	[0.0540,0.2660]	1.4535
3	0.2115	[0.0870,0.3130]	1.6986
4	0.4038	[0.2680,0.5320]	2.9728
5	0.2308	[0.1050,0.3350]	1.8349
6	0.0385	[0.0000,0.0900]	0.7465

confidence interval $\left[\hat{\theta}(x_{th}, n)_{-}, \hat{\theta}(x_{th}, n)_{+}\right]$ in (2.41). Table 2.7 lists the concrete values of $\hat{\theta}(x_{th}, n)$ and the confidence intervals $\left[\hat{\theta}(x_{th}, n)_{-}, \hat{\theta}(x_{th}, n)_{+}\right]$.

In order to valid confidence intervals, the above calculations are repeated for 100 Monte Carlo simulations. As shown in Fig. 2.16, the numbers of simulations that an actual PMF value lies inside its 95% confidence interval are respectively equal to 95, 94, 96, 97, 100 for $T_h = 2, 3, \cdots, 6$. The above results clearly support the validness of Bayesian estimates and their confidence intervals.

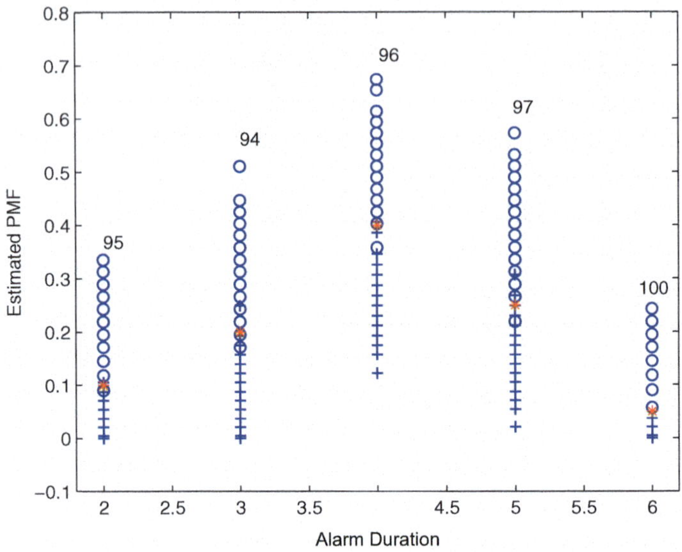

Fig. 2.16 Upper (blue circle) and lower (blue cross) bounds of confidence intervals of Bayesian estimates in 100 Monte Carlo simulations and actual PMF values (red star), reprinted from Ref. [30], copyright 2022, with permission from Elsevier

2.2.4.2 Industrial Examples

The proposed method is applied to two process variables at a large-scale power plant to remove false and missed alarms. The first is IID and the second is non-IID.

The first process variable with tagname 4U94AIO1 is the generator frequency x_1. The generator frequency should be synchronous to the power grid frequency so that x_1 is monitored at a high-alarm threshold $x_{1,th} = 50.3$ Hz. There are no data on missed alarms so that only false alarms are considered here. A 24-hour data samples of x_1 with sampling period 0.2 sec are available, where the number of false alarms in the current alarm variable is 16416. A section of the data samples and its alarm data are shown in Fig. 2.17. Based on operational requirements, the alarm threshold is allowed to adjust in the range [50.30, 50.33] with a measurement resolution 0.001 Hz. The proposed method is applied on the basis of the 24-hour data. Different pairs of n and $x_{1,th}$ are obtained from (2.33) and (2.34). Table 2.8 lists the detailed results, and Fig. 2.18 presents the loss function $J(x_{1,th}, n)$. Equation (2.31) yields the optimal design parameters $x_{1,th} = 50.327$ and $n = 48$ with $J(50.327, 48) = 0.0001$. As a comparison, the existing method in [2, 33] is also applied to the same 24-hour data, and Fig. 2.19 shows the loss function $J_{iid}(x_{1,th}, n)$ in (2.47). The design results $x_{1,th} = 50.327$ and $n = 48$ are with the minimum loss function $J_{iid}(50.327, 48) = 0.0000828$, and are the same as the counterparts in the proposed method. These designed parameters are applied on the next 48-hours data samples and the results are given in Table 2.9 and Fig. 2.20. The same results are achieved by both the

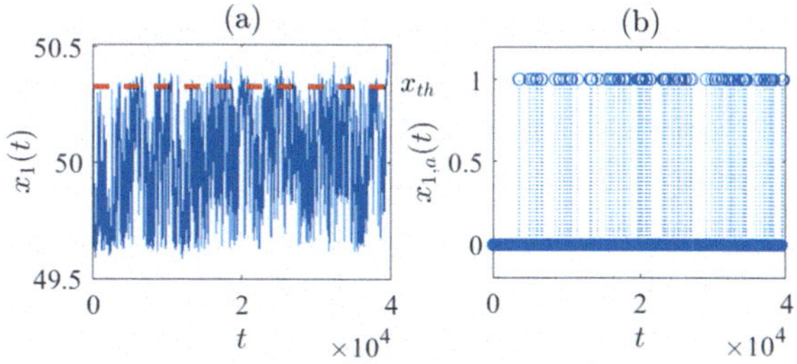

Fig. 2.17 **a** The generator frequency x_1 in the normal condition, **b** alarm variable corresponding to x_1, reprinted from Ref. [15], copyright 2023, with permission from Elsevier

Table 2.8 Different pairs of $x_{1,th}$ and n, as well as probability $F(x_{1,th}, n)$ in (2.32) with their distances $J(x_{1,th}, n)$ to the desired value $F_o = 0.05$, reprinted from Ref. [15], copyright 2023, with permission from Elsevier

$x_{1,th}$	n	$F(x_{1,th}, n)$	$J(x_{1,th}, n)$
50.33	47	0.04905	0.00095
50.331	47	0.0494	0.00058
50.326	48	0.0510	0.00156
50.327	48	0.499	0.0001
50.328	48	0.0384	0.0116
50.323	49	0.0700	0.002
50.324	49	0.04909	0.000909
50.325	49	0.0639	0.0139

proposed and existing methods. This is attributed to the fact that the process variable is IID which is correctly assumed and handled by both methods.

The second process variable with tagname 4U29FT02A02 is the inlet flow rate x_2 for a feedwater pump. A schematic diagram of the boiler pump system is shown in Fig. 2.21. One electric pump and two steam pumps A and B supply water to the boiler drum, where the water is heated into high-pressure high-temperature steam. If the pumps are in abnormal conditions, there may be insufficient or too much water entering the drum such that the required water level cannot be achieved. A too high or low water level will trigger an emergency shutdown, so that it is important to monitor inlet flow rates of the pumps. The inlet flow rate x_2 of the steam pump A is monitored by setting a high-alarm threshold at $x_{2,th} = 560\ m^3/s$. A 24-hour data samples of x_2 are obtained, where some parts of data samples are presented in Fig. 2.22. The numbers of false and missed alarms in the current alarm variable respectively are 11560 and 8467. According to some process knowledge, the alarm threshold can be adjusted in the range [560, 565] with a measurement resolution of the flow rate as $0.01\ m^3/s$.

The proposed method is applied to the 24-hour historical data to jointly design the alarm threshold and the delay factor in order to make both the FAR and MAR

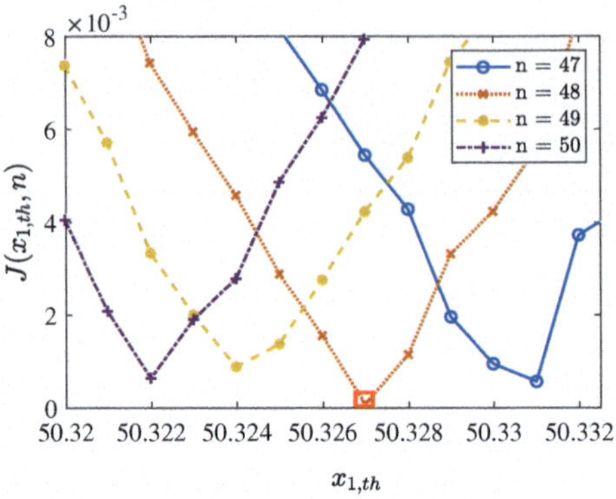

Fig. 2.18 The loss function $J(x_{1,th}, n)$ for the generator frequency, reprinted from Ref. [15], copyright 2023, with permission from Elsevier

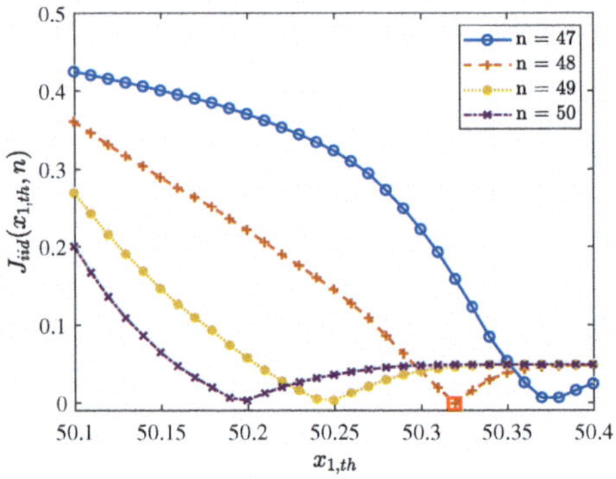

Fig. 2.19 The loss function $J_{iid}(x_{1,th}, n)$ for the generator frequency, reprinted from Ref. [15], copyright 2023, with permission from Elsevier

Table 2.9 The number N_f of false alarms for the current alarm variable, the number $N_{f,opt}$ of false alarms as well as the FAR F_{opt} from the proposed method by using $x_{1,th} = 50.327$ and $n = 48$, together with the counterparts $N_{f,iid}$ and F_{iid} from the existing method by using $x_{1,th} = 50.327$ and $n = 48$ for the generate frequency x_1, reprinted from Ref. [15], copyright 2023, with permission from Elsevier

N	$N_{f,opt}$	F_{opt}	N_{iid}	$F_{f,iid}$
226	11	0.0487	11	0.0487

close to a desired value 0.05. Different pairs of n and $x_{2,th}$ are formulated according to (2.33) and (2.34). Table 2.10 lists the values of $F(x_{2,th}, n)$ and $M(x_{2,th}, n)$. Figure 2.23 presents the loss function $J(x_{2,th}, n)$ in (2.32). The alarm threshold $x_{2,th} = 561.5$ and the parameter $n = 74$ of an alarm delay timer are selected as the optimal design parameters.

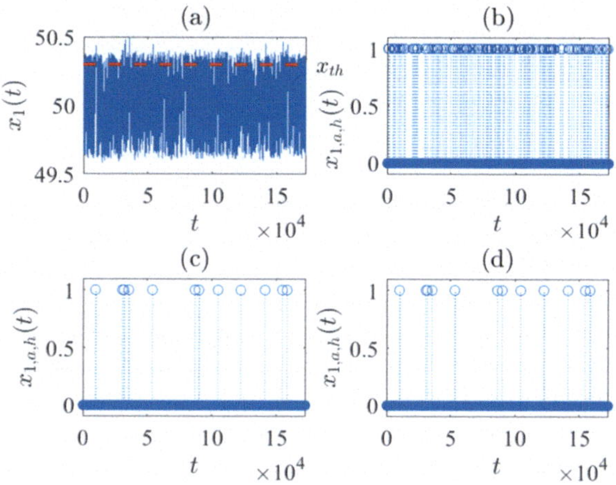

Fig. 2.20 **a** The generator frequency in a normal condition, **b** alarm variable for the basic alarm generator, **c** alarm variable for the proposed method and **d** alarm variable for the existing method, reprinted from Ref. [15], copyright 2023, with permission from Elsevier

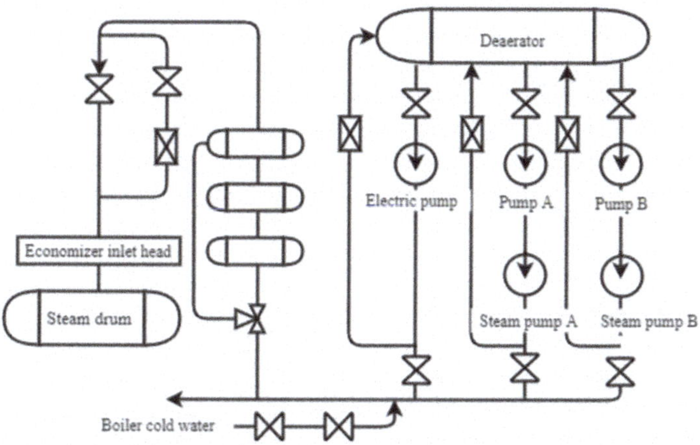

Fig. 2.21 Schematic diagram of a boiler pump system, reprinted from Ref. [15], copyright 2023, with permission from Elsevier

For a comparison, $x_{2,th}$ and n are redesigned based on the same 24-hour data via an existing method in [2, 33]. The loss function $J_{iid}(x_{2,th}, n)$ in (2.47) is shown in Fig. 2.24 where the minimum value $J_{iid}(561.5, 2) = 0.016$ is achieved at $x_{2,th} = 561.5$ and $n = 2$. Two designed alarm systems are applied to data samples for 8 next consecutive weeks. As shown by the results in Table 2.11 as well as Figs. 2.25 and 2.26, the FAR and MAR from the proposed method are much closer to 0.05 than the counterparts from the existing method. In particular, both methods take the same alarm threshold $x_{2,th} = 561.5$, but very different values of n for the alarm delay timer.

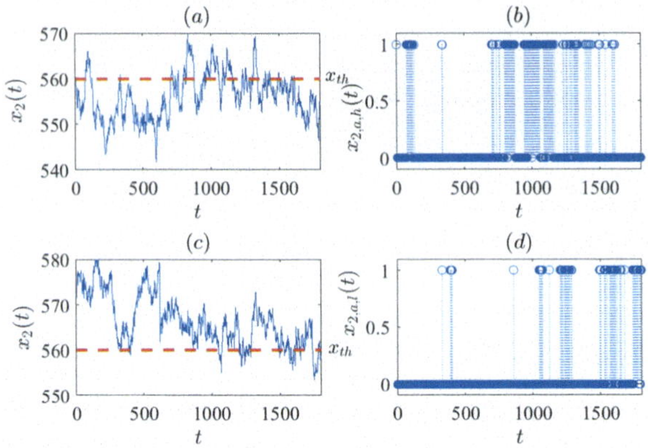

Fig. 2.22 **a** The inlet flow rate x_2 in the normal condition, **b** false alarms, **c** the inlet flow rate x_2 in the abnormal condition, and **d** missed alarms, reprinted from Ref. [15], copyright 2023, with permission from Elsevier

Table 2.10 Different pairs of $x_{2,th}$ and n, as well as probabilities $F(x_{2,th}, n)$ and $M(x_{2,th}, n)$ in (2.32) with their distances to the desired values $F_o = 0.05$ and $M_o = 0.05$ for the inlet flow rate x_2, reprinted from Ref. [15], copyright 2023, with permission from Elsevier

$x_{2,th}$	n	$F(x_{2,th}, n)$	$\lvert F(x_{2,th}, n) - F_o\rvert$	$M(x_{2,th}, n)$	$\lvert M(x_{2,th}, n) - M_o\rvert$	$J(x_{2,th}, n)$
560.5	72	0.069	0.019	0.034	0.016	0.018
561	72	0.059	0.009	0.046	0.004	0.0065
561.21	72	0.055	0.005	0.055	0.005	0.005
561.35	72	0.054	0.004	0.058	0.008	0.006
561.5	72	0.053	0.0025	0.061	0.011	0.007
560.5	73	0.067	0.017	0.031	0.019	0.018
561	73	0.057	0.007	0.044	0.011	0.009
561.25	73	0.053	0.003	0.053	0.003	0.003
561.5	73	0.049	0.001	0.055	0.0046	0.0028
561.5	73	0.048	0.002	0.061	0.005	0.0035
560.5	74	0.068	0.018	0.039	0.011	0.015
561	74	0.056	0.006	0.044	0.006	0.003
561.5	74	0.049	0.001	0.052	0.0044	0.0027
561.2	74	0.045	0.005	0.06	0.01	0.008
561.26	74	0.0519	0.0019	0.0519	0.0019	0.0019
560.5	75	0.086	0.036	0.0182	0.032	0.034
561	75	0.058	0.008	0.0425	0.025	0.0165
561.5	75	0.066	0.016	0.053	0.003	0.0095
562	75	0.045	0.005	0.068	0.018	0.034
562	75	0.017	0.033	0.024	0.026	0.03

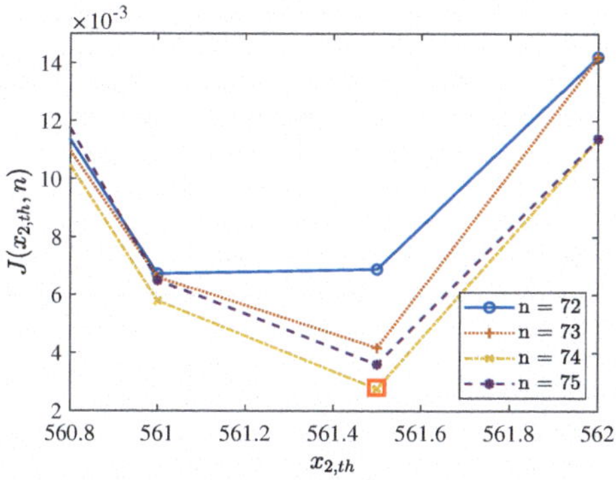

Fig. 2.23 The loss function $J(x_{2,th}, n)$ for the second process variable, reprinted from Ref. [15], copyright 2023, with permission from Elsevier

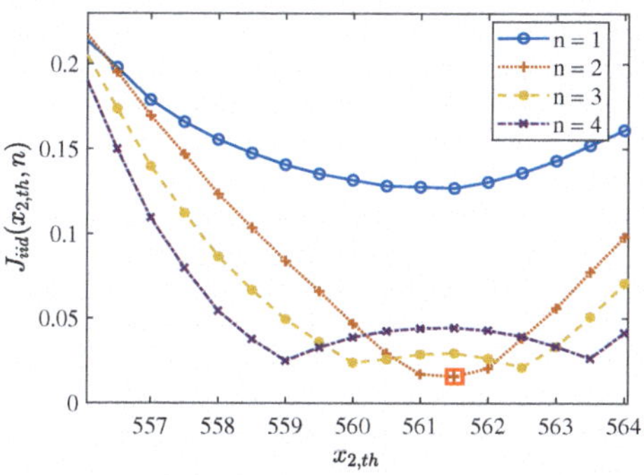

Fig. 2.24 The loss function $J_{iid}(x_{2,th}, n)$ for the second process variable, reprinted from Ref. [15], copyright 2023, with permission from Elsevier

This is due to a fact that the flow rate is clearly non-IID as revealed in Figs. 2.22, 2.25 and 2.26.

2.3 Alarm Deadbands for Non-IID Process Variables

This section formulates an index to determine whether alarm deadbands or alarm delay timers are suitable for removing false alarms in given process variables [32], and designs an optimal deadband width to achieve desired performance indices [31]. First, the deadband index is defined from two metrics of normalized alarm durations and deviations. If the index is larger than a threshold, then alarm deadbands are suitable to remove false alarms associated with small alarm deviations and large alarm durations. Second, an optimal value of the deadband width is determined based on the cumulative probability of alarm deviations. Bayesian estimation rule is used to evaluate uncertainties of the estimated cumulative probability. Unlike existing meth-

Table 2.11 The numbers N_f and N_m of false and missed alarms in the current alarm variable, the number $N_{f,opt}$ of false alarms and the number $N_{m,opt}$ of missed alarms as well as the FAR F_{opt} and MAR M_{opt} from the proposed method by using $x_{2,th} = 561.5$ and $n = 74$, together with the counterparts $N_{f,iid}$, $N_{m,iid}$, F_{iid} and M_{iid} from the existing method by using $x_{2,th} = 561.5$ and $n = 2$ for the inlet flow rate x_2, reprinted from Ref. [15], copyright 2023, with permission from Elsevier

Week	N_f	$N_{f,opt}$	F_{opt}	$N_{f,iid}$	F_{iid}	N_m	$N_{m,opt}$	M_{opt}	$N_{m,iid}$	M_{iid}
1	250	12	0.048	23	0.091	218	10	0.049	21	0.096
2	345	17	0.049	31	0.089	256	12	0.046	25	0.097
3	233	11	0.047	22	0.093	199	9	0.045	22	0.110
4	510	22	0.043	44	0.087	224	11	0.049	24	0.106
5	219	11	0.050	20	0.095	189	9	0.048	18	0.094
6	453	21	0.046	46	0.101	221	11	0.049	21	0.094
7	392	19	0.048	36	0.091	250	12	0.046	22	0.087
8	321	14	0.044	34	0.105	248	12	0.048	27	0.109

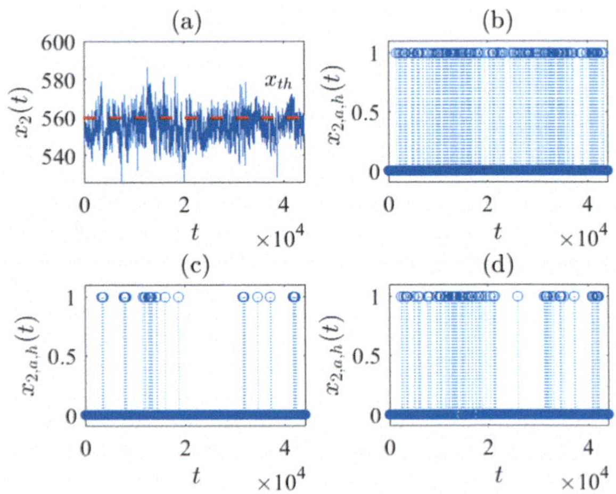

Fig. 2.25 **a** The inlet flow rate x_2 in the normal condition, **b** false alarms for the basic alarm generation mechanism, **c** false alarms for the proposed method and **d** false alarms for the existing method, reprinted from Ref. [15], copyright 2023, with permission from Elsevier

ods being limited to IID process variables, the proposed method extracts information from alarm deviations for designing alarm deadbands, and has no discriminations to IID and non-IID process variables.

2.3.1 *Problem Description*

Alarm deadbands are powerful tools that are frequently used in practice to remove nuisance alarms [18]. For a high alarm threshold, an alarm deadband with the width d raises an alarm event if $x(t)$ is larger than or equal to $x_{th} + d$, and clears an alarm event if $x(t)$ is smaller than x_{th}, i.e.,

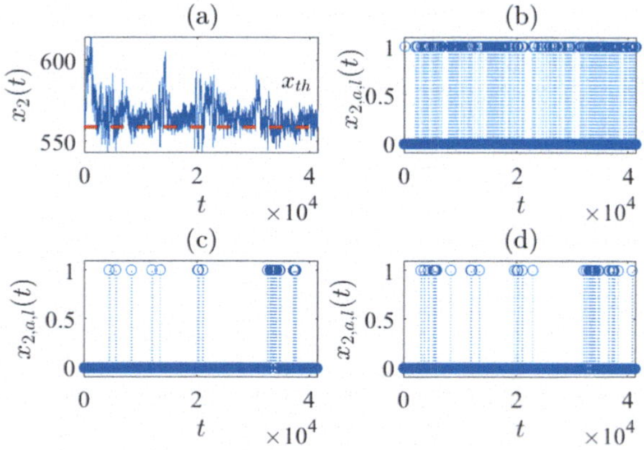

Fig. 2.26 **a** The inlet flow rate x_2 in the abnormal condition, **b** missed alarms for the basic alarm generation mechanism, **c** missed alarms for the proposed method and **d** missed alarms for the existing method, reprinted from Ref. [15], copyright 2023, with permission from Elsevier

$$x_a(t; d) = \begin{cases} 1, & \text{if } x(t) \geq x_{th} + d, \\ 0, & \text{if } x(t) < x_{th}, \\ x_a(t-1; d), & \text{otherwise.} \end{cases} \tag{2.49}$$

Thus, two different thresholds are used to raise and clear alarm events. Analogously, an alarm deadband is defined for a low alarm threshold,

$$x_a(t; d) = \begin{cases} 1, & \text{if } x(t) \leq x_{th} - d, \\ 0, & \text{if } x(t) > x_{th}, \\ x_a(t-1; d), & \text{otherwise.} \end{cases} \tag{2.50}$$

The alarm deadband is commonly implemented as an embedded function in distributed control systems, e.g., the High Low Alarm (HLALM) function depicted in Fig. 2.27. The parameters HDB and LDB in the function are the counterparts of d in (2.49) and (2.50), respectively.

There is another definition of alarm deadbands in practice [13]. If an alarm deadband of width d is configured for the high-alarm threshold, then an alarm occurs when $x(t)$ is larger than a higher-alarm threshold $x_{th} + d$; however, the occurring alarm is not cleared until $x(t)$ is smaller than a lower-alarm threshold $x_{th} - d$ [13]. Mathematically, the corresponding alarm variable is

$$\tilde{x}_a(t; d) = \begin{cases} 1, & \text{if } x(t) \geq \tilde{x}_{th} + \tilde{d}, \\ 0, & \text{if } x(t) < \tilde{x}_{th} - \tilde{d}, \\ \tilde{x}_a(t-1; d), & \text{otherwise.} \end{cases} \tag{2.51}$$

The two sets of definitions are indeed equivalent: (2.49) and (2.51) are the same by taking the parameters as

$$x_{th} = \tilde{x}_{th} - \tilde{d}, \; d = 2\tilde{d}.$$

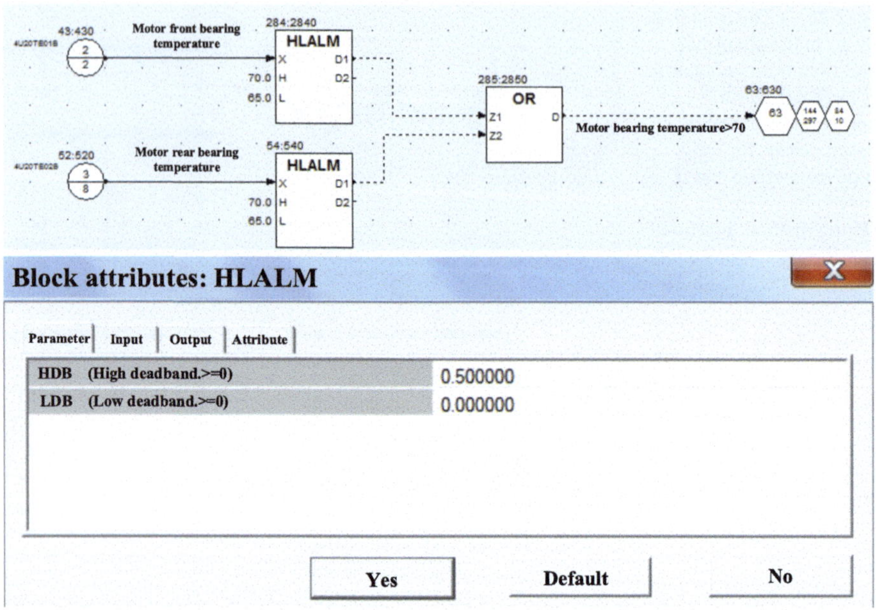

Fig. 2.27 An embedded alarm deadband function in distributed control systems, reprinted from Ref. [31], copyright 2023, with permission from IEEE

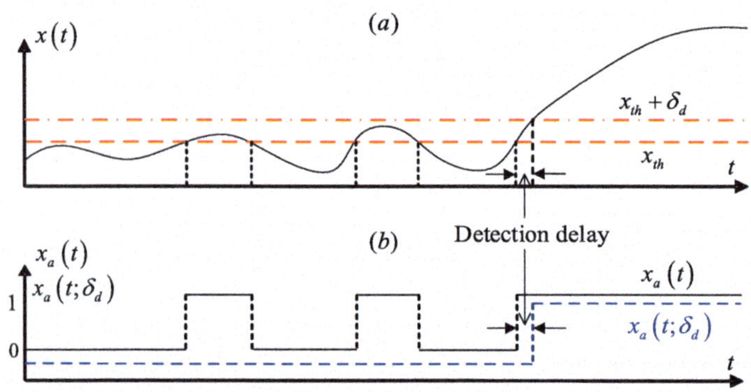

Fig. 2.28 **a** $x(t)$ (black solid), the high-alarm threshold x_{th} (red dash) and $x_{th} + d$ (red dot-dash), **b** $x_a(t)$ (black solid), $x_a(t; d)$ (blue dash, shifted downwards for 0.1), reprinted from Ref. [31], copyright 2023, with permission from IEEE

In the sequel, the first definition of alarm deadbands in (2.49) for a high alarm threshold is considered without the loss of generality.

The functionality of an alarm deadband is further illustrated in Fig. 2.28. Some samples of $x(t)$ are larger than a high alarm threshold x_{th}, but are smaller than $x_{th} + d$. As a result, some false alarms are present in $x_a(t)$, but are absent in $x_a(t; d)$. Clearly, more false alarms can be removed for a larger value of d; however, the price is the increment of a detection delay between the time instant of $x_a(t)$ running to the alarm state and that of $x_a(t; d)$ switching from '0' to '1', both due to the same abnormal condition. Hence, d should be chosen to make a good balance between the detection delay and the number of false alarms to be removed. Note that the deadband width d in (2.49) is irrelevant to missed alarms; in other words, if both false and missed alarms are considered, then the alternative definition (2.51) should be used.

Detection delays cannot be explicitly calculated without historical samples of $x(t)$ in abnormal conditions, because the time instants of $x_a(t)$ and $x_a(t; d)$ running into alarm states are dependent on when the samples of $x(t)$ in abnormal conditions go beyond x_{th} and $x_{th} + d$, respectively. However, historical samples in abnormal conditions are hardly available sometimes, because many process variables in practice stay at normal conditions for most of time instances and run into abnormal conditions occasionally. Hence, this section designs alarm deadbands solely based on historical samples in normal conditions. The deadband width affects two performance indices of the false alarm rate and detection delay [1]: when the deadband width increases, the false alarm rate decreases and the detection delay increases accordingly. Hence, the detection delay can be implicitly considered by selecting a proper value of the deadband width to control the number of false alarms to be removed, instead of removing all false alarms in order to avoid excessive detection delays.

Suppose that some historical data samples $\{x(t)\}_{t=1}^{T}$ are known to be in normal conditions according to process knowledge, and false alarms are observed in the corresponding alarm variable $x_a(t)$ in (2.1). In this section, the first problem to be solved is to determine whether an alarm deadband is suitable for removing false alarms for a given process variable. If the answer is positive, then the second problem to be solved is to design an optimal value of the deadband width d. The optimality lies at choosing d such that the percentage of false alarms in $x_a(t; d)$ with respect to the counterpart in $x_a(t)$ is closest to a desired value.

2.3.2 Formulation of a Deadband Index

This subsection formulates an index to determine whether an alarm deadband is suitable to remove nuisance alarms for given process variables.

First, two basic metrics are introduced. The first metric is the alarm duration in (2.27), denoted as $T_a(k)$ representing the k-th alarm duration. The second metric is the alarm deviation as the maximum value of the deviation between $x(t)$ and x_{th} during the time period from an alarm occurrence until its clearance. Mathematically, the k-th alarm deviation is

$$D_a(k) = \max\{(x(t_1(k)) - x_{th}), \cdots, (x(t_0(k)) - x_{th})\}, \tag{2.52}$$

where $t_1(k)$ and $t_0(k)$ are respectively the sampling indices of the k-th alarm occurrence and clearance defined in (2.27). Figure 2.29 illustrates the metrics $T_a(k)$ and $D_a(k)$ for a process variable $x(t)$ configured with a high-alarm threshold x_{th}.

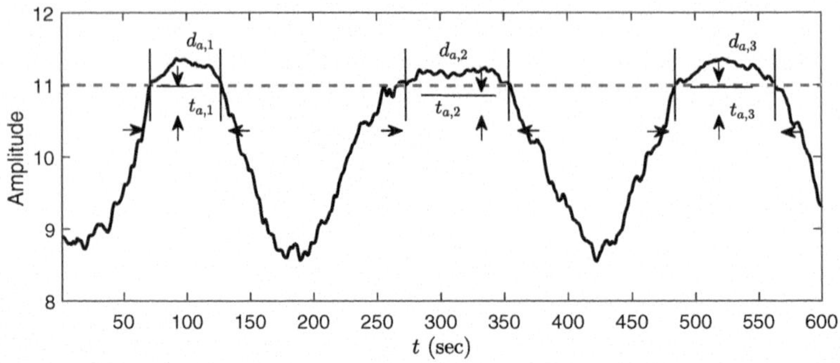

Fig. 2.29 Illustrations of $T_a(k)$ and $D_a(k)$ for a process variable $x(t)$ (black solid) configured with a high-alarm threshold x_{th} (red dash), reprinted from Ref. [32], copyright 2019, with permission from IEEE

Table 2.12 Default values of $T_{a,max}$ and p for $D_{a,max}$, reprinted from Ref. [32], copyright 2019, with permission from IEEE

Variable Type	$T_{a,max}$ (sec)	p (%)
Flow	15	5
Level	60	5
Pressure	15	2
Temperature	60	1

Alarm durations and deviations are normalized so that the two metrics in different units are comparable. The normalized alarm duration is

$$T_{a,n}(k) = \frac{T_a(k)}{T_{a,max}}, \tag{2.53}$$

where $T_{a,max}$ is the acceptable maximum value of alarm durations. The normalized alarm deviation is

$$D_{a,n}(k) = \frac{D_a(k)}{D_{a,max}}, \tag{2.54}$$

where $D_{a,max}$ is the acceptable maximum value of alarm deviations. The physical meaning of $T_{a,max}$ ($D_{a,max}$) is the acceptable maximum value of alarm durations (alarm deviations) for users to treating occurring alarms as nuisance ones. In other words, if an occurring alarm has the alarm duration (deviation) larger than $T_{a,max}$ ($D_{a,max}$), then it could be a true alarm, instead of a nuisance alarm, so that users have to take a further investigation. The recommended values for alarm delay timers in the EEMUA guideline can be taken as the default values of $T_{a,max}$, as shown in Table 2.12. A reasonable choice of $D_{a,max}$ is proportional to the alarm threshold x_{th}, i.e.,

$$D_{a,max} = p \cdot x_{th}. \tag{2.55}$$

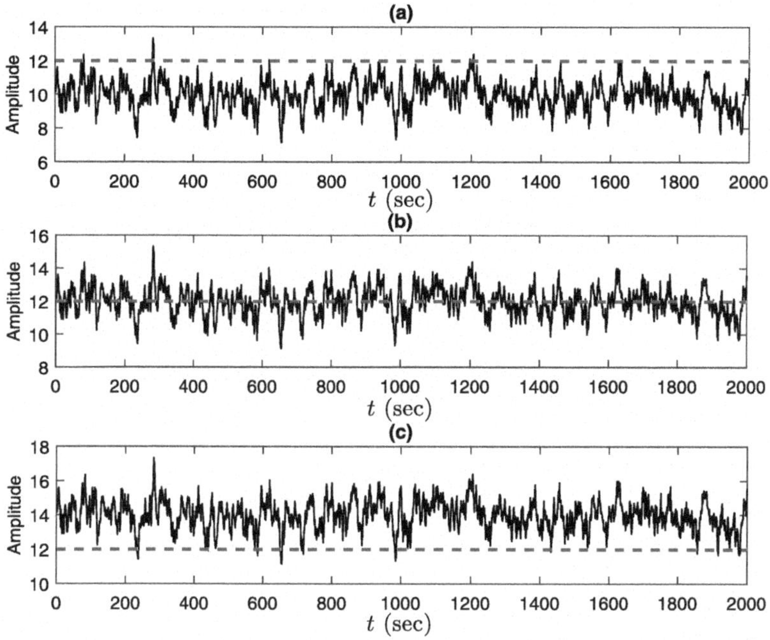

Fig. 2.30 **a** data samples of $x_1(t)$ (black solid) with the high-alarm threshold $x_{th} = 12$ (red dash); **b** data samples of $x_2(t)$ (black solid) with $x_{th} = 12$ (red dash); **c** data samples of $x_3(t)$ (black solid) with $x_{th} = 12$ (red dash), reprinted from Ref. [32], copyright 2019, with permission from IEEE

Table 2.12 gives default values of the ratio $p \in (0, 1)$ for different types of process variables, following the recommended values for alarm deadbands in the EEMUA guideline [13].

A method is proposed here to determine $T_{a,max}$ and $D_{a,max}$ for types of process variables that are not listed in Table 2.12. Three variables $x_1(t)$, $x_2(t)$ and $x_3(t)$ are presented in Fig. 2.30 to explain the rationale of this method. The variables are generated as

$$x_1(t) = -0.85 \cdot x_1(t-1) + e(t) + 10,$$
$$x_2(t) = -0.85 \cdot x_2(t-1) + e(t) + 12,$$
$$x_2(t) = -0.85 \cdot x_2(t-1) + e(t) + 14,$$

where $e(t)$ is Gaussian white noise with zero mean and variance 0.25. The three variables have the same high-alarm threshold $x_{th} = 12$. It is obvious that the alarms occurring in Fig. 2.30a, c are very likely to be nuisance (true) alarms, because most data samples of $x_1(t)$ ($x_3(t)$) are below (above) x_{th}. As stated earlier, the physical meaning of $T_{a,max}$ ($D_{a,max}$) is the acceptable maximum value of alarm durations (alarm deviations) for nuisance alarms. Hence, it is reasonable to design $T_{a,max}$ and $D_{a,max}$ according to alarm durations and deviations of $x_2(t)$ in Fig. 2.30b, where x_{th} is equal to the mean value of $x_2(t)$.

The proposed method consists of the following steps.

- Step-1: A new alarm variable $x'_a(t)$ is generated by setting the alarm threshold x_{th} to the sample mean of $x(t)$, i.e.,

$$x'_a(t) = \begin{cases} 1, & \text{if } x(t) \geq x_{th}, \\ 0, & \text{otherwise,} \end{cases} \tag{2.56}$$

where $x_{th} = \frac{1}{T}\sum_{t=1}^{T} x(t)$. Alarm durations and deviations are calculated as (2.27) and (2.52), respectively; they are sorted in the ascending order as $\{T'_a(k)\}_{k=1}^{K}$ and $\{D'_a(k)\}_{k=1}^{K}$.
- Step-2: $T_{a,max}$ and $D_{a,max}$ are determined respectively as the 95-th percentiles of $\{T'_a(k)\}_{k=1}^{K}$ and $\{D'_a(k)\}_{k=1}^{K}$ as

$$T_{a,max} = T'_a\left(\left\lceil 0.95 \times K \right\rceil\right), \tag{2.57}$$

$$D_{a,max} = D'_a\left(\left\lceil 0.95 \times K \right\rceil\right). \tag{2.58}$$

Here $\lceil \cdot \rceil$ is the ceil function, whose value is the smallest integer larger than the operand. The 95-th percentile is used to avoid effects of erroneous large values of T'_a and D'_a originating from outliers in the measurements of $x(t)$, and has probability 5% of Type-I errors, i.e., 5% of nuisance alarms will not be included in the estimation of $T_{a,max}$ and $D_{a,max}$.
- Step-3: A sufficiently large number of alarm occurrences is necessary to obtain reliable values of $T_{a,max}$ in (2.57) and $D_{a,max}$ in (2.58). A variational coefficient [19] is used,

$$c_v(T_{a,max}) = \sigma(T_{a,max})/\mu(T_{a,max}), \tag{2.59}$$

where

$$\mu(T_{a,max}) = \frac{\sum_{K=K_1}^{K_2} T_{a,max}(K)}{K_2 - K_1 + 1},$$

$$\sigma(T_{a,max}) = \sqrt{\frac{\sum_{K=K_1}^{K_2}\left(T_{a,max}(K) - \mu(T_{a,max})\right)^2}{K_2 - K_1}}.$$

If $c_v(T_{a,max})$ is no longer more than a small positive real value, e.g., $c_v(T_{a,max}) < 0.1$, then the number of alarm durations is sufficiently large so that $T_{a,max}$ is a reliable value. Analogously, a reliable estimate of $D_{a,max}$ can be determined.

Second, an index is proposed on the basis of normalized alarm durations and deviations. It is readily observed that alarm deadbands are suitable to deal with nuisance alarms with small alarm deviations and large alarm durations. High prices must be paid by applying alarm deadbands to handle nuisance alarms with large alarm deviations and small alarm durations. Based on such an observation, a deadband index $\theta \in (0°, 90°)$ is defined as

$$\theta = \arctan(s). \tag{2.60}$$

It is the angle of a straight line describing the relationship between the normalized alarm deviation $D_{a,n}$ and the normalized alarm duration $T_{a,n}$, i.e.,

$$T_{a,n} = s \cdot D_{a,n}. \tag{2.61}$$

Note that the deadbands index can be equivalently defined as the slope s, owing to the relation in (2.60). The ranges of θ and s are $(0°,\ 90°)$ and $(0,\ +\infty)$, respectively; to determine whether alarm deadbands are suitable, θ and s are respectively compared to the thresholds 45° and 1 (to be clarified at the end of this section). Thus, it is numerically more convenient to use θ with a finite range than s with an infinite range as the deadband index, although there is no special physical advantage of using θ other than s.

The slope s can be estimated from $\left\{D_{a,n}(k), T_{a,n}(k)\right\}_{k=1}^{K}$ through the linear least squares method,

$$\hat{s} = \frac{\sum_{k=1}^{K} D_{a,n}(k) \cdot T_{a,n}(k)}{\sum_{k=1}^{K} D_{a,n}^2(k)},$$

so that the estimate of θ is

$$\hat{\theta} = \arctan(\hat{s}). \tag{2.62}$$

According to the central limit theorem, $\hat{s}$ is Gaussian distributed. The 95% confidence interval of s is

$$\left[s_-, s_+\right] := \left[\hat{s} - 1.96\sigma_{\hat{s}}, \hat{s} + 1.96\sigma_{\hat{s}}\right],$$

where

$$\sigma_{\hat{s}} = \sqrt{\frac{\frac{1}{K-2}\sum_{k=1}^{K}\left(T_{a,n}(k) - \hat{s}D_{a,n}(k)\right)^2}{\sum_{k=1}^{K}\left(D_{a,n}(k) - \bar{D}_{a,n}\right)^2}}$$

and $\bar{D}_{a,n}$ is the sample mean value of $D_{a,n}$, i.e., $\bar{D}_{a,n} = \frac{1}{K}\sum_{k=1}^{K} D_{a,n}(k)$. Since the inverse tangent is a monotonic function, $\hat{\theta}$ is also Gaussian distributed. Analogously, the 95% confidence interval of θ is

$$\left[\theta_-, \theta_+\right] := \left[\hat{\theta} - 1.96\sigma_{\hat{\theta}}, \hat{\theta} + 1.96\sigma_{\hat{\theta}}\right], \tag{2.63}$$

where $\sigma_{\hat{\theta}} = \arctan\left(\sigma_{\hat{s}}\right)$.

Finally, it is ready to tell whether alarm deadbands are suitable for $x(t)$ based on the index θ. As $D_{a,n}$ and $T_{a,n}$ are normalized, they become comparable metrics. If $D_{a,n}$ is small and $T_{a,n}$ is large, then alarm deadbands are suitable to remove the corresponding nuisance alarms. Thus, the threshold of θ is associated with an intermittent situation $T_{a,n} = D_{a,n}$, which leads to $s = 1$ and $\theta = 45°$ from (2.60) and (2.61). Hence, if θ is larger than 45°, then $x(t)$ is suitable for using alarm deadbands; otherwise, alarm delay timers should be used. In other words, alarm deadbands and alarm delay timers are commonly-used candidates to reduce the number of nuisance alarms have different application scenarios: alarm deadbands are suitable to deal with nuisance alarms when $x(t)$ deviates away not too far away from x_{th} for a long time, while alarm delay timers are good for nuisance alarms when $x(t)$ traverses x_{th} quickly.

To tolerate uncertainties in the estimation of θ, a hypothesis test is introduced,

$$H_0 : \theta = 45°, \ H_1 : \theta \neq 45°. \tag{2.64}$$

Under the null hypothesis H_0, the acceptance region for the significance level 0.05 is

$$\left[45 - 1.96\sigma_{\hat{\theta}}, 45 + 1.96\sigma_{\hat{\theta}}\right]. \tag{2.65}$$

Therefore, if $\hat{\theta} \geqslant \left(45 + 1.96\sigma_{\hat{\theta}}\right)$, then it is safe to reject H_0 with a small probability of Type-I errors, so that alarm deadbands are suitable to remove nuisance alarms of $x(t)$. Otherwise, H_0 cannot be rejected, and alarm deadbands should not be used.

2.3.3 Design of Alarm Deadband Widths

This subsection proposes a method to design an optimal deadband width based on historical samples of process variables in normal conditions. The method is based on an observation that if the k-th alarm deviation $D_a(k)$ is less than the deadband width d, then the k-th alarm occurrence in $x_a(t)$ will not appear in $x_a(t; d)$. Hence, d can be designed based on certain information of $\{D_a(k)\}_{k=1}^{K}$.

First, an optimization function is formulated. If $\{x(t)\}_{t=1}^{N}$ is in the normal condition, the number of false alarms for $x_a(t; d)$ in (2.49) is

$$F\left(\{x_a(t; d)\}_{t=1}^{N}\right) = \sum_{t=1}^{N} x_{a,o}(t; d). \tag{2.66}$$

Here $x_{a,o}(t; d)$ takes the value '1' only at the time instant when an alarm event appears, i.e.,

$$x_{a,o}(t; d) = \begin{cases} 1, & \text{if } x_a(t-1; d) = 0 \ \& \ x_a(t; d) = 1, \\ 0, & \text{otherwise.} \end{cases}$$

The effectiveness of an alarm deadband is measured by the ratio of the number of false alarms in $\{x_a(t)\}_{t=1}^{T}$ from (2.1) with respect to the counterpart in $\{x_a(t; d)\}_{t=1}^{T}$,

$$\eta(d) = \frac{F\left(\{x_a(t;d)\}_{t=1}^{T}\right)}{F\left(\{x_a(t)\}_{t=1}^{T}\right)}. \tag{2.67}$$

The ratio $\eta(d)$ is in the range $[0, 1]$. If $\eta(d)$ is closer to zero, then more false alarms in $x_a(t)$ are removed. As discussed in Sect. 2.3.1, d cannot be too large in order to avoid an excessive value of the abnormality detection delay. An optimal value of d is chosen as the one to make $\eta(d)$ in (2.67) closest to a user-selected desired value η_0,

$$d_{opt} = \min\{D_{a,max}, \arg\min_{d} |\eta(d) - \eta_0|\}. \tag{2.68}$$

Here η_0 has a similar physical meaning as the probability of Type-I errors in Statistics [11], because false alarms are the alarms arisen in normal conditions and belong to Type-I errors. The probability of Type-I errors is usually chosen as 1% or 5% [11]. Here a default value of η_0 is taken as 5%. As discussed in Sect. 2.3.2, the parameter $D_{a,max}$ in (2.68) is an upper bound of deadband widths, in order to avoid missing correct alarms.

The alarm deviation $D_a(k)$ can be regarded as a realization of a continuous random variable Δ with the probability density function (PDF) $f_\Delta(\delta)$. It is reasonable to assume that alarm occurrences are mutually independent, so that $\eta(d)$ in (2.67) is equivalent to the area covered by $f_\Delta(\delta)$ above the value d, i.e.,

$$\eta(d) = \int_{d}^{+\infty} f_\Delta(\delta)\,\mathrm{d}\delta. \tag{2.69}$$

Thus, determining d_{opt} in (2.68) is equivalent to calculating the cumulative probability in (2.69). By introducing a symbol

$$\Theta(d) := \int_{d}^{+\infty} f_\Delta(\delta)\,\mathrm{d}\delta, \tag{2.70}$$

d_{opt} in (2.68) becomes

$$d_{opt} = \min\{D_{a,max}, \arg\min_{d} |\Theta(d) - \eta_0|\}. \tag{2.71}$$

The main concern is how to obtain a reliable estimate of $\Theta(d)$. Note that the probability $\Theta(d)$ is required here; such a requirement is much less demanding than the estimation of the PDF $f_\Delta(\delta)$. Note that the PDF of $x(t)$ is not required so that whether x is IID or non-IID is irrelevant.

Second, a reliable estimate of the cumulative probability $\Theta(d)$ in (2.70) is obtained by the Bayesian estimation approach. The probability that $D_a(k)$ falls inside an interval $[d, +\infty]$ for a certain value d is approximately equal to

$$\Theta(d) = \frac{c(d)}{K}, \tag{2.72}$$

where $c(d)$ is the number of samples in $\{D_a(k)\}_{k=1}^{K}$ to satisfy $D_a(k) \geq d$. For the Bayesian estimation approach [28], $\Theta(d)$ is not treated as an unknown constant in the range (0, 1), but is regarded as a continuous random variable $\Theta(\delta)$ taking $\theta(\delta)$ as a realization. If no extra information is available, the prior PDF of $\Theta(\delta)$ is usually taken as a continuous uniform distribution,

$$f_{\Theta(\delta)}(\theta(\delta)) = \begin{cases} 1, & \text{if } 0 < \theta(\delta) < 1, \\ 0, & \text{otherwise.} \end{cases} \tag{2.73}$$

Let us take the number of elements in $\{D_a(k)\}_{k=1}^{K}$ falling inside $[d, +\infty]$ as a realization $c(d)$ of a discrete random variable $C(d)$. Since maximum amplitude deviations are mutually independent, $C(d)$ takes the binomial distribution with the probability of success as $\Theta(\delta)$ among K independent trials with two outcomes of success and failure [28]. The conditional PDF of $C(d)$ based on $\Theta(\delta)$ is

$$f_{C(d)|\Theta(\delta)}(c(d)|\theta(\delta)) = \frac{K!}{c(d)!(K-c(d))!} \cdot (\theta(\delta))^{c(d)} \cdot (1-\theta(\delta))^{K-c(d)}. \tag{2.74}$$

The Bayesian formula yields the posterior PDF of $\Theta(\delta)$ based on the realization $c(d)$ of $C(d)$,

$$f_{\Theta(\delta)|C(d)}(\theta(\delta)|c(d)) = \frac{f_{C(d),\Theta(\delta)}(c(d), \theta(\delta))}{f_{C(d)}(c(d))}, \tag{2.75}$$

where the joint PDF of $C(d)$ and $\Theta(\delta)$ is

$$f_{C(d),\Theta(\delta)}(c(d), \theta(\delta)) = f_{C(d)|\Theta(\delta)}(c(d)|\theta(\delta)) \cdot f_{\Theta(\delta)}(\theta(\delta)), \tag{2.76}$$

and the marginal PDF of $C(d)$ is

$$f_{C(d)}(c(d)) = \int f_{C(d),\Theta(\delta)}(c(d), \theta(\delta)) \,\mathrm{d}\theta(\delta). \tag{2.77}$$

The Bayesian estimate of $\theta(\delta)$ is often taken as the conditional mean,

$$\hat{\theta}(\delta) = \int \theta(\delta) \cdot f_{\Theta(\delta)|C(d)}(\theta(\delta)|c(d)) \,\mathrm{d}\theta(\delta). \tag{2.78}$$

It is ready to obtain the $(1-\alpha)$ confidence interval of $\theta(\delta)$ from (2.75) as the narrowest interval $\left[\hat{\theta}(\delta)_-, \hat{\theta}(\delta)_+\right]$ satisfying the equality,

$$\int_{\hat{\theta}(\delta)_-}^{\hat{\theta}(\delta)_+} f_{\Theta(\delta_0)|C(d)}(\theta(\delta)|c(d)) \,\mathrm{d}\theta(\delta) = 1-\alpha. \tag{2.79}$$

A default value $\alpha = 0.05$ is used for the 95% confidence level. If the difference between $\hat{\theta}(\delta)$ and $\hat{\theta}(\delta)_-$ and that between $\hat{\theta}(\delta)$ and $\hat{\theta}(\delta)_+$ are small with respect to $\hat{\theta}(\delta)$, e.g.,

$$r\left(\hat{\theta}(\delta)\right) := \frac{\hat{\theta}(\delta)}{\max\left(\hat{\theta}(\delta) - \hat{\theta}(\delta)_-, \hat{\theta}(\delta)_+ - \hat{\theta}(\delta)\right)} \geq \beta, \tag{2.80}$$

then $\hat{\theta}(\delta)$ is regarded as a reliable estimate of $\Theta(\delta)$ in (2.72).

As the Bayesian estimation approach yields a consistent estimate [28], the confidence interval $\left[\hat{\theta}(\delta)_-, \hat{\theta}(\delta)_+\right]$ in (2.79) will be narrower when the sample size K of $\{D_a(k)\}_{k=1}^K$ is larger. Hence, it is expected that $r\left(\hat{\theta}(\delta)\right)$ in (2.80) gets larger as K increases. The parameter β is a threshold of $r\left(\hat{\theta}(\delta)\right)$ as the ratio of the estimate $\hat{\theta}(\delta)$ to its estimation uncertainty. A sensitivity analysis of β given later at Sect. 2.3.4.1 shows that if β is not less than 1, then the designed deadband width is not sensitive to β. Hence, a default value $\beta = 1$ is used to say that $\hat{\theta}(\delta)$ should be equal to or larger than its estimation uncertainty.

Finally, the alarm deadband is designed based on historical data $\{x(t)\}_{t=1}^N$ configured with the alarm threshold x_{th} in normal conditions as follows:

- Step 1: Samples of the alarm deviation $D_a(k)$ are collected from $\{x(t)\}_{t=1}^T$ as $\{D_a(k)\}_{k=1}^K$, and $D_{a,max}$ in (2.58) is obtained;
- Step 2: The Bayesian estimate $\hat{\theta}(\delta)$ in (2.78) and its confidence interval $\left[\hat{\theta}(\delta)_-, \hat{\theta}(\delta)_+\right]$ in (2.79) are calculated on the basis of $\{D_a(k)\}_{k=1}^K$. If $r\left(\hat{\theta}(\delta)\right)$ in (2.80) is large enough, then proceed to the next step; otherwise, above two steps are repeated for more samples of the alarm deviation.
- Step 3: The optimal deadband width is $d_{opt} = \delta$ as the solution to the problem in (2.68) or equivalently that in (2.71).

For Step 2, if there are not enough samples of the alarm deviation, the proposed method can still proceed to Step 3; however, the optimal deadband width designed in Step 3 may not be reliable to achieve the desired value η_0 for new samples of $x(t)$. The integral to calculate $\hat{\theta}(\delta)$ in (2.78) is implemented as a numerical integration by discretizing $\theta(\delta)$ and δ. In particular, $\theta(\delta)$ is discretized in the range (0, 1) by a small step size, e.g., 0.001, since $\theta(\delta)$ represents a possible value of the probability of false alarms; δ is discretized in the range $(0, D_{a,max})$ by a proper step size, e.g., $0.01 \cdot x_{th}$, since δ stands for a possible value of the deadband width.

As said by the deadband index in Sect. 2.3.2, alarm deadbands are very effective for process variables that are lingering around the alarm threshold with small amplitude deviations possibly for several consecutive samples. As depicted in Fig. 2.28, these kinds of process variables are not IID. Hence, most of design methods based on the IID assumption of process variables are suffering from such an invalid assumption. For non-IID process variables, time-series models are built to describe $x(t)$ [17], and

Kalman filters or particle filters are exploited to design d [27]; however, it is a well-known fact that building time-series models is not an easy task in general. By contrast, the proposed method requires a cumulative probability of the alarm deviation $D_a(k)$, instead of the PDFs or time-series models of $x(t)$, so that the proposed method is easy for implement and has no discrimination to IID and non-IID process variables.

2.3.4 Numerical and Industrial Examples

This section presents two numerical examples respectively to support the Bayesian estimation and investigate effects of the sample size, as well as an industrial example to illustrate the deadband index and the proposed method to design optimal deadband widths.

2.3.4.1 Numerical Example 1

This numerical example validates the estimated cumulative probability of alarm deviations, and compares the proposed method with existing ones based on the IID assumption of process variables. Consider a moving average process

$$x(t) = \frac{1}{4}\sum_{i=1}^{4} e(t-i), \tag{2.81}$$

where $e(t)$ is a Gaussian white noise with zero mean and standard deviation 4. Figure 2.31a shows some samples of $x(t)$, together with a high-alarm threshold $x_{th} = 1$. By assuming that $x(t)$ is in the normal condition, many false alarms occur in the alarm variable $x_a(t)$ in (2.1), as shown in Fig. 2.31b. An alarm deadband is designed to remove 95% of false alarms, that is, the number of false alarms in $x_a(t; d)$ from the alarm deadband should be about 5% of false alarms in $x_a(t)$.

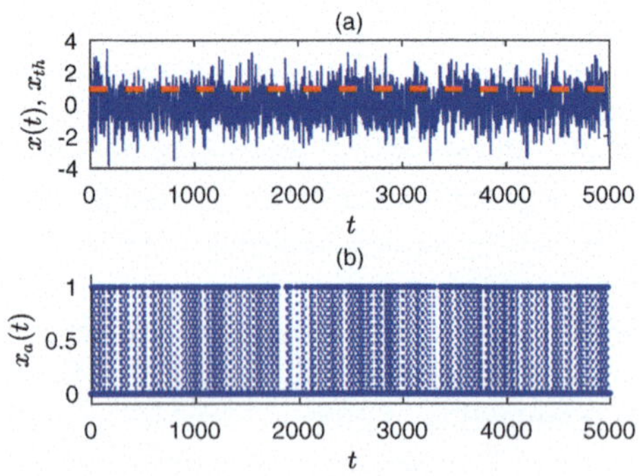

Fig. 2.31 **a** Samples of $x(t)$ (blue solid) and a high-alarm threshold x_{th} (red dash), **b** samples of $x_a(t)$, reprinted from Ref. [31], copyright 2023, with permission from IEEE

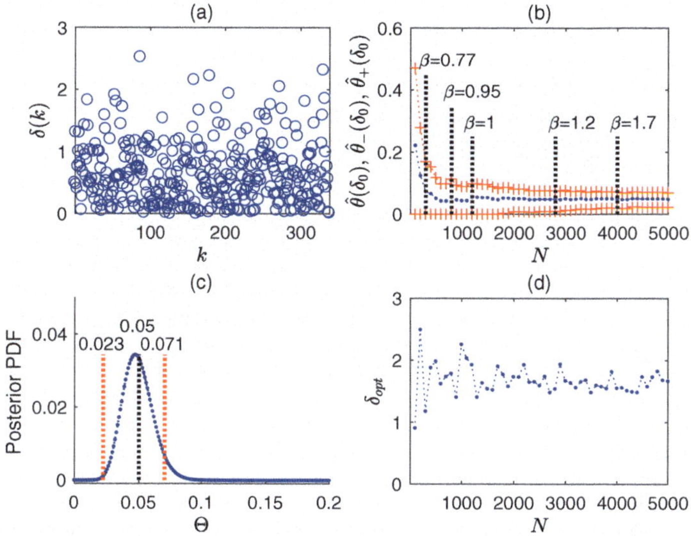

Fig. 2.32 **a** Alarm deviations $\{D_a(k)\}_{k=1}^{K=338}$ for $N = 5000$, **b** estimated cumulative probability $\hat{\theta}(\delta)$ (blue dot) and its confidence interval $\left[\hat{\theta}(\delta)_-, \hat{\theta}(\delta)_+\right]$ (red plus) as a function of T, **c** posterior PDF $f_{\Theta(\delta)|C(d)}$ for $N = 5000$, **d** optimal deadband width d_{opt} as a function of T, reprinted from Ref. [31], copyright 2023, with permission from IEEE

The proposed method is applied to $\{x(t)\}_{t=1}^{T}$ for different values of T. The upper bound of deadband widths in (2.58) is calculated as $D_{a,max} = 2.633$. All samples of the maximum amplitude deviation for $T = 5000$ are given as $\{D_a(k)\}_{k=1}^{K=338}$ in Fig. 2.32a. As T and K increase, the confidence interval $\left[\hat{\theta}(\delta)_-, \hat{\theta}(\delta)_+\right]$ in Fig. 2.32b becomes narrower, so that the estimated cumulative probability $\hat{\theta}(\delta)$ gets more reliable. As shown in Fig. 2.32d, the optimal deadband width $d_{opt} = \delta = 1.685$ is found to be the one such that $\hat{\theta}(\delta) = 0.05$ is closest to $\eta_0 = 0.05$. Figure 2.32c shows the corresponding posterior PDF $f_{\Theta(\delta)|C(d)}$. By taking the confidence level 95%, the confidence interval $\left[\hat{\theta}(\delta)_-, \hat{\theta}(\delta)_+\right]$ is [0.023, 0.071], and the ratio $r\left(\hat{\theta}(\delta)\right)$ in (2.80) is equal to 1.8519, being larger than $\beta = 1$. Hence, the estimated probability $\hat{\theta}(\delta) = 0.05$ is reliable, so that $d_{opt} = 1.685$ is trustworthy. In other words, the ratio $\eta(d_{opt})$ is expected to be 0.05 in the confidence interval [0.023, 0.071] with a confidence level 95%.

As a validation, 100 Monte Carlo simulations are performed for $T = 5000$. Figure 2.33a, b present samples of $x(t)$ and $x_a(t; d_{opt})$ in one typical simulation. By using $d_{opt} = 1.685$, $\eta(\delta_{opt})$'s in 100 simulations are given as blue dots in Fig. 2.33d. For 97 simulations, $\eta(d_{opt})$'s are inside the confidence interval [0.023, 0.071], which is consistent with expectations.

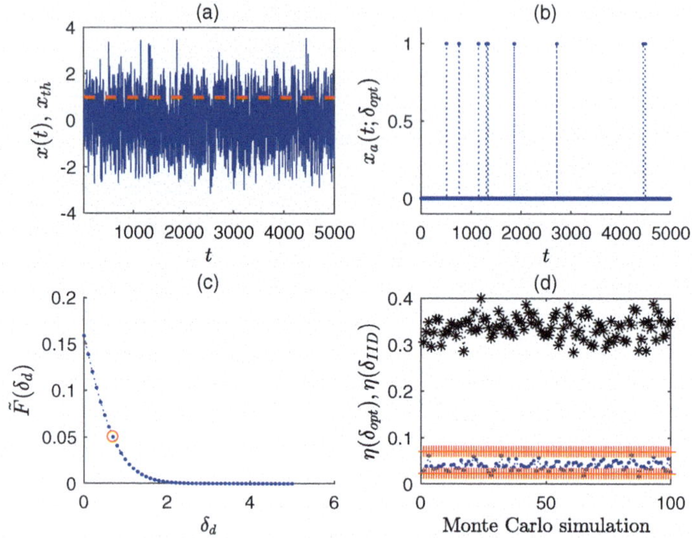

Fig. 2.33 **a** Samples of x (blue solid) and a high-alarm threshold x_{th} (red dash), **b** samples of $x_a(t; d_{opt})$, **c** $\tilde{F}(d)$ in (2.83), **d** $\eta(d_{opt})$ (blue dot) with the confidence interval [0.023, 0.071] (red plus) and $\eta(d_{iid})$ (black star), reprinted from Ref. [31], copyright 2023, with permission from IEEE

Table 2.13 Sensitivity analysis of β, reprinted from Ref. [31], copyright 2023, with permission from IEEE

β	d_{opt}	$r\left(\hat{\theta}\left(d_{opt}\right)\right)$
0.7	1.9870	0.7574
0.9	1.7893	0.9506
1	1.6699	1.8519
2	1.6614	2.1739
3	1.6298	3.5714

A sensitivity analysis of β in (2.80) is performed here. Figure 2.32b marks the positions of $\hat{\theta}(\delta)$'s for five values of β, and Table 2.13 lists the corresponding values of d_{opt}. As shown in Fig. 2.32b and Table 2.13, $\hat{\theta}(\delta)$ and its confidence interval $\left[\hat{\theta}(\delta)_-, \hat{\theta}(\delta)_+\right]$ begin to change slowly at about $\beta = 1$, and the differences among d_{opt}'s for $\beta = 1, 2, 3$ are quite small. Hence, $\beta = 1$ is chosen as the default value, saying that $\hat{\theta}(\delta)$ should be equal to or larger than its estimation uncertainty as $\hat{\theta}(\delta) - \hat{\theta}(\delta)_-$ or $\hat{\theta}(\delta)_+ - \hat{\theta}(\delta)$.

For a comparison, existing design methods are applied based on the PDF of the process variable $x(t)$. If $x(t)$ is IID, then $x(t)$ is equivalent to a random variable x with the PDF $q(x)$, and the transitions between alarm states and non-alarm states for an alarm deadband can be described by a Markov process depicted in Fig. 2.34 [1].

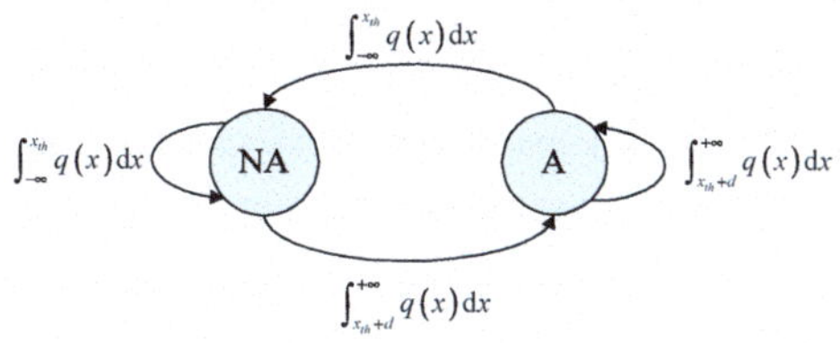

Fig. 2.34 A Markov process to describe transitions between the alarm state (A) and non-alarm state (NA)

If a false alarm rate (FAR) is defined as the percentage of samples in $\{x_a(t;d)\}_{t=1}^{T}$ in the alarm state when $\{x(t)\}_{t=1}^{T}$ is in the normal condition, i.e.,

$$\tilde{F}(x_a(t;d)) = \frac{\sum_{t=1}^{T} x_a(t;d)}{T}, \tag{2.82}$$

then it is ready to derive that the FAR for an alarm deadband in (2.49) has a theoretical expression

$$\tilde{F}(d) = \frac{\int_{x_{th}+d}^{+\infty} q(x)\,\mathrm{d}x}{\int_{x_{th}+d}^{+\infty} q(x)\,\mathrm{d}x + \int_{-\infty}^{x_{th}} q(x)\,\mathrm{d}x}. \tag{2.83}$$

Analogously to (2.68), an optimal deadband width can be designed as

$$d_{iid} = \min\{D_{a,max}, \arg\min_{d} \left|\tilde{F}(d) - \tilde{F}_0\right|\}, \tag{2.84}$$

where $\tilde{F}_0$ is a desired FAR value. Figure 2.33c presents the FAR function $\tilde{F}(d)$ in (2.83), from which $d_{iid} = 0.7$ is obtained from (2.84). For the 100 Monte Carlo simulations, too many false alarms are left by using $d_{iid} = 0.7$, and $\eta(d_{iid})$'s in Fig. 2.33d are about 0.35, much than the desired value 0.05. Since the process variable $x(t)$ is not IID, the Markov process in Fig. 2.34 and $\tilde{F}(d)$ in (2.83) are invalid. Hence, the unsatisfactory performance of d_{iid} is essentially due to the invalid assumption that $x(t)$ is IID.

2.3.4.2 Numerical Example 2

A numerical investigation is performed to tell a recommended value of the sample size K such that a reliable estimate $\hat{\theta}(\delta)$ in (2.78) is reached. Let us consider three types of $D_a(k)$ respectively taking the standard Gaussian distribution with zero mean and unit variance, the continuous-uniform distribution in the range [0, 1] and the exponential distribution with unit mean. Such a numerical investigation based on different probability distributions is commonly adopted in Statistics, e.g., robust mean estimators [10] (pages 257–258 therein). If the recommend value is insensitive to different probability distributions, then it could be applicable to general cases.

Figure 2.35a presents 1000 random realizations of $D_a(k)$ taking the standard Gaussian distribution, from which $\delta = 1.6449$ is found to be associated with $\theta(\delta) = \eta_0 = 0.05$. By using sample sizes varying from $K = 10$ up to $K = 1000$ in the step 10, $\hat{\theta}(\delta)$ and its confidence interval $\left[\hat{\theta}(\delta)_-, \hat{\theta}(\delta)_+\right]$ are obtained in Fig. 2.35b, and the ratio $r\left(\hat{\theta}(\delta)\right)$ in (2.80) is given in Fig. 2.35c.

As expected, the confidence interval becomes narrower and the ratio gets larger with the increment of K. By sorting the number $c(d)$ in (2.72), the relation between $r\left(\hat{\theta}(\delta)\right)$ and $c(d)$ demonstrates a monotonically increasing trend as shown in Fig. 2.35d. Figures 2.36 and 2.37 are the counterparts of Fig. 2.35 for the uniform and exponential distributions, respectively. Despite of different distributions, the subplots (b), (c) and (d) in Figs. 2.35, 2.36 and 2.37 are very similar. If a default value $\beta = 1$ in (2.80) is used, then $c(d) = 5$ is the corresponding threshold from Fig. 2.35d. Figures 2.36d or 2.37d, so that $K = c(d)/\eta_0 = 5/\eta_0$ is the recommended minimum sample size.

Such a sample size implies that the proposed method can be further simplified by omitting the step of Bayesian estimation, to ease the design of alarm deadbands in practice. If more than $K = 5/\eta_0$ false alarms have been observed, the width d for an alarm deadband to remove $100 \cdot \eta_0\%$ of false alarms is designed as

$$d_{opt} = \min\{D_{a,max}, \arg\min_{d} \left|\frac{c(d)}{K} - \eta_0\right|\}, \tag{2.85}$$

where $c(d)$ is the one in (2.72). The numerical investigation here indicates that $c(d)$ is a reliable estimate of $\theta(\delta)$ if the sample size K is larger than $5/\eta_0$. For instance, if $\eta_0 = 0.05$, then a default value of K is 100; in words, about 100 false alarms occurred in the past are sufficient to design a reliable alarm deadband via the simple method in (2.85).

2.3.4.3 Industrial Example

This example presents the results for a process variable $x(t)$ in the unit Nm^3/h with tagname X71NAO084 as the inlet air flow of an ammonia-air mixer in a large-scale thermal power plant. It is a common flow rate variable, being configured with a low-alarm threshold $x_{th} = 1800$ Nm^3/h. Figure 2.38a shows 10-hour samples of $x(t)$ with a sample period 1 sec. Industrial plant operators confirm that $x(t)$ in the ten hours are indeed in the normal condition, so that all occurred alarms in Fig. 2.38b are false ones.

First, normalized alarm durations and deviations are respectively calculated from (2.53) and (2.54). Their scatter plot is given in Fig. 2.39. Taking values for flow signals from Table 2.12, the acceptable maximum values of alarm deviations and alarm durations are $T_{a,max} = 15$ sec and $D_{a,max} = 0.05x_{th} = 90$ Nm3/h from (2.55). The deadband indices are calculated from (2.62) as $\hat{\theta} = 87.81°$ with the confidence

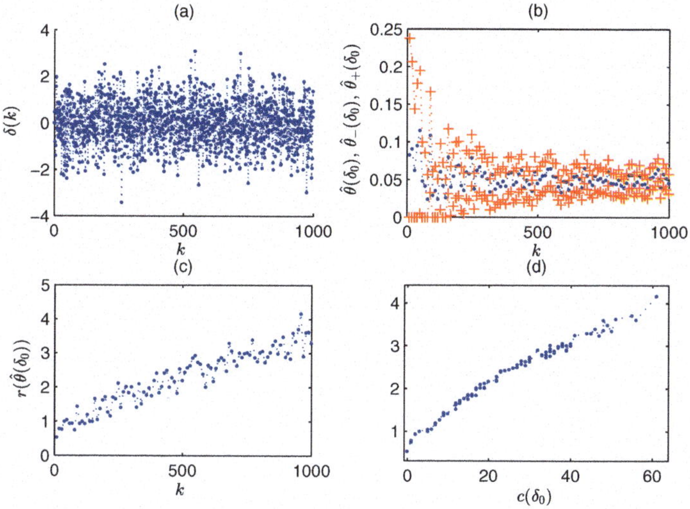

Fig. 2.35 **a** Samples of $D_a(k)$ with Gaussian distribution, **b** estimated probability $\hat{\theta}\ (\delta)$ (blue dot) and its confidence interval $\left[\hat{\theta}\ (\delta)_-\ ,\ \hat{\theta}\ (\delta)_+\right]$ (red plus), **c** ratio $r\left(\hat{\theta}\ (\delta)\right)$, (d) scatter plot of $r\left(\hat{\theta}\ (\delta)\right)$ and $c(d)$

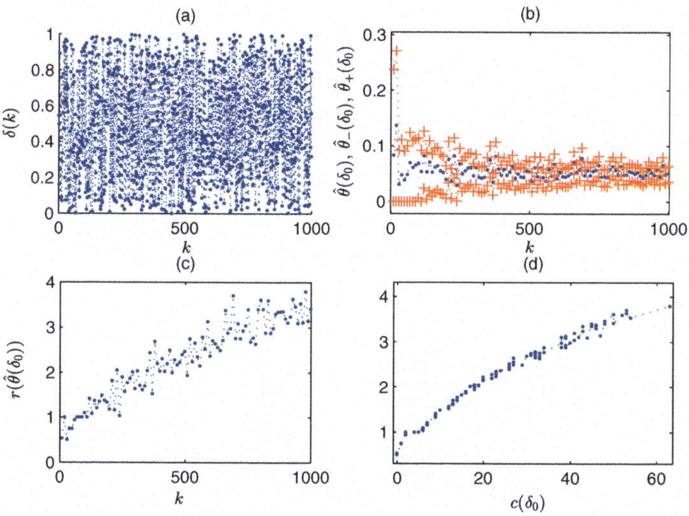

Fig. 2.36 **a** Samples of $D_a(k)$ with uniform distribution, **b** estimated probability $\hat{\theta}\ (\delta)$ (blue dot) and its confidence interval $\left[\hat{\theta}\ (\delta)_-\ ,\ \hat{\theta}\ (\delta)_+\right]$ (red plus), **c** ratio $r\left(\hat{\theta}\ (\delta)\right)$, **d** scatter plot of $r\left(\hat{\theta}\ (\delta)\right)$ and $c(d)$

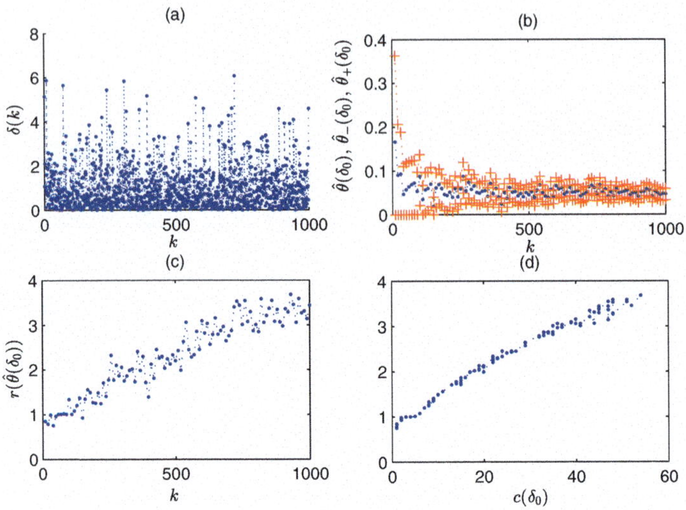

Fig. 2.37 **a** Samples of $D_a(k)$ with exponential distribution, **b** estimated probability $\hat{\theta}(\delta)$ (blue dot) and its confidence interval $\left[\hat{\theta}(\delta)_-, \hat{\theta}(\delta)_+\right]$ (red plus), (c) ratio $r\left(\hat{\theta}(\delta)\right)$, **d** scatter plot of $r\left(\hat{\theta}(\delta)\right)$ and $c(d)$

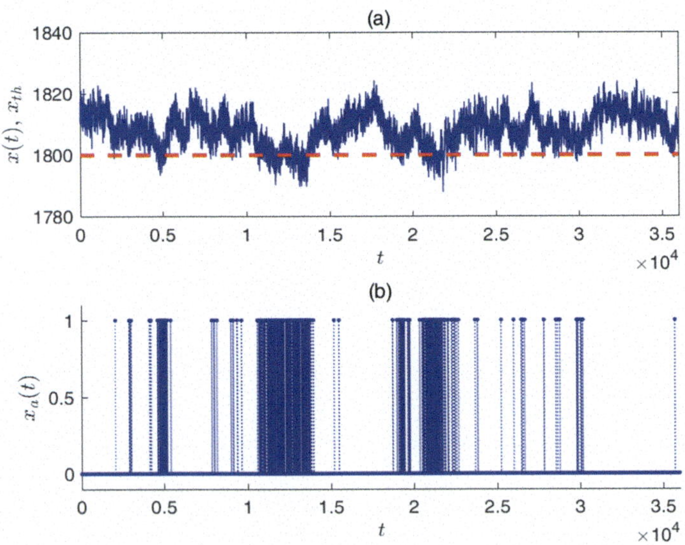

Fig. 2.38 **a** Samples of $x(t)$ (blue solid) and a low-alarm threshold x_{th} (red dash), **b** samples of $x_a(t)$, reprinted from Ref. [31], copyright 2023, with permission from IEEE

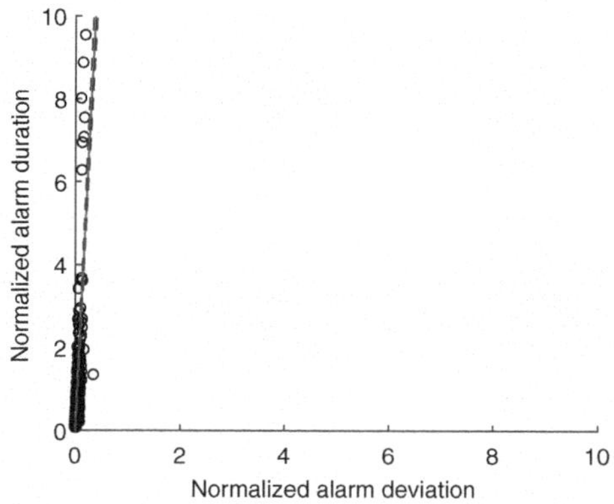

Fig. 2.39 Scatter plot of normalized alarm deviations and durations (black dot), estimated line $\hat{T}_{a,n} = \hat{s} \cdot D_{a,n}$ (red solid) and its 95% confidence interval $\hat{T}_{a,n,\pm} = \left(\hat{s} \pm 1.96\sigma_{\hat{s}}\right) \cdot D_{a,n}$ (blue dash) for $x_1(t)$, reprinted from Ref. [32], copyright 2019, with permission from IEEE

interval in (2.63) as [87.62°, 88.00°]. The acceptance region in (2.65) is [44.81°, 45.19°]. Thus, the null hypothesis H_0 is rejected, and it is claimed that $\hat{\theta} > 45°$. In fact, $\hat{\theta}$ is far from the acceptance region. Alarm deadbands are suitable to deal with false alarms of $x(t)$ here.

The default values of $T_{a,max}$ and $D_{a,max}$ in Table 2.12 are easy to determine; however, the corresponding normalization may have lower qualities, because the default values are given for all process variables in the same type, not for one particular process variable. By looking at Fig. 2.38a, the variation ranges of $x(t)$ are much smaller than $D_{a,max} = 90$ Nm^3/h, so that the default value is less appropriate. Instead, $T_{a,max}$ in (2.57) and $D_{a,max}$ in (2.58) are respectively equal to 28 sec and 9.7 Nm^3/h; the corresponding deadband index is re-calculated as $\hat{\theta} = 56.50°$ with the confidence interval as [55.53°, 57.47°], also supporting the usage of alarm deadbands.

Second, the proposed method is applied to design an optimal deadband width based on historical samples under normal conditions for removing 95% of false alarms in upcoming time periods. Figure 2.40a presents all samples $\{D_a(k)\}_{k=1}^{K=427}$ of the alarm deviation in the ten hours. By increasing the data length T with a step of 3600 samples from the first hour till the tenth, the proposed method yields the estimated cumulative probability $\hat{\theta}(\delta)$ with its confidence interval $\left[\hat{\theta}(\delta)_-, \hat{\theta}(\delta)_+\right]$ as a function of T in Fig. 2.40b, as well as the corresponding optimal deadband width d_{opt} as a function of T in Fig. 2.40d. As T gets larger, the confidence interval becomes narrower. For the entire samples $\{D_a(k)\}_{k=1}^{K=427}$ in the ten hours, $\delta = 6.3223$ is the one such that $\hat{\theta}(\delta) = 0.04662$ is the closest to $\eta_0 = 0.05$. The corresponding posterior PDF $f_{\Theta(\delta)|C(\delta)}$ is given in Fig. 2.40c. The confidence interval $\left[\hat{\theta}(\delta)_-, \hat{\theta}(\delta)_+\right]$ is [0.023, 0.065] for the confidence level 95%, and the ratio $r\left(\hat{\theta}(\delta)\right) = 1.9737$ is larger than the threshold $\beta = 1$. Hence, $\hat{\theta}(\delta)$ is reliable so that $d_{opt} = 6.3223$ is trustworthy.

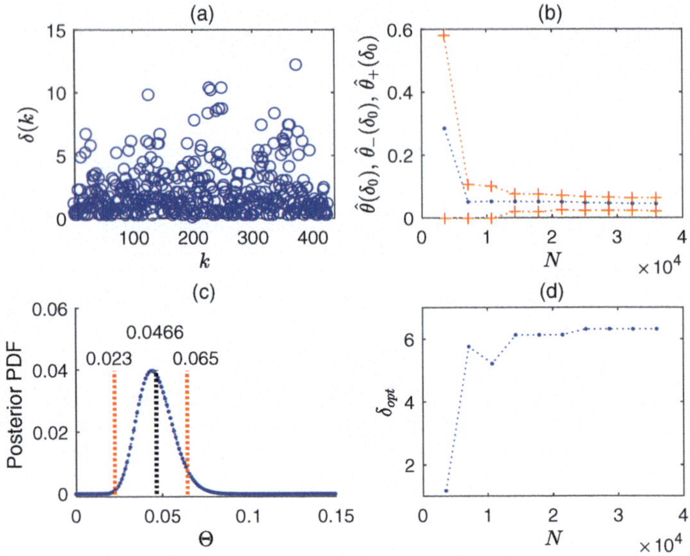

Fig. 2.40 **a** Maximum amplitude deviations $\{D_a(k)\}_{k=1}^{K=427}$, **b** estimated cumulative probability $\hat{\theta}(\delta)$ (blue dot) and its confidence interval $\left[\hat{\theta}(\delta)_-, \hat{\theta}(\delta)_+\right]$ (red plus) as a function of T, **c** posterior PDF $f_{\Theta(\delta)|C(d)}$ based on all maximum amplitude deviations, **d** optimal deadband width d_{opt} as a function of T, reprinted from Ref. [31], copyright 2023, with permission from IEEE

Next, the existing method in (2.84) is applied for a comparison. For the 10-hour samples of $x(t)$, the PDF of x is estimated via a standard kernel PDF estimation method, as shown in Fig. 2.41a. Based on the estimated PDF of x, the FAR function $\tilde{F}(d)$ is calculated as given in Fig. 2.41b, leading to $d_{iid} = 1.2$.

Finally, a validation test is taken by applying the two deadband widths $d_{opt} = 6.322$ and $d_{iid} = 1.2$ to a new set of 8-hour samples of $x(t)$. Figure 2.42 depicts the samples of $x(t)$ as well as the alarm variables $x(t; d_{opt})$ and $x(t; d_{iid})$. The unremoved false alarm ratio $\eta(d_{opt})$ is equal to 0.0483, which is close to the desired value 0.05. By contrast, $\eta(d_{iid}) = 0.4897$ is much larger than the design specification 0.05. Only half of false alarms in $x_a(t)$ are removed by using d_{iid}. The failure of $d_{iid} = 1.2$ is due to the invalid assumption that $x(t)$ is IID. As shown in Figs. 2.38a and 2.42a, $x(t)$ is clearly not IID.

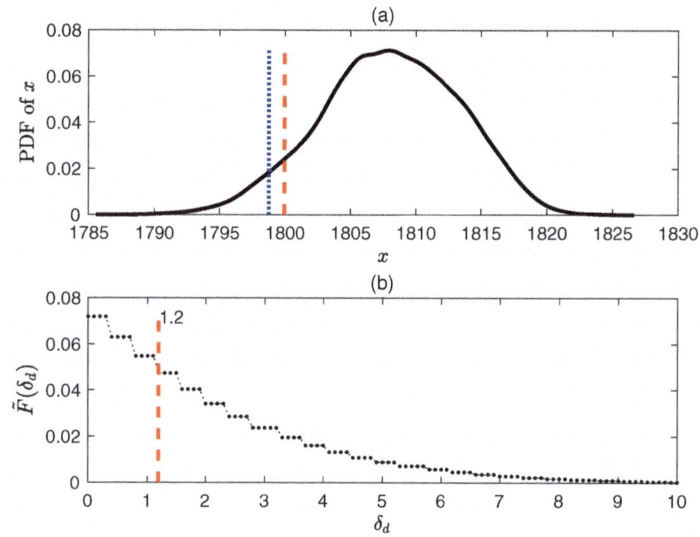

Fig. 2.41 **a** Estimated PDF of $x(t)$ (black dot), a low-alarm threshold x_{th} (red dash) and $x_{th} - d_{iid}$ (blue solid), **b** $\tilde{F}(d)$ in (2.83) (black dot) and d_{iid} (red dash), reprinted from Ref. [31], copyright 2023, with permission from IEEE

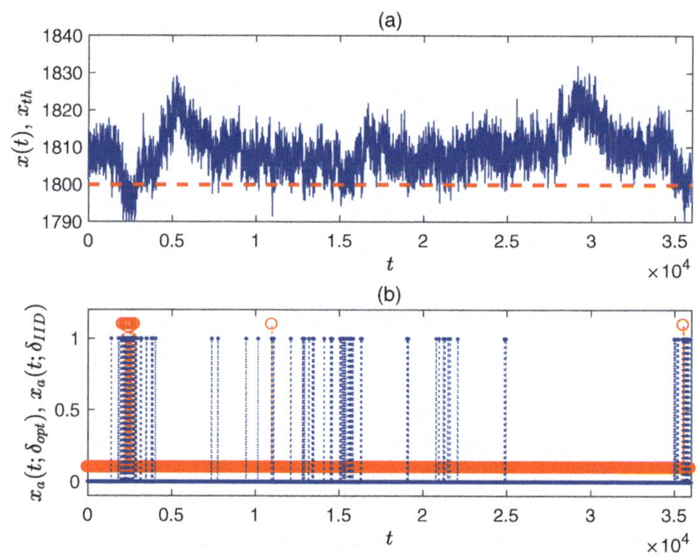

Fig. 2.42 **a** 8-hour samples of $x(t)$ (blue solid) and a low-alarm threshold x_{th} (red dash), **b** samples of $x_a(t; d_{opt})$ (red circle, shifted upwards for 0.1) and $x_a(t; d_{iid})$ (blue dot), reprinted from Ref. [31], copyright 2023, with permission from IEEE

2.4 Alarm Thresholds Based on Alarm Probability Plots

This section proposes a method to design optimal alarm thresholds for non-IID process variables based on alarm probability plots [34]. An alarm probability plot involves four statistics, namely, the transition probability and the averaged transition step towards an alarm threshold, the transition probability and the averaged transition step away from an alarm threshold. Theoretical expressions of the four statistics are developed based on Markov chain theory. The statistics are estimated with a proper determination of quantification intervals transferring analog process variables to discrete state variables. An objective function composed of the four statistics is used to determine an optimal choice of alarm thresholds.

2.4.1 Problem Description

Alarm thresholds, also known as alarm limits or alarm trippoints, are indispensable parameters in industrial alarm systems. Alarm thresholds are usually determined based on process knowledge and operation requirement; however, they are sometimes subject to design within certain ranges. For instance, if a bearing temperature for a coal grinding mill is higher than 55 °C, a protective logic will shut down the mill automatically; thus, an alarm threshold is allowed to vary in a range up to 55.

This section is motivated by an idea of designing alarm thresholds based on a so-called alarm probability plot formulated in a public technical report [9]. The alarm probability plot is composed of two curves involving four statistics:

- The probability and the amount of time that a process variable will take to go into alarms, after reaching a certain value not overpassing the alarm threshold.
- The probability and the amount of time that a process variable will take to attain a specified value beyond the alarm threshold, once an alarm has occurred.

An example of the alarm probability plot, shown in Fig. 2.43 (from the numerical example in Sect. 2.4.4.2), says that a high alarm threshold can be decreased in order to have more amount of time for industrial plant operators to deal with abnormal conditions, without increasing the number of false alarms. That is, the current alarm threshold being associated with the state $i = 8$ in Fig. 2.43 should be decreased to the one associated with the state $i = 5$. Hence, the alarm probability plot is very useful for the alarm threshold design, but it requires a systematic study. In particular, the two curves in the alarm probability plot have not been mathematically defined and numerically estimated in the public technical report [9].

Given data samples $\{x(t)\}_{t=1}^{T}$ of a process variable x, two problems are going to be solved. First, mathematically definitions and theoretical expressions of four statistics to formulate the alarm probability plot need to be established, and the four statistics need to be estimated from data samples. The four statistics are the transition probability and the averaged transition step towards an alarm threshold, the transition probability and the averaged transition step away from an alarm threshold. Second,

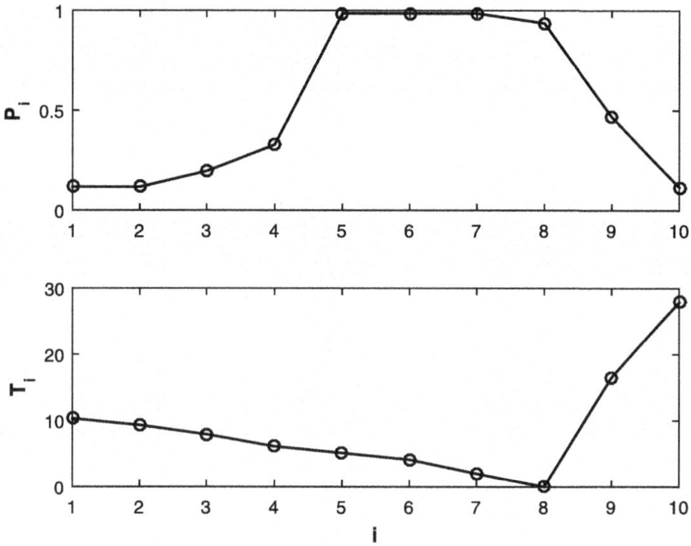

Fig. 2.43 An alarm probability plot from the numerical example in Sect. 2.4.4.2, reprinted from Ref. [34], copyright 2017, with permission from IEEE

an optimal choice of alarm thresholds should be determined from a discriminant function composed of the four estimated statistics.

2.4.2 *Theoretical Expressions of Four Statistics*

This subsection presents the definitions and theoretical expressions of the four statistics in alarm probability plots.

Consider the alarm variable $x_a(t)$ in (2.1) to be generated by comparing a process variable $x(t)$ to a high alarm threshold x_{th}. For the purpose of designing a proper value of x_{th}, a state variable $x_s(t)$ is defined by quantifying $x(t)$ into a number of discrete states. Let $[x_i, x_{i+1})$ be the quantification interval for the state i; thus, $x_s(t) = i$ if $x(t) \in [x_i, x_{i+1})$. The symbol a is specified for the first alarm state, i.e., $x_s(t) = a$ if $x(t) \in [x_a, x_{a+1})$ with $x_a = x_{th}$. In words, when the process variable $x(t)$ reaches the alarm threshold x_{th}, the state variable $x_s(t)$ is in the first alarm state a. The states taking values smaller and larger than a are referred to as the non-alarm states and subsequential alarm states, respectively. The determination of quantification intervals for all states will be discussed later in Sect. 2.4.3.

With the state variable $x_s(t)$, the four statistics in alarm probability plots are defined as:

- P_{ka} and T_{ka} respectively represent the transition probability and the averaged transition step that $x(t)$ transfers from a non-alarm state k, where $x(t)$ takes a value smaller than x_{th}, to the first alarm state a in a non-decreasing manner;

- P_{am} and T_{am} respectively represent the transition probability and the averaged transition step that $x(t)$ transfers from the first alarm state a to a subsequential alarm state m, where $x(t)$ takes a value larger than x_{th}, without going beneath the state a.

The state variable $x_s(t)$ is assumed to satisfy Markov property, which is a mild assumption for industrial process variables and can be verified via some statistical tests [8]. The Markov chain of $x_s(t)$ is described by its one-step transition probability $p_{ij}^{(1)}$ that the state i transfers to the next state j in one step, i.e.,

$$p_{ij}^{(1)} = Pr[x_s(t+1) = j | x_s(t) = i], \ i, j \in V. \tag{2.86}$$

Here $V := \{1, \cdots, v\}$, for some positive integer v, is a state space containing all finite states. Theoretical expressions of the four statistics can be developed based on the one-step transition probabilities as given in the next subsections.

2.4.2.1 Theoretical Expressions of P_{ka} and T_{ka}

First, let $P_{i(i+1)}$ be the transition probability that $x(t)$ transfers from the state i to the state $i+1$ in a non-decreasing manner. If the transfer is completed in l steps, then the l-step transition probability $p_{i(i+1)}^{(l)}$ is

$$p_{i(i+1)}^{(l)} = \left(p_{ii}^{(1)}\right)^{l-1} p_{i(i+1)}^{(1)}.$$

Since the transfer is non-decreasing, the former $l-1$ steps stay at the state i, and the last step transfers from the state i to the state $i+1$. Hence, $P_{i(i+1)}$ is the sum of all l-step transition probabilities, i.e.,

$$P_{i(i+1)} = \sum_{l=1}^{\infty} p_{i(i+1)}^{(l)} = \sum_{l=1}^{\infty} \left(p_{ii}^{(1)}\right)^{l-1} p_{i(i+1)}^{(1)} = \frac{p_{i(i+1)}^{(1)}}{1 - p_{ii}^{(1)}}.$$

Owing to the Markov property, $P_{i(i+1)}$ is independent to $P_{(i-1)(i)}$ and $P_{(i+1)(i+2)}$ for all values of i. Thus, the transition probability P_{ka} is a multiplication of all involved transition probabilities $P_{i(i+1)}$ from $i = k$ to $i = a - 1$,

$$P_{ka} = \prod_{i=k}^{a-1} \frac{p_{i(i+1)}^{(1)}}{1 - p_{ii}^{(1)}}. \tag{2.87}$$

Second, recall some related concepts in Markov chain theory [5] as follows. The states of a Markov chain can be divided into two classes, namely, transient states and recurrent states. A state is recurrent if it always goes back to itself after the transfer of finite steps, and a state is transient if it is not recurrent. In addition, every recurrent state, denoted by s here, is absorbing, if two conditions are satisfied [5],

$$\begin{cases} p_{ss}^{(1)} & = 1, \\ p_{sj}^{(1)} & = 0 \text{ for all } j \neq s. \end{cases}$$

Denote the symbol T_{is} as the expected value of transition steps from any state i to the absorbing state s. Then, T_{is} is the unique solution to the following equations [5],

$$\begin{cases} T_{is} & = 0, \text{ for all absorbing states } i, \\ T_{is} & = 1 + \sum\limits_{j \in V} p_{ij}^{(1)} T_{js}, \text{ for all transient states } i. \end{cases} \tag{2.88}$$

Here V is a state space containing all finite states.

Next, let $T_{i(i+1)}$ represent the averaged transition step that $x(t)$ transfers from the state i to the state $i+1$ in a non-decreasing manner, and the transfer graph from the state i to the state $i+1$ is given in Fig. 2.44. For the state transition graph in Fig. 2.44, (2.88) leads to

$$\begin{pmatrix} 1 & 0 & 0 \\ p_{i(i-1)}^{(1)} & p_{ii}^{(1)} - 1 & p_{i(i+1)}^{(1)} \\ 0 & 0 & 1 \end{pmatrix} \begin{pmatrix} T_{(i-1)(i+1)} \\ T_{i(i+1)} \\ T_{(i+1)(i+1)} \end{pmatrix} = \begin{pmatrix} 0 \\ -1 \\ 0 \end{pmatrix}. \tag{2.89}$$

The averaged transition step $T_{i(i+1)}$ is obtained from (2.89) as

$$T_{i(i+1)} = \frac{1}{1 - p_{ii}^{(1)}}.$$

Owing to the Markov property, the averaged transition step T_{ka} is a summation of all involved averaged transition steps $T_{i(i+1)}$'s from $i = k$ to $i = a - 1$,

$$T_{ka} = \sum_{i=k}^{a-1} \frac{1}{1 - p_{ii}^{(1)}}. \tag{2.90}$$

2.4.2.2 Theoretical Expressions of P_{am} and T_{am}

First, recall that the probability P_{is} of eventually reaching an absorbing state s, starting from the state i, is the unique solution to following equations [5],

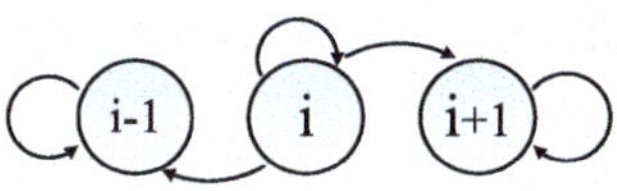

Fig. 2.44 The state transition graph from the state i to the state $i+1$, reprinted from Ref. [34], copyright 2017, with permission from IEEE

$$\begin{cases} P_{ss} &= 1, \\ P_{is} &= 0, \text{ for all absorbing states } i \neq s, \\ P_{is} &= \sum\limits_{j \in V} p_{ij}^{(1)} P_{js}, \text{ for all transient states } i. \end{cases} \tag{2.91}$$

Here V is a state space containing all finite states.

Second, the definition of P_{am} says that the associated transfer of states needs not to be in a non-decreasing manner. Figure 2.45 depicts the state transition graph from the first alarm state a to a subsequential alarm state m. For the state transition graph in Fig. 2.45, (2.91) leads to

$$PA = C, \tag{2.92}$$

where

$$A := \begin{bmatrix} P_{(a-1)m} \\ P_{am} \\ P_{(a+1)m} \\ \vdots \\ P_{(m-1)m} \\ P_{mm} \end{bmatrix}, \quad C := \begin{bmatrix} 0 \\ 0 \\ 0 \\ \vdots \\ 0 \\ 1 \end{bmatrix},$$

and

$$P := \begin{bmatrix} 1 & 0 & 0 & 0 & 0 \cdots 0 & 0 & 0 & 0 \\ p_{a(a-1)}^{(1)} & p_{aa}^{(1)} - 1 & p_{a(a+1)}^{(1)} & 0 & 0 \cdots 0 & 0 & 0 & 0 \\ 0 & p_{(a+1)a}^{(1)} & p_{(a+1)(a+1)}^{(1)} - 1 & p_{(a+1)(a+2)}^{(1)} & 0 \cdots 0 & 0 & 0 & 0 \\ \vdots & \vdots & \vdots & \vdots & \vdots \ddots \vdots & \vdots & \vdots & \vdots \\ 0 & 0 & 0 & 0 & 0 \cdots 0 & p_{(m-1)(m-2)}^{(1)} & p_{(m-1)(m-1)}^{(1)} - 1 & p_{(m-1)m}^{(1)} \\ 0 & 0 & 0 & 0 & 0 \cdots 0 & 0 & 0 & 1 \end{bmatrix}.$$

Based on (2.92), the transition probability P_{am} is obtained as

$$P_{am} = (P^{-1}C)_2, \tag{2.93}$$

where the subscript '2' denotes the second element of the vector $P^{-1}C$. Note that P_{aa} is the same as the one-step transition probability $p_{aa}^{(1)}$.

Fig. 2.45 The state transition graph from the first alarm state a to a subsequential alarm state m, reprinted from Ref. [34], copyright 2017, with permission from IEEE

Next, for the state transition graph in Fig. 2.45, (2.88) leads to

$$PB = D, \tag{2.94}$$

where P is the same as that in (2.92), and

$$B := \begin{pmatrix} T_{(a-1)m} \\ T_{am} \\ \vdots \\ T_{(m-1)m} \\ T_{mm} \end{pmatrix}, \quad D := \begin{pmatrix} 0 \\ -1 \\ \vdots \\ -1 \\ 0 \end{pmatrix}.$$

Thus, the averaged transition step T_{am} from the first alarm state a to the absorbing subsequential alarm state m is obtained from (2.94) as

$$T_{am} = (P^{-1}D)_2, \tag{2.95}$$

where the subscript '2' denotes the second element of the vector $P^{-1}D$.

2.4.2.3 Numerical Verification of Four Statistics

This subsection validates the theoretical expressions of the four statistics via numerical studies. Given a historical data set $\{x(t)\}_{t=1}^{T}$, there are basically two approaches to estimate the four statistics. The first approach is to estimate the one-step transition probability $p_{ij}^{(1)}$ in (2.86), and to calculate P_{ka} in (2.87), T_{ka} in (2.90), P_{am} in (2.93) and T_{am} in (2.95). The one-step transition probability $p_{ij}^{(1)}$ is estimated as

$$\hat{p}_{ij}^{(1)} = \frac{n_{ij}}{n_i}, \quad j = i-1, i, i+1, \tag{2.96}$$

where n_i is the number of times that the state i occurs, and n_{ij} is the number of times that the state i is followed by the state j in the next sample.

The second approach is to estimate the four statistics directly by the classical probability as follows:

- The estimate of P_{ka} is

$$\hat{P}_{ka} = \frac{n_{ka}}{n_k}, \tag{2.97}$$

and the estimate of T_{ka} is

$$\hat{T}_{ka} = \frac{\sum_i t_{i,ka}}{n_{ka}}. \tag{2.98}$$

Here n_{ka} is the number of times that a non-alarm state k transfers to the first alarm state a in a non-decreasing manner, n_k is the number of times that the state k occurs, and $t_{i,ka}$ is the number of steps taken by the transfer from the state k to the state a in a non-decreasing manner for the i-th time.
- The estimate of P_{am} is

$$\hat{P}_{am} = \frac{n_{am}}{n_a}, \tag{2.99}$$

and the estimate of T_{am} is

$$\hat{T}_{am} = \frac{\sum_i t_{i,am}}{n_{am}}. \tag{2.100}$$

Here n_{am} is the number of times that the first alarm state a transfers to a subsequential alarm state m without going beneath the state a, n_a is the number of times that the state a occurs, and $t_{i,am}$ is the number of steps taken by the transfer from the state a to the state m without going beneath the state a for the i-th time.

If the data length T goes to infinity, by the law of large numbers, the estimated statistics by the second approach converge to the true values of the four statistics. Thus, if the estimates from two approaches are close for a large data length T, then the theoretical expressions of the four statistics are validated. An example will be provided later at Sect. 2.4.4.1 to numerically verify the theoretical expression of the four statistics.

2.4.3 Design of Optimal Alarm Thresholds

This subsection presents a method to design alarm thresholds based on alarm probability plots. In order to estimate the four statistics in a reliable manner, a prerequisite step is a proper determination of quantification intervals in order to obtain the discrete state variable $x_s(t)$ from the process variable $x(t)$. Assume that the samples below the high alarm threshold x_{th} are divided into N_1 non-alarm states, while the others are into N_2 alarm states. Let us determine the values of N_1 and N_2 and the quantification intervals as follows.

First, the estimate $\hat{p}_{ij}^{(1)}$ in (2.96) may be statistically unreliable if the sample size of each state is too small, which will make the estimates of the four statistics inaccurate; meanwhile, the number of states should be large enough so that the resolution to choose a proper value for alarm thresholds is satisfactory. A natural idea of determining the quantification intervals for states is that the sample size of the states in each interval should not be too small and the width of a quantification interval should be narrower when the values of $x(t)$ in the interval are dense. Algorithm 1 implements this idea by using the smallest sample size z of all states as the tuning parameter. A recommended value of z is determined in the next example.

Algorithm 1 Determine the quantification intervals

Input Argument #1: $x(t)$, the process variable;
Input Argument #2: x_{th}, the high alarm threshold;
Input Argument #3: z, the smallest sample size of all the states;
L_1=sum($x(t)< x_{th}$)-1; L_2=sum($x(t) \geq x_{th}$)+1;
N_1=floor(L_1/z); N_2=floor(L_2/z)
$x_o(t)$=sort($x(t)$);
$t = 1$;
for $t = 1$ to $L_1 + L_2$ **do**
  for $k = 1$ to $N_1 - 1$ **do**
    if $x_o(1) \leq x(t) \leq x_o(L_1 - (N_1 - 1)z)$ **then**
      $x_s(t) = 1$;
    end if
    if $x_o(L_1 - (N_1 - k)z + 1) \leq x(t) \leq x_o(L_1 - (N_1 - 1 - k)z)$ **then**
      $x_s(t) = k + 1$;
    end if
  end for
  for $k = 1$ to $N_2 - 1$ **do**
    if $x_o((k-1)z + 1 + L_1) \leq x(t) \leq x_o(kz + L_1)$ **then**
      $x_s(t) = k + N_1$;
    end if
    if $x_o((N_2 - 1)z + 1 + L_1) \leq x(t) \leq x_o(L_1 + L_2)$ **then**
      $x_s(t) = N_1 + N_2$;
    end if
  end for
end for

Second, it is worthy to note that if the value of P_{ka} is too small, the state k hardly ever transfers to the first alarm state a in a non-decreasing manner. Hence, a threshold γ_0 can be set as the lower bound of P_{ka}. A default value $\gamma_0 = 0.001$ is used, because an event is generally regarded as a small probability event when its probability of occurrence is smaller than 0.001. As implied by (2.87), P_{ka} is a non-decreasing function of the state k, so that the minimum of P_{ka} is P_{1a}. In other words, P_{1a} is set to be no smaller than γ_0. This is achieved by adjusting the value of N_1 as presented in Algorithm 2.

Now it is ready to propose a method to design an optimal alarm threshold. Given an alarm probability plot for a specified value of the alarm threshold x_{th}, a natural

Algorithm 2 Transform $x(t)$ to $x_s(t)$

1) Set the initial value of $N_1 = n_0$ to achieve the smallest sample size $z = 600$ for all states, and determine the quantification intervals by Algorithm 1;
2) Calculate the value of P_{1a}. If it is smaller than γ_0, then $N_1 = N_1 - 1$;
3) Repeat the steps 1 and 2 till the inequality $P_{1a} \geq \gamma_0$ holds.

idea is to formulate a discriminant function as

$$F(P_i, T_i) = \omega P_i + (1 - \omega)\frac{T_i}{\max(T_i)}. \tag{2.101}$$

Here P_i and T_i respectively represent the transition probability P_{ia} or P_{ai} and the averaged transition step T_{ia} or T_{ai} associated with the state i, and $0 \leq \omega \leq 1$ is a real-valued weighting factor. A default value is $\omega = 0.5$, saying that P_i and T_i are equally important. A larger value of P_i implies that the state i is more likely to move towards x_{th} to raise an alarm, or to go beyond x_{th} after an alarm has occurred. In addition, a larger value of T_i implies that industrial plant operators have more time to handle the upcoming alarm or the occurring one. Thus, the optimal value of x_{th} is the one associated with the state maximizing the discriminant function in (2.101), i.e.,

$$x_{th,opt} = x_{i^{(opt)}}, \tag{2.102}$$

where $x_{i^{(opt)}}$ is the lower bound of the quantification interval $[x_{i^{(opt)}}, x_{i^{(opt)}+1})$ to achieve the maximum value of $F(P_i, T_i)$,

$$i^{(opt)} = \underset{i}{\text{argmax}}\, F(P_i, T_i).$$

The definitions of the four statistics reveal that P_i and T_i vary for different alarm thresholds. Thus, for different alarm thresholds, the discriminant function in (2.101) is updated as

$$F(P_{i,x_{th}}, T_{i,x_{th}}) = \alpha P_{i,x_{th}} + (1 - \alpha)\frac{T_{i,x_{th}}}{T_{max}}. \tag{2.103}$$

Here $P_{i,x_{th}}$ and $T_{i,x_{th}}$ respectively are the counterparts of P_i and T_i for the alarm threshold x_{th}, and T_{max} is the maximum of all $T_{i,x_{th}}$'s, i.e., $T_{max} := \max_{i,x_{th}}\left(T_{i,x_{th}}\right)$. It is possible that the maximum points of $F(P_{i,x_{th}}, T_{i,x_{th}})$ for different alarm thresholds are close to each others. To handle this condition, the average of $F(P_{i,x_{th}}, T_{i,x_{th}})$'s with respect to a fixed value of x_{th} is compared to choose the optimal value of alarm thresholds, i.e.,

$$x_{th,opt} = \underset{x_{th}}{\text{argmax}}\, \overline{F(P_{i,x_{th}}, T_{i,x_{th}})}, \tag{2.104}$$

where $\overline{F(P_{i,x_{th}}, T_{i,x_{th}})} = \frac{\sum_i F(P_{i,x_{th}}, T_{i,x_{th}})}{N_{1,x_{th}}+N_{2,x_{th}}}$ with $N_{1,x_{th}}$ ($N_{2,x_{th}}$) as the number of non-alarm (alarm) states in $x_s(t)$ associated with x_{th}.

Given a historical data set $\{x(t)\}_{t=1}^{T}$, the steps of the proposed method are summarized as follows.

1. Choose a specified value of alarm threshold x_{th}, and quantify $x(t)$ to obtain the state variable $x_s(t)$ via Algorithms 1 and 2;
2. Smooth $x_s(t)$, when necessary, to guarantee that the states in $x_s(t)$ transfer only among adjacent states: if $x_s(t) = i$ and $x_s(n+1) \leq i-1$, then $x_s(n+1)$ is set to the state $i-1$; if $x_s(t) = i$ and $x_s(n+1) \geq i+1$, then $x_s(n+1)$ is set to $i+1$;
3. Estimate the one-step transition probability $p_{ij}^{(1)}$ as that in (2.96);
4. Calculate the four statistics P_{ka} in (2.87), T_{ka} in (2.90), P_{am} in (2.93) and T_{am} in (2.95) to formulate the alarm probability plot;
5. Select the optimal value of alarm thresholds as $x_{th,opt}$ in (2.102), or repeat the above steps 1-4 for multiple alarm thresholds and select the optimal one as $x_{th,opt}$ in (2.104).

The proposed method provides a solution to design alarm thresholds for non-IID analog process variables. By contrast, many existing methods to design alarm thresholds for univariate alarm systems require $x(t)$ to be IID [20, 33]. If they are applied to more general process variables, then erroneous results may be obtained. For instance, if $x(t)$ is IID, then it is correct to estimate the false alarm rate by countering the numbers of all samples of $x(t)$ larger than the high alarm threshold x_{th} for the case that $x(t)$ is in the normal condition. However, doing so will over-estimate the false alarm rate if $x(t)$ is non-IID, because several consecutive samples could be larger than x_{th} and should be counted as one single false alarm occurrence instead of multiple ones.

2.4.4 Numerical and Industrial Examples

This section presents two numerical examples and one industrial example, mainly for verifying theoretical expressions of four statistics in the alarm probability plot and designing an optimal alarm threshold for an industrial process variable.

2.4.4.1 Numerical Example

The process variable $x(t)$ is generated as a first-order auto-regressive model, $x(t) = 4.5 + 0.1x(t-1) + e(t)$, where $e(t)$ is the Gaussian white noise with zero mean and variance 4. The samples are quantified into 10 states, namely, $x_s(t) = 1$ for $x(t) < 1$, $x_s(t) = 2$ for $1 \leq x(t) < 2$, $\cdots$, $x_s(t) = 9$ for $8 \leq x(t) < 9$, and $x_s(t) = 10$ for $x(t) \geq 9$. Without loss of generality, the state 8 is selected as the first alarm state.

Table 2.14 The estimates of four statistics via two approaches, reprinted from Ref. [34], copyright 2017, with permission from IEEE

State	The first approach		The second approach	
i	$\overline{\hat{P}_i}$	$\overline{\hat{T}_i}$	$\overline{\hat{P}_i}$	$\overline{\hat{T}_i}$
1	0.033	8.086	0.031	8.105
2	0.033	7.084	0.032	7.232
3	0.039	6.038	0.037	6.287
4	0.042	4.852	0.043	4.847
5	0.055	3.660	0.053	3.718
6	0.087	2.442	0.084	2.422
7	0.217	1.199	0.219	1.205
8	0.197	0	0.200	0
9	0.083	1.107	0.087	1.073
10	0.003	2.292	0.003	2.171

100 independent realizations of $\{x(t)\}_{t=1}^{T}$ are generated for different data lengths T. Tuning T actually adjusts the smallest sample size z of all states. Table 2.14 presents the sample means of the estimates of P_{ka}, P_{am}, T_{ka} and T_{am} from two approaches in 100 realizations with $z = 1500$. For the ease of notations, the estimates of P_{ia} and P_{ai} are represented by $\hat{P}_i$ associated with the state i. Similarly, the estimates of T_{ia} and T_{ai} are represented by $\hat{T}_i$. The two groups of sample means are close to each other, which says that the theoretical expressions of the four statistics are correct.

Comparing with the second approach, the first approach requires a smaller sample size to achieve the same level of estimation accuracy. Figures 2.46 and 2.47 provide the sample variances of the estimated statistics for the states 6, 7, 9 and 10 close to the first alarm state 8 for different sample sizes, where 100 independent realizations are generated for each sample size. The sample variances of the estimates from the first approach are smaller than the counterparts from the second approach. The difference is larger for a smaller sample size, and becomes smaller for a larger sample size as expected. Thus, the first approach should be used in practice, especially when the sample size z is small.

A recommended value of z is determined. Based on 100 independent realizations of $\{x(t)\}_{t=1}^{T}$, Fig. 2.48 presents the estimates of $P_{48}/E(P_{48})$ and $T_{48}/E(T_{48})$ in the 100 realizations, where $E(\cdot)$ is the expectation operation, for different values of z (achieved by varying the data length T). It is observed that the estimates are fluctuating around their sample means within the range of $\pm 5\%$ once z is no less than 500.

To verify the above conclusion for all states, Fig. 2.49 presents the estimated alarm probability plots for 100 independent realizations of $\{x(t)\}_{t=1}^{T}$, where z is set to 500. In Fig. 2.49, all the estimated alarm probability plots in the 100 realizations are almost overlapped. Therefore, it is reliable to estimate the four statistics from historical data samples as long as z is no less than 500.

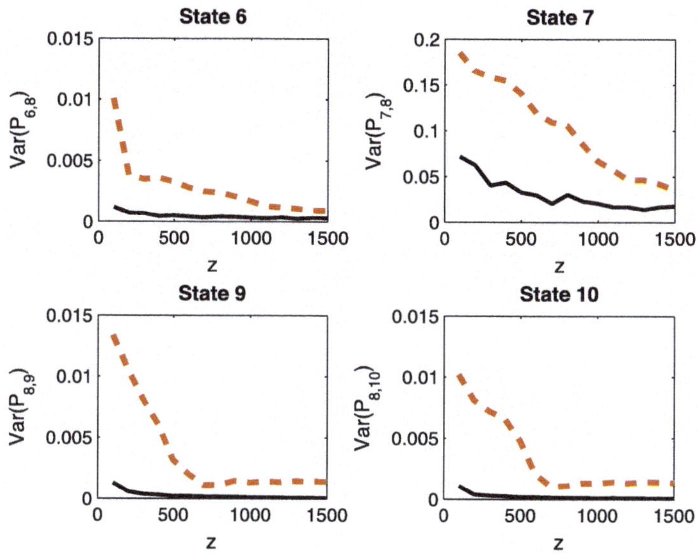

Fig. 2.46 The sample variances of the estimated transition probabilities for different sample sizes z from the first (solid) and second (dash) approaches, reprinted from Ref. [34], copyright 2017, with permission from IEEE

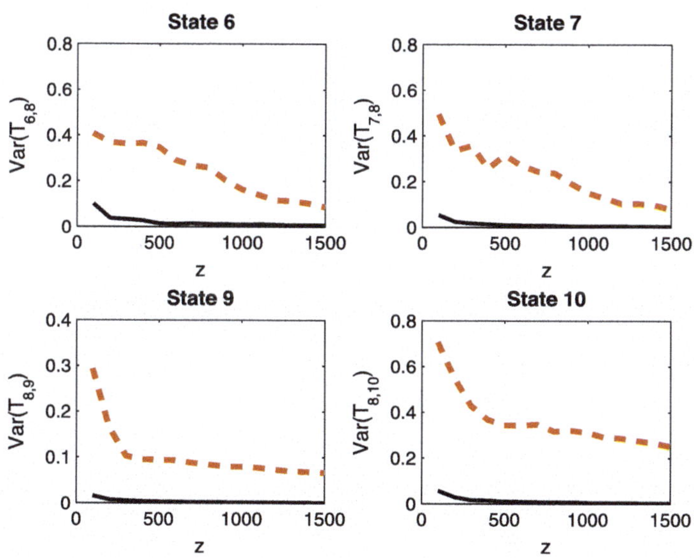

Fig. 2.47 The sample variances of the estimated averaged transition steps for different sample sizes z from the first (solid) and second (dash) approaches, reprinted from Ref. [34], copyright 2017, with permission from IEEE

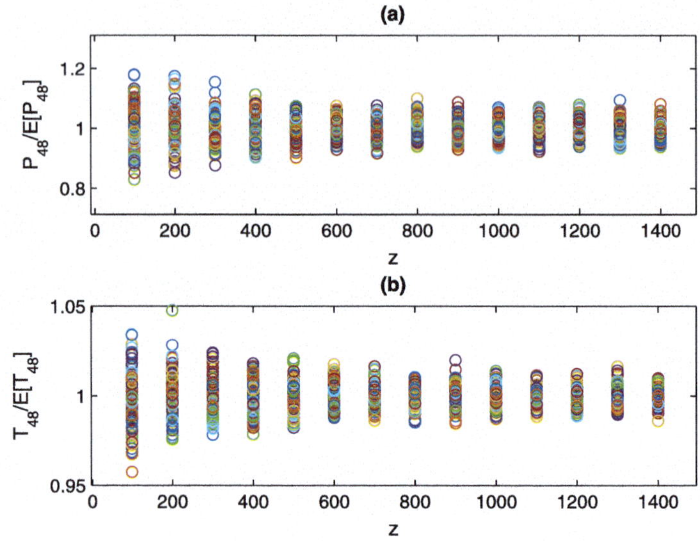

Fig. 2.48 The transition probabilities **a** and the averaged transition steps **b** for different values of *z*, reprinted from Ref. [34], copyright 2017, with permission from IEEE

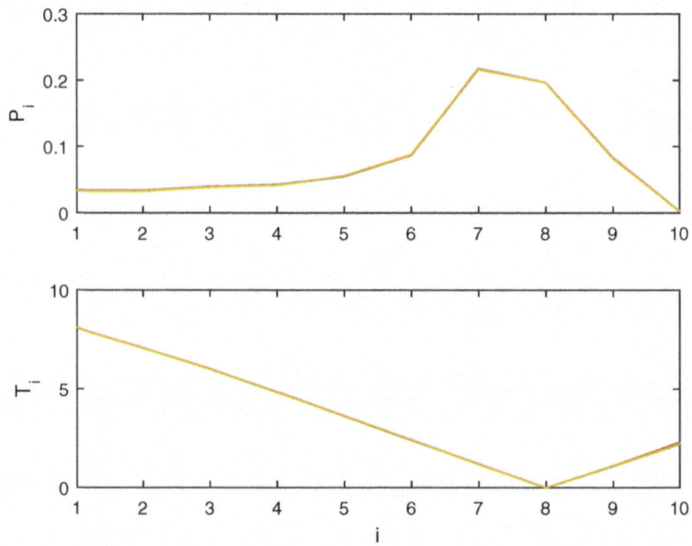

Fig. 2.49 The alarm probability plots in 100 independent realizations, reprinted from Ref. [34], copyright 2017, with permission from IEEE

Table 2.15 The values of P_i, T_i and $F(P_i, T_i)$ for all states, reprinted from Ref. [34], copyright 2017, with permission from IEEE

State i	P_i	T_i	$F(P_i, T_i)$
1	0.12	10.33	0.24
2	0.12	9.33	0.23
3	0.20	7.93	0.24
4	0.33	6.13	0.27
5	0.98	5.13	0.58
6	0.98	4.13	0.57
7	0.98	2.01	0.53
8	0.95	0	0.47
9	0.47	16.55	0.53
10	0.12	27.91	0.56

2.4.4.2 Numerical Example 2

This example reproduces an illustrative example in the public technical report [9], which provides an alarm probability plot shown in Fig. 2.43. The state 8 is the first alarm state. Table 2.15 lists the values of the transition probability P_i, the averaged transition step T_i and the discriminant function $F(P_i, T_i)$ in (2.101) for all states.

Eq. (2.102) gives the optimal alarm threshold to be associated with the state 5. This is consistent with the observation from Fig. 2.43 as follows. If the high alarm threshold x_{th} is set to be associated with the state 4, then the process variable $x(t)$ is likely (with probability 0.67) to return to non-alarm states on its own after arriving at the first alarm state 4; as a result, the occurring alarm is a false one, because no corrective actions are taken to clear the alarm. By contrast, the state 5 has a large transition probability $P_5 = 0.98$, saying that once $x(t)$ reaches the state 5, it is very likely (with probability 0.98) to keep increasing to higher values. If x_{th} is adjusted to make the state 5 as the first alarm state, then there is a small probability 0.02 of generating false alarms in terms that $x(t)$ returns to non-alarm states $i < 5$ on its own after arriving at the first alarm state 5. The same situation applies to the states 6 and 7 owing to the same transition probabilities $P_6 = P_7 = 0.98$. Setting the first alarm state at the state 5 will increase few false alarms. In addition, the state 5 takes more averaged transition steps than the states 6 and 7 for the transfer to the state 8 in a non-decreasing manner. Setting x_{th} to be associated with the state 5 yields more time for industrial plant operators to response to alarms than the states 6 and 7. Therefore, x_{th} could be decreased from the state 8 to the state 5 in order to have more amount of time for operators to deal with abnormal conditions, with few increments of false alarms.

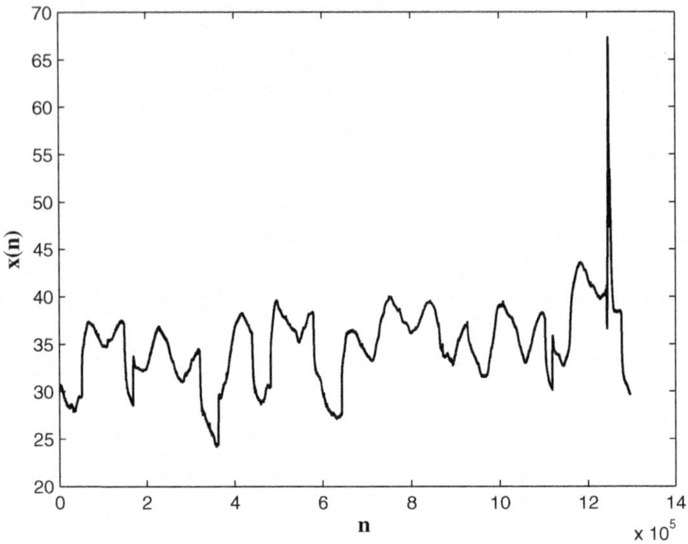

Fig. 2.50 The time series plot of the bearing temperature, reprinted from Ref. [34], copyright 2017, with permission from IEEE

2.4.4.3 Industrial Example

In a large-scale thermal power plant, a coal grinding mill is associated with several measurable process variables. One process variable $x(t)$ is the bearing temperature in the non-driving end of the mill. If $x(t)$ goes beyond the value of 55 Celsius degree, then the mill will be shut down automatically to prevent a potential mill damage from the abnormal high temperature in $x(t)$. Thus, it is desirable to design a proper high alarm threshold x_{th} for $x(t)$ for operators to detect and handle such an abnormal condition in a prompt manner. The time series plot of $x(t)$ for 15 days is given in Fig. 2.50 with the sampling period $h = 1$ sec. Note that $x(t)$ in Fig. 2.50 is clearly not IID so that the methods in literature requiring $x(t)$ to be IID cannot be applied here to design x_{th}. If the value of x_{th} is too small, then $x(t)$ will return to the values smaller than x_{th} on its own after an alarm occurs so that the occurring alarm is a false one. On the other hand, if the value of x_{th} is too large, then there would be less time for plant operators to handle the abnormal condition.

Owing to the space limitation, detailed results are presented for three alarm threshold candidates, namely, $x_{th} = 45, 47, 51$. The corresponding alarm probability plots are presented in Figs. 2.51, 2.52 and 2.53. Tables 2.16, 2.17 and 2.18 list the estimates of the transition probability $P_{i,x_{th}}$, the averaged transition step $T_{i,x_{th}}$ and the discriminant function $F(P_{i,x_{th}}, T_{i,x_{th}})$ (abbreviated as F_i) for $x_{th} = 45, 47, 51$, respectively. As revealed from Tables 2.16, 2.17 and 2.18, the maximum values of $F(P_{i,x_{th}}, T_{i,x_{th}})$ for the three alarm thresholds are very close to each other. Hence, the averaged values

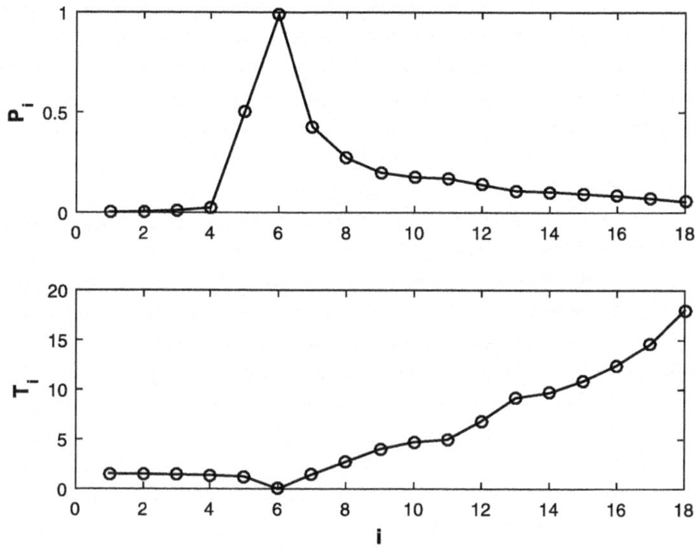

Fig. 2.51 The alarm probability plot for $x_{th} = 45$, reprinted from Ref. [34], copyright 2017, with permission from IEEE

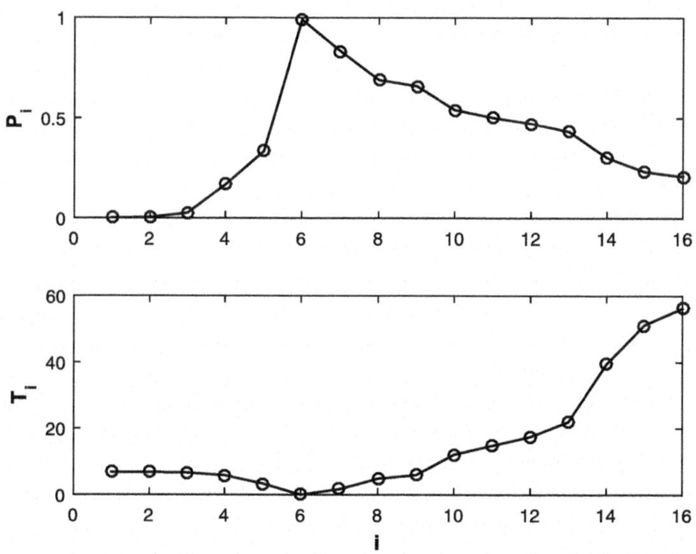

Fig. 2.52 The alarm probability plot for $x_{th} = 47$, reprinted from Ref. [34], copyright 2017, with permission from IEEE

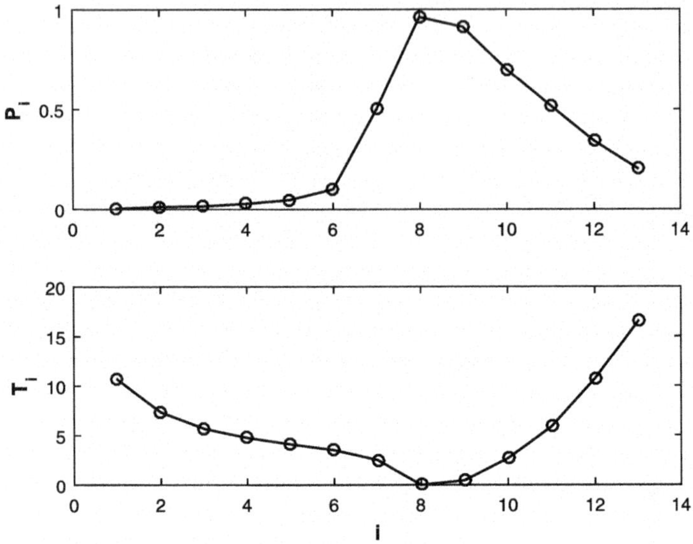

Fig. 2.53 The alarm probability plot for $x_{th} = 51$, reprinted from Ref. [34], copyright 2017, with permission from IEEE

Table 2.16 The estimated statistics for $x_{th} = 45$, reprinted from Ref. [34], copyright 2017, with permission from IEEE

State	Interval	$P_{i,x_{th}}$	$T_{i,x_{th}}/T_{max}$	F_i
1	[43.51, 43.53)	0.0021	0.0075	0.0048
2	[43.53, 43.55)	0.0048	0.0073	0.0061
3	[43.55, 43.58)	0.0115	0.0071	0.0093
4	[43.58, 43.86)	0.0247	0.0068	0.0157
5	[43.86, 45.00)	0.5000	0.0061	0.2531
6	[45.00, 46.13)	0.9883	0.0000	0.4941
7	[46.13, 47.34)	0.4286	0.0071	0.2178
8	[47.34, 48.10)	0.2727	0.0135	0.1431
9	[48.10, 48.76)	0.2000	0.0197	0.1099
10	[48.76, 49.22)	0.1765	0.0232	0.0998
11	[49.22, 50.33)	0.1707	0.0244	0.0976
12	[50.33, 51.47)	0.1391	0.0336	0.0863
13	[51.47, 52.24)	0.1088	0.0450	0.0769
14	[52.24, 53.03)	0.1038	0.0475	0.0756
15	[53.03, 54.66)	0.0940	0.0534	0.0737
16	[54.66, 57.93)	0.0836	0.0612	0.0724
17	[57.93, 62.46)	0.0716	0.0718	0.0717
18	[62.46, 67.35)	0.0557	0.0886	0.0722

Table 2.17 The estimated statistics for $x_{th} = 47$, reprinted from Ref. [34], copyright 2017, with permission from IEEE

State	Interval	$P_{i,x_{th}}$	$T_{i,x_{th}}/T_{max}$	F_i
1	[43.54, 43.57)	0.0014	0.0334	0.0174
2	[43.57, 43.61)	0.0034	0.0331	0.0183
3	[43.61, 44.71)	0.0256	0.0324	0.0290
4	[44.71, 45.89)	0.1667	0.0287	0.0977
5	[45.89, 47.00)	0.3333	0.0164	0.1748
6	[47.00, 47.91)	0.9900	0.0000	0.4950
7	[47.91, 48.57)	0.8333	0.0083	0.4208
8	[48.57, 49.15)	0.6897	0.0238	0.3567
9	[49.15, 49.98)	0.6593	0.0292	0.3442
10	[49.98, 51.10)	0.5405	0.0594	0.2999
11	[51.10, 52.16)	0.5018	0.0727	0.2873
12	[52.16, 52.77)	0.4722	0.0858	0.2790
13	[52.77, 53.96)	0.4314	0.1081	0.2698
14	[53.96, 57.03)	0.3014	0.1941	0.2478
15	[57.03, 60.98)	0.2316	0.2516	0.2416
16	[60.98, 67.35)	0.2076	0.2765	0.2420

Table 2.18 The estimated statistics for $x_{th} = 51$, reprinted from Ref. [34], copyright 2017, with permission from IEEE

State	Interval	$P_{i,x_{th}}$	$T_{i,x_{th}}/T_{max}$	F_i
1	[44.40, 45.59)	0.0043	0.0524	0.0284
2	[45.59, 46.69)	0.0128	0.0360	0.0244
3	[46.69, 47.76)	0.0154	0.0278	0.0216
4	[47.76, 48.41)	0.0282	0.0234	0.0258
5	[48.41, 48.65)	0.0471	0.0201	0.0336
6	[48.65, 49.03)	0.1000	0.0172	0.0586
7	[49.03, 51.00)	0.5000	0.0123	0.2561
8	[51.00, 52.10)	0.9815	0.0000	0.4907
9	[52.10, 52.54)	0.9130	0.0022	0.4576
10	[52.54, 53.41)	0.7000	0.0132	0.3566
11	[53.41, 56.14)	0.5185	0.0289	0.2737
12	[56.14, 59.76)	0.3415	0.0526	0.1971
13	[59.76, 67.35)	0.2029	0.0812	0.1420

of $F(P_{i,x_{th}}, T_{i,x_{th}})$ are calculated as 0.1481, 0.3414 and 0.2172 for $x_{th} = 45, 47, 51$, respectively, and (2.104) gives the optimal choice $x_{th,opt} = 47$.

Let us verify the optimal choice $x_{th} = 47$ via the alarm probability plots. As revealed from Fig. 2.51, once $x(t)$ reaches the value $x_{th} = 45$, it keeps increasing with a small probability $P_{i=7,x_{th}=45} = 0.4286$. In other words, $x(t)$ will go back to the values smaller than $x_{th} = 45$ instead of increasing to higher values. Thus, setting $x_{th} = 45$ is expected to result in many false alarms. In Figs. 2.52 or 2.53, when x_{th} takes the value of 47 or 51, $x(t)$ will keep increasing beyond x_{th} with a large probability $P_{i=7,x_{th}=47} = 0.8333$ or $P_{i=9,x_{th}=51} = 0.9130$ once $x(t)$ reaches the corresponding alarm threshold. In addition, setting $x_{th} = 47$ clearly leads to more time for plant operators to handle the alarms. The averaged values of $F(P_{i,x_{th}}, T_{i,x_{th}})$ for $x_{th} = 47$ is larger than the counterpart for $x_{th} = 51$. Therefore, it is reasonable to choose the optimal alarm threshold as $x_{th,opt} = 47$.

References

1. Adnan, N. A., Izadi, I., & Chen, T. (2011). On expected detection delays for alarm systems with deadbands and delay-timers. *Journal of Process Control, 21*, 1318–1331.
2. Afzal, M. S., & Chen, T. (2017). Analysis and design of multimode delay-timers. *Chemical Engineering Research and Design, 120*, 179–193.
3. Alrowaie, F., Gopaluni, R.B., & Kwok, K.E. (2014). Alarm design for nonlinear stochastic systems. In *The 11th World Congress on Intelligent Control and Automation* (pp. 472–479). June 29-July 4, 2014, Shenyang, China.
4. Basseville, M., & Nikiforov, I. V. (1993). *Detection of abrupt changes: Theory and application*. Englewood Cliffs, NJ, USA: Prentice Hall.
5. Bertsekas, D. P., & Tsitsiklis, J. N. (2002). *Introduction to probability*. USA, MA: Athena Scientific Belmont.
6. Bransby, M.L., & Jenkinson, J. (1998). *The management of alarm systems*. Health and Safety Executive.
7. Chen, J., & Gupta, A. K. (2000). *Parametric statistical change point analysis*. Boston, USA: Birkhauser.
8. Chen, B., & Hong, Y. (2012). Testing for the Markov property in time series. *Econometric Theory, 28*, 130–178.
9. Control Arts Inc. (2011). *Alarm system engineering*, www.controlartsinc.com
10. Devore, J.L. (2004). *Probability and statistics for engineering and the sciences* (5th edn.). Thomson Learning.
11. Doan, A. E. (2005). Type I and Type II errors. *Encyclopedia of Social Measurement, 3*, 883–888.
12. Douglas, C. M., & George, C. R. (2011). *Applied statistics and probability for engineers* (5th ed.). New Jersey, USA: John Wiley and Sons.
13. EEMUA. (2013). *EEMUA-191: Alarm systems-A guide to design*. Management and Procurement: Engineering Equipment and Materials Users' Association.
14. Gulhar, M., Kibria, B. G., Albatineh, A. N., & Ahmed, N. U. (2012). A comparison of some confidence intervals for estimating the population coefficient of variation: a simulation study. *SORT-Statistics and Operations Research Transactions, 36*, 45–68.
15. Gyasi, P., & Wang, J. (2023). Design of alarm thresholds and delay timers for non-IID process variables based on alarm durations. *Process Safety and Environmental Protection, 170*, 1173–1187.
16. Hautus, M. J., Macmillan, N. A., & Creelman, C. D. (2021). *Detection theory: A user's guide* (3rd ed.). New York, USA: Routledge.

17. Hugo, A.J. (2009). Estimation of alarm deadbands. In *The 7th IFAC Symposium on Fault Detection, Supervision and Safety of Technical Processes* (pp. 663–667). June 30–July 3, 2009, Barcelona, Spain.
18. ISA (2009). *ANSI/ISA-18.2: Management of alarm systems for the process industries*. International Society of Automation.
19. Lehmann, E. L. (1986). *Testing statistical hypothesis* (2nd ed.). Wiley: New York.
20. Naghoosi, E., Izadi, I., & Chen, T. (2011). Estimation of alarm chattering. *Journal of Process Control, 21*, 1243–1249.
21. Papoulis, A., & Pillai, S. U. (2002). *Probability, random variables and stochastic processes* (4th ed.). New York, USA: McGraw-Hill.
22. Pettitt, A. N. (1979). A non-parametric approach to the change-point problem. *Applied Statistics, 28*, 126–135.
23. Pishro-Nik, H. (2016). *Introduction to probability, statistics, and random processes*. USA: Kappa Research LLC.
24. Rice, J. A. (2006). *Mathematical statistics and data analysis* (2nd ed.). Patparganj, India: Cengage Learning.
25. Rothenberg, D. (2009). *Alarm management for process control*. Momentum Press.
26. Sharma, K. K., & Krishna, H. (1994). Asymptotic sampling distribution of inverse coefficient-of-variation and its applications. *IEEE Transactions on Reliability, 43*, 630–633.
27. Tulsyan, A., & Gopaluni, R. B. (2019). Univariate model-based deadband alarm design for nonlinear processes. *Industrial & Engineering Chemistry Research, 58*, 11295–11302.
28. Walpole, R. E., Myers, R. H., Myers, S. L., & Ye, K. (2011). *Probability and statistics for engineers and scientists* (9th ed.). Prentice Hall.
29. Wang, J., & Chen, T. (2014). An online method to remove chattering and repeating alarms based on alarm durations and intervals. *Computers and Chemical Engineering, 67*, 43–52.
30. Wang, J., Wang, Z., Zhou, X., & Yang, F. (2022). Design of delay timers based on estimated probability mass functions of alarm durations. *Journal of Process Control, 110*, 154–165.
31. Wang, J., Sun, S., Wang, Z., Zhou, X., & Gyasi, P. (2023). Alarm deadband design based on maximum amplitude deviations and Bayesian estimation. *IEEE Transactions on Control Systems Technology, 31*, 1941–1948.
32. Wang, Z., Bai, X., Wang, J., & Yang, Z. (2019). Indexing and designing deadbands for industrial alarm signals. *IEEE Transactions on Industrial Electronics, 66*, 8093–8103.
33. Xu, J., Wang, J., Izadi, I., & Chen, T. (2012). Performance assessment and design for univariate alarm systems based on FAR, MAR and AAD. *IEEE Transactions on Automation Science & Engineering, 9*, 296–307.
34. Yu, Y., Wang, J., & Yang, Z. (2017). Design of alarm trippoints for univariate analog process variables based on alarm probability plots. *IEEE Transactions on Industrial Electronics, 64*, 6496–6505.

Chapter 3
Optimal Design of Multivariate Alarm Systems

Abstract This chapter is about the optimal design of multivariate alarm systems. The essence of designing multivariate alarm systems is to exploit some features of multiple related process variables to distinguish normal and abnormal conditions. The features are diverse, being general for many industrial processes or being specific for particular industrial devices. First, two general methods are presented to respectively exploit normal operating zones and variation directions of multiple variables as the features being applicable to many industrial processes. Second, two specific methods are provided for two types of industrial devices, electrical pumps and condensers, based on their own unique features. Numerical and industrial examples are presented to illustrate the proposed methods.

3.1 Normal Operating Zone-Based Multivariate Alarm Systems

This section proposes a method to design multivariate alarm systems based on the so-called normal operating zones (NOZs) [43, 50]. NOZs are the geometric spaces formed by variation ranges of multiple related variables to be allowed under normal operating conditions. Mathematical models such as convex hulls and search cones are established from historical data for NOZs in the convex or non-convex shapes. Dynamic alarm thresholds of process variables are obtained based on the established NOZ models.

3.1.1 Problem Description

Consider a multivariate process $\tilde{X}$ composed by n process variables $\tilde{x}_1, \tilde{x}_2, \cdots, \tilde{x}_n$. The selection of n process variables and their variation ranges can be determined based on the physical meanings and normal operating requirements. Historical data samples in normal conditions are given as $\left\{\tilde{X}(t) \mid t \in [1,\ T]\right\}$, with $\tilde{X}(t) := [\tilde{x}_1(t),$

J. Wang et al., *Intelligent Industrial Alarm Systems*,
https://doi.org/10.1007/978-981-97-6516-4_3

$\tilde{x}_2(t), \cdots, \tilde{x}_n(t)]$. Here t and T are the sampling time index and the number of data points, respectively. These data samples formulate a normal operating zone (NOZ), so that $\tilde{X}(t)$ in normal (abnormal) conditions should lie inside (outside) the NOZ. It is assumed that $\tilde{x}_1(t), \tilde{x}_2(t), \cdots, \tilde{x}_n(t)$ are stationary processes. In other words, new data samples of $\tilde{x}_1, \tilde{x}_2, \cdots, \tilde{x}_n$ in normal conditions should be distributed in a consistent way as these historical data samples.

The problem to be solved is to design multivariate alarm systems based on mathematical models of the NOZs established from historical data samples $\left\{\tilde{X}(t) \mid t \in [1, T]\right\}$ in normal conditions to generate dynamic alarm thresholds of each process variable for alarm monitoring. For a new data point $\tilde{X}(t)$, each process variable $\tilde{x}_j$ should have its own dynamic low alarm threshold $\tilde{l}_j(t)$ and high alarm threshold $\tilde{h}_j(t)$, which are consistent with the NOZs. A comparison between $\tilde{x}_j(t)$ and $\tilde{l}_j(t)$ or $\tilde{h}_j(t)$ should tell whether $\tilde{X}(t)$ is in the normal condition. If all process variables are inside their own dynamic alarm thresholds, the process is in the normal condition and no alarms are raised; otherwise, the process is in the abnormal condition and alarms will occur. An alarm variable $x_a(t)$ is defined as

$$x_a(t) = \begin{cases} 0, & \text{if } \tilde{l}_j(t) \le \tilde{x}_j(t) \le \tilde{h}_j(t), \ \forall j = 1, 2, \ldots, n, \\ 1, & \text{otherwise.} \end{cases} \tag{3.1}$$

Here the word 'dynamic' has two meanings that $\tilde{l}_j(t)$ and $\tilde{h}_j(t)$ change with t, and are related to other process variables $\tilde{x}_{j'}$ for $j' \neq j$. Figure 3.1 presents a schematic illustration of the NOZ and dynamic alarm thresholds for a bivariate process formulated by variables x_1 and x_2. When the bivariate process changes from an operating point 'A' to another point 'B' along a trajectory as the blue dashed line in Fig. 3.1a, x_1 and x_2 have their own dynamic alarm thresholds as the red solid lines in Fig. 3.1b.

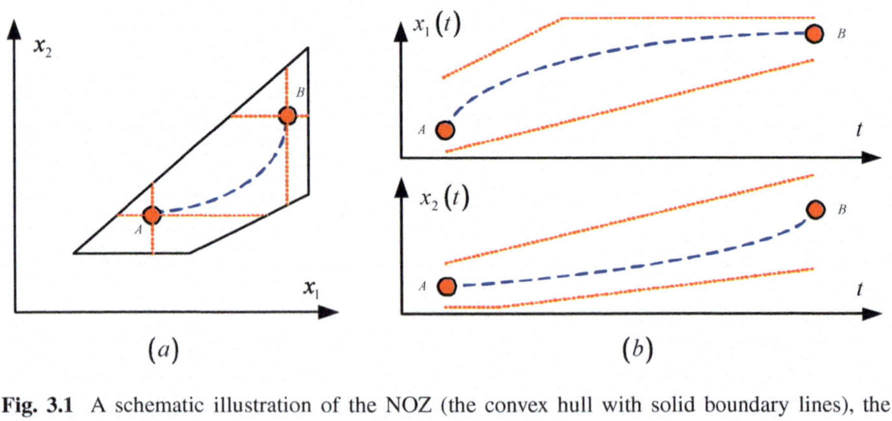

Fig. 3.1 A schematic illustration of the NOZ (the convex hull with solid boundary lines), the trajectory of operating points (red circles and blue dash curve) and dynamic alarm thresholds (red dashed lines)

It is worthy to point out that the NOZs and dynamic alarm thresholds are complementary to each other. The NOZs reveal the relations among multiple process variables in a transparent manner. By looking at the geometric positions of data points with respect to the NOZs, it is ready to tell whether an abnormal condition occurs. However, the time information is absent in the NOZs, and industrial plant operators are used to the standard ways of comparing process variables with their alarm thresholds for alarm monitoring. As a complement, dynamic alarm thresholds are introduced to generate alarms when the abnormal condition occurs.

The designed multivariate alarm system can avoid false and missed alarms that occurred if univariate alarm systems with static alarm thresholds were used. This is owing to a fact that dynamic alarm thresholds of one process variable are related to other process variables, while static alarm thresholds ignore the relations among process variables. In addition, each process variable has its own dynamic alarm thresholds that have the same physical meaning as the process variable.

3.1.2 Multivariate Alarm Systems for Convex NOZs

This section establishes a convex hull as the NOZ model from historical data points in normal conditions based on a formulated fitness index; afterwards, dynamic alarm thresholds of each process variable are developed based on the established NOZ. If the NOZ is in the shape of other geometric models such as hyper-ellipsoids, the proposed method is equally applicable by replacing the convex hull by other geometric models.

3.1.2.1 Establishment of NOZ Models Via Convex Hulls

This section builds a convex hull as the mathematical model to describe the NOZ and measures the fitness between the convex hull model and historical data points.

First, the mathematical model of a convex hull is obtained from $\{\tilde{X}(t)\}_{t=1}^{T}$. It should be noted that $\tilde{X}(t)$ for $t = 1, 2, \cdots, T$ is collected on the condition that the process works normally without reported abnormalities, and data points with too small/large values are removed as outliers in advance [15]. In order to avoid that some process variables with large values erroneously play dominate roles, the raw data points are pre-processed by normalizing values of each process variable to have zero means and unit variances. For a raw data point $\tilde{X}(t)$, the corresponding normalized data point is

$$X(t) := \left[x_1(t),\ x_2(t),\ \cdots,\ x_n(t) \right],$$

where

$$x_j(t) = [\tilde{x}_j(t) - \mu_{\tilde{x}_j}]/\sigma_{\tilde{x}_j},\ j = 1, 2, \cdots, n. \tag{3.2}$$

Here $\mu_{\tilde{x}_j}$ and $\sigma_{\tilde{x}_j}$ indicate the sample mean and standard deviation of $\tilde{x}_j$, respectively.

A convex hull with m facets is associated with m hyperplanes, and the i-th facet $f^{(i)}$ is on the i-th hyperplane $p^{(i)}$. Thus, the mathematical expression of a convex hull is

$$AX'_{ch}(t) - B \leq \mathbf{0}, \tag{3.3}$$

where $X_{ch}(t)$ is a data point inside the convex hull or on a facet of the convex hull, the subscript 'ch' stands for the convex hull, the superscript '′' is the transpose operation, and

$$A = [a^{(1)}, a^{(2)}, \cdots, a^{(m)}]', \; B = [b^{(1)}, b^{(2)}, \cdots, b^{(m)}]', \mathbf{0} = [0, 0, \cdots, 0]'.$$

For $i \in \{1, 2, \cdots, m\}$, $a^{(i)} = [a_1^{(i)}, a_2^{(i)}, \cdots, a_n^{(i)}]$ is the unit normal vector of $p^{(i)}$, and $b^{(i)}$ is the distance of $p^{(i)}$ away from the origin point. The hyperplane function of $p^{(i)}$ is

$$a^{(i)}X'_p(t) - b^{(i)} = 0, \tag{3.4}$$

where $X_p(t)$ is a data point on $p^{(i)}$.

The values of $a^{(i)}$ and $b^{(i)}$ are estimated from $\{\bar{X}(t)\}_{t=1}^{T}$ as follows. Because the i-th hyperplane $p^{(i)}$ passes through n vertices of $f^{(i)}$, namely, $X(I_{i,1})$, $X(I_{i,2})$, $\cdots$, and $X(I_{i,n})$. These vertices are obtained by the quick-hull algorithm [4] that returns a m-by-n matrix I with m as the number of facets of the convex hull. The element $I_{i,j}$ on the i-th row and j-th column of I is the j-th vertex of the i-th facet of the convex hull, and is associated with the data point $X(I_{i,j})$. Substituting these n vertices into (3.4) leads to

$$a^{(i)}X'(I_{i,k}) - b^{(i)} = 0, \; k = 1, 2, \cdots, n. \tag{3.5}$$

Because $X(I_{i,k}) = [x_1(I_{i,k}), x_2(I_{i,k}), \cdots, x_n(I_{i,k})]$, (3.5) can be rewritten as

$$x_1(I_{i,k})a_1^{(i)} + x_2(I_{i,k})a_2^{(i)} + \cdots + x_n(I_{i,k})a_n^{(i)} - b^{(i)} = 0, \tag{3.6}$$

where $k = 1, \; 2, \; \cdots, \; n$. Moreover, $a^{(i)}$ is an unit vector, i.e.,

$$a_1^{(i)} \cdot a_1^{(i)} + a_2^{(i)} \cdot a_2^{(i)} + \cdots + a_n^{(i)} \cdot a_n^{(i)} = 1. \tag{3.7}$$

Equations (3.6) and (3.7) lead to the unique solutions of $a_1^{(i)}, a_2^{(i)}, \cdots, a_n^{(i)}$ and $b^{(i)}$. By setting the value of i to $1, \; 2, \; \cdots, \; m$, all hyperplane functions of the convex hull are obtained, so that the convex hull model in (3.3) is established.

Second, a fitness index is formulated to measure the fitness between the data points and the convex hull. The range $[r_j^{(l)}, r_j^{(h)}]$ for each normalized process variable x_j is computed as

$$r_j^{(l)} = \min(x_j) \text{ and } r_j^{(h)} = \max(x_j),$$

where $\min(x_j)$ and $\max(x_j)$ denote the minimum and maximum values of x_j, respectively. The convex hull is divided into equal-sized grids, each of which is represented by the set

$$\{[r_j^l + (k_j - 1)\delta_j, r_j^l + k_j\delta_j]\}_{j=1}^n.$$

Here $1 \le k_j \le \lfloor (r_j^{(h)} - r_j^{(l)})/\delta_j \rfloor$, and δ_j is the minimal interval of x_j to be determined by operators based on some prior process knowledge. The floor function $\lfloor \cdot \rfloor$ returns the largest integer that is not greater than the operand. The center point of each grid is

$$C(k_1, k_2, \cdots, k_n) = [c_1(k_1), c_2(k_2), \cdots, c_n(k_n)],$$

where $c_j(k_j) = r_j^{(l)} + (k_j - 0.5)\delta_j$. If a center point locates inside the convex hull (i.e., (3.3) is satisfied), it is defined as an inside point. The number of inside points is denoted as n_i. For an inside point $(k_1', k_2', \cdots, k_n')$, if a data point can be found in the corresponding grid, i.e.,

$$\exists t \in [1, N], \ |c_j(k_j') - x_j(t)| \le 0.5\delta_j, \ \forall j = 1, 2, \cdots, n,$$

the inside point is determined as a counting point. The number of counting points is denoted as n_c. A fitness index η is formulated as

$$\eta = n_c/n_i. \tag{3.8}$$

A smaller η means that there are lots of regions in the convex hull that have not been occupied by historical data points in normal conditions. Thus, the value of η measures the fitness between historical data points and the convex hull as a model of the NOZ.

A simple example is given in Fig. 3.2 to illustrate η. The variables x_1 and x_2 are with the same range $[-0.5, 0.5]$ and the same minimal interval 0.2. In Fig. 3.2, the big blue dots are the historical data points after normalization to build the NOZ, the small black dots are center points, and the red solid lines are boundaries of the convex hull. There are 25 center points, and the center points (3, 1), (2, 2), (3, 2), (4, 2), (1, 3), (2, 3), (3, 3), (4, 3), (5, 3), (2, 4), (3, 4), (4, 4) and (3, 5) are inside points. In addition, the center points (3, 1), (2, 2), (3, 2), (4, 2), (1, 3), (5, 3), (2, 4), (4, 4) and (3, 5) are counting points. The fitness index of the convex hull in Fig. 3.2 is $\eta = 9/13$.

An algorithm is summarized as follows to build the NOZ from historical data points $\{\bar{X}(t)\}_{t=1}^N$ in normal conditions with $\bar{X}(t) = [\bar{x}_1(t), \bar{x}_2(t), \cdots, \bar{x}_n(t)]$ based on the fitness index η.

Algorithm 1:

1. Normalize $\{\tilde{X}(t)\}_{t=1}^T$ by (3.2) so that each process variable has zero mean and unit variance, and the normalized historical data points in normal conditions are denoted as $\{X(t)\}_{t=1}^T$.

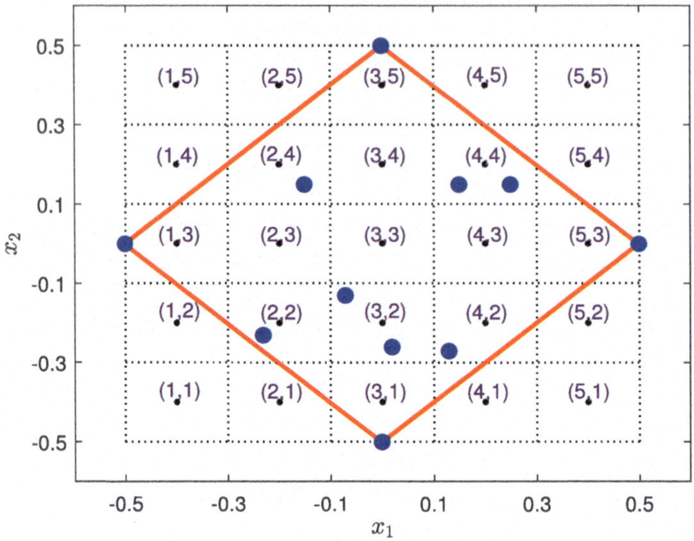

Fig. 3.2 A simple example to illustrate the fitness index η, reprinted from Ref. [50], copyright 2020, with permission from IEEE

2. Divide $\{X(t)\}_{t=1}^{T}$ into subsets as $\{X(t)\}_{t=1}^{T_s}$, $\{X(t)\}_{t=1}^{2T_s}$, $\cdots$, $\{X(t)\}_{t=1}^{dT_s}$, where T_s is the number of elements in the first subset, and $d = \lfloor T/T_s \rfloor$ is the number of subsets.
3. Select r data points to be included in the NOZ from $\{X(t)\}_{t=1}^{kT_s}$ using a distance-based clustering algorithm [27]. The initial value of k is 1, $r = \lfloor kT_s - \beta kT_s \rfloor$, and $\beta \in (0,\ 1)$ is the false alarm rate (FAR).
4. Find the vertices of a convex hull to represent the NOZ by the quick-hull algorithm [4] based on the r data points obtained in Step 3, estimate parameters of the convex hull model via (3.6) and (3.7), and calculate the fitness index η in (3.8).
5. If the value of η tends to be stable and its stable value is greater than a given threshold η_t, the algorithm is terminated to yield the NOZ established in Step 4 as the final result. Otherwise, k is updated to $k + 1$, and the algorithm repeats the iteration from Steps 3 to 5.

Not all data points are used in Step 3 to build the NOZ. This is due to an observed fact that data points far away from the NOZ (cluster) center are sparsely distributed, and they may be contaminated by noises. Thus, it is a reasonable choice that the NOZ encloses data points in normal conditions closer to the NOZ center, instead of all data points. Those data points that are not included in the NOZ are treated as false alarms. Under the assumption that statistical distributions of noises are the same for historical and current data points in normal conditions, the FAR is equal to β, since β in Step 3 is the ratio between the number of data points not included in the NOZ and the number of total data points.

3.1.2.2 Development of Dynamic Alarm Thresholds

This section develops dynamic alarm thresholds for each process variable based on the NOZ described by a convex hull model.

First, dynamic alarm thresholds of each process variable are defined. The most common way in raising alarms is to compare the value of a process variable x_i to two constant values l_i and h_i as the low and high alarm thresholds respectively, i.e., an alarm occurs if $x_i(t) < l_i$ and/or $x_i(t) > h_i$ (see, e.g., [11, 17]). Here dynamic alarm thresholds of each involved process variable are designed based on the NOZ.

If a new data point $\tilde{X}(t)$ satisfies (3.3), then it is a normal data point; its projections on facets of the convex hull in (3.3) are defined as its dynamic alarm thresholds. In this way, dynamic alarm thresholds of $\tilde{X}(t)$ reflect the distances of $\tilde{X}(t)$ away from the boundaries of the NOZ. When some process variables are close to their dynamic alarm thresholds, industrial plant operators should pay more attentions to the multivariate process and may take proactive actions for preventing potential abnormalities. For an abnormal data point $\tilde{X}(t)$ that locates outside the NOZ, i.e., (3.3) is not satisfied, dynamic alarm thresholds of a corresponding normal data point $\tilde{X}_c(t)$ are defined as dynamic alarm thresholds of $\tilde{X}(t)$. Here $\tilde{X}_c(t)$ is the data point on a facet of the convex hull being closest to $\tilde{X}(t)$. Dynamic alarm thresholds of an abnormal data point can tell the distances of $\tilde{X}(t)$ deviating away from the nearest boundary of the NOZ.

An example is given in Fig. 3.3a to illustrate dynamic alarm thresholds for a normal data point $\tilde{X}(t)$. Based on projections of $\tilde{X}(t)$ on facets of the convex hull, the dynamic alarm thresholds of $\tilde{x}_1$ and $\tilde{x}_2$ are calculated as $\tilde{l}_1(t) = -0.80, \tilde{h}_1(t) = 0.80$, and $\tilde{l}_2(t) = -0.70, \tilde{h}_2(t) = 0.70$. Figure 3.3b shows an example for an abnormal data point $\tilde{X}(t)$. Based on projections of the corresponding normal data point of $\tilde{X}(t)$ on facets of the convex hull, the dynamic alarm thresholds of $\tilde{x}_1$ and $\tilde{x}_2$ are calculated as $\tilde{l}_1(t) = -0.60, \tilde{h}_1(t) = 0.60$, and $\tilde{l}_2(t) = -0.40, \tilde{h}_2(t) = 0.40$.

Second, theoretical expressions of dynamic alarm thresholds are derived. For a normal data point $\tilde{X}(t) = [\tilde{x}_1(t), \tilde{x}_2(t), \cdots, \tilde{x}_n(t)]$, its corresponding normalized data point is $X(t) = [x_1(t), x_2(t), \cdots, x_n(t)]$. $X(t)$ is separated into three parts, namely, $x_j(t)$,

$$v_1 = [x_1(t), x_2(t), \cdots, x_{j-1}(t)]$$

and

$$v_2 = [x_{j+1}(t), x_{j+2}(t), \cdots, x_n(t)].$$

Equation (3.4) becomes

$$\left[u_1\ a_j^{(i)}\ u_2\right]\left[v_1\ x_j(t_0)\ v_2\right]' - b^{(i)} = 0, \tag{3.9}$$

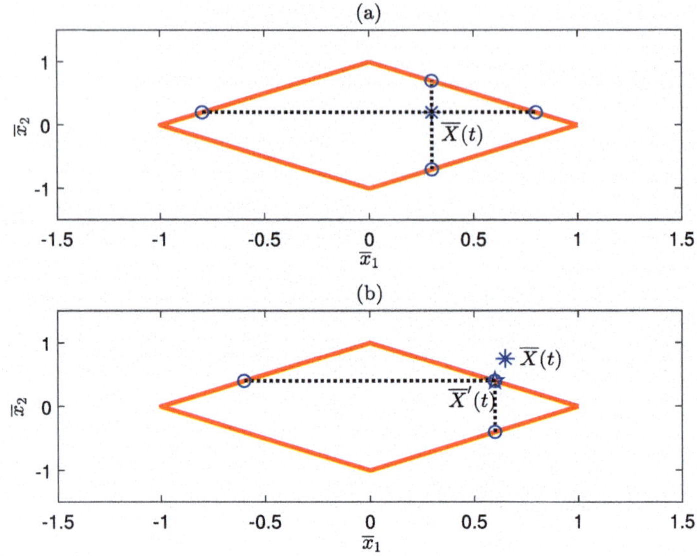

Fig. 3.3 **a** A normal data point $\tilde{X}(t)$ (blue star), and its projections on facets of the convex hull (blue circles), **b** an abnormal data point $\tilde{X}(t)$ (blue star), its corresponding normal data point $\tilde{X}_c(t)$ (blue pentagram), and projections of $\tilde{X}_c(t)$ on facets of the convex hull (blue circles), reprinted from Ref. [50], copyright 2020, with permission from IEEE

where

$$\begin{cases} u_1 = [a_1^{(i)}, a_2^{(i)}, \cdots, a_{j-1}^{(i)}], \\ u_2 = [a_{j+1}^{(i)}, a_{j+2}^{(i)}, \cdots, a_n^{(i)}]. \end{cases}$$

Equation (3.9) is a matrix equation of $x_j(t)$,

$$a_j^{(i)} x_j(t) + u_1 v_1 + u_2 v_2 - b^{(i)} = 0. \tag{3.10}$$

The solution of (3.10) is

$$s_j^{(i)}(t) = -[u_1 v_1 + u_2 v_2 - b^{(i)}]/a_j^{(i)}. \tag{3.11}$$

Hence, the projection of $X(t)$ on the hyperplane $p^{(i)}$ along the j-th axis is

$$X_{p,j}^{(i)}(t) = [x_1(t), \cdots, s_j^{(i)}(t), \cdots, x_n(t)]. \tag{3.12}$$

If $X_{p,j}^{(i)}(t)$ meets (3.3), $X_{p,j}^{(i)}(t)$ is defined as a projection of $X(t)$ on the facet $f^{(i)}$ along the j-th axis.

Third, the next algorithm gives the steps to calculate dynamic alarm thresholds for the current normal data point $\tilde{X}(t)$ based on the convex hull:

Algorithm 2:

1. Compute the projection of $X(t)$ on $p^{(i)}$ along the j-th axis by (3.12), where $i = 1, 2, \cdots, m$. The result is recorded as $X_{p,j}^{(i)}(t)$ and a null set is initialized as z. If $X_{p,j}^{(i)}(t)$ meets (3.3), add $s_j^{(i)}(t)$ in (3.11) to z.
2. Calculate the low and high dynamic alarm thresholds of $X(t)$ for x_j as

$$l_j(t) = \min(z),\ h_j(t) = \max(z), \tag{3.13}$$

where $\min(z)$ and $\max(z)$ are defined as the minimum and maximum values of the elements in z, respectively. Equations (3.2) and (3.13) lead to the low and high dynamic alarm thresholds of $\tilde{X}(t)$ for $\tilde{x}_j$ as

$$\tilde{l}_j(t) = \sigma_{\tilde{x}_j} l_j(t) + \mu_{\tilde{x}_j},\ \tilde{h}_j(t) = \sigma_{\tilde{x}_j} h_j(t) + \mu_{\tilde{x}_j}. \tag{3.14}$$

3. Set the value of j to 1, 2, $\cdots$, n, and calculate dynamic alarm thresholds of $\tilde{X}(t)$ for each process variable from (3.14).

If $X(t)$ is an abnormal data point, its dynamic alarm thresholds are the same as those of a corresponding normal data point $X_c(t)$ that is the data point on a facet of the convex hull being closest to $X(t)$. The determination of $X_c(t)$ is equivalent to solving the optimization problem

$$\begin{aligned} &\text{minimize} \quad \| X_c(t) - X(t) \|^2, \\ &\text{subject to} \quad AX_c'(t) - B \leq \mathbf{0}, \end{aligned} \tag{3.15}$$

where $\| \cdot \|$ represents the Euclidean-norm of the operand. Both the objective function $\| X_c(t) - X(t) \|^2$ and constraint function $AX_c'(t) - B \leq \mathbf{0}$ in (3.15) are convex functions. Thus, (3.15) is a convex quadratic optimization problem and can be solved in polynomial time. In particular, a function 'quadprog' in MATLAB can solve (3.15) efficiently.

After obtaining dynamic alarm thresholds for all process variables $\tilde{x}_1, \tilde{x}_2, \cdots, \tilde{x}_n$, it is ready to define an alarm variable X_a that takes the value 1 (0) when $\tilde{X}(t)$ is inside (outside) the NOZ, i.e.,

$$X_a(t) = \begin{cases} 0, \ \tilde{l}_j(t) \leq \tilde{x}_j(t) \leq \tilde{h}_j(t), \ \forall j = 1, 2, \cdots, n, \\ 1, \ \text{otherwise}. \end{cases} \tag{3.16}$$

In words, if all process variables in $\tilde{X}(t)$ are inside their own dynamic alarm thresholds, no alarms occur; otherwise, an alarm is announced.

The proposed method is composed by the offline and online stages. In the offline stage, Algorithm 1 includes three main steps, namely, the selection of historical data points in normal conditions based on a clustering algorithm, the determination of the convex hull vertices by the quick-hull algorithm, and the parameters estimation of the convex hull model through the Gaussian elimination algorithm. In the online stage, dynamic alarm thresholds for the current data point are calculated based on Algorithm 2.

The computation complexities of Algorithms 1 and 2 are analyzed as follows. The computation complexities of the three main steps in Algorithm 1 are $O(Tlog(T))$, $O(rlog(r))$, and $O(n^3)$, respectively. Here T is the number of historical data points, r is the number of historical data points included by the NOZ, and n is the number of process variables. Thus, the total computation complexity for Algorithm 1 is $O(Tlog(T) + n^3)$, which is acceptable for the offline implementation. The computation complexity of Algorithm 2 is $O(m)$, where m is the number of facets of the convex hull. It imposes no difficulties in the online application. In a typical example, the offline stage takes 12.64 sec at a personal computer with Intel i5 2.5GHz CPU and 8.0 GB memory, to establish the NOZ from 21600 data points (6 hours with sampling period 1 sec) of four process variables, and the online stage spends 0.24 sec in calculating dynamic alarm thresholds for a new data point.

3.1.3 Multivariate Alarm Systems for Non-Convex NOZs

This section extends NOZ-based multivariate alarm systems from convex NOZs to non-convex ones. First, mathematical models are established by exploiting search cones from historical data points in normal conditions to describe non-convex NOZs of multiple related process variables. Second, dynamic alarm thresholds of each process variable are designed for new data points being inside or outside the NOZs.

3.1.3.1 Establishment of NOZ Models Via Search Cones

This section builds NOZ models based on search cones from historical data points in normal conditions. Data points of each process variable are normalized to have zero means and unit variances. As described in Sect. 3.1.2.1, the normalized data point is $X(t) := [x_1(t), x_2(t), \cdots, x_n(t)]$.

The search cone is defined as an isosceles cone in a high-dimensional space, and is easier to be described in the n-sphere coordinate system than the Cartesian coordinate system. The n-sphere coordinate consists of a radial coordinate r, and n-1 angular coordinates $\varphi_1, \varphi_2, \cdots, \varphi_{n-1}$, where the angles $\varphi_1, \varphi_2, \cdots, \varphi_{n-2}$ range over $[0, \pi]$ radians and φ_{n-1} ranges over $[0, 2\pi)$ radians. If $x_1, \cdots, x_n$ of a data point is in the Cartesian coordinate system, then the coordinates $x_1, \cdots, x_n$ can be computed from $r, \varphi_1, \cdots$, and φ_{n-1} [6],

$$\begin{cases} x_1 = r\cos(\varphi_1), \\ x_2 = r\sin(\varphi_1)\cos(\varphi_2), \\ \quad\vdots \\ x_{n-1} = r\sin(\varphi_1)\cdots\sin(\varphi_{n-2})\cos(\varphi_{n-1}), \\ x_n = r\sin(\varphi_1)\cdots\sin(\varphi_{n-2})\sin(\varphi_{n-1}). \end{cases} \tag{3.17}$$

The inverse transformation is

$$\begin{cases} r = \sqrt{x_n^2 + x_{n-1}^2 + \cdots + x_2^2 + x_1^2}, \\ \varphi_i = \arccos \dfrac{x_i}{\sqrt{x_n^2 + x_{n-1}^2 + \cdots + x_i^2}}, \text{ for } i \in [1, n-1], \\ \varphi_{n-1} = \begin{cases} \arccos \frac{x_{n-1}}{\sqrt{x_n^2 + x_{n-1}^2}}, & \text{if } x_n \geq 0, \\ 2\pi - \arccos \frac{x_{n-1}}{\sqrt{x_n^2 + x_{n-1}^2}}, & \text{if } x_n < 0. \end{cases} \end{cases} \tag{3.18}$$

If $x_j \neq 0$ for some j but all of $x_{j+1}, \cdots, x_n$ are zero, then $\varphi_j = 0$ when $x_j > 0$, and $\varphi_j = \pi$ when $x_j < 0$. The coordinate transformation in a 3-dimensional Euclidean space is shown in Fig. 3.4, where the angular coordinates are φ_1, φ_2 and the radial coordinate is r. Figure 3.5a shows an example of the search cone (an isosceles triangle with vertices O, A, and B). Figure 3.5b shows a non-convex NOZ model built by search cones in the 2-dimensional space (consisting of 72 search cones). Here α is the step angle standing for the change of angular coordinates in search cones. As shown in Fig. 3.5, it is evident that search cones can be used as basic elements to describe non-convex NOZs.

If a data point falls in the k-th search cone, then the angular coordinates φ_1, φ_2, $\cdots$, φ_{n-1} of this point should satisfy

$$\varphi_j \in [\varphi_{l_j}^{(k)}, \varphi_{u_j}^{(k)}), \; j \in [1, n-1]. \tag{3.19}$$

Here $\varphi_{l_j}^{(k)}$ and $\varphi_{u_j}^{(k)}$ respectively are the lower and upper bounds of the j-th angular coordinate in the k-th search cone. Let α be the step angle of search cones,

$$\alpha = \varphi_{u_j}^{(k)} - \varphi_{l_j}^{(k)}, \quad \forall j = 1, 2, \cdots, n-1. \tag{3.20}$$

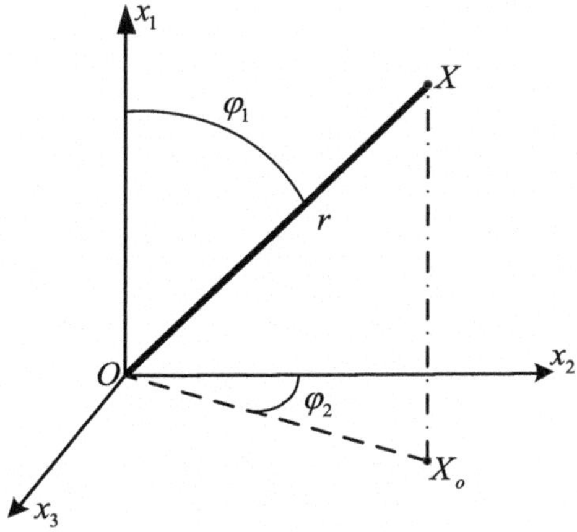

Fig. 3.4 Coordinate transformation in the 3-dimensional space, reprinted from Ref. [43], copyright 2022, with permission from IEEE

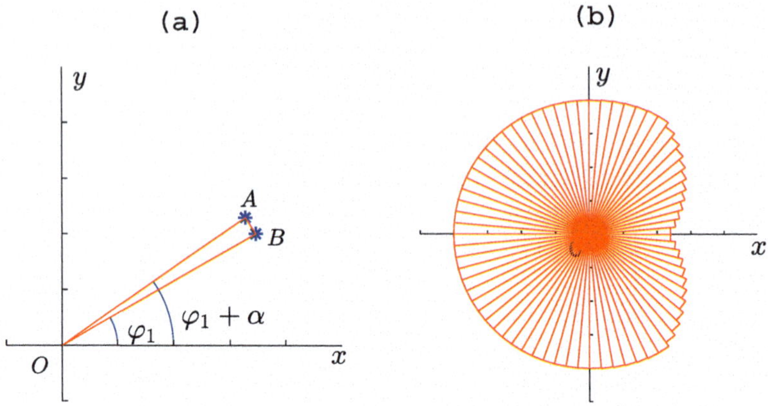

Fig. 3.5 **a** The search cone. **b** Non-convex NOZ model built by 72 search cones in the 2-dimensional space, reprinted from Ref. [43], copyright 2022, with permission from IEEE

The n-sphere coordinate for a data point $X(t) = [x_1(t), \cdots, x_j(t), \cdots, x_n(t)]$ is defined as

$$X_\phi(t) := [\varphi_1(t), \cdots, \varphi_j(t), \cdots, \varphi_{n-1}(t), r(t)],$$

where $r(t)$ is r in (3.18) for $X(t)$.

The n-dimensional space NOZ is composed of many search cones, and each search cone can be represented by a convex hull in a high-dimensional space, as the search cone is convex. The vertex set of each search cone is composed of the origin point and vertices of hypercubes at the bottom of the search cone. There are 2^{n-1} vertices and $2^{n-2}(n-1)$ edges in the hypercube at the bottom of the k-th search cone. Thus, the k-th search cone consists of $2^{n-2}(n-1)+1$ facets. According to the expression of a convex hull model, the mathematical expression of the k-th search cone is

$$A^{(k)} X'_{ch}(t) - B^{(k)} \leq 0, \tag{3.21}$$

where $X_{ch}(t)$ is a data point in the k-th search cone, and

$$\begin{cases} A^{(k)} = \left[a_1^{(k)}, \cdots, a_m^{(k)}, \cdots, a_M^{(k)}\right]', \\ B^{(k)} = \left[b_1^{(k)}, \cdots, b_m^{(k)}, \cdots, b_M^{(k)}\right]', \\ \quad 0 = [0, 0, \cdots, 0]'. \end{cases} \tag{3.22}$$

For $m \in \{1, 2, \cdots, M\}$ with $M = 2^{n-2}(n-1)+1$, $a_m^{(k)} = [a_{m,1}^{(k)}, a_{m,2}^{(k)}, \cdots, a_{m,n}^{(k)}]$ is the unit normal vector of the m-th facet p_m and $b_m^{(k)}$ is the offset distance from the m-th facet p_m to the origin point. The m-th facet in the k-th search cone can be described by a hyperplane as

$$a_m^{(k)} X'_p(t) - b_m^{(k)} = 0, \tag{3.23}$$

where $X_p(t)$ is the data point in the m-th hyperplane.

The theoretical expressions of $a_m^{(k)}$ and $b_m^{(k)}$ for the k-th search cone are derived as follow. First, the data point farthest from the origin point in the k-th search cone is $X(t_{far}^{(k)}) = [x_1(t_{far}^{(k)}), \cdots, x_j(t_{far}^{(k)}), \cdots, x_n(t_{far}^{(k)})]$. The unit normal vector $a_{m,n}^{(k)}$ of the bottom hyperplane p_M is

$$a_{m,n}^{(k)} = -\frac{[x_1(t_m), \cdots, x_j(t_m), \cdots, x_n(t_m)]}{\sqrt{\sum_{j=1}^{n} x_j^2(t_m)}}. \tag{3.24}$$

$X(t_m) = [x_1(t_m), \cdots, x_j(t_m), \cdots, x_j(t_m)]$ takes the corresponding n-sphere coordinate as

$$X_\phi(t_m) = [\varphi_1(t_m), \cdots, \varphi_j(t_m), \cdots, \varphi_{n-1}(t_m), 1],$$

and $\varphi_j(t_m) = (\varphi_{d_j}^{(k)} + \varphi_{u_j}^{(k)})/2$ for $j \in [1, n-1]$. $b_M^{(k)}$ in (3.22) is

$$b_M^{(k)} = \sum_{j=1}^{M} a_{m,j}^{(k)} x_j(t_{far}^{(k)}). \tag{3.25}$$

Second, each vertex of the bottom hyperplane p_M in the k-th search cone is defined as

$$X_{\phi,v}(t) := [\varphi_{v_1}^{(k)}(t), \cdots, \varphi_{v_j}^{(k)}(t), \cdots, \varphi_{v_{n-1}}^{(k)}(t), r^{(k)}(t)]. \tag{3.26}$$

where $v_j = l_j$ or u_j stands for the lower or upper bound in (3.19). Then, each adjacent vertex of $X_{\phi,v}(t)$ is

$$X_{\phi,v'_j}(t) := [\varphi_{v_1}^{(k)}(t), \cdots, \varphi_{v'_j}^{(k)}(t), \cdots, \varphi_{v_{n-1}}^{(k)}(t), r^{(k)}(t)], \tag{3.27}$$

where

$$v'_j = \begin{cases} l_j, & \text{if } v_j = u_j, \\ u_j, & \text{if } v_j = l_j. \end{cases}$$

For a vertex $X_{\phi,v'_j}(t_k)$ in (3.26), the corresponding $a_m^{(k)}$ is obtained by solving

$$\begin{cases} I_{a_m^{(k)}} (a_m^{(k)})' = 0, \\ a_m^{(k)} (a_m^{(k)})' = 1, \\ a_m^{(k)} X(t_m) \leq 0. \end{cases} \tag{3.28}$$

Here $I_{a_m^{(k)}}$ is a $(n-1) \times n$ matrix in (3.29),

$$\begin{bmatrix} X_{v_1'}(t_k)-X_{v_1}(t_k) \\ \vdots \\ X_{v_j'}(t_k)-X_{v_j}(t_k) \\ \vdots \\ X_{v_{n-1}'}(t_k)-X_{v_{n-1}}(t_k) \end{bmatrix}, \tag{3.29}$$

where n is the dimension of data points. The j-th row element in the matrix $I_{a_m^{(k)}}$ represents a vector on one side hyperplane, and any two rows in the matrix represent two non-collinear vectors. The above steps will obtain a pair of repeated unit normal vectors, as each unit normal vector $a_m^{(k)}$ is calculated twice. Thus, if $a_m^{(k)}$ has been obtained in the previous results for $[1, m-1]$, delete $a_m^{(k)}$ and choose another vertex in (3.26) for (3.28) until all unit normal vectors $\{a_m^{(k)}\}_{m=1}^{M-1}$ have been obtained. As the side hyperplanes in the k-th search cone will cross the origin point, it is ready to have

$$b_m^{(k)} = 0, \ m = 1, 2, \cdots, M-1. \tag{3.30}$$

Mathematical expression of the k-th search cone (3.21) can be obtained by solving $a_m^{(k)}$ and $b_m^{(k)}$ of all hyperplanes in (3.24), (3.25), (3.28), and (3.30) for m from 1 to $M = 2^{n-2}(n-1)+1$. If there is no data point inside the k-th search cone, the k-th search cone is assumed to have the same size as the (k-1)-th search cone. Then, the parameters $a_m^{(k)}$ and $b_m^{(k)}$ are calculated in (3.24), (3.25), (3.28), and (3.30) with the farthest point being denoted as $X_\phi(t_{far}^{(k)}) = [\varphi_1(t_{far}^{(k)}) + \alpha/2, \cdots, \varphi_j(t_{far}^{(k)}), \cdots, \varphi_{n-1}(t_{far}^{(k)}), r(t_{far}^{(k)})]$ for (3.25).

Given the optimal step angle α, the number of search cones K for building the NOZ model can be determined

$$K = \lceil \frac{2\pi}{\alpha_{opt}} \rceil \lceil \frac{\pi}{\alpha_{opt}} \rceil^{n-2}. \tag{3.31}$$

Here $\lceil \cdot \rceil$ is the ceiling function taking the largest integer no larger than the operand. Thus, the NOZ model is composed by K search cones as

$$\text{NOZ} := \{A^{(k)}X'(t) - B^{(k)} \leq 0\}_{k=1}^{K}. \tag{3.32}$$

Here the parameters $\{a_m^{(k)}\}_{m=1}^{M}$ in $A^{(k)}$ are calculated by (3.24) and (3.28), and $\{b_m^{(k)}\}_{m=1}^{M}$ in $B^{(k)}$ are calculated by (3.25) and (3.30).

The optimal step angle α_{opt} is determined as follows. The boundary of an ideal NOZ should be as smooth as possible. That is, the size difference between adjacent search cones in an ideal NOZ should be small. Thus, a lateral edge difference index η is formulated to measure the smoothness of the NOZ boundary,

$$\eta(\alpha) = \sum_{k=2}^{K} \left| r^{(k)}(t) - r^{(k-1)}(t) \right|, \tag{3.33}$$

where K is the number of search cones in NOZ and $r^{(k)}(t)$ in (3.26) is the lateral edge length of the k-th search cone with a step angle α. When historical data points are not enough, data in the NOZ will be sparse. It is easy for search cones to fall into sparse data spaces or empty data spaces when the step angle is too small, which leads to the overfitting of the NOZ model. The lateral edge difference between two adjacent search cones will be too large in this condition.

The plot of $\eta(\alpha)$ is the "L" shaped curve (e.g., Fig. 3.13 given later in Sect. 3.1.4.2), as $\eta(\alpha)$ decreases along with the increment of α. The decreasing speed of $\eta(\alpha)$ is fast at first and then gradually slows down. Thus, it is "L" shaped. Being inspired by a method in [52] for selecting the optimal number of segments in polynomial fitting, the "knee" of the curve $\eta(\alpha)$ is obtained as the optimal step α_{opt},

$$\alpha_{opt} = \arg\max \frac{\theta_1 - \theta_2}{1 + \theta_1\theta_2}, \tag{3.34}$$

where

$$\theta_1 = \frac{\eta(\alpha(i-1)) - \eta(\alpha(i))}{\alpha(i) - \alpha(i-1)},$$
$$\theta_2 = \frac{\eta(\alpha(i)) - \eta(\alpha(i+1))}{\alpha(i+1) - \alpha(i)}.$$

As shown in Fig. 3.6, α_{opt} is the maximal intersection angle between two lines $\eta(\alpha(i-1))$-$\eta(\alpha(i))$ and $\eta(\alpha(i))$-$\eta(\alpha(i+1))$.

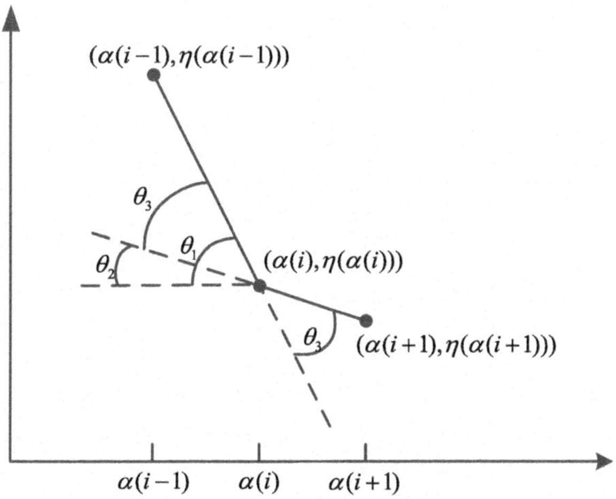

Fig. 3.6 The intersection angle between two lines $\eta(\alpha(i-1))$-$\eta(\alpha(i))$ and $\eta(\alpha(i))$-$\eta(\alpha(i+1))$, reprinted from Ref. [43], copyright 2022, with permission from IEEE

3.1.3.2 Development of Dynamic Alarm Thresholds

The development of dynamic alarm thresholds for each process variable is illustrated in this part based on the NOZ model described by search cones. For a data point $X(t)$, the dynamic alarm thresholds mean that the j-th process variable $x_j(t)$ has its own high and low alarm thresholds at time instant t for $j = 1, 2, \cdots, n$. The dynamic alarm thresholds show the variation range of $x_j(t)$ when $x_j(t)$ is inside NOZ. That is, it explains how much margin the current operating state has for deviating away from the NOZ. If $x_j(t)$ is outside the NOZ, the dynamic alarm thresholds can show how far $x_j(t)$ is away from the NOZ. As illustrated in Fig. 3.7, the low and high dynamic alarm thresholds $\tilde{l}_j(t), \tilde{h}_j(t)$ of $\tilde{X}(t)$ for $\tilde{x}_j(t)$ can be obtained in an analogous manner as the counterparts in Sect. 3.1.2.2 with some extensions for non-convex NOZs. In particular, (3.35)–(3.43) for calculating projections on facets of NOZs for new data points are the same as the counterparts in Sect. 3.1.2.2.

First, if a data point $\tilde{X}(t) = [\tilde{x}_1(t), \tilde{x}_2(t), \cdots, \tilde{x}_n(t)]$ is inside the NOZ in (3.32) as illustrated in Fig. 3.7a, its corresponding normalized data point is $X(t) = [x_1(t), x_2(t), \cdots, x_n(t)]$. $X(t)$ is separated into three parts, namely, $x_j(t)$,

$$v_1 = \left[x_1(t), x_2(t), \cdots, x_{j-1}(t)\right], \tag{3.35}$$

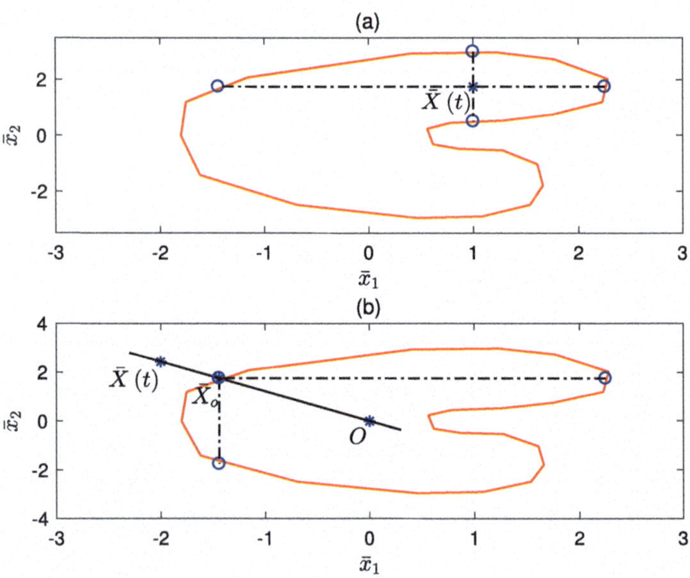

Fig. 3.7 **a** A data point (blue star) inside the NOZ with its thresholds (blue circles). **b** A data point (blue star) outside the NOZ and projection point $X_o(t)$ with its thresholds (blue circles), reprinted from Ref. [43], copyright 2022, with permission from IEEE

and

$$v_2 = \left[x_{j+1}(t), x_{j+2}(t), \cdots, x_n(t)\right]. \quad (3.36)$$

In order to reduce the computational complexity, the boundary of the NOZ model is used here to calculate dynamic alarm thresholds.

$$A_M X'(t) - B_M \leq 0, \quad (3.37)$$

where

$$\begin{cases} A_M = \left[a_M^{(1)}, \cdots, a_M^{(k)}, \cdots, a_M^{(K)}\right]', \\ B_M = \left[b_M^{(1)}, \cdots, b_M^{(k)}, \cdots, b_M^{(K)}\right]', \\ 0 = [0, 0, \cdots, 0]'. \end{cases} \quad (3.38)$$

Similar to (3.23), the k-th facet in (3.37) is described as

$$a_M^{(k)} X_p'(t) - b_M^{(k)} = 0, \quad (3.39)$$

where $k \in [1, K]$ and K takes the value in (3.31). Then, (3.39) becomes

$$\left[u_1 \; a_{M,j}^{(k)} \; u_2\right]' \left[v_1 \; x_j(t) \; v_2\right] - b_M^{(k)} = 0, \quad (3.40)$$

where

$$\begin{cases} u_1 = \left[a_{M,1}^{(k)}, a_{M,2}^{(k)}, \ldots, a_{M,j-1}^{(k)}\right], \\ u_2 = \left[a_{M,j+1}^{(k)}, a_{M,j+2}^{(k)}, \ldots, a_{M,n}^{(k)}\right]. \end{cases}$$

Equation (3.40) is a matrix equation of $x_j(t)$,

$$a_{M,j}^{(k)} x_j(t) + u_1' v_1 + u_2' v_2 - b_M^{(k)} = 0. \quad (3.41)$$

The solution of (3.41) is

$$s_{M,j}^{(k)}(t) = -\left[u_1' v_1 + u_2' v_2 - b_M^{(k)}\right] / a_{M,j}^{(k)}. \quad (3.42)$$

Hence, the projection of $X(t)$ on the k-th hyperplane along the j-th axis is

$$X_{M,j}^{(k)}(t) = \left[x_1(t), \ldots, s_{M,j}^{(k)}(t), \ldots, x_n(t)\right]. \quad (3.43)$$

If $X_{m,j}^{(k)}(t)$ meets (3.37), it is regarded as a projection of $X(t)$. If $X_{m,j}^{(k)}(t)$ satisfies $X_{m,j}^{(k)}(t) \leq x_j(t)$ and the point $[x_1(t), \cdots, s_{M,j}^{(k)} - \varepsilon, \cdots, x_n(t)]$ does not meet (3.37), where ε is an arbitrarily small positive number, $X_{m,j}^{(k)}(t)$ is added to the set $Z_j^{(l)}$.

If $X_{m,j}^{(k)}(t)$ satisfies $X_{m,j}^{(k)}(t) \geq x_j(t)$ and the point $[x_1(t), \cdots, s_{M,j}^{(k)} + \varepsilon, \cdots, x_n(t)]$ does not meet (3.37), $X_{m,j}^{(k)}(t)$ is added to the set $Z_j^{(h)}$. After applying the above steps for all K facets, the dynamic high and low alarm thresholds for $x_j(t)$ in $X(t)$ are obtained as

$$l_j(t) = \max(Z_j^{(l)}), \quad h_j(t) = \min(Z_j^{(h)}). \tag{3.44}$$

Second, if the data point $X(t)$ is outside the NOZ as illustrated in Fig. 3.7b, its dynamic high and low alarm thresholds are defined the same as the projection point $X_o(t)$ of $X(t)$. The projection point $X_o(t)$ is on the NOZ boundary and satisfies

$$\begin{aligned} &\min \|X_o(t) - X(t)\|^2, \\ &\text{s.t. } A^{(k)} X_o'(t) - B^{(k)} \leq \mathbf{0}. \end{aligned} \tag{3.45}$$

The dynamic alarm thresholds $l_j(t)$ and $h_j(t)$ for x_j of $X_o(t)$ can be obtained by (3.35)–(3.44). The low and high dynamic alarm thresholds $\tilde{l}_j(t)$, $\tilde{h}_j(t)$ of $\tilde{X}(t)$ for $\tilde{x}(j)$ are calculated according to (3.2) as

$$\begin{aligned} \tilde{l}_j(t) &= \sigma_{\tilde{x}_j} l_j(t) + \mu_{\tilde{x}_j}, \\ \tilde{h}_j(t) &= \sigma_{\tilde{x}_j} h_j(t) + \mu_{\tilde{x}_j}. \end{aligned} \tag{3.46}$$

Finally, the alarm variable $x_a(t)$ is generated in (3.1).

In summary, the proposed method consists of the following steps:

Step 1: Calculate the lateral edge length $r^{(k)}(t)$ in (3.26) of each search cone and the lateral edge difference index $\eta(\alpha)$ in (3.33) for a step angle α.

Step 2: Select the optimal step angle α_{opt} in (3.34).

Step 3: Compute $A^{(k)}$ and $B^{(k)}$ in the mathematical expression (3.21) of search cones in (3.24), (3.25), (3.28), and (3.30) based on the optimal step angle α_{opt}.

Step 4: Obtain the NOZ model via search cones in (3.32).

Step 5: Design the dynamic alarm thresholds $\tilde{l}_j(t)$, $\tilde{h}_j(t)$ in (3.46) for online data points.

The proposed method is composed of the offline and online parts. The offline part includes Steps 1-4 to obtain the NOZ model. In the online part, dynamic alarm thresholds for the current data point are calculated based on Step 5. The computation complexities for the offline and online parts are $O(n2^n)$ and $O(2^n)$, respectively. Thus, the proposed method is feasible for implementation if there are not too many process variables. For instance, the numerical example in Sect. 3.4.4 is associated with four process variables. The offline stage takes about 38.720 sec at a personal computer with Intel i5 2.4-GHz CPU and 16.0-GB memory to build the NOZ model of four process variables. Moreover, the online stage spends about 0.302 sec in calculating dynamic alarm thresholds for a new data point.

3.1.4 Numerical and Industrial Examples

This section illustrates the effectiveness of the proposed methods via two examples, one with a convex NOZ and another with a non-convex one.

3.1.4.1 Industrial Example

A bivariate industrial process from a large-scale power plant is composed by the inlet flow rate $\tilde{x}_1$ of a feedwater pump and the electrical current $\tilde{x}_2$ of a booster pump. Figure 3.8 presents 2-hour historical data points $\{\tilde{X}(t)\}_{t=1}^{7200}$ in normal conditions, with the sampling period 1 second. In the contemporary industrial alarm system, the process variable $\tilde{x}_1$ is configured with a high alarm threshold 560 t/h, as shown by the black dash-dotted line in Fig. 3.8a. A process knowledge provided by operators is that if $\tilde{x}_1$ and $\tilde{x}_2$ change in a synchronized manner, the process is in normal conditions.

The convex-hull-based NOZ method is implemented as follows. First, a convex hull is established through Algorithm 1 as the NOZ model. In particular, $\{\tilde{X}(t)\}_{t=1}^{7200}$ is normalized by (3.2) to obtain the normalized data points $\{X(t)\}_{t=1}^{7200}$. The data set $\{X(t)\}_{t=1}^{7200}$ is divided into 20 subsets as $\{X(t)\}_{t=1}^{360}$, $\{X(t)\}_{t=1}^{720}$, $\cdots$, $\{X(t)\}_{t=1}^{7200}$. For

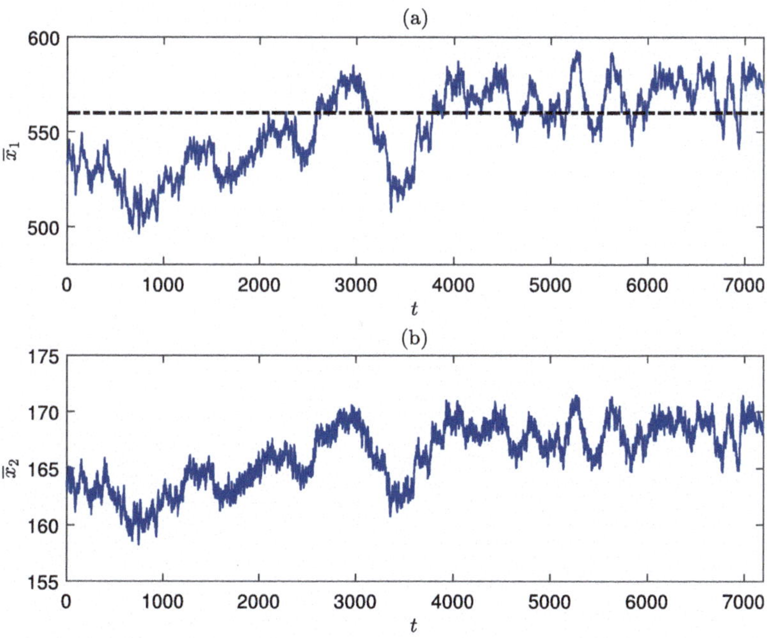

Fig. 3.8 Time sequence plots of process variables: **a** the inlet flow rate $\tilde{x}_1$ (blue solid) with the static alarm threshold (black dash), **b** the electrical current $\tilde{x}_2$, reprinted from Ref. [50], copyright 2020, with permission from IEEE

each subset, the clustering algorithm [27] is utilized to determine the normalized data points to be included in the NOZ with a given false alarm rate $\beta = 0.01$. The quick-hull algorithm [4] is employed to find vertices of the convex hull to represent the NOZ. The parameters of the convex hull are estimated via (3.6) and (3.7). The fitness index η of the NOZ for each subset is calculated by (3.8), where the minimal intervals to determine the equal-sized grids for x_1 and x_2 are respectively chosen as 0.04 and 0.02 based on the process knowledge. The value of η changes with the subset index k, as shown in Fig. 3.9. The established NOZ is from the 15-th subset $\{X(t)\}_{t=1}^{5400}$, because the value of η is larger than a given threshold 0.90 and tends to be stable after $k = 15$. The scatter plot of normalized process variables x_1 and x_2 is given in Fig. 3.10a, where the boundaries of the NOZ containing 99% data points in $\{X(t)\}_{t=1}^{5400}$ are also presented.

Second, dynamic alarm thresholds of each process variable are calculated based on the NOZ model. The data points $\{\tilde{X}(t)\}_{t=1}^{600}$ are new data points, their dynamic alarm thresholds are computed through (3.14), as given in Fig. 3.10b, c. Because $\tilde{x}_1$ and $\tilde{x}_2$ in $\tilde{X}(t)$ for $t \in [1,\ 600]$ are inside their own dynamic alarm thresholds, these data points are in the normal condition and no alarms occur.

Third, the NOZ method is compared with the conventional method based on $\{\tilde{X}(t)\}_{t=1}^{600}$. As presented in Fig. 3.10b, the data points of $\tilde{x}_1(t)$ for $t \in [39,\ 580]$ are larger than the static alarm threshold 560 t/h, so that alarms occur. However, the relationship between $\tilde{x}_1$ and $\tilde{x}_2$ is unchanged during the alarm occurrences, as presented by the scatter plot of $\tilde{x}_1(t)$ and $\tilde{x}_2(t)$ in Fig. 3.10a. Thus, all the 542 alarms are indeed false alarms. The conventional method uses the static alarm threshold for $\tilde{x}_1$ that is isolated from the related process variable $\tilde{x}_2$; such an isolation results in the presence of false alarms.

Finally, the NOZ method is compared with the principle component analysis (PCA) based method. In Fig. 3.11, all values of the T^2 statistic for $\{\tilde{X}(t)\}_{t=1}^{600}$ are smaller than the control limit of the T^2 statistic. Thus, the NOZ method and the

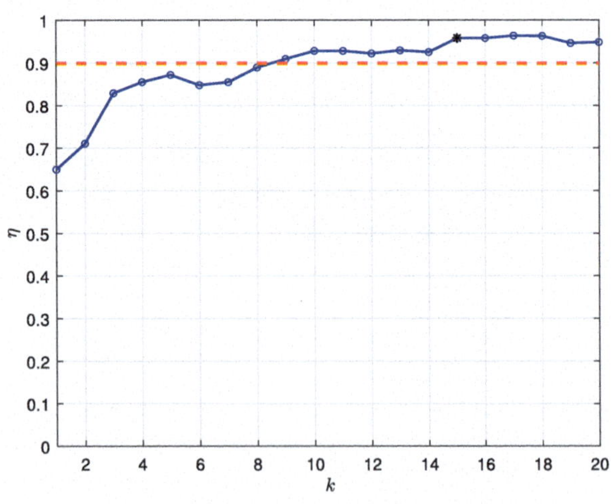

Fig. 3.9 The change of the fitness index η with the subset index k, reprinted from Ref. [50], copyright 2020, with permission from IEEE

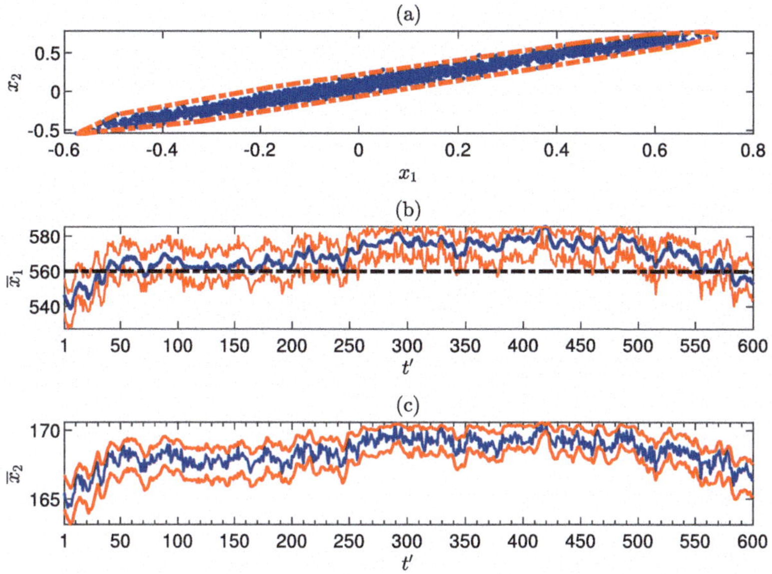

Fig. 3.10 **a** The scatter plot (blue dot) of x_1 and x_2, and the established convex hull (red dash), **b** the time sequence plot of $\bar{x}_1$ (blue solid) and its dynamic alarm thresholds (red solid), **c** the time sequence plot of $\bar{x}_2$ (blue solid) and its dynamic alarm thresholds (red solid), reprinted from Ref. [50], copyright 2020, with permission from IEEE

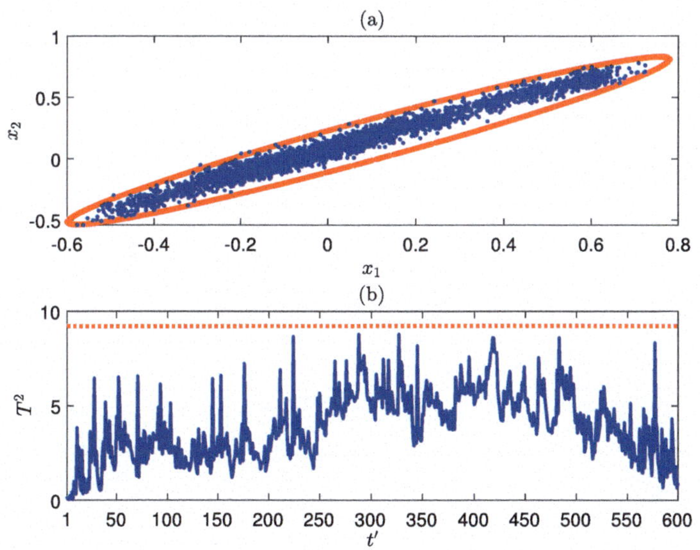

Fig. 3.11 **a** The scatter plot (blue dot) of x_1 and x_2 and the estimated ellipsoid from the PCA-based method (red solid), **b** the time sequence plot of T^2 statistic (blue solid) and its control limit (red dot), reprinted from Ref. [50], copyright 2020, with permission from IEEE

PCA-based method have the same performance for this example in terms that no false alarms are raised, even though the two methods respectively use a convex hull and a hyper-ellipsoid to represent the NOZ. However, the two methods are very different in the way of monitoring whether a data point is in the normal condition. For the NOZ method, the value of each process variable is compared with its dynamic alarm thresholds, i.e., $\tilde{x}_1(t)$ is compared with $\bar{l}_1(t)$ and $\bar{h}_1(t)$, and $\bar{x}_2(t)$ is compared with $\bar{l}_2(t)$ and $\bar{h}_2(t)$. The dynamic alarm thresholds have the same physical meanings as the corresponding process variables. By contrast, the PCA-based method compares the T^2 statistic with its control limit associated with a confidence level 0.99. Here $T^2 = 315.46x_1^2 - 614.74x_1x_2 + 319.20x_2^2$, where x_1 and x_2 are respectively the normalized counterparts of $\bar{x}_1$ and $\bar{x}_2$. The T^2 statistic does not have the same physical meanings of $\bar{x}_1$ and $\bar{x}_2$, so that it is difficult for industrial plant operators to accept.

3.1.4.2 Numerical Example

This example is about the continuous stirred tank reactor (CSTR) with a cooling jacket, which is a well-established benchmark for process monitoring [23]. As shown in Fig. 3.12, a reactant A (cyclopentadiene) is fed in the reactor with the concentration C_{A0}, flow q, and temperature T_f. There are three reactions in this reactor: a main reaction from A to B with reaction velocity k_1, a follow-up reaction from B to C with reaction velocity k_2, and a side reaction from $2A$ to D with reaction velocity

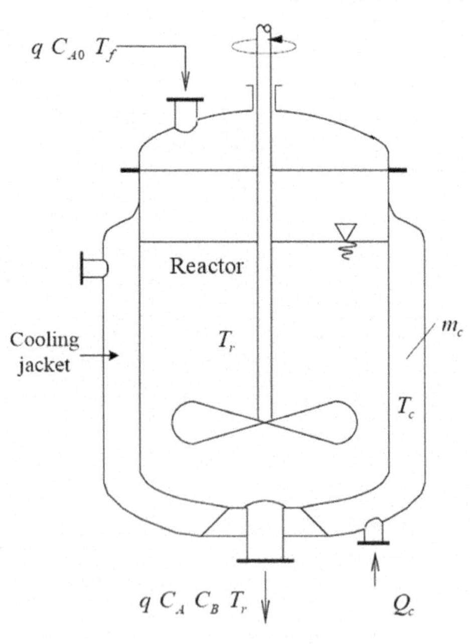

Fig. 3.12 The continuous stirred tank reactor, reprinted from Ref. [43], copyright 2022, with permission from IEEE

k_3. Here B (cyclopentene) is the desired product, and C (cyclopentanone) and D (dicyclopentadiene) are undesirable byproducts [8].

The dynamic model of the CSTR is obtained based on the material balance of concentrations C_A for the reactant A and C_B for the desired product B, as well as the energy balance for the reactor and the cooling jacket with temperatures T_r and T_c [8],

$$\begin{aligned}
\frac{\mathrm{d}C_A(t)}{\mathrm{d}t} &= \frac{q(t)}{V}\left(C_{A0}(t) - C_A(t)\right) - k_1(T_r(t))C_A(t) - k_3(T_r(t))C_A^2(t),\\
\frac{\mathrm{d}C_B(t)}{\mathrm{d}t} &= -\frac{q(t)}{V}C_B(t) + k_1(T_r(t))C_A(t) - k_2(T_r(t))C_B(t),\\
\frac{\mathrm{d}T_r(t)}{\mathrm{d}t} &= \frac{q(t)}{V}\left(T_f(t) - T_r(t)\right) + \frac{k_w A_R}{\rho V C_p}\left(T_c(t) - T_r(t)\right) + k_2(T_r(t))C_B(t)\Delta H_{BC}\\
&\quad - \frac{1}{\rho C_p}\left(k_1(T_r(t))C_A(t)\Delta H_{AB} + k_3(T_r(t))C_A^2(t)\Delta H_{AD}\right),\\
\frac{\mathrm{d}T_c(t)}{\mathrm{d}t} &= \frac{1}{m_c C_{pc}}\left(-Q_c(t) + k_w A_R\left(T_r(t) - T_c(t)\right)\right).
\end{aligned} \tag{3.47}$$

Here $k_i(T_r(t))$ is the reaction velocity via Arrhenius functions of the temperature (in °C):

$$k_i(T_r(t)) = k_{i0}e^{\frac{-E_i}{(T_r(t)+273.15)}},\ i = 1, 2, 3. \tag{3.48}$$

The constant model parameters in (3.47) and (3.48) are listed in Table 3.1.

The input variables are q, C_{A0}, Q_c and T_f, and the output variables are C_B, C_A, T_c and T_r . The input variables are set as $T_f = 104.9$ °C, $Q_c = 1000:10:4000\ \frac{\mathrm{kJ}}{\mathrm{h}}$, $q = 100:400\ \frac{\mathrm{l}}{\mathrm{h}}$, and $C_{A0} = 4.5:0.01:5.7\ \frac{\mathrm{mol}}{\mathrm{l}}$ according to process knowledge. Data points of output variables in normal operating conditions can be generated based on the CSTR model (3.47) by using the above values of input variables. Based on these data points, the search-cone-based NOZ model is built as described in (3.32). The lateral edge difference index $\eta(\alpha)$ is shown in Fig. 3.13, and the step angle is selected as $\alpha_{opt} = 1.4°$ by (3.34). As the 4-dimensional NOZ cannot be visualized in the Cartesian coordinate, the scatter plot of data points of $\tilde{C}_A$ and $\tilde{T}_r$ is shown in Fig. 3.14a and the corresponding NOZ of $\tilde{C}_A$ and $\tilde{T}_r$ is given in Fig. 3.14b (red solid line). The NOZ is non-convex, as clearly revealed in Fig. 3.14.

1800 new data points in Fig. 3.15 are generated from (3.47) as online data for alarm monitoring. An abnormality is introduced as a rapid increment of Q_c starting at $t = 1313$. The proposed method is applied to yield the dynamic alarm thresholds of C_A and T_r, respectively shown in Fig. 3.16b, c (red dashed lines). The black solid line in Fig. 3.16a means that these data points are in the abnormal condition, which can be detected promptly when C_A crosses its dynamic high alarm threshold and T_r goes below its dynamic low alarm threshold at $t = 1395$.

Table 3.1 CSTR model parameters, reprinted from Ref. [43], copyright 2022, with permission from IEEE

Model parameter	Symbol	Value
Collision factor for reaction k_1	k_{10}	$1.287 \cdot 10^{12}\ \text{h}^{-1}$
Collision factor for reaction k_2	k_{20}	$1.287 \cdot 10^{12}\ \text{h}^{-1}$
Collision factor for reaction k_3	k_{30}	$9.043 \cdot 10^{9}\ \frac{1}{\text{mol}\cdot\text{h}}$
Activation energy for reaction k_1	E_1	9758.3 K
Activation energy for reaction k_2	E_2	9758.3 K
Activation energy for reaction k_3	E_3	8560 K
Enthalpy of reaction k_1	ΔH_{AB}	$4.2\ \frac{\text{kJ}}{\text{mol}}$
Enthalpy of reaction k_2	ΔH_{BC}	$-11.0\ \frac{\text{kJ}}{\text{mol}}$
Enthalpy of reaction k_3	ΔH_{AD}	$-41.85\ \frac{\text{kJ}}{\text{mol}}$
Density of reactant	ρ	$0.9342\ \frac{\text{kg}}{\text{l}}$
Reactant heat capacity	C_p	$3.01\ \frac{\text{kJ}}{\text{kg}\cdot\text{K}}$
Coolant heat capacity	C_{pc}	$2.0\ \frac{\text{kJ}}{\text{kg}\cdot\text{K}}$
Jacket heat transfer coefficient	k_w	$4032\ \frac{\text{kJ}}{\text{h}\cdot\text{m}^2\cdot\text{K}}$
Cooling jacket surface	A_R	$0.215\ \text{m}^2$
Reactor volume	V	$0.01\ \text{m}^3$
Coolant mass	m_c	5.0 kg

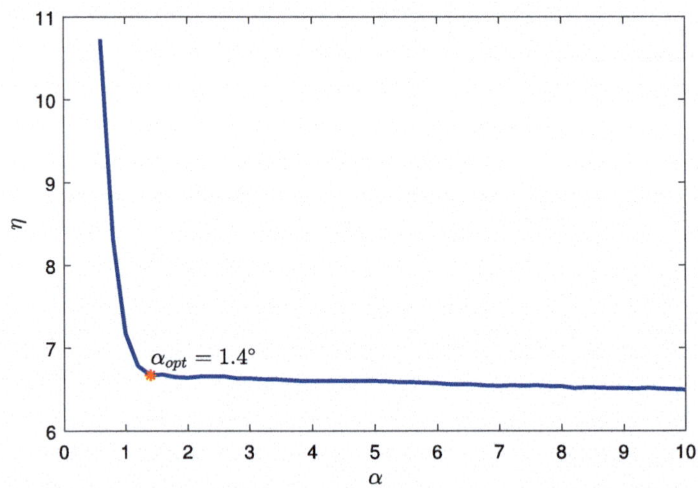

Fig. 3.13 The lateral edge difference index $\eta(\alpha)$ (blue solid line) and the optimal step angle α_{opt} (red star), reprinted from Ref. [43], copyright 2022, with permission from IEEE

As a comparison, the convex-hull-based NOZ model in Fig. 3.16a (green dashed line) is also built as the model describing the NOZ of $\tilde{C}_A$ and $\tilde{T}_r$. Some areas outside the real normal operation zone are erroneously included, because a convex hull

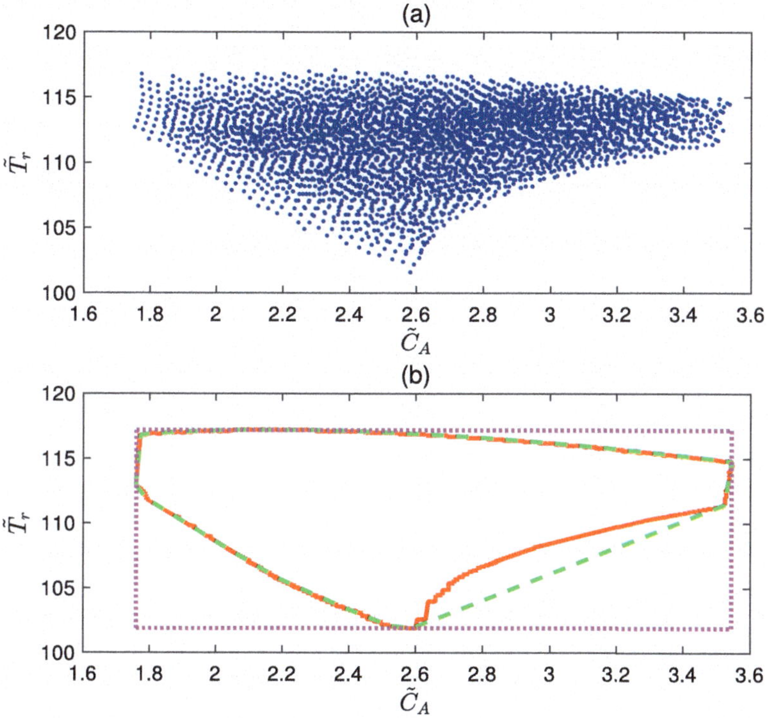

Fig. 3.14 **a** Scatter plot (blue dotted line) of $\tilde{C}_A$ and $\tilde{T}_r$, **b** the search-cone-based NOZ model (red solid line), the convex-hull-based NOZ model (green dashed line), and the univariate alarm thresholds (pink dotted line) of $\tilde{C}_A$ and $\tilde{T}_r$, reprinted from Ref. [43], copyright 2022, with permission from IEEE

cannot describe the actual non-convex NOZ. Therefore, the abnormal data points in this area will not trigger alarms, even if the CSTR is already in the abnormal operation zone. The corresponding dynamic alarm thresholds from the convex-hull-based NOZ model are shown in Fig. 3.16b, c (green dot-dashed lines), the alarm variable $x_a(t)$ changes to '1' at $t = 1417$. Hence, the convex-hull-based NOZ model for such a non-convex NOZ has a detection delay of 22 sampling periods later than the search-cone-based NOZ model, due to the modeling error in using a convex hull to describe an non-convex NOZ.

The univariate alarm thresholds are designed for $\tilde{C}_A$ and $\tilde{T}_r$ as a comparison. The high (low) univariate alarm threshold is defined as the maximum (minimum) value of $\tilde{C}_A$ and $\tilde{T}_r$ in the NOZ. These thresholds can cover all normal conditions, ensuring that normal data points will not trigger alarms. As shown in Fig. 3.16, the high and low univariate alarm thresholds (pink dotted lines) are 3.5448 and 1.75 for $\tilde{C}_A$, 101.9 and 117.2 for $\tilde{T}_r$, respectively. No alarm occurs as all online data points do not cross the high or low univariate alarm thresholds. Univariate alarm thresholds fail to detect

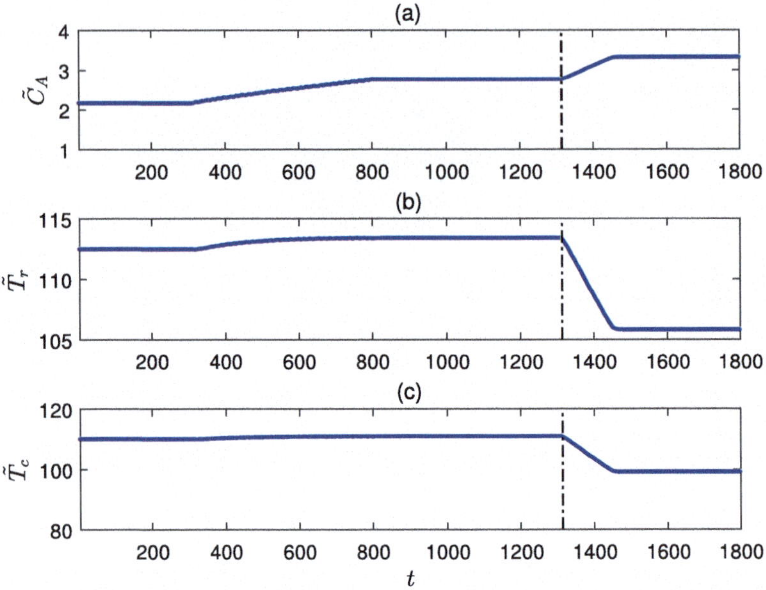

Fig. 3.15 The online data (blue solid line) of $\tilde{C}_A$, $\tilde{T}_r$, and $\tilde{T}_c$ with the abnormal condition occurring at $t = 1313$ (black dash-dotted line), reprinted from Ref. [43], copyright 2022, with permission from IEEE

the abnormal condition in this example, since they do not consider the relationship among these related process variables.

3.2 Variation Direction-Based Multivariate Alarm Systems

This section proposes a method to design a multivariate alarm system based on variation directions of involved process variables [9]. The main idea is to raise alarms when process variables are changing in directions being inconsistent with a prior-known relationship in normal conditions. By using adaptive time scales, time gradients of process variables are extracted to calculate variation directions in an online manner. Adaptive time scales are determined from a function between time scales and volatilities of process variables. Alarms arise at the moment that a nominal relationship of variation directions among process variables is invalidated.

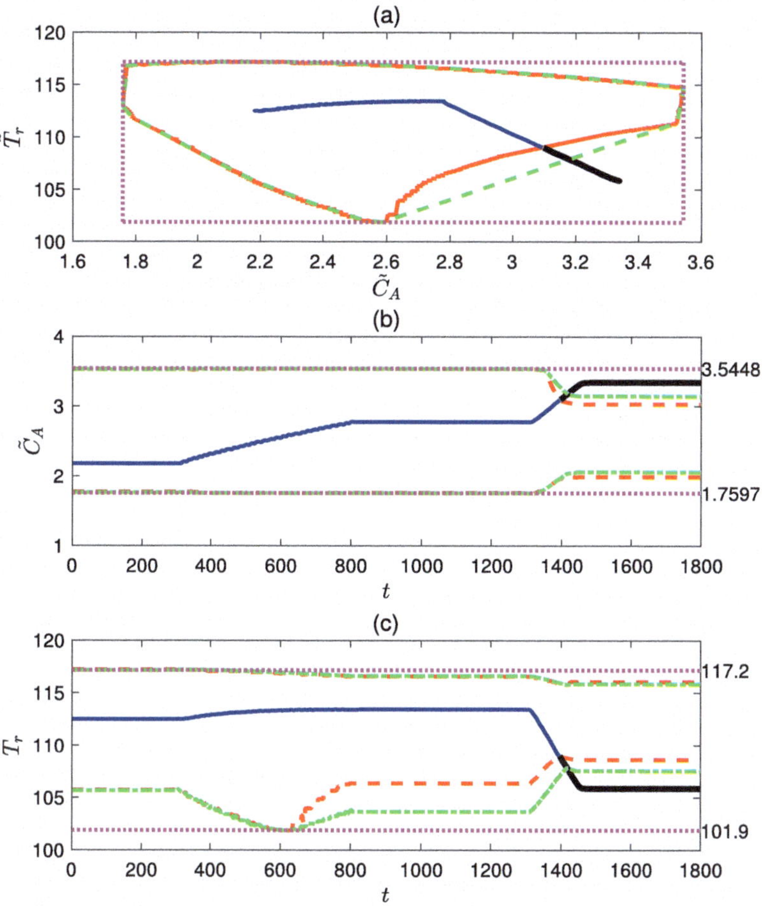

Fig. 3.16 **a** The search-cone-based NOZ model (red solid line), the convex-hull-based NOZ model (green dashed line), the univariate alarm thresholds (pink dotted line), and the test data points in the normal (blue solid line) or abnormal (black solid line) condition, **b** time sequence of $\tilde{C}_A$ and its corresponding dynamic alarm thresholds from the search-cone-based NOZ model (red dashed line), the convex-hull-based NOZ model (green dot-dashed line), and univariate alarm thresholds (pink dotted line), **c** time sequence of $\tilde{T}_r$ and its corresponding dynamic alarm thresholds from the search-cone-based NOZ model (red dashed line), the convex-hull-based NOZ model (green dot-dashed line), and univariate alarm thresholds (pink dotted line), reprinted from Ref. [43], copyright 2022, with permission from IEEE

3.2.1 Problem Description

Variation directions of related process variables are exploited here as the features to detect abnormal conditions. Let us take a variable frequency pump as an example. It can be easily observed that the outlet flow rate is always increasing or decreasing in a synchronized manner with the pump speed, when the pump works in normal conditions. This observation is guaranteed by physical laws of process variables involved the pump operation. Hence, if the outlet flow and speed of a variable frequency pump are found moving in opposite directions, then there must be something wrong in the pump operation. Thus, by looking at the directions of variations in the outlet flow rate and the pump speed, industrial plant operators are able to judge whether the pump is in abnormal conditions or not.

Consider a multivariate alarm system involving multiple process variables, $X := [x_1, x_2, \cdots, x_n]$. The variation direction of X_i is defined as

$$\operatorname{sign}\left(T_{x_i}\right) := \begin{cases} +1 & \text{if } T_{x_i} \in (\mu, +\infty), \\ 0 & \text{if } T_{x_i} \in [-\mu, \mu], \\ -1 & \text{if } T_{x_i} \in (-\infty, -\mu), \end{cases} \tag{3.49}$$

where T_{x_i} is the time gradient of x_i. Here μ is the significance threshold to be determined later in (3.66). In words, when $\operatorname{sign}\left(T_{x_i}\right)$ equals to $+1, 0$ and -1, x_i is increasing, unchanged (or having no evident trends) and decreasing in a statistical sense, respectively.

If X is in normal conditions, then variation directions of $x_1, x_2, \cdots, x_n$ are assumed to have a fixed relationship. Mathematically, the nominal relationship is represented by

$$f\left(\operatorname{sign}\left(T_{x_1}\right), \operatorname{sign}\left(T_{x_2}\right), \cdots, \operatorname{sign}\left(T_{x_n}\right)\right) = 0. \tag{3.50}$$

By contrast, when X runs into abnormal conditions, the relationship in (3.50) is assumed to be no longer valid, i.e.,

$$f\left(\operatorname{sign}\left(T_{x_1}\right), \operatorname{sign}\left(T_{x_2}\right), \cdots, \operatorname{sign}\left(T_{x_n}\right)\right) \neq 0.$$

The nominal relationship in (3.50) is usually obtained based on some process knowledge of X.

Under the above assumptions, the problem to be solved is to calculate time gradients and variation directions of X and to raise an alarm for X being in abnormal conditions in an online manner.

3.2.2 *Online Calculation of Time Gradients*

In the literature, variation directions are often obtained by qualitative trend analysis (QTA) methods that extract the increasing, decreasing or steady trends of process variables. Time sequences of process variables are approximated by several piecewise linear segments [7, 21] or high-order polynomials [29, 36]. Splines methods and wavelet methods decompose time sequences into different scale bases [12, 13, 38]. Clustering methods use a fuzzy C-means clustering technique to identify different trends [30]. Primitives methods represent time sequences by a combination of primitives in series whose local trends are pre-defined [19, 34]. However, the above-mentioned methods may have two limitations:

1. Some online methods modify the extracted trends in the past after the most recent trends are updated, so that they are not implemented in a real online sense (see e.g., [7, 21, 29]). Taking the sliding window method in [21] as an example, this method fits a time sequence using a linear segment growing with the time until the fitting error becomes larger than a threshold, and then updates the linear segment by restarting the fitting procedure.

2. Few methods are able to adapt extracted trends to the co-existed sharp and smooth variations in process variables. For sharp variations, narrow time windows or equivalently smaller time scales should be used in order to detect the changes of trends promptly; by contrast, for smooth variations, larger time windows or larger time scales should be exploited in order to reduce negative effects of noises on the trend extraction. As an illustration, the example in Sect. 3.2.4.1 later illustrates this limitation for a widely used method, the Savitzky-Golay (S-G) filter [36].

Due to the above two limitations, the existing QTA methods are not adopted here to calculate the variation directions. Instead, a so-called adaptive time gradient (ATG) approach is formulated here to calculate the time gradients and the variation directions by using adaptive time scales. The key step of the ATG approach is to determine adaptive time scales from a function between time scales and volatilities of time sequences. The function parameters are estimated from training data sets. Compared to the existing QTA methods, the ATG approach is adaptive in time scales and is online in a real sense.

First, time gradients of a discrete-time process variable $x(t)$ are defined, where $t \in \mathbb{Z}$ is the sampling time index and $\mathbb{Z}$ is the set of non-negative integers. To calculate the time gradient in an online manner, the current time index is denoted as j. The time gradient at $t = j$ is defined as the slope $k(j)$ of a local linear regression model on the time interval $t \in [j - L + 1, j]$, where L is the time window size, i.e.,

$$x(t) = k(j)t + b(j), t \in [j - L + 1, j]. \tag{3.51}$$

The loss function for the local regression model is

$$G = \sum_{t=j-L+1}^{j} K_\lambda(j, t)[x(t) - k(j)t - b(j)]^2, \tag{3.52}$$

where $K_\lambda(j, t)$ is the kernel function,

$$K_\lambda(j, t) = \lambda^{j-t}. \tag{3.53}$$

Here $\lambda \in (0, 1)$ is a constant forgetting factor so that λ^{j-t} is no larger than 1 and decreases with t.

A time scale C is defined as the time duration that the weighting term λ^{j-t} decreases from 1 to a small positive real number η,

$$\lambda^{C} = \eta. \tag{3.54}$$

A default value of η is 1%. The physical meaning of C is more transparent than the counterpart of λ, even though the two are closely related. A suitable range of C will be analyzed later. The resolution of a trend is closely related to the concept of time scales, i.e., if a trend takes a higher resolution, then a greater local change rate of the trend can be observed and vice versa [3]. As a consequence, a proper range of time scales is required in analyzing qualitative trends of time sequences. This is achieved as follows.

Setting the partial derivatives of G with respect to $k\,(j)$ and $b\,(j)$ to zeros and solving the corresponding equations yield the optimal solutions to minimize G as

$$\hat{k}\,(j) = \frac{\sum\limits_{t} tx\,(t)\,\lambda^{j-t}\sum\limits_{t}\lambda^{j-t} - \sum\limits_{t} x\,(t)\,\lambda^{j-t}\sum\limits_{t} t\lambda^{j-t}}{\sum\limits_{t} t^{2}\lambda^{j-t}\sum\limits_{t}\lambda^{j-t} - \left(\sum\limits_{t} t\lambda^{j-t}\right)^{2}}, \tag{3.55}$$

$$\hat{b}\,(j) = \frac{\sum\limits_{t} x\,(t)\,\lambda^{j-t}\sum\limits_{t} t^{2}\lambda^{j-t} - \sum\limits_{t} tx\,(t)\,\lambda^{j-t}\sum\limits_{t} t\lambda^{j-t}}{\sum\limits_{t} t^{2}\lambda^{j-t}\sum\limits_{t}\lambda^{j-t} - \left(\sum\limits_{t} t\lambda^{j-t}\right)^{2}},$$

where the index t in the summations ranges from $j - L + 1$ to j. The optimal estimate of $x\,(j)$ is $\hat{x}\,(j) = \hat{k}\,(j)\,t + \hat{b}\,(j)$.

It is apparent that the time gradient $k\,(j)$ is highly dependent on the forgetting factor λ. A higher value of λ, or equivalently a larger time scale C, implies that $k\,(j)$ takes account of the information from more historical data samples. Hence, at the current time index j, the time gradient T_x of $x\,(t)$ is defined as the time gradient $k(j)$ associated with a proper time scale C,

$$T_x\,(j) = k_C\,(j)\,. \tag{3.56}$$

Second, a function is first established to connect time scales with volatilities. There are two basic assumptions as follows:

1. Signals are composed of different variations in various time scales. The smaller time scale corresponds to sharp variations, while the larger time scale leads to variation lasting for a longer period of time.
2. In order to adapt to time-varying variations in time sequences, adaptive time scales have to be used, and the current time scale $C\,(j)$ can be determined from the corresponding volatility $V\,(j)$,

$$C(j) = g(V(j)). \tag{3.57}$$

Owing to the mapping relationship between λ and C in (3.54), an equivalence of (3.57) is

$$\lambda(j) = f(V(j)). \tag{3.58}$$

The volatility is defined as the range of time gradients in the current time scale, i.e.,

$$V(j) = \max_{t\in[j-C(j)+1,j]} k(t) - \min_{t\in[j-C(j)+1,j]} k(t). \tag{3.59}$$

A proper range $[C_{\min}, C_{\max}]$ of the time scale C for a process variable could be determined based on empirical knowledge as follows. The minimum time scale $C_{\min}$ is oriented from a common sense that if an abnormality arises, it should last at least for a while before being regarded as an actual abnormality, instead of being treated as short-time random variations due to noises. Thus, $C_{\min}$ is a boundary parameter in time to distinguish noises and trends in the finest resolution. The rationale of $C_{\min}$ is similar to alarm delay timers [11, 17], whose role is to raise an alarm when several consecutive samples of a process variable overpass an alarm threshold. A default value of $C_{\min}$ here is taken as $C_{\min} = 20$, by taking the recommended value of 20 seconds for alarm delay timers [39]. The maximum time scale $C_{\max}$ is associated with trends in the roughest resolution. It is oriented from the influence of some events, at the time duration of a specific trend. Usually $C_{\max}$ can be obtained from historical data sequences as the maximum duration of observed trends.

When $V(j)$ is small, $\lambda(j)$ approaches to $\eta^{\frac{1}{C_{\max}}}$ and $C(j)$ is forced to be no larger than $C_{\max}$. In the opposite side, a higher value of $V(j)$ corresponds to $\lambda(j)$ approaching $\eta^{\frac{1}{C_{\min}}}$ and $C(j)$ is constrained to be no less than $C_{\min}$. It is a reasonable fact that a shorter (longer) time scale would be more suitable for a higher (lower) volatility case. Hence, (3.58) is formulated as

$$\lambda(j) = \max\left(-\beta_0 \cdot \max(0, V(j) - V_0) + \eta^{\frac{1}{C_{\max}}}, \eta^{\frac{1}{C_{\min}}}\right), \tag{3.60}$$

where β_0 and V_0 are two constants to be determined.

Finally, parameters of the established function between time scales and volatilities are determined from training data sets. In particular, the parameters to be determined are β_0 and V_0 in (3.60) and μ in (3.49).

A data segment $\{s(t)\}_{t=1}^{l} := [s(1), s(2), \cdots, s(l)]$ is selected as the training data set with $l \geq L$. The function (3.60) describes a quantitative relationship between time scales and volatilities. The user-selected range $[C_{min}, C_{max}]$ of time scales has been defined for the variations in process variables of interest. Thus, fitting the function (3.60) is mainly for capturing the volatilities induced by noises. Hence, a requirement of $s(t)$ is that the data segment has no notable variations on time scales less than $C_{\min}$ in the view of industrial plant operators. A mild assumption should be mentioned that the variations on time scales less than $C_{\min}$ are assumed to be absent, and the variations induced by noises in the training data set are expected to be persistent in the subsequent online data sequences.

The volatility of the training data set is calculated to fit the function (3.60). The time scale is defined as $C_p = C_{\min} + p - 1$ for $p = 1, 2, \cdots, C_{\max} - C_{\min} + 1$, and $\lambda_p = \eta^{\frac{1}{C_p}}$. The time gradient $k_{C_p}(j)$ is computed for each data sample $s(j)$ with $j = L, L+1, \cdots, l$ as that in (3.55),

$$k_{C_p}(j) = \frac{\sum_t ts(t)\lambda_p^{j-t}\sum_t \lambda_p^{j-t} - \sum_t s(t)\lambda_p^{j-t}\sum_t t\lambda_p^{j-t}}{\sum_t t^2\lambda_p^{j-t}\sum_t \lambda_p^{j-t} - \left(\sum_t t\lambda_p^{j-t}\right)^2}, \tag{3.61}$$

where the index t ranges from $j - L + 1$ to j. Since the maximum meaningful time scale is $C_{\max}$, the window size L of the local regression model in (3.51) should be no larger than $C_{\max}$. Because the forgetting factor λ_p is changing adaptively, L can be simply chosen as $L = C_{\max}$. Next, the volatility $V_{C_p}(j)$ for $s(j)$ on the time scale C_p is obtained as

$$V_{C_p}(j) = \max_{t \in [j-C_p+1, j]} k_{C_p}(t) - \min_{t \in [j-C_p+1, j]} k_{C_p}(t). \tag{3.62}$$

The calculations of $k_{C_p}(j)$ and $V_{C_p}(j)$ are repeated for all values of p and j, so that both $k_{C_p}(j)$ and $V_{C_p}(j)$ are two dimensional matrices with j as the row index and p as the column index.

Figure 3.17 plots each row of $V_{C_p}(j)$ versus C_p or λ_p as the dot lines, based on an industrial data set of a pump speed appeared later in Sect. 3.2.4.3. The envelope curve of all the dot lines marks the maximum volatility $V_p^{(Env)}$ corresponding to a specific time scale $C_p^{(Env)}$ or $\lambda_p^{(Env)}$. Any data samples with volatility less than $V_p^{(Env)}$ on the specific time scale $C_p^{(Env)}$ should be associated with a smoother trend so that a longer time scale larger than $C_p^{(Env)}$ should be chosen. Then only the data samples with the highest volatility on each time scale are expected to be on the function curve of (3.60). Hence, the envelope curve denoted as $\{\lambda_p^{(Env)}, V_p^{(Env)}\}$ or $\{C_p^{(Env)}, V_p^{(Env)}\}$ is calculated from

$$\begin{cases} \lambda_p^{(Env)} := \eta^{\frac{1}{C_{\min}+p-1}}, \\ C_p^{(Env)} := C_{\min} + p - 1, \\ V_p^{(Env)} = \max_{j \in [C_{\max}, l]} \left(V_{C_p^{(Env)}}(j)\right), \end{cases} \tag{3.63}$$

where $p = 1, 2, \cdots, C_{\max} - C_{\min} + 1$. The least-squares method is used to fit the function (3.60) to the envelope curve $\{\lambda_p^{(Env)}, V_p^{(Env)}\}$ with the optimal solution as

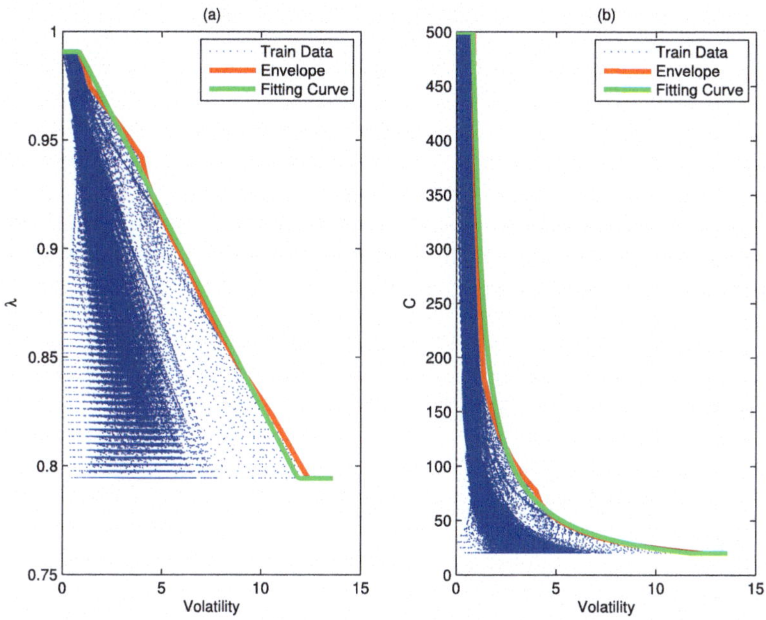

Fig. 3.17 **a** The volatility $V_{C_p}(j)$ versus λ_p, **b** $V_{C_p}(j)$ versus C_p, reprinted from Ref. [9], copyright 2017, with permission from Elsevier

$$\begin{cases} V_0' = V_{C_{\max}-C_{\min}+1}^{(Env)}, \\ \beta_0' = \arg\min_{\beta_0} \sum_{p=1}^{C_{\max}-C_{\min}+1} \left[f\left(V_p^{(Env)}, \beta_0, V_0'\right) - \lambda_p^{(Env)} \right]^2. \end{cases} \tag{3.64}$$

A goodness of fit is defined as

$$\rho = 1 - \frac{\sum_{p=1}^{C_{\max}-C_{\min}+1} \left(\lambda_p^{(Env)} - \widehat{\lambda}_p^{(Env)}\right)^2}{\sum_{p=1}^{C_{\max}-C_{\min}+1} \left(\lambda_p^{(Env)} - \overline{\lambda}_p^{(Env)}\right)^2}, \tag{3.65}$$

where

$$\widehat{\lambda}_p^{(Env)} = f\left(V_p^{(Env)}, \beta_0', V_0'\right), \ \overline{\lambda}_p^{(Env)} = \frac{1}{C_{\max} - C_{\min} + 1} \sum_{i=1}^{C_{\max}-C_{\min}+1} \lambda_p^{(Env)}.$$

A large value of ρ close to 1 indicates that the quality of the fitting function is satisfactory.

The parameter μ in (3.60) is a threshold to tolerate small values of time gradients induced by noises. The variation direction is unchanged if the time gradient is inside the range $[-\mu, \mu]$. The unchanged variation direction means that the volatility is low. The minimum volatility corresponds to the maximum time scale according to (3.60). Hence, the significance threshold μ in (3.49) is determined as

$$\mu = 2\underset{j}{\mathrm{std}}\left(k_{C_{\max}}(j)\right), \tag{3.66}$$

where $k_{C_{\max}}(j)$ is from (3.61) and $j \in [C_{\max}, l]$. Equation (3.66) is developed from a hypothesis test on mean values with the null hypothesis of zero time gradients by using a significance level 0.05.

3.2.3 Steps of the Proposed Method

For the multivariate system $X = [x_1, x_2, \cdots, x_n]$, the time gradients of x_i for $i = 1, 2, \cdots, n$ are calculated by the proposed ATG approach in two steps, namely, the offline training of the unknown parameters $\{V_0, \beta_0, \mu\}$ and the online calculating of time gradient $T_x(j)$. Here the subscript x refers to anyone of $x_1, x_2, \cdots, x_n$.

In the offline training part, a time scale range $[C_{\min}, C_{\max}]$ is determined and a training data set $s(t)$ is selected as described in Sect. 3.2.2. The unknown parameters $\{V_0, \beta_0, \mu\}$ are obtained from (3.64) and (3.66). If ρ in (3.65) is less than a threshold (e.g., $\rho < 0.8$), another training data set is chosen and this part of offline training is repeated.

For the online calculating part, the following steps are implemented when a new data sample of X is available.

Step 1: Calculate the time gradient $T_x(j)$ as that in (3.55) by using the time scale $C(j-1)$ or equivalently the forgetting factor $\lambda(j-1)$.

Step 2: Update the time scale as

$$C(j) = \min\left\{C_{\max}, \frac{\log(\eta)}{\log(\lambda(j-1))}\right\}.$$

Step 3: Calculate the volatility for the updated time scale $C(j)$,

$$V(j) = \max_{t \in [j-C(j)+1, j]} k(t) - \min_{t \in [j-C(j)+1, j]} k(t).$$

Step 4: Update the forgetting factor,

$$\lambda(j) = \max\left(-\beta_0' \cdot \max\left(0, V(j) - V_0'\right) + \eta^{\frac{1}{C_{\max}}}, \eta^{\frac{1}{C_{\min}}}\right).$$

By using the time gradients, the variation direction of x_i is obtained from (3.49). The multivariate alarm system is designed as follows.

To compare the variation directions with the relationship in (3.50) in a systematic manner, a matrix R with size $r \times n$ is introduced as

$$R := \begin{bmatrix} \text{sign}\left(T_{1,x_1}\right) & \text{sign}\left(T_{1,x_2}\right) & \cdots & \text{sign}\left(T_{1,x_n}\right) \\ \text{sign}\left(T_{2,x_1}\right) & \text{sign}\left(T_{2,x_2}\right) & \cdots & \text{sign}\left(T_{2,x_n}\right) \\ \vdots & \vdots & \ddots & \vdots \\ \text{sign}\left(T_{r,x_1}\right) & \text{sign}\left(T_{r,x_2}\right) & \cdots & \text{sign}\left(T_{r,x_n}\right) \end{bmatrix}. \tag{3.67}$$

Each row of R represents one rule against the relationship in (3.50) under a specific circumstance. For example, if $X := [x_1, x_2]$ is in normal conditions when both variables are increasing, decreasing or unchanged simultaneously, then the relationship matrix R is

$$R = \begin{bmatrix} 1 & -1 \\ -1 & 1 \end{bmatrix}. \tag{3.68}$$

The variation directions of all process variables in the current time index j are represented by a vector

$$\text{sign}\left(T_X\left(j\right)\right) = \left[\text{sign}\left(T_{x_1}\left(j\right)\right), \text{sign}\left(T_{x_2}\left(j\right)\right), \cdots, \text{sign}\left(T_{x_n}\left(j\right)\right)\right]. \tag{3.69}$$

The variation direction vector $\text{sign}\left(T_X\left(j\right)\right)$ is compared with the relationship matrix R to determine whether $X(j)$ is in normal or abnormal conditions. In particular, if the variation direction vector is the same as one row of R, then $X\left(j\right)$ is in abnormal conditions, and the alarm variable $x_a\left(j\right) = 1$; otherwise, $X\left(j\right)$ is regarded as being in normal conditions, and no alarm arises, i.e., $x_a\left(j\right) = 0$.

3.2.4 Numerical and Industrial Examples

This subsection provides three examples to illustrate the effectiveness of the ATG approach, and design multivariate alarm systems for a simulated water tank system and an industrial variable-frequency pump.

3.2.4.1 Numerical Example #1

This numerical example illustrates the effectiveness of the ATG approach for calculating time gradients. Consider a process variable

$$x(t) = \begin{cases} e(t), \text{ for } t \in [1, 300), \\ -1 \cdot (t - 300) + e(t), \text{ for } t \in [300, 1000], \end{cases}$$

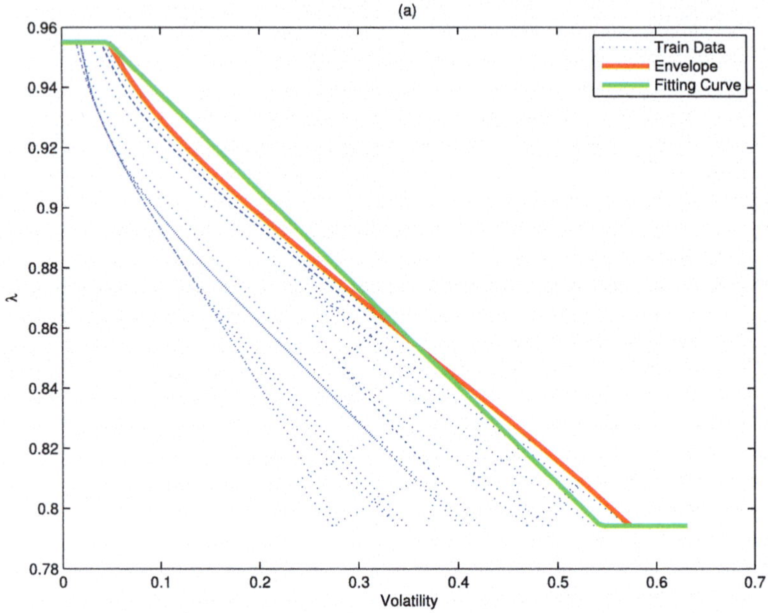

Fig. 3.18 The volatility $V_C(j)$ versus λ, reprinted from Ref. [9], copyright 2017, with permission from Elsevier

where $e(t)$ is the Gaussian white noise with zero mean and unit variance. Figure 3.18 shows the result of the offline training part with $C_{\min} = 20$, $C_{\max} = 100$, and $s(t) = x(1 : 200)$. The unknown parameters β_0 and V_0 in (3.60) and μ in (3.49) and the goodness of fit ρ in (3.65) are determined,

$$V_0 = 0.0481,\ \beta_0 = 0.3242,\ \mu = 0.0073,\ \rho = 0.9843.$$

In particular, $\rho = 0.9843$ says that the quality of the fitting function is satisfactory. Figure 3.19 presents the time sequence plots of the process variable $x(t)$, the time gradient $k_C(t)$, and the time scale $C(t)$. The notable change occurring on $t = 300$ is well captured by the ATG approach, so that the time scale $C(t)$ decreases to near $C_{\min}$. As $x(t)$ keeps a steady descent trend after $t = 300$, $C(t)$ grows to near $C_{\max}$ and the time gradient $k_C(t)$ becomes smoother.

The results in Fig. 3.19 are benchmarked by the noise-free counterpart in Fig. 3.20. The S-G filter [36] is applied as a comparison. Since the S-G filter is associated with a fixed time window for computation of time gradients, the lower bound $C_{\min}$ is used as the time window size. It can be found that even though both the S-G filter and the ATG approach response to the change of $x(t)$ at $t = 300$ immediately, the ATG approach gives a much smoother and more accurate time gradient than the S-G filter. This observation is also supported by the boxplots of the estimation errors in Fig. 3.21. The superior performance of the ATG approach is owing to the adaptive

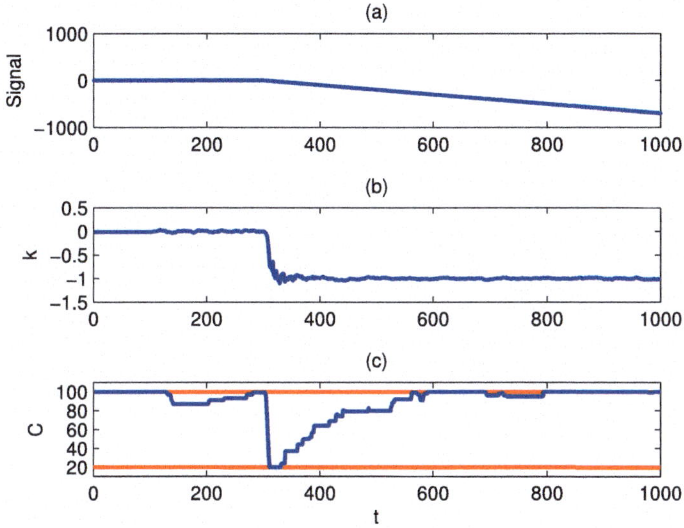

Fig. 3.19 The time sequence plots of variables: **a** $x(t)$ (dot), **b** the time gradient $k_C(t)$, and **c** the time scale $C(t)$, reprinted from Ref. [9], copyright 2017, with permission from Elsevier

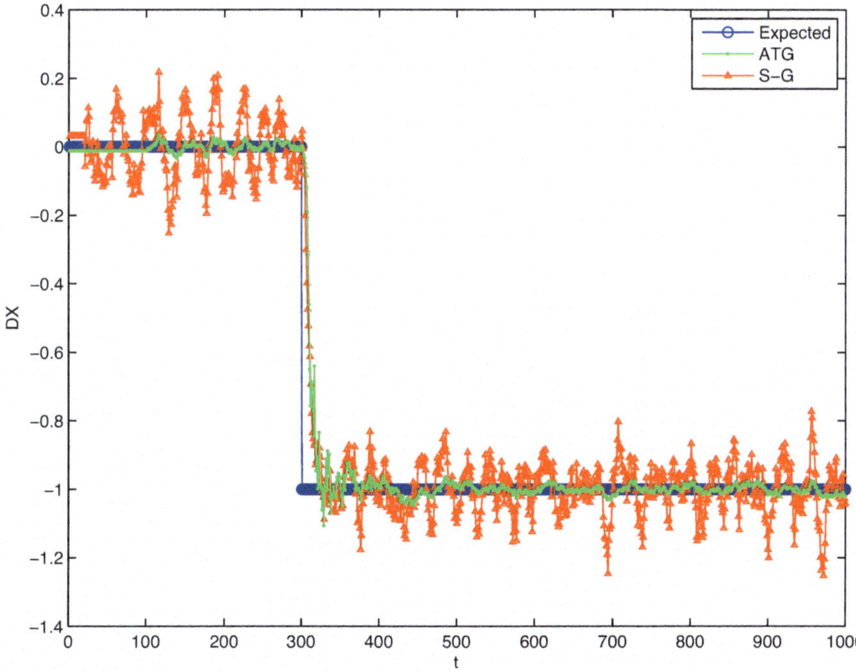

Fig. 3.20 The time sequence plots of the expected time gradient (circle), the estimated time gradient from the ATG approach (dot) and the counterpart from the S-G filter (triangle), reprinted from Ref. [9], copyright 2017, with permission from Elsevier

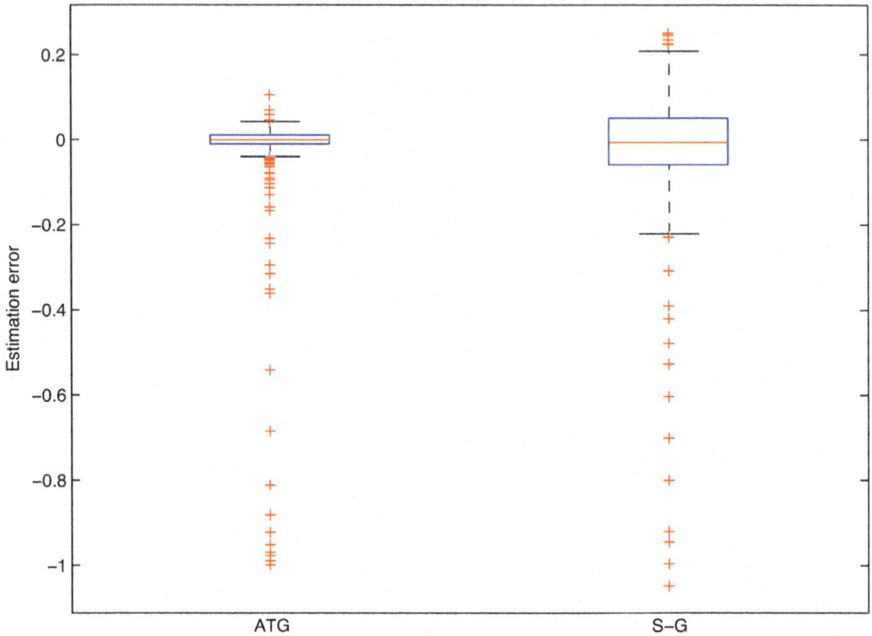

Fig. 3.21 The boxplots of the estimation errors from the ATG approach and the counterpart from the S-G filter, reprinted from Ref. [9], copyright 2017, with permission from Elsevier

capability of time scales. In other words, the ATG approach is able to smooth (track) time sequences on a larger (smaller) time scale than the S-G filter when a trend with low (high) volatility is detected.

3.2.4.2 Numerical Example #2

This numerical example designs a multivariate alarm system for a simulated water tank system. As schematically shown in Fig. 3.22, a water tank system can be described by a dynamic model

$$\begin{cases} S_1 \frac{\mathrm{d}H(t)}{dt} = F_i(t) - F_o(t), \\ F_o(t) = S_2 \cdot \sqrt{2gH(t)} + e(t). \end{cases} \tag{3.70}$$

Here F_i and F_o respectively denote the inlet and outlet flows passing through two valves, H is the liquid level, S_1 and S_2 respectively are the sectional areas of the tank and the outlet valve, g represents the gravity constant and e is the system noise. By adjusting the inlet flow F_i via a control valve, the tank level H is adjusted to a desired position, while the outlet flow F_o varies with H. Thus, (3.70) implies that F_o and H increase or decrease simultaneously when the system is in normal conditions.

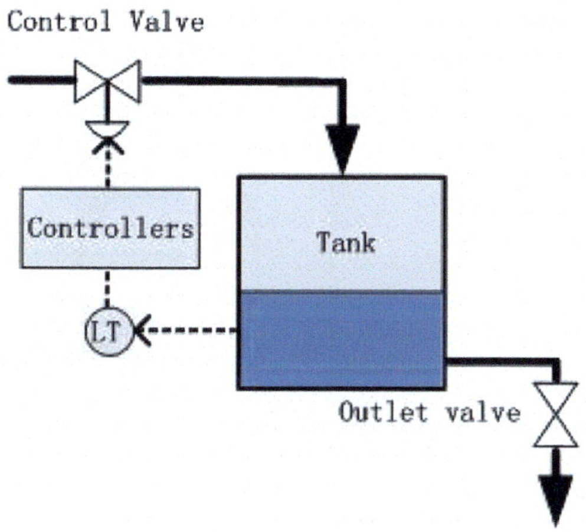

Fig. 3.22 Schematic diagram of a water tank system in a laboratory, reprinted from Ref. [9], copyright 2017, with permission from Elsevier

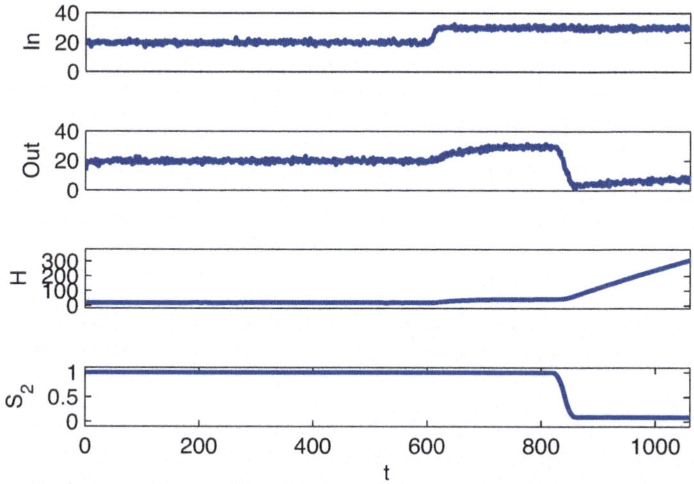

Fig. 3.23 Time sequence plots of variables: **a** F_i, **b** F_o, **c** H and **d** S_2, reprinted from Ref. [9], copyright 2017, with permission from Elsevier

Figure 3.23 shows the time sequence plots of F_i, F_o and S_2. The inlet flow F_i is increased on $t = 600$, so that both F_o and H are increasing. A fault occurs such that S_2 rapidly decays by manually blocking the outlet valve from $t = 821$ to $t = 860$. As a result, an abnormality occurs, because F_o decreases while H increases. Hence F_o and H are in abnormal conditions in the time period $t \in [821, 860]$.

The proposed method is applied with $C_{\min} = 20$, $C_{\max} = 100$, $s(t) = X(1:400) := \{X(1), X(2), \cdots, X(400)\}$), and the relationship matrix R in (3.68).

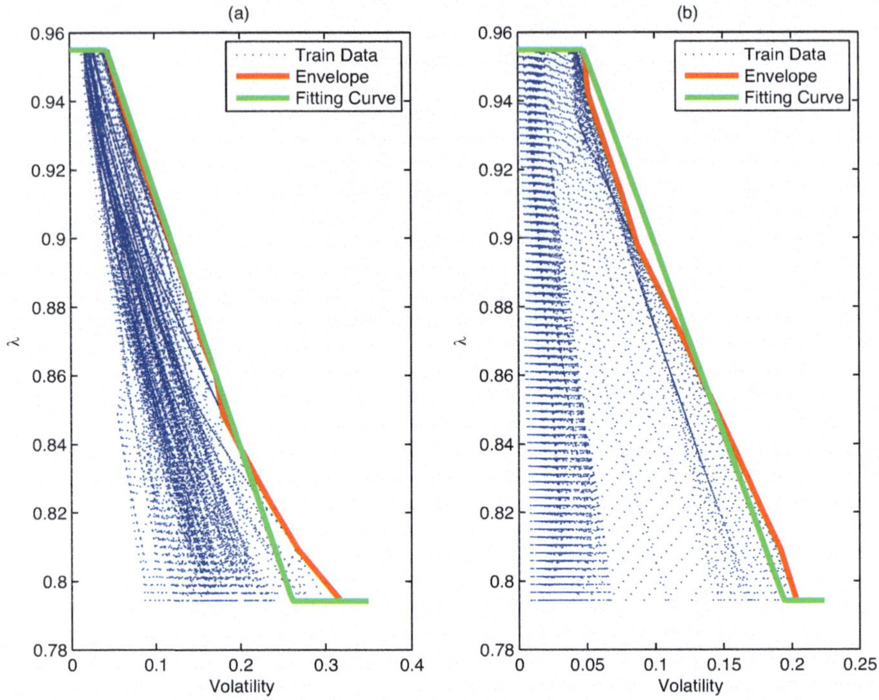

Fig. 3.24 **a** The volatility $V_C(j)$ versus λ for F_o, and **b** the volatility $V_C(j)$ versus λ for H, reprinted from Ref. [9], copyright 2017, with permission from Elsevier

Figure 3.24 shows the results of the offline data training for F_o and H, respectively. The two groups of unknown parameters β_0 and V_0 in (3.60) and μ in (3.49) and the goodness of fit ρ in (3.65) for F_o and H are respectively

$$\begin{cases} V_{0,F_o} = 0.0444 \\ \beta_{0,F_o} = 0.7425 \\ \mu_{F_o} = 0.0134 \\ \rho_{F_o} = 0.9919 \end{cases}, \quad \begin{cases} V_{0,H} = 0.0481 \\ \beta_{0,H} = 1.0961 \\ \mu_H = 0.0161 \\ \rho_H = 0.9678 \end{cases}.$$

The fitness metrics ρ_{F_o} and ρ_H are close to 1, saying that the qualities of corresponding functions in (3.60) are satisfactory. The proposed method yields the time gradients $k_{F_o}(t)$ and $k_H(t)$ and time scales $C_{F_o}(t)$ and $C_H(t)$ as shown in Fig. 3.25. The red horizontal lines in the middle and bottom subplots in Fig. 3.25 denote the significance thresholds $[-\mu, \mu]$ in (3.49) and the range $[C_{min}, C_{max}]$ of time scales, respectively. As expected, the time scales $C_{F_o}(t)$ and $C_H(t)$ take the maximum values when $F_o(t)$ ad $H(t)$ are steady, and decrease quickly when $F_o(t)$ ad $H(t)$ experience sharp variations. As presented in Fig. 3.26, the alarm system raises alarms in the time period $t = [832, 875]$, which is close to the actual time interval $t \in [821, 860]$ when the

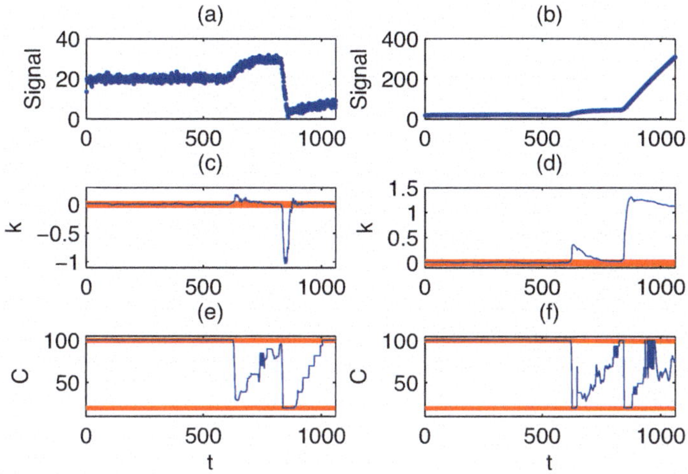

Fig. 3.25 The time sequence plots of variables: **a** $F_o(t)$, **b** $H(t)$, **c** the time gradient $k_{F_o}(t)$, **d** the time gradient $k_H(t)$, **e** the time scale $C_{F_o}(t)$ and **f** the time scale $C_H(t)$, reprinted from Ref. [9], copyright 2017, with permission from Elsevier

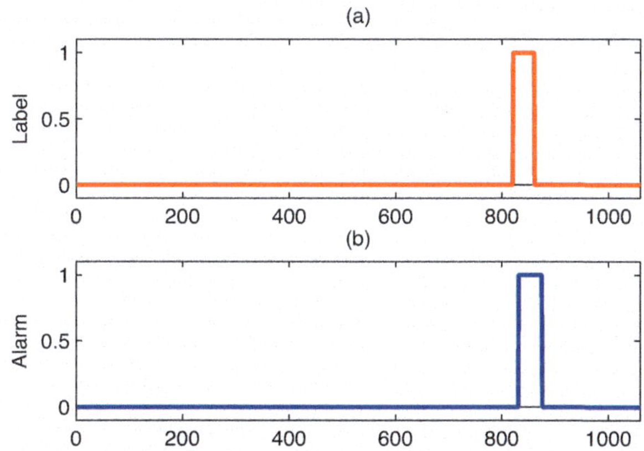

Fig. 3.26 The actual alarm variable **a** and the alarm variable $X_a(t)$ from the designed multivariate alarm system **b**, reprinted from Ref. [9], copyright 2017, with permission from Elsevier

abnormality presents. Thus, with an expected detection delay between the detected and actual abnormalities less than $C_{\min} = 20$, the designed multivariate alarm system successfully detects the abnormality against the nominal relationship between F_0 and H.

3.2.4.3 Industrial Example

This industrial example designs a multivariate alarm system for a variable-frequency feedwater pump in a large-scale power plant. The current alarm system to monitor the operation of the feedwater pump raises an alarm when the inlet flow rate is larger than a high alarm threshold 560 t/h. Such a univariate alarm system sometimes results in false alarms, when the inlet flow rate of the feedwater pump increases beyond 560 t/h to supply more water to downstream devices. As shown in Fig. 3.27, even though the inlet flow rate overpasses the high alarm threshold 560 t/h, the relationship between the inlet flow rate and pre-pump electrical current are unchanged. Thus, the feedwater pump is in normal conditions, and the occurred alarms from the univariate alarm system are nuisance ones. Furthermore, the univariate alarm system may also lead to missed alarms, to be illustrated by the data sequences in Fig. 3.28 later.

To avoid false and missed alarms from the current univariate alarm system, a three-variable alarm system $X := [x_1, x_2, x_3]$ is formulated for the pump, where x_1, x_2 and x_3 are the inlet flow rate, the pump speed and the pre-pump electrical current, respectively. Based on physical laws of pumps, the three process variables are expected to increase or decrease synchronously in normal conditions, so that the relationship matrix R is

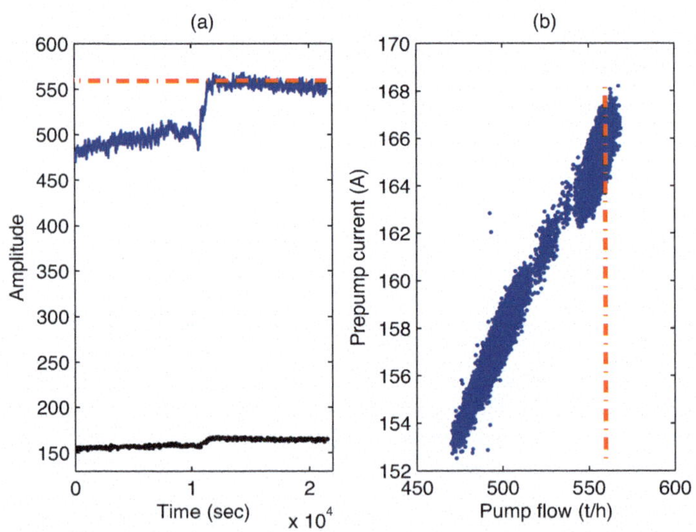

Fig. 3.27 **a** The time sequence plots of the inlet flow rate (the upper one) and pre-pump electrical current (the lower one), **b** the scatter plot of the inlet flow rate and pre-pump electrical current with the high alarm threshold (red dash-dotted), reprinted from Ref. [9], copyright 2017, with permission from Elsevier

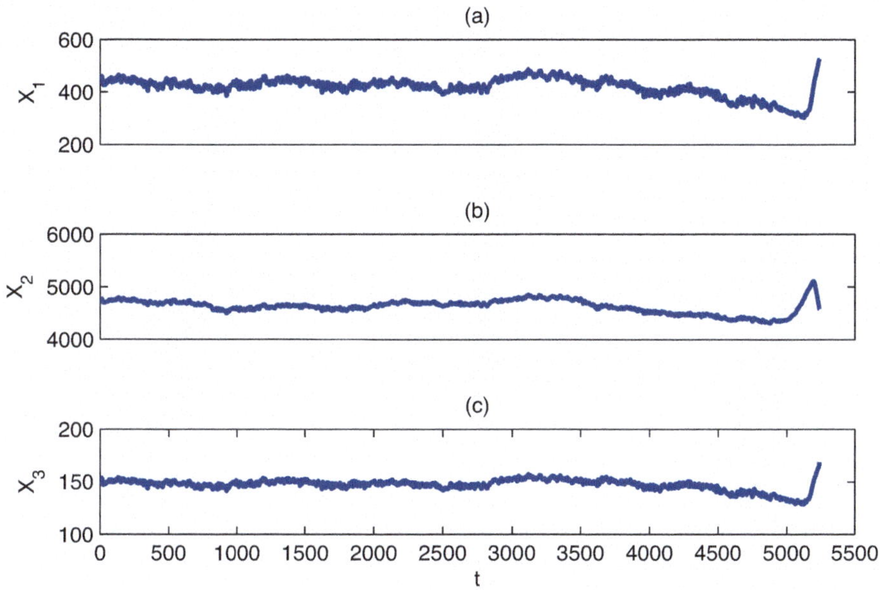

Fig. 3.28 The time sequence plots of three process variables: **a** x_1, **b** x_2 and **c** x_3, reprinted from Ref. [9], copyright 2017, with permission from Elsevier

$$R = \begin{bmatrix} 1 & 1 & 1 & 1 & 1 & 0 & 0 & -1 & -1 & -1 & -1 & -1 \\ -1 & -1 & 1 & -1 & 0 & 1 & -1 & 1 & 1 & -1 & 0 & 1 \\ -1 & 1 & -1 & 0 & -1 & -1 & 1 & 1 & -1 & 1 & 1 & 0 \end{bmatrix}'.$$

Figure 3.28 presents the time sequence plots of the three process variables collected with the sampling period 1 sec. By visualization, X is in normal conditions before $t = 5035$. After that, there are two intervals with obvious abnormalities: x_2 increases while x_1 and x_3 keep decreasing during the interval $t \in [5035, 5160]$, and x_2 decreases while x_1 and x_3 keep increasing during the interval $t \in [5195, 5239]$. This is an offline detection of normal and abnormal conditions by visualization. Let us do so in an online manner via the proposed method. Note that the univariate alarm system currently used in the power plant is not able to detect such abnormalities, since x_1 is always smaller than the high alarm threshold 560 t/h; thus, missed alarms are present.

The proposed method is applied by setting $C_{\min} = 20$, $C_{\max} = 500$ and $s(t) = X(1:2000)$. Figure 3.29 shows the offline training results. The unknown parameters and the goodness of fit for x_1, x_2 and x_3 are respectively

$$\begin{cases} V_{0,x_1} = 0.2096 \\ \beta_{0,x_1} = 0.0629 \\ \mu_{x_1} = 0.1093 \\ \rho_{x_1} = 0.9886 \end{cases}, \begin{cases} V_{0,x_2} = 0.8608 \\ \beta_{0,x_2} = 0.0178 \\ \mu_{x_2} = 0.4496 \\ \rho_{x_2} = 0.9962 \end{cases}, \begin{cases} V_{0,x_3} = 0.0319 \\ \beta_{0,x_3} = 0.3711 \\ \mu_{x_3} = 0.0173 \\ \rho_{x_3} = 0.9877 \end{cases}. \quad (3.71)$$

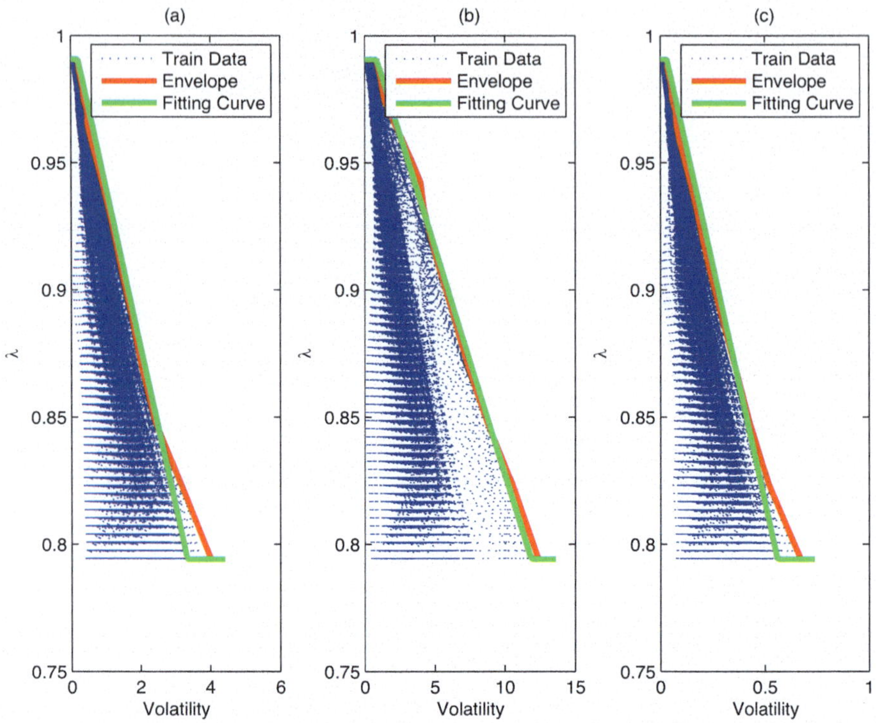

Fig. 3.29 **a** The volatility $V_C(j)$ versus λ for x_1, **b** the volatility $V_C(j)$ versus λ for x_2, and **c** the volatility $V_C(j)$ versus λ for x_3, reprinted from Ref. [9], copyright 2017, with permission from Elsevier

The fitness metrics ρ_{x_1}, ρ_{x_2} and ρ_{x_3} are close to 1, saying that the qualities of corresponding functions in (3.60) are satisfactory. Figure 3.30 presents the time gradients and time scales for x_1, x_2 and x_3. The red horizontal lines in the middle and bottom subplots in Fig. 3.30 denote the significance thresholds $[-\mu, \mu]$ in (3.49) and the range $[C_{min}, C_{max}]$ of time scales, respectively. Before $t = 5000$, the volatilities of the pump system are low so that large time scales are used to calculate the time gradients. The time scales decay rapidly when some sharp variations are detected around $t = 5080$. The designed multivariate alarm system yields the alarm variable $x_a(t)$ in Fig. 3.31. The alarms arise at the time instants $t = [5051, 5178]$ and $t = [5215, 5239]$, which are close to the offline detection of abnormal conditions.

The designed multivariate alarm system is also able to avoid the false alarms illustrated in Fig. 3.27. By using the offline training results in (3.71), the proposed method yields the time gradients and adaptive time scales as shown in Fig. 3.32. The increment of x_1 over the original alarm threshold 560 t/h is accompanied by synchronized increments in x_2 and x_3 around $t = 2750$. Thus, no alarms occur for such a normal condition. Note that the time scales in Fig. 3.32 are able to adapt to the sharp

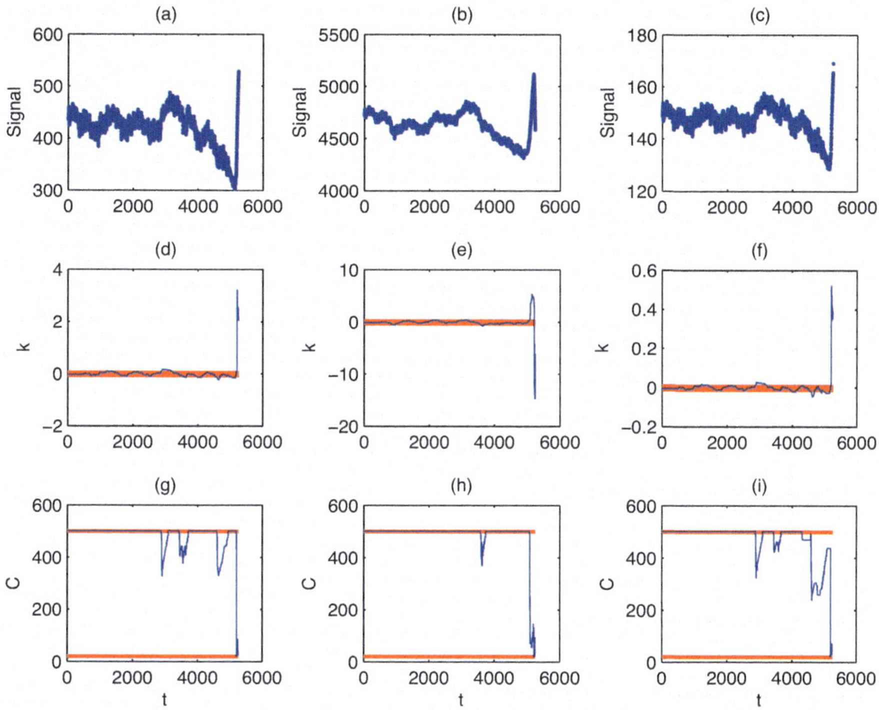

Fig. 3.30 The time sequence plots of variables: **a** $x_1(t)$, **b** $x_2(t)$, **c** $x_3(t)$, **d** the time gradient $k_{x_1}(t)$, **e** $k_{x_2}(t)$, **f** $k_{x_3}(t)$, **g** the time scale $C_{x_1}(t)$, **h** $C_{x_2}(t)$ and **i** $C_{x_3}(t)$, reprinted from Ref. [9], copyright 2017, with permission from Elsevier

variations in variables around $t = 2750$, and to the smooth variations in other time periods.

3.3 Multivariate Alarm Systems for Electrical Pumps

This section is about abnormality detection for condensate-water electrical pumps in thermal power plants [46]. A method is proposed to design multivariate alarm systems for electrical pumps that are in fact time-varying processes. A novel feature to distinguish normal and abnormal conditions is observed on the variation rates of parameters in a multiple linear regression model. A model estimator based on Bayesian filters is formulated to track the variations of model parameters in normal conditions, and not to do so in abnormal conditions so that absolute cumulative modeling errors are large enough to raise alarms.

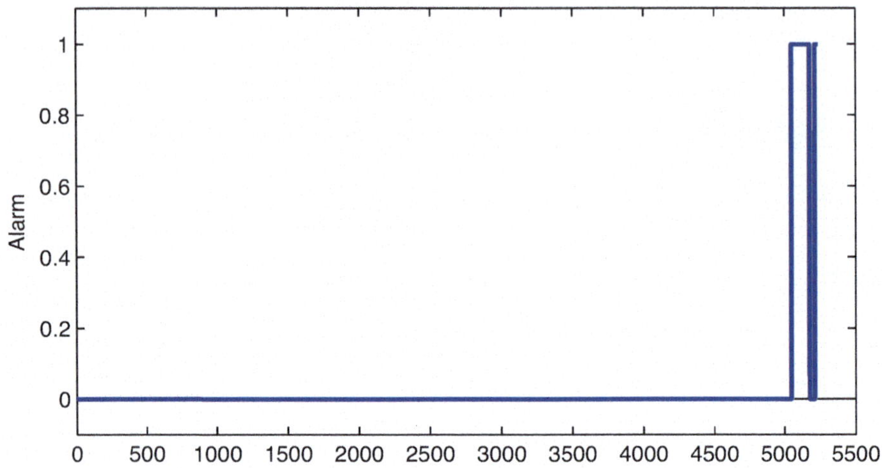

Fig. 3.31 The alarm variable $x_a(t)$ for process variables in Fig. 3.30, reprinted from Ref. [9], copyright 2017, with permission from Elsevier

3.3.1 Problem Description

This section presents a novel feature to distinguish normal and abnormal conditions of electrical pumps, and describes the problem to be solved.

Abnormality detection for condensate-water electrical pumps in thermal power plants reveals a novel feature to distinguish normal and abnormal conditions. The feature says that if a time-varying process is in normal (abnormal) conditions, then the variation rates of model parameters of the process are small (large).

Consider a condensate-water electrical pump with a variable-frequency drive from a large-scale thermal power plant. Figure 3.33 presents the time sequences of three process variables of the pump from January 15th to June 1st, namely, the differential pressure $\Delta_p(t)$ between the inlet and outlet fluids, the rotational speed $r(t)$ and the fluid flow rate $q(t)$. Here $t \in \mathbb{Z}$ is the sampling time index, with $\mathbb{Z}$ representing the set of non-negative integers. In order to remove the effects of different amplitudes, all data samples of the three variables have been normalized into the range [0, 1]. Three accidents of the pump occurred at 10:40:00 on January 24, 23:35:00 on March 25 and 13:00:00 on May 25. In these accidents, the pump was forced to shut down in emergency as indicated by the abrupt changes of $\Delta_p(t)$ from nonzero-values to zero. Once the pump was off, a back-up pump was started automatically; because the two pumps shared the same transducers for $r(t)$ and $q(t)$, the values of $r(t)$ and $q(t)$ were not equal to zero in the accidents. Let us take a further look at process variables of the pump in two cases.

In the first case, the time sequences of $\Delta_p(t)$, $r(t)$ and $q(t)$ from 11:34:00 to 12:57:40 on March 15 are shown in Fig. 3.34a, where the pump is known to be in normal conditions throughout the entire time period. The sampling period is 1 second.

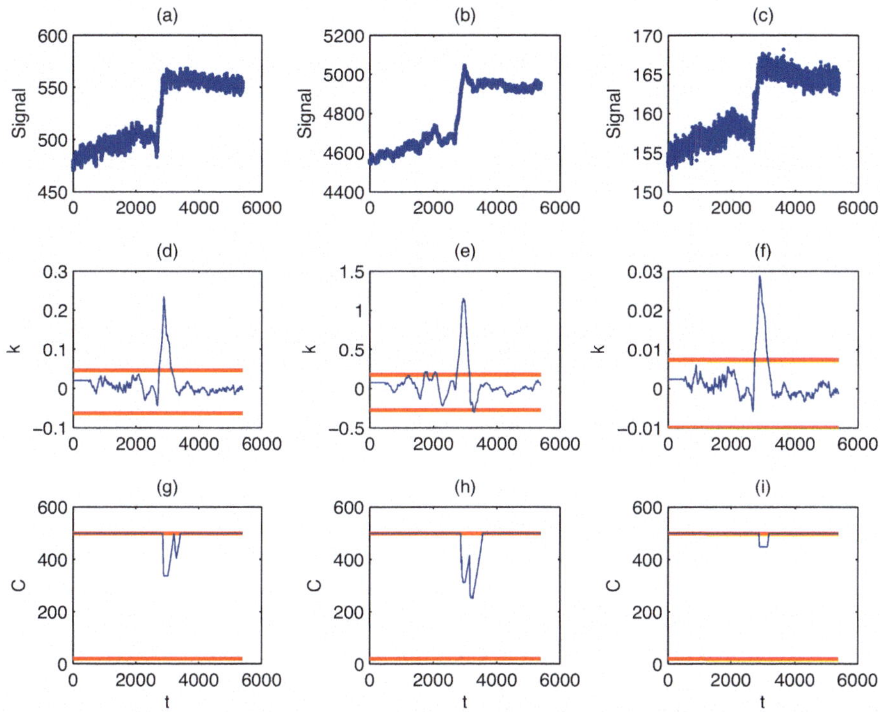

Fig. 3.32 The time sequence plots of variables: **a** $x_1(t)$, **b** $x_2(t)$, **c** $x_3(t)$, **d** the time gradient $k_{x_1}(t)$, **e** $k_{x_2}(t)$, **f** $k_{x_3}(t)$, **g** the time scale $C_{x_1}(t)$, **h** $C_{x_2}(t)$ and **i** $C_{x_3}(t)$, reprinted from Ref. [9], copyright 2017, with permission from Elsevier

The 3D scatter plots of the three variables from different visual angles are showed in Fig. 3.34b, c. Two clusters are formulated, namely, the blue star points in the first 3750 seconds and the purple dot points in the last 1043 seconds. The remaining 207 black triangle points in the middle positions are in the transition stage between the two clusters. The clusters represent two different normal operating conditions of the pump, which can be numerically confirmed as follows. Based on some physical principles, the pump can be described by a multiple linear regression model with time-varying parameters [18],

$$y(t) = \boldsymbol{x}'(t)\boldsymbol{\beta}(t), \tag{3.72}$$

where

$$y(t) := \Delta_p(t), \boldsymbol{\beta}(t) := [\beta_0(t), \beta_1(t), \beta_2(t), \beta_3(t)]',$$
$$\boldsymbol{x}(t) := [1, x_1(t), x_2(t), x_3(t)]' = [1, r^2(t), q(t)r(t), q^2(t)]'.$$

The parameter vector $\beta(t)$ is estimated via the ordinary least-squares (OLS) method as $\hat{\beta}(t) = (\boldsymbol{X}'\boldsymbol{X})^{-1}\boldsymbol{X}'\boldsymbol{Y}$, where $\boldsymbol{X} := [\boldsymbol{x}(1), \cdots, \boldsymbol{x}(t)]'$ and $\boldsymbol{Y} := [y(1), \cdots, y(t)]'$. If the blue star points are used, the estimated parameters are $\hat{\beta}(t) = [-0.3649, 2.2064, -0.9629, 0.3658]'$, and the corresponding estimate $\hat{\Delta}_p(t) = \boldsymbol{x}'(t)\hat{\beta}(t), q(t)$ and $r(t)$ formulate the light blue surface in Fig. 3.34b. Under mild conditions, the estimated parameter vector $\hat{\beta}(t)$ asymptotically approaches to Gaussian distribution $\mathcal{N}\left(\beta, \hat{\sigma}^2(\boldsymbol{X}'\boldsymbol{X})^{-1}\right)$, where $\hat{\sigma}^2$ is the sample variance of the model residual $y(t) - \boldsymbol{x}'(t)\hat{\beta}(t)$ [24]. Thus, the 95% confidence interval of $\beta(t)$ can be obtained as $[(-0.3708, 0.3590), (2.1876, 2.2252), (-0.9684, -0.9574), (0.3656, 0.3660)]'$. Analogously, if the purple dot points are used, $\hat{\beta}(t) = [-0.9759, 2.4453, -0.4928, 0.1804]'$ and the corresponding surface is the light purple one in Fig. 3.34b. Clearly, the two sets of $\hat{\beta}(t)$'s are different, because $\hat{\beta}(t)$ for the purple dot points is not inside the 95% confidence interval of $\beta(t)$ for the blue star points; it is also transparent to see the difference by comparing the light blue and purple surfaces. Thus, the pump in the first 3750 seconds is in a normal operating condition different from the condition in the last 1043 seconds.

The second case is about the accident of the pump at 23:35:00 on March 25. The 5000 data points from 20:31:40 to 21:55:00 before the accident are enlarged in Fig. 3.35a. The 3D scatter plots of $\Delta_p(t)$, $r(t)$ and $q(t)$ with different visual angles are presented in Fig. 3.35b, c. Obviously, there are two clusters, namely, the blue star points in the first 3670 seconds and the red dot points in the last 1327 seconds. Between the two clusters, there are only 3 data points (black triangle points) in the middle. Analogously to the first case, the parameters in (3.72) are estimated as $\hat{\beta}(t) = [-0.7937, 2.7290, -1.1682, 0.4070]'$ with the 95% confidence interval as $[(-2.0732, 0.4858), (-1.0252, 6.4853), (-1.8579, -0.4785), (0.4069, 0.4071)]'$, by applying the OLS method to the blue star data points, and the corresponding estimate $\hat{\Delta}_p(t)$, $q(t)$ and $r(t)$ formulate the blue surface in Fig. 3.35b. If the red dot points are used, $\hat{\beta}(t) = [-4.7213, -5.0859, 1.7412, 0.2953]'$ and a red surface is obtained. Clearly, the two clusters represent two different operating conditions of the pump. The abnormality can be observed from Fig. 3.35a: both $\Delta_p(t)$ and $r(t)$ decrease after the first 3670 seconds, while $q(t)$ moves in an opposite direction with a short large increment. This phenomenon implies the presence of some abnormalities, because $\Delta_p(t)$, $r(t)$ and $q(t)$ are expected to change in the same direction when the pump is in normal conditions, according to process knowledge of the pump.

The measured fluid flow $q_m(t)$ (before normalization) was originally configured with a univariate alarm system with a high alarm threshold 560 t/h. For the normalized variable $q(t)$ in Fig. 3.33, an alarm will be arisen for $q(t) \geq 0.77$, i.e.,

$$q_a(t) = \begin{cases} 1, & \text{if } q(t) \geq 0.77, \\ 0, & \text{otherwise.} \end{cases} \tag{3.73}$$

Here $q_a(t)$ denotes the alarm variable associated with $q(t)$. Let us investigate the performance of the univariate alarm system for the above two cases. Figure 3.36a, b

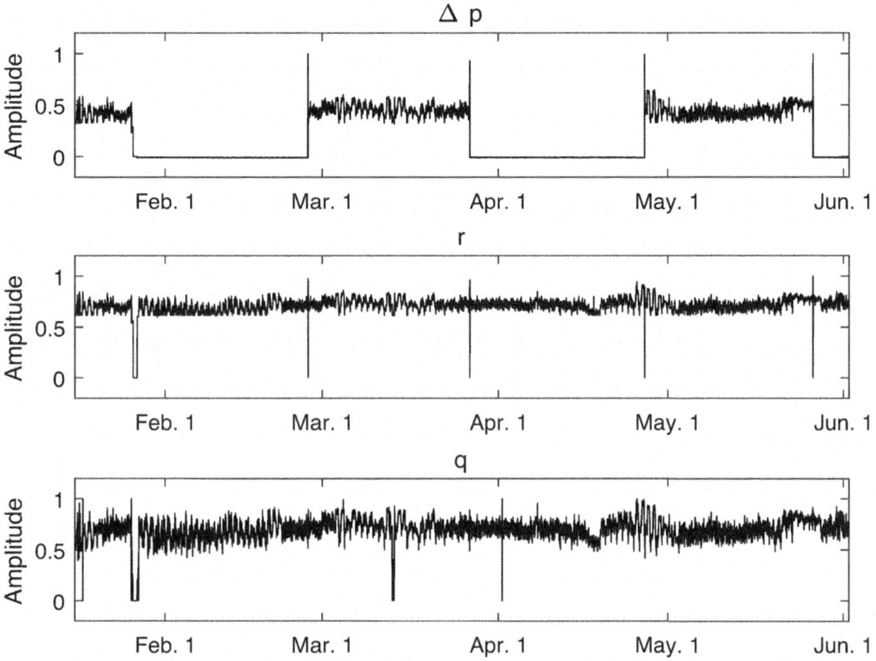

Fig. 3.33 The time sequence plots of $\Delta_p(t)$, $r(t)$ and $q(t)$, reprinted from Ref. [46], copyright 2018, with permission from IEEE

show the time sequence plots of $q(t)$ with the alarm threshold (red dash) for the two cases; Fig. 3.36c, d present the corresponding alarm variables $q_a(t)$ in (3.73). For the first case, the pump is in normal conditions for the entire time period; however, there are some false alarms in Fig. 3.36c. For the second case, the pump experiences abnormalities; however, no alarms occur in Fig. 3.36d, so that $q_a(t)$ has a poor performance in terms of having missing alarms. Therefore, the univariate alarm system leads to both false and missing alarms, which motivates us to design multivariate alarm systems for electrical pumps. Note that redesigning alarm thresholds or using alarm strategies such as alarm delay-timers and dead-bands [40, 41] cannot resolve the issue of having false and missing alarms for the univariate alarm system. First, increasing the alarm threshold 0.77 can remove the false alarms in Fig. 3.36c, but cannot avoid the missing alarms in Fig. 3.36d. By contrast, decreasing the alarm threshold 0.77 may alleviate missing alarms in Fig. 3.36d; however, doing so results in more false alarms in Fig. 3.36c, because the maximum value of $q(t)$ in Fig. 3.36a is larger than the counterpart in Fig. 3.36b. Second, using alarm delay timers may remove the false alarms in Fig. 3.36c, but has no positive effects on missing alarms in Fig. 3.36d. The alarm deadbands do not work, due to the same reason for redesigning alarm thresholds.

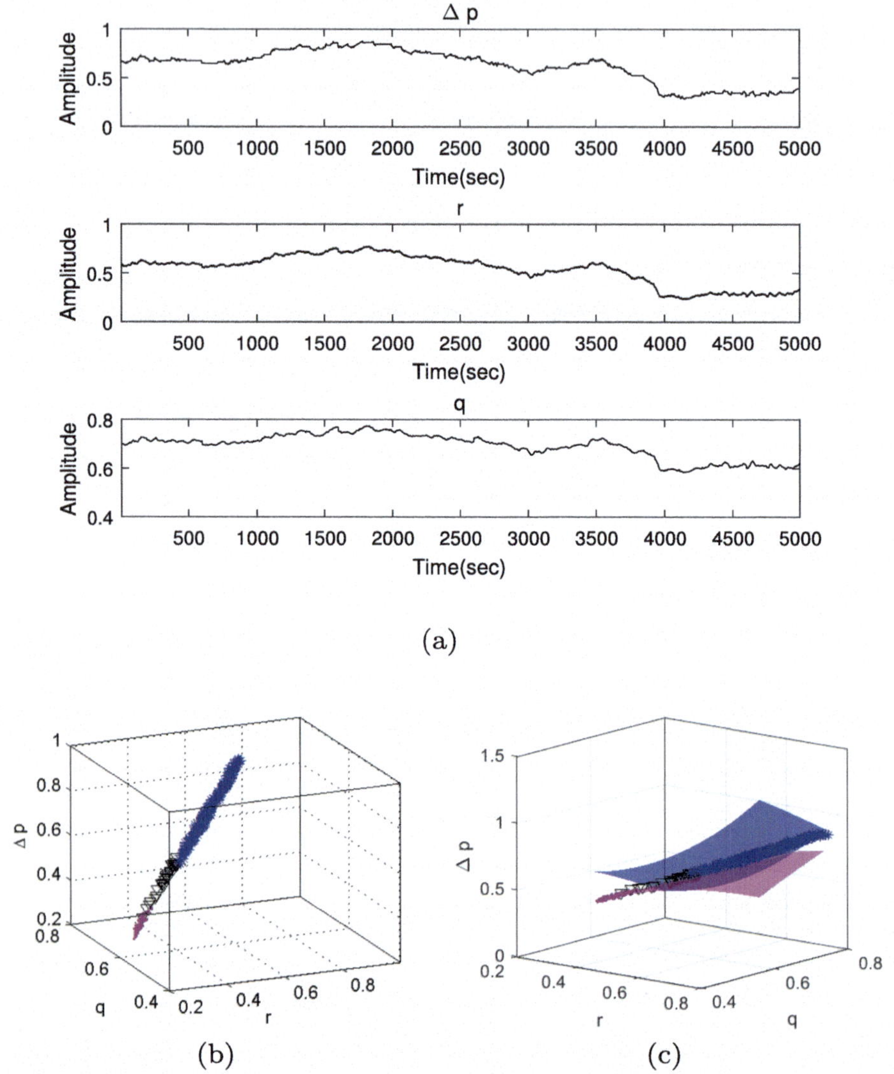

Fig. 3.34 **a** The time sequence plots of $\Delta_p(t)$, $r(t)$ and $q(t)$, and **b**, **c** the 3D scatter plots of $q(t)$, $r(t)$ and $\Delta_p(t)$ from different visual angles in the first case, reprinted from Ref. [46], copyright 2018, with permission from IEEE

The problem to be solved is to design a multivariate alarm system for time-varying processes described by a linear regression model based on (3.72),

$$y(t) = \boldsymbol{x}'(t)\boldsymbol{\beta}(t) + \varepsilon(t), \tag{3.74}$$

where

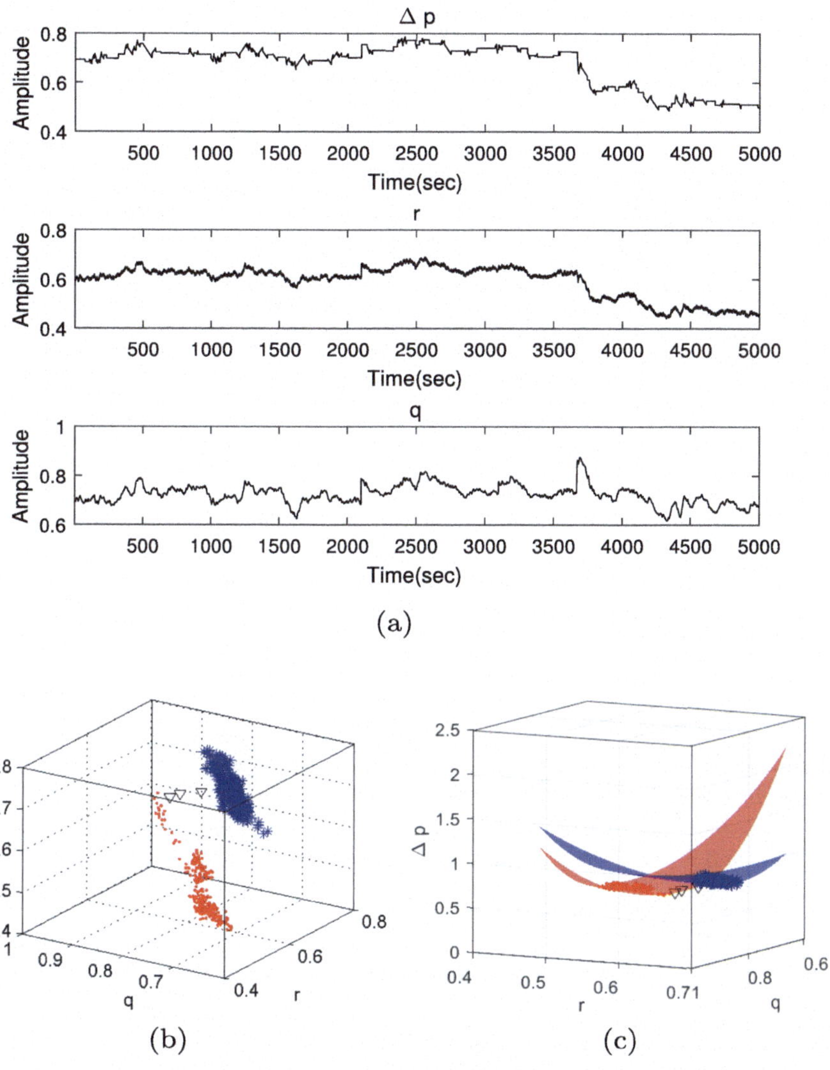

Fig. 3.35 **a** Time sequence plots of $\Delta_p(t)$, $r(t)$ and $q(t)$, **b**, **c** the 3D scatter plots of $q(t)$, $r(t)$ and $\Delta_p(t)$ from different visual angles in the second case, reprinted from Ref. [46], copyright 2018, with permission from IEEE

$$\boldsymbol{x}(t) := [1, x_1(t), x_2(t), \cdots, x_M(t)]',$$
$$\boldsymbol{\beta}(t) := [\beta_0(t), \beta_1(t), \beta_2(t), \cdots, \beta_M(t)]'.$$

Here M is a positive integer and $\varepsilon(t)$ is the modeling error. By comparing the above two cases, it is observed that the transition between adjacent operating conditions in the second case is much shorter than the counterpart in the first case. This observation

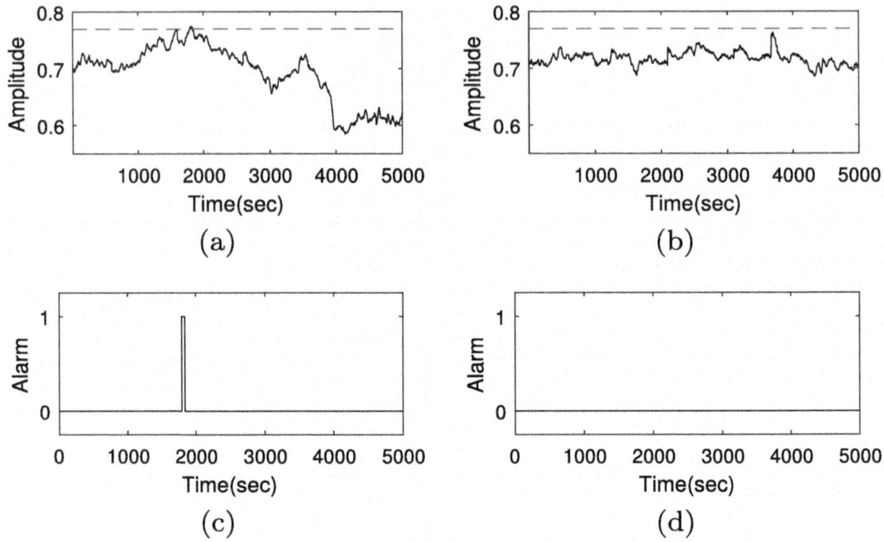

Fig. 3.36 The time sequence plots of $q(t)$ (solid) with a high alarm threshold (dash) and $q_a(t)$ in (3.73): **a**, **c** the first case, **b**, **d** the second case, reprinted from Ref. [46], copyright 2018, with permission from IEEE

leads to a feature to distinguish normal and abnormal conditions: if the process is in normal (abnormal) conditions, then the variation rates of $\beta(t)$ in (3.74) are small (large). If a model estimator is exploited to only track $\beta(t)$ having small variation rates, then absolute cumulative modeling errors would be large when variation rates are large, and alarms would be raised for such abnormalities.

3.3.2 Steps of the Proposed Method

This section proposes a method to design multivariate alarm systems for time-varying processes. First, the main idea of the proposed method is introduced. Second, a model estimator based on Bayesian filters is formulated. Finally, the steps of the proposed method are summarized.

The main idea goes as follows. Define $\varphi(t)$ as the absolute cumulative modeling error,

$$\varphi(t) = \left| \sum_{l=t-N}^{t} \varepsilon(l) \right|, \tag{3.75}$$

where $\varepsilon(l)$ is the modeling error in (3.74) and the positive integer N stands for the window length. A model estimator is designed on the basis of Bayesian filters to synthesize the prior estimate and the information in the current data point. Based on the feature discussed in Sect. 3.3.1, a filter should be designed to track the time-varying parameters when the variation rates of these parameters are small and not to do so when the rates are large. In the latter case, the transitions of time-varying parameters cannot be tracked so that $\varphi(t)$ in (3.75) is large. As a result, there will be a distinct difference between the value of $\varphi(t)$ under normal conditions and the counterpart under abnormal conditions. Such a distinct difference is exploited to raise an alarm for a large value of $\varphi(t)$, i.e.,

$$\varphi_a(t) = \begin{cases} 1, & \text{if } \varphi(t) > \varphi_{th}, \\ 0, & \text{otherwise}, \end{cases} \tag{3.76}$$

where φ_{th} is the alarm threshold to be determined later.

Second, the Bayesian model estimator is formulated. Bayesian filters, including Kalman filters and particle filters as special cases, are perhaps the most effective techniques for parameter estimation of time-varying processes [10]. Thus, a model estimator based on Bayesian filters is formulated to track the variations of time-varying model parameters in normal conditions, and not to do so in abnormal conditions so that absolute cumulative modeling errors are large enough to raise alarms.

A sliding data window is defined, with N as the window length and j as the serial number of the window. The j-th window includes the data points of $\boldsymbol{x}(t)$ and $y(t)$ from $[N(j-1)+1]$-th sampling instant to Nj-th sampling instant. Use y_j, $\boldsymbol{x}_j$, $\boldsymbol{\beta}_j$ and ε_j to represent y, $\boldsymbol{x}$, $\boldsymbol{\beta}$ and ε in the j-th window, respectively. Thus, the output variable can be estimated as $\hat{y}_j = \boldsymbol{x}'_j\boldsymbol{\beta}_j$, and the modeling error is

$$\varepsilon_j = y_j - \hat{y}_j. \tag{3.77}$$

According to Bayesian theorem, $\boldsymbol{\beta}_j$ is estimated by maximizing the posterior distribution function as

$$\boldsymbol{p}(\boldsymbol{\beta}_j|\boldsymbol{Y}_j) = \frac{\boldsymbol{p}(\boldsymbol{Y}_j|\boldsymbol{\beta}_j)\cdot \boldsymbol{p}(\boldsymbol{\beta}_j)}{\boldsymbol{p}(\boldsymbol{Y}_j)} \propto \boldsymbol{p}(\boldsymbol{Y}_j|\boldsymbol{\beta}_j)\cdot \boldsymbol{p}(\boldsymbol{\beta}_j),$$

where

$$\boldsymbol{Y}_j = \left[y(N(j-1)+1)\ y(N(j-1)+2)\ \cdots\ y(Nj)\right]'.$$

Here $\boldsymbol{p}$ stands for the probability density function. If ε_j is assumed to be Gaussian distributed with zero mean and variance σ_ε^2, then an analytic expression of β_j can be obtained as follows. This assumption is often ensured by the central limit theorem, and is shown later in Sect. 3.3.4 to be valid for the condensate-water electrical pump. Under this assumption, it is ready to see that $\boldsymbol{\beta}_j$ and $\boldsymbol{\beta}_{j-1}$ are Gaussian distributed, i.e., $\boldsymbol{\beta}_j \sim \mathcal{N}(\boldsymbol{\mu}_{\beta_j}$ and $\boldsymbol{\Sigma}_{\beta_j})$, $\boldsymbol{\beta}_{j-1} \sim \mathcal{N}(\boldsymbol{\mu}_{\beta_{j-1}}, \boldsymbol{\Sigma}_{\beta_{j-1}})$, where $\boldsymbol{\mu}_{\beta_j}$ ($\boldsymbol{\mu}_{\beta_{j-1}}$) and $\boldsymbol{\Sigma}_{\beta_j}$ ($\boldsymbol{\Sigma}_{\beta_{j-1}}$) are respectively the mean vector and the covariance matrix of $\boldsymbol{\beta}_j$ ($\boldsymbol{\beta}_{j-1}$), and σ_ε^2 is the variance of ε_j. Under normal conditions, the variation rate of $\boldsymbol{\beta}(t)$ is small, so that it is reasonable to take $\boldsymbol{\beta}_{j-1}$ as the prior estimate of $\boldsymbol{\beta}_j$. The posterior distribution of $\boldsymbol{\beta}_j$ is

$$\begin{aligned}
\boldsymbol{p}(\boldsymbol{\beta}_j|\boldsymbol{Y}_j) &\propto \exp\left[-\frac{1}{2}\left(\boldsymbol{Y}_j - \boldsymbol{X}_j\boldsymbol{\beta}_j\right)' \sigma_\varepsilon^{-2} \left(\boldsymbol{Y}_j - \boldsymbol{X}_j\boldsymbol{\beta}_j\right)\right] \\
&\cdot \exp\left[-\frac{1}{2}\left(\boldsymbol{\beta}_j - \boldsymbol{\mu}_{\beta_{j-1}}\right)' \boldsymbol{\Sigma}_{\beta_{j-1}}^{-1} \left(\boldsymbol{\beta}_j - \boldsymbol{\mu}_{\beta_{j-1}}\right)\right] \\
&\propto -\boldsymbol{\beta}_j' \left(\sigma_\varepsilon^{-2}\boldsymbol{X}_j'\boldsymbol{X}_j + \boldsymbol{\Sigma}_{\beta_{j-1}}^{-1}\right)\boldsymbol{\beta}_j - 2\left(\sigma_\varepsilon^{-1}\boldsymbol{Y}_j'\boldsymbol{X}_j + \boldsymbol{\mu}_{\beta_{j-1}}\boldsymbol{\Sigma}_{\beta_{j-1}}^{-1}\right)\boldsymbol{\beta}_j,
\end{aligned} \tag{3.78}$$

where

$$\boldsymbol{X}_j = \begin{bmatrix} 1 & \cdots & 1 \\ x_1(Nj - N + 1) & \cdots & x_1(Nj) \\ \vdots & \ddots & \vdots \\ x_M(Nj - N + 1) & \cdots & x_M(Nj) \end{bmatrix}'.$$

For the model in (3.72), the components in X_j are mutually correlated. In order to reduce the negative effects of collinearity on parameter estimation, the idea of ridge regression is borrowed to impose a penalty term $k_r\boldsymbol{\beta}_j'\boldsymbol{\beta}_j\sigma_\varepsilon^{-2}$ to the posterior distribution, where $k_r \geq 0$ is a weighting parameter for the penalty term. In addition, a forgetting factor $0 < k_p < 1$ is introduced to control the credibility of the prior information. Thus, the right hand side of (3.78) becomes

$$\begin{aligned}
&\boldsymbol{\beta}_j'\left(\sigma_\varepsilon^{-2}\boldsymbol{X}_j'\boldsymbol{X}_j + k_p\boldsymbol{\Sigma}_{\beta_{j-1}}^{-1}\right)\boldsymbol{\beta}_j + \left(\sigma_\varepsilon^{-2}\boldsymbol{Y}_j'\boldsymbol{X}_j + k_p\boldsymbol{\mu}_{\beta_{j-1}}\boldsymbol{\Sigma}_{\beta_{j-1}}^{-1}\right)\boldsymbol{\beta}_j \\
&+ k_r\boldsymbol{\beta}_j'\boldsymbol{\beta}_j\sigma_\varepsilon^{-2} = \boldsymbol{\beta}_j'\left(\sigma_\varepsilon^{-2}\boldsymbol{X}_j'\boldsymbol{X}_j + k_r\boldsymbol{I}\sigma_\varepsilon^{-2} + k_p\boldsymbol{\Sigma}_{\beta_{j-1}}^{-1}\right)\boldsymbol{\beta}_j \\
&+ 2\left(\sigma_\varepsilon^{-2}\boldsymbol{Y}_j'\boldsymbol{X}_j + k_p\boldsymbol{\mu}_{\beta_{j-1}}\boldsymbol{\Sigma}_{\beta_{j-1}}^{-1}\right)\boldsymbol{\beta}_j,
\end{aligned}$$

where $\boldsymbol{I}$ is the identity matrix. Define $\boldsymbol{Z}_1 := \sigma_\varepsilon^{-2}\boldsymbol{X}_j'\boldsymbol{X}_j + k_r\boldsymbol{I}\sigma_\varepsilon^{-2} + k_p\boldsymbol{\Sigma}_{\beta_{j-1}}^{-1}$ and $\boldsymbol{Z}_2 := \sigma_\varepsilon^{-2}\boldsymbol{Y}_j'\boldsymbol{X}_j + k_p\boldsymbol{\mu}_{\beta_{j-1}}\boldsymbol{\Sigma}_{\beta_{j-1}}^{-1}$. It is ready to reach the penalized posterior distribution of $\boldsymbol{\beta}_j$,

$$\boldsymbol{p}(\boldsymbol{\beta}_j|\boldsymbol{Y}_j) \propto -\left(\boldsymbol{\beta}_j - \boldsymbol{Z}_1^{-1}\boldsymbol{Z}_2\right)'\boldsymbol{Z}_1\left(\boldsymbol{\beta}_j - \boldsymbol{Z}_1^{-1}\boldsymbol{Z}_2\right). \tag{3.79}$$

By maximizing $\boldsymbol{p}(\boldsymbol{\beta}_j|\boldsymbol{Y}_j)$ in (3.79), $\boldsymbol{\beta}_j$ is obtained as the mean vector, i.e.,

$$\begin{aligned}
\boldsymbol{\beta}_j &:= \boldsymbol{\mu}_{\beta_j} = \boldsymbol{Z}_1^{-1}\boldsymbol{Z}_2 \\
&= \left[\sigma_\varepsilon^{-2}(\boldsymbol{X}_j'\boldsymbol{X}_j + k_r\boldsymbol{I}) + k_p\boldsymbol{\Sigma}_{\beta_{j-1}}^{-1}\right]^{-1} \cdot \left(\sigma_\varepsilon^{-2}\boldsymbol{X}_j'\boldsymbol{Y}_j + k_p\boldsymbol{\mu}_{\beta_{j-1}}\boldsymbol{\Sigma}_{j-1}^{-1}\right),
\end{aligned} \tag{3.80}$$

with the covariance matrix

$$\boldsymbol{\Sigma}_j = \boldsymbol{Z}_1^{-1} = \left[\sigma_\varepsilon^{-2}(\boldsymbol{X}_j'\boldsymbol{X}_j + k_r\boldsymbol{I}) + k_p\boldsymbol{\Sigma}_{\beta_{j-1}}^{-1}\right]^{-1}. \tag{3.81}$$

Here $\sigma_{\varepsilon}^{-2}(\boldsymbol{X}'_j\boldsymbol{X}_j + k_r\boldsymbol{I})$ is the reliability measures of the information in the j-th window and $k_p\boldsymbol{\Sigma}_{\boldsymbol{\beta}_{j-1}}^{-1}$ is the prior information $\boldsymbol{\beta}_{j-1}$. Owing to the forgetting factor $k_p \in (0, 1)$, the prior distribution of $\boldsymbol{\beta}_{j-1}$ at the first step $j = 1$ has little importance, since its effects on $\boldsymbol{\beta}_j$ decay quickly as j increases. Hence, the mean vector and covariance matrix of $\boldsymbol{\beta}_j$ at $j = 1$ are obtained from (3.80) and (3.81) by ignoring the prior information of $\boldsymbol{\beta}_{j-1}$ as

$$\begin{cases} \boldsymbol{\mu}_{\beta_j} = \left(\boldsymbol{X}'_j\boldsymbol{X}_j + k_r\boldsymbol{I}\right)^{-1}\boldsymbol{X}'_j\boldsymbol{Y}_j \\ \boldsymbol{\Sigma}_{\beta_j} = \sigma_{\varepsilon}^2\left(\boldsymbol{X}'_j\boldsymbol{X}_j + k_r\boldsymbol{I}\right)^{-1} \end{cases}, \quad \text{for } j = 1.$$

Finally, the proposed method is composed of the following steps:

Step 1: Determine whether an alarm should be arisen for a new set of data points. At the current sampling instant, the data points $x_1(t), x_2(t), \cdots, x_M(t)$ and $y(t)$ for the model (3.74) are collected. The absolute cumulative modeling error $\varphi(t)$ in (3.75) is calculated based on the model parameter $\boldsymbol{\beta}_{j-1}$ in the previous data window. Then, the alarm variable $\varphi_a(t)$ in (3.76) is generated based on the alarm threshold φ_{th} given later in (3.88).

Step 2: Perform a hypothesis test to decide whether the model should be updated. At the Nj-th sampling instant, a hypothesis test is introduced to judge whether the process in the j-th window is in normal or abnormal conditions, and to decide whether the model parameters should be updated. First, the number of alarms in the j-th window is counted as A_j. If the j-th window is in normal conditions, then the occurred alarms are often caused by random noises so that they are independent and identically distributed. Thus, A_j follows the binomial distribution, i.e.,

$$A_j \sim \mathcal{B}(N, \text{FAR}). \tag{3.82}$$

Here FAR is the false alarm rate of the designed alarm system. Next, with a significance level α, e.g., $\alpha = 0.05$, a hypothesis threshold A_α is the one to satisfy the equality,

$$\Pr(A_j \leq A_\alpha) = 1 - \alpha. \tag{3.83}$$

The model parameter vector is updated based on the hypothesis test as

$$\begin{cases} \beta_j = \beta_{j-1}, & \text{if } A_j > A_\alpha, \\ \beta_j \text{ in (3.80)}, & \text{if } A_j \leq A_\alpha. \end{cases} \tag{3.84}$$

If $A_j > A_\alpha$, then the j-th window is against the null hypothesis (3.82) and thus is regarded as being abnormal. Hence, the model parameter should not be updated and $\boldsymbol{\beta}_{j-1}$ is preserved. If $A_j \leq A_\alpha$, then the j-th window is in normal conditions, and (3.80) is used to update the model parameters. If the process in the j-th window is judged to be in normal conditions, then occurring alarms in the j-th window are the false alarms; otherwise, alarm clearances for $\varphi_a(t)$ in (3.76) taking the value of 0 in the j-th window are the missing alarms. By counting the frequencies of false and missing alarms for past data windows, the false alarm rate and the missing alarm rate of the designed alarm system can be obtained.

Step 3: Repeat Steps 1 and 2 to update the alarm variable $\varphi_a(t)$ *in* (3.76) *at each sampling instant, and the model parameter vector* β_j *in* (3.84) *for every N samples.* The two online steps involve simple algebraic computations, which can be completed very quickly. Hence, the configuration effort of the proposed method is acceptable for online implementation in practice.

3.3.3 Determination of Tuning Parameters

The subsection determines three tuning parameters in (3.80), namely, the window length N, the forgetting factor k_p and the penalty factor k_r.

First, a small value of N is obviously helpful in tracking the variations of time-varying model parameters. However, the number of rows in $\boldsymbol{X}_j$ is usually required to be larger than the number of columns in $\boldsymbol{X}_j$ to yield a non-singular matrix $\boldsymbol{X}'_j\boldsymbol{X}_j$. Hence, the smallest possible value of N is used for the model (3.74),

$$N = M + 2. \tag{3.85}$$

Second, the forgetting factor k_p is a critical tuning parameter. Two historical data sets, referred to as the normal and abnormal data sets, are required to determine a proper value of k_p. The normal data set does not contain abnormal conditions, while the abnormal data set does. Denote $\varphi_1(t)$ and $\varphi_2(t)$ as the values of $\varphi(t)$ in (3.75) for the normal and abnormal data sets, respectively. A fitness between $y(t)$ and its estimate $\hat{y}(t)$ is calculated to measure the model quality [24],

$$F := \max\left(0, 1 - \frac{\left\|y(t) - \hat{y}(t)\right\|_2}{\left\|y(t) - \overline{y(t)}\right\|_2}\right). \tag{3.86}$$

Here $\hat{y}(t)$ in the j-th data window is the same as $\hat{y}_j$ in (3.77), $\|\cdot\|_2$ denotes the 2-norm of the operand, and $\overline{y(t)}$ is the sample mean of $y(t)$. The effects of k_p are investigated based on the first and second cases from Sect. 3.3.1 as follows. Figure 3.37a, b, c show $\varphi_1(t)$ obtained by using the normal data set in the first case, for three different forgetting factors $k_p = 0$, $k_p = 0.3$ and $k_p = 0.9$, respectively. Table 3.2 shows the fitness values under different values of k_p for the first case. As expected, the fitness decreases with the increment of k_p, because a heavier weight of the prior information reduces the traceability of the model estimator. Figure 3.37d, e, f present $\varphi_2(t)$ based on the abnormal data set in the second case for $k_p = 0$, $k_p = 0.3$ and $k_p = 0.9$, respectively. A heavier weight of the prior information leads to a larger modeling error at the time when the abnormal condition occurs. Hence, a proper value of k_p is

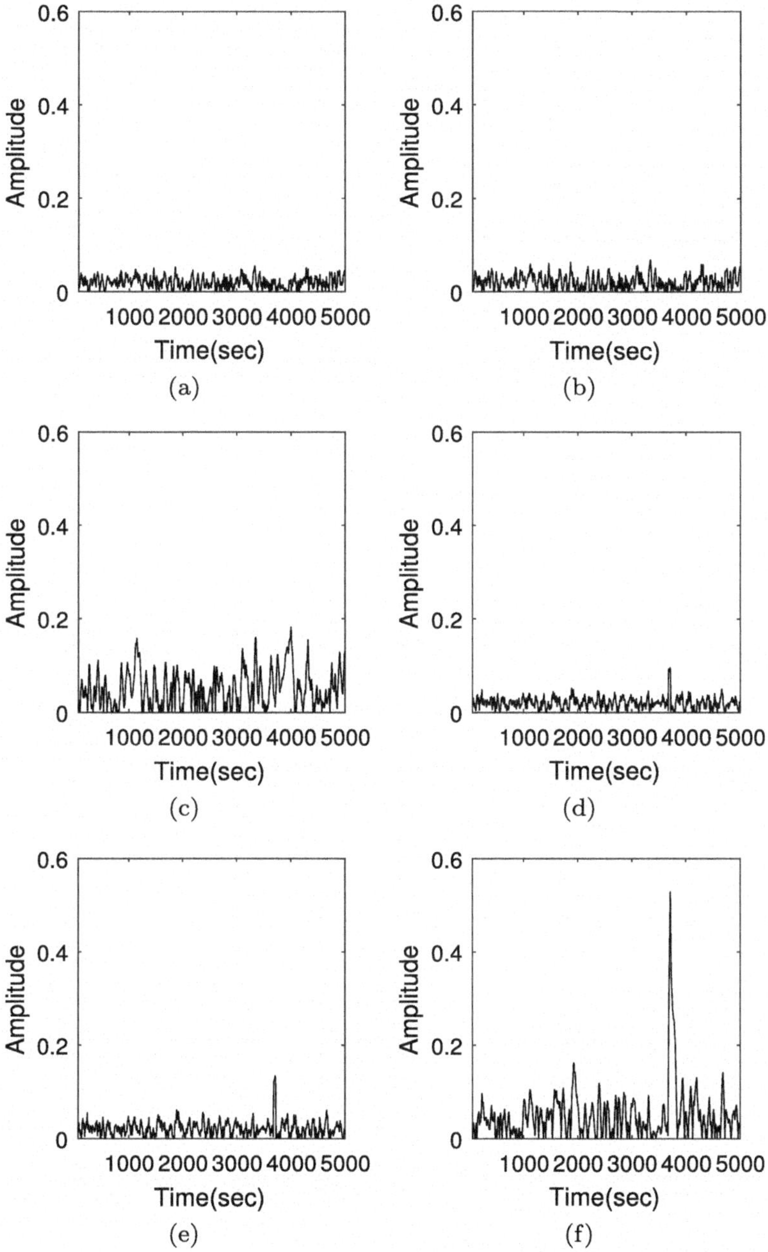

Fig. 3.37 The influence of k_p on $\varphi(t)$: **a** $k_p = 0$, **b** $k_p = 0.3$ and **c** $k_p = 0.9$ for the first case, and **d** $k_p = 0$, **e** $k_p = 0.3$ and **f** $k_p = 0.9$ for the second case, reprinted from Ref. [46], copyright 2018, with permission from IEEE

Table 3.2 The finesses values under different values of k_p for the first case, reprinted from Ref. [46], copyright 2018, with permission from IEEE

	$k_p = 0$	$k_p = 0.3$	$k_p = 0.9$
F	96.29%	96.17%	94.42%

expected to achieve a balance between the ability to detect abnormalities and that to track parameter variations in normal conditions. Here k_p is determined as the one to maximize the difference between $\varphi_1(t)$ and $\varphi_2(t)$ subject to the requirement on the fitness,

$$k_p^{(\text{opt})} = \arg\max_{k_p} \quad (\max(\varphi_2(t)) - \max(\varphi_1(t))) \\ \text{s.t.} \quad F \geq F_0, \tag{3.87}$$

where F_0 is a user-defined threshold as the lower limit of the fitness, e.g., $F_0 = 95\%$. The alarm threshold φ_{th} in (3.76) is set as

$$\varphi_{th} = \frac{\max(\varphi_1(t)) + \max(\varphi_2(t))}{2}, \tag{3.88}$$

in order to be consistent with the choice of k_p in (3.87). Recall that $\varphi_1(t)$ and $\varphi_2(t)$ are the values of $\varphi(t)$ in (3.75) for the normal and abnormal data sets, respectively.

Third, the value of k_r involves a tradeoff between the modeling error ε_j in (3.77) and the variability of $\boldsymbol{\beta}_j$ in (3.80) due to collinearity. A variance inflation factor (VIF), which measures the variability of an estimated regression coefficient, is used to choose a suitable value of k_r [16]. In the j-th window of the normal historical data set, VIF_j can be calculated as

$$\text{VIF}_j = \left(\boldsymbol{X}'_j\boldsymbol{X}_j + k_r\boldsymbol{I}\right)^{-1} \boldsymbol{X}'_j\boldsymbol{X}_j \left(\boldsymbol{X}'_j\boldsymbol{X}_j + k_r\boldsymbol{I}\right)^{-1}.$$

The average of VIF_j's for J windows in total is defined as VIF, i.e.,

$$\text{VIF} = \frac{1}{J}\sum_{j=1}^{J}\text{VIF}_j.$$

The optimal value of k_r is chosen as the maximum one to make each element in VIF less than 0.1, i.e.,

$$k_r^{(\text{opt})} = \max\left(k_r|\text{VIF} \prec 0.1\right). \tag{3.89}$$

Finally, by using the data sets in the first and second cases in Sect. 3.3.1 as the normal and abnormal historical data sets, the tuning parameters k_p in (3.87) and k_r in (3.89), the alarm threshold φ_{th} in (3.88) and the hypothesis threshold A_α in (3.83) are designed.

Table 3.3 The elements of VIF under different values of k_r for the first case, reprinted from Ref. [46], copyright 2018, with permission from IEEE

VIF	
$k_r = 0.006$	$[0.1184, 0.0487, 0.1010, 0.1371]'$
$k_r = 0.009$	$[0.0619, 0.0305, 0.0534, 0.0988]'$
$k_r = 0.012$	$[0.0422, 0.0242, 0.0368, 0.0853]'$
$k_r = 0.015$	$[0.0330, 0.0212, 0.0290, 0.0790]'$

Table 3.4 The values of $(\max(\varphi_2(t)) - \max(\varphi_1(t)))$ and the fitness values under different values of k_p for the first case, reprinted from Ref. [46], copyright 2018, with permission from IEEE

$\max(\varphi_2(t)) - \max(\varphi_1(t))$		F
$k_p = 0.74$	$0.3216 - 0.1225 = 0.1902$	95.52%
$k_p = 0.79$	$0.3612 - 0.1331 = 0.2281$	95.32%
$k_p = 0.84$	$0.4245 - 0.1503 = 0.2742$	95.04%
$k_p = 0.89$	$0.5086 - 0.1758 = 0.3328$	94.56%

Table 3.3 presents the elements of VIF under different values of k_r for the first case; clearly, $k_r^{(\mathrm{opt})}$ in (3.89) is designed as 0.009 to make each element of VIF no larger than 0.1. Next, Table 3.4 lists the values of $(\max(\varphi_2(t)) - \max(\varphi_1(t)))$ and the values of the fitness F in (3.86) under different values of k_p for the first case. The optimal choice $k_p^{(\mathrm{opt})}$ in (3.87) is obtained as 0.84 to maximize the difference between $\varphi_1(t)$ and $\varphi_2(t))$ and to meet the fitness requirement $F \geq 95\%$. Thus, the alarm threshold φ_{th} in (3.88) is chosen as $\varphi_{th} = \frac{0.4245+0.1503}{2} = 0.2874$. The sample variance of $\varphi_1(t)$ is estimated as $\hat{\sigma}^2_{\varphi_1} = 0.1503$ so that A_α in (3.83) is determined as $A_\alpha = 1$ for $\alpha = 0.05$.

The Kolmogorov-Smirnov test is exploited to verify whether the model error $\varepsilon(t)$ follows the Gaussian distribution [28]. Figure 3.38 compares the standard Gaussian cumulative density function (CDF) with the estimated CDF of $(\varepsilon(t) - \mu_\varepsilon)/\sigma_\varepsilon$ for the first case in Sect. 3.3.1, where μ_ε and σ_ε are respectively the sample mean and standard deviation of $\varepsilon(t)$. The two CDFs are almost overlapped so that the Kolmogorov-Smirnov test is passed.

3.3.4 *Industrial Examples*

This section provides industrial examples for the condensate-water electrical pump to illustrate the effectiveness of the proposed method, and compares with existing methods for monitoring the electrical pump.

First, by following the steps in Sect. 3.3.2 and using the tuning parameters in Sect. 3.3.3, the multivariate alarm system is designed. It is applied to two new data

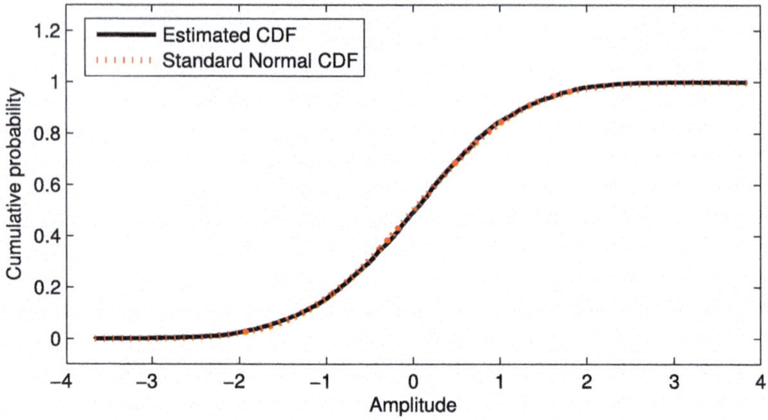

Fig. 3.38 The estimated CDF (black solid) of $(\varepsilon(t) - \mu_\varepsilon)/\sigma_\varepsilon$ and the standard Gaussian CDF (red dash), reprinted from Ref. [46], copyright 2018, with permission from IEEE

sets, referred to as the third and fourth cases, which are different from the first and second cases in Sect. 3.3.1.

In the third case, the pump is in normal conditions for the entire time period: Fig. 3.39 presents 5000 data points of the three variables $\Delta_p(t)$, $q(t)$ and $r(t)$ from 9:32:10 to 10:23:20 on May 17. The four parameters in $\boldsymbol{\beta}_j(t)$ are presented in Fig. 3.40; all the parameters are clearly not constant, saying that the pump indeed is a time-varying process. Figure 3.41 gives the absolute cumulative modeling error $\varphi(t)$ with the designed alarm threshold φ_{th}. The model estimator is able to follow the variations of the time-varying parameters so that $\varphi(t) < \varphi_{th}$, $\forall t$ and no alarms occur.

In the fourth case, the pump experiences abnormal conditions: Fig. 3.42 shows 5000 data points of the three variables $\Delta_p(t)$, $q(t)$ and $r(t)$ from 11:29:40 to 12:53:00 on May 25. Some abnormalities clearly occurred around the 3200-th data point, and a few minutes later, the pump was shut down in emergency at 13:00:00. The designed alarm system is applied to the 5000 data points in Fig. 3.42. Figures 3.43 and 3.44 present the model parameters in $\boldsymbol{\beta}_j(t)$ and the absolute cumulative modeling error $\varphi(t)$, respectively. As shown in Fig. 3.44b, an alarm occurs at the 3218-th sampling instant where $\varphi(t) > \varphi_{th}$ and the model parameters in $\boldsymbol{\beta}_j(t)$ have abrupt changes. Thus, the alarm variable $\varphi_a(t)$ stays at the alarm state in the rest sampling instants, and the model parameters are no longer updated afterwards.

Comparing the model parameters in Fig. 3.40 with those in Fig. 3.43 reveals that the transition of $\boldsymbol{\beta}_j(t)$ at $t = 3218$ in Fig. 3.43 are much quicker than the counterparts in Fig. 3.40. The designed multivariate alarm system is able to follow the variations of model parameters for the third case in Fig. 3.40, and is not able to follow the abrupt change of model parameters for the fourth case in Fig. 3.43. Thus, the designed multivariate alarm system achieves a satisfactory performance, i.e., no alarms occur in Fig. 3.41 for the third case where abnormalities are absent, while an alarm occurs

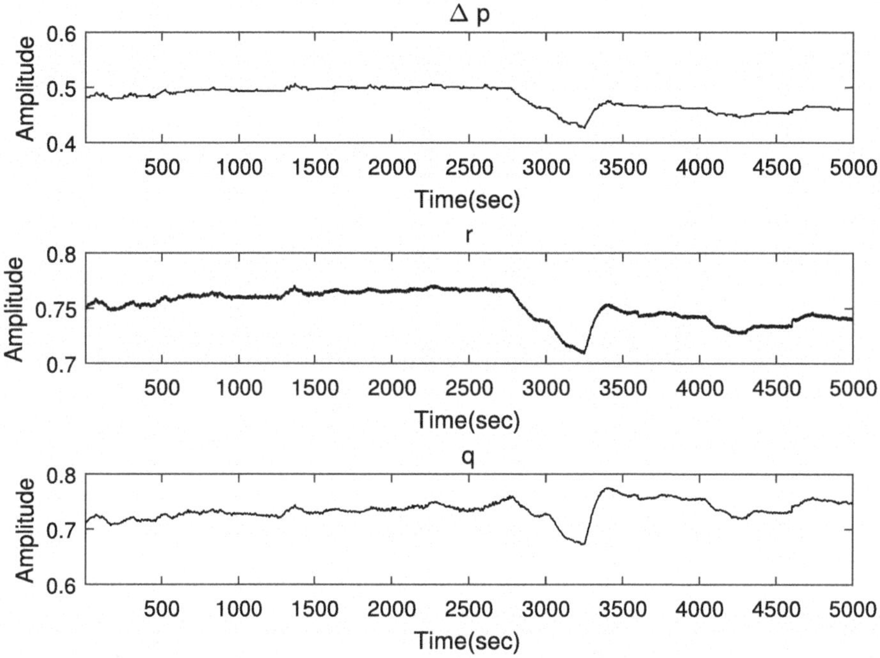

Fig. 3.39 The time sequence plots of $\Delta_p(t)$, $r(t)$ and $q(t)$ for the third case, reprinted from Ref. [46], copyright 2018, with permission from IEEE

in Fig. 3.44b and the associated abnormal condition in the fourth case is successfully detected.

Second, the performance of the univariate alarm system in (3.73) is investigated. For the third and fourth cases, the time sequence plots of $q(t)$ are respectively given in Fig. 3.45a, b, together with the alarm threshold 0.77 (red dash). The corresponding alarm variables $q_a(t)$ in (3.73) for the third and four cases are presented in Fig. 3.45c, d, respectively. Although the pump is in normal conditions for the entire time period in the third case, some false alarms occur in $q_a(t)$ as shown in Fig. 3.45c. For the fourth case, $q_a(t)$ in Fig. 3.45d is erroneously in the alarm state, when the pump is in normal conditions before the presence of abnormalities; when the pump runs into abnormal conditions, missing alarms are clearly presented in $q_a(t)$. Therefore, the univariate alarm system has a poor performance in terms of having too many false and missing alarms.

Finally, the technique of principle component analysis (PCA) has been widely used in multivariate process monitoring. The PCA-based method is used here for monitoring the operating status of electrical pumps. A PCA model is obtained by taking $[\Delta_p(t), r(t), q(t)]$ in the first case in Sect. 3.3.1 as the normal training data set. Define the 99% confidence intervals of T-squared and SPE statistics associated with the PCA model as two alarm thresholds (see, e.g., [45] for the definitions of

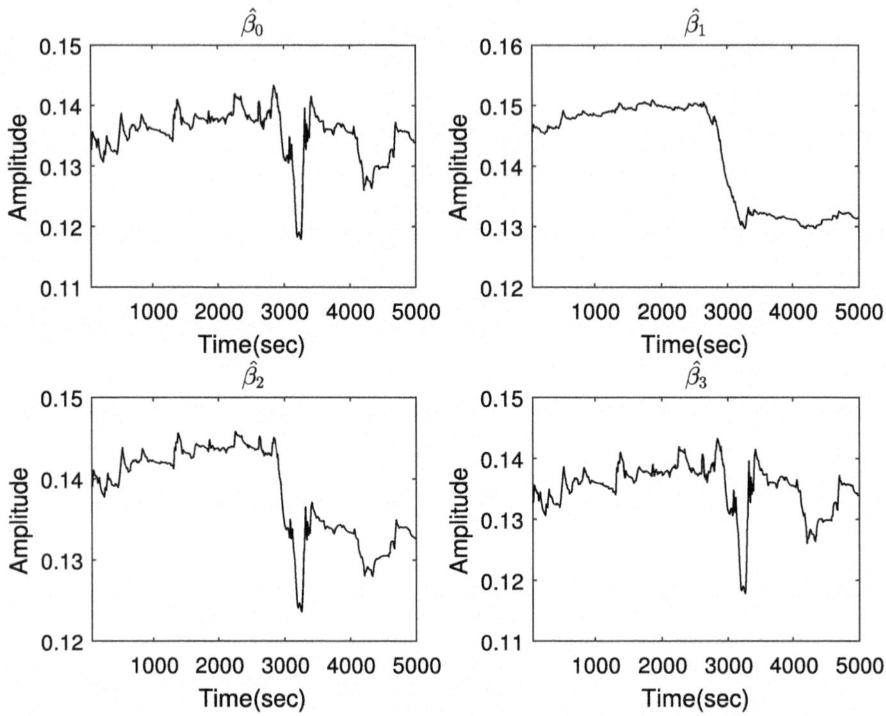

Fig. 3.40 The values of $\boldsymbol{\beta}_j(t)$ for the third case, reprinted from Ref. [46], copyright 2018, with permission from IEEE

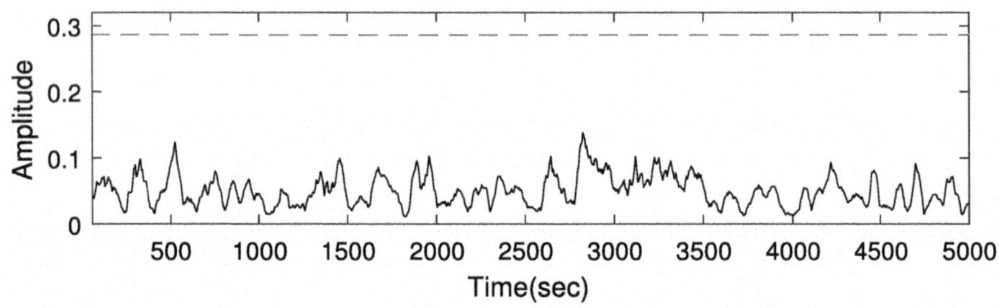

Fig. 3.41 The values of $\varphi(t)$ (solid) and the alarm threshold φ_{th} (dash) for the third case, reprinted from Ref. [46], copyright 2018, with permission from IEEE

T-squared and SPE statistics). Thus, an alarm will occur if either T-squared statistic or the SPE statistic is larger than the corresponding 99% confidence interval.

The PCA model is applied to the data sets in the third and fourth cases. Figure. 3.46a, b respectively present T-squared and SPE statistics, with the corresponding 99% confidence intervals (dash) for the third case. Figure 3.46c, d show

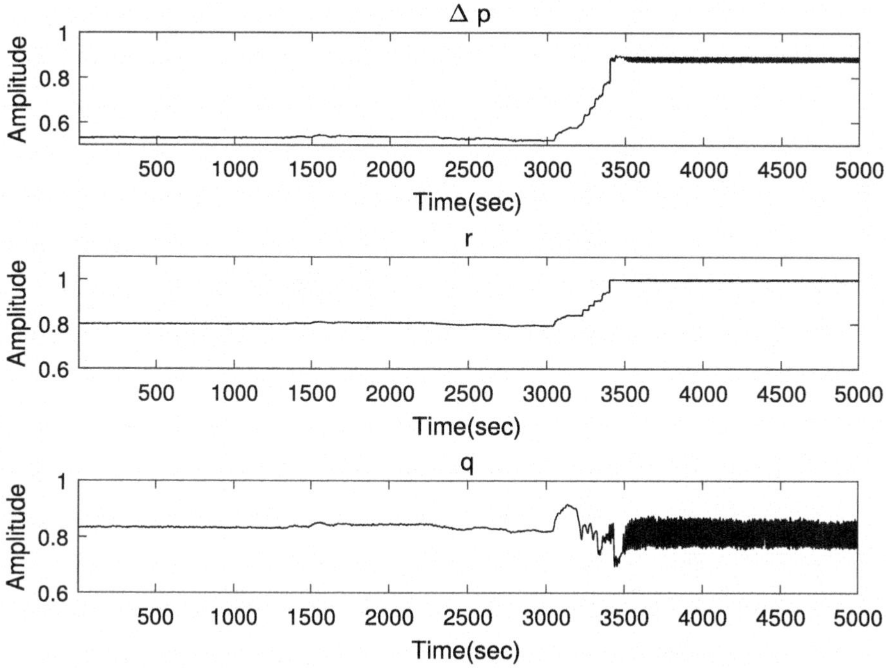

Fig. 3.42 The time sequence plots of $\Delta_p(t)$, $r(t)$ and $q(t)$ for the fourth case, reprinted from Ref. [46], copyright 2018, with permission from IEEE

the corresponding alarm sequences based on T-squared and SPE statistics, respectively. The two alarm sequences are erroneously in the alarm state for most of the time instants, whereas the pump is in normal conditions throughout the entire time period. Thus, false alarms persist for a long time period. Figure 3.47 is the counterpart of Fig. 3.46 for the fourth case. False alarms occur in the alarm sequences in Fig. 3.47c, d, because the alarm sequences are erroneously in the alarm state for the first 3200 seconds when the pump is indeed in normal conditions. Missing alarms appear in the alarm sequences in Fig. 3.47c, d, for the remaining time period after some abnormalities occur in the pump. In Fig. 3.46a, b, both T-squared statistic and SPE statistic are quite larger than their corresponding 99% confidence intervals. Thus, the PCA model from the first case cannot capture the relationships among the three process variables in the third case, even though the pump is in normal conditions for both cases. This is essentially due to the fact that the pump is a time-varying process.

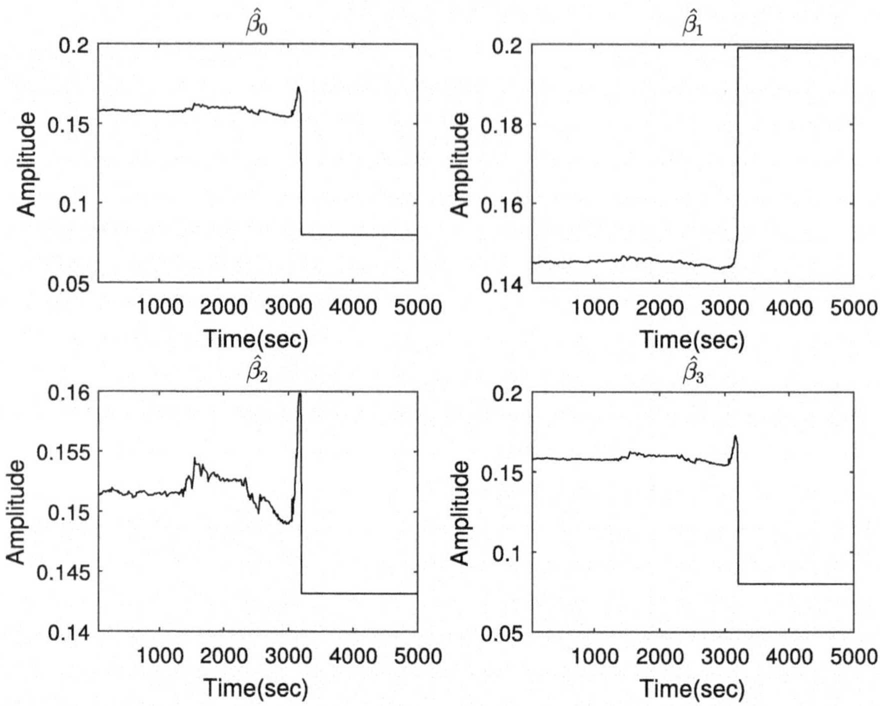

Fig. 3.43 The values of $\beta_j(t)$ for the fourth case, reprinted from Ref. [46], copyright 2018, with permission from IEEE

3.4 Multivariate Alarm Systems for Condensers

This section is about an optimal multivariate alarm system for safe operation of condensers in industrial power plants [44]. A condenser physical model is developed from physical mechanisms, and model uncertainties are obtained from multiple sets of model parameters being estimated from a large amount of condenser historical data. Operating zones are formulated as variation ranges of four related variables of condensers under normal operating conditions. A mathematical model is established to describe a final choice of condenser operating zones by achieving an optimal trade-off between the false alarm rate (FAR) and missed alarm rate (MAR) caused by condenser model uncertainties. Theoretical results for the upper bounds of FAR and MAR are developed based on the Bayesian estimation rule.

3.4.1 Problem Description

Condensers are critical assets constituting thermal power plants and are crucial auxiliary equipments of steam turbines [22, 49]. Their main task is to condense the

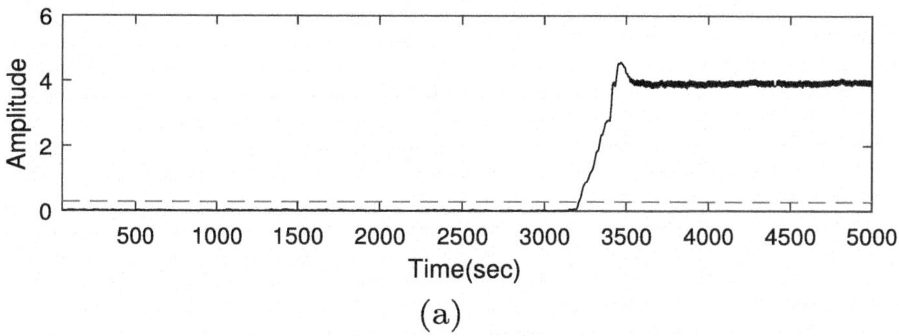

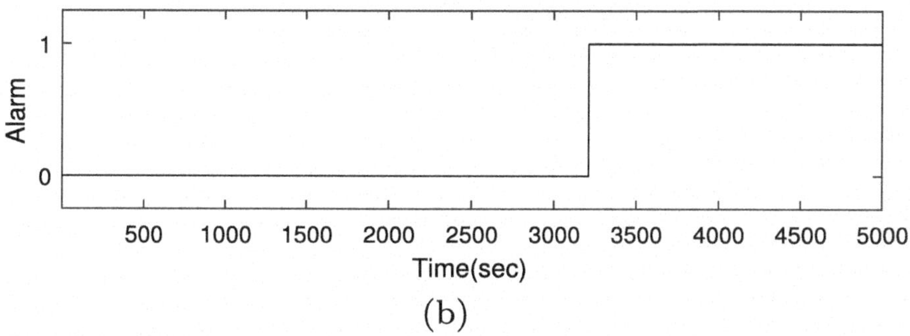

Fig. 3.44 **a** The values of $\varphi(t)$ (solid) with the alarm threshold φ_{th} (dash) and **b** the alarm sequence for the fourth case, reprinted from Ref. [46], copyright 2018, with permission from IEEE

turbine exhaust steam into water and maintain a certain degree of vacuum at the steam exhaust port of a steam turbine [32]. Figure 3.48 shows a schematic diagram of a condenser in thermal power plants. The condenser is a water-cooled shell and tube heat exchanger located downstream of the turbine to condense exhaust steam from a turbine in thermal power plants [42]. The turbine is a device that converts the heat in superheated steam into mechanical energy and drives a generator to generate electricity. Differences between the heat of steam per unit mass at the inlet and outlet of turbines represent the heat converted to mechanical power. Therefore, the amount of heat converted into mechanical energy per unit mass of steam in the turbine will influence the power plant's operational safety and efficiency. The exhaust steam of a turbine can be converted from its gaseous state to its liquid state through heat exchange with cooling water in the cooling water pipes of the condenser. The condenser pressure can be maintained during this condensation process, increasing the amount of heat available for conversion to mechanical power. Thus, a condenser is one of the most important auxiliary equipment, and its performance will directly affect the safe operation and thermal economy of power plants [14, 33, 48]. Hence, the operation monitoring of condensers has attracted growing attentions from industry and academic communities [2, 26].

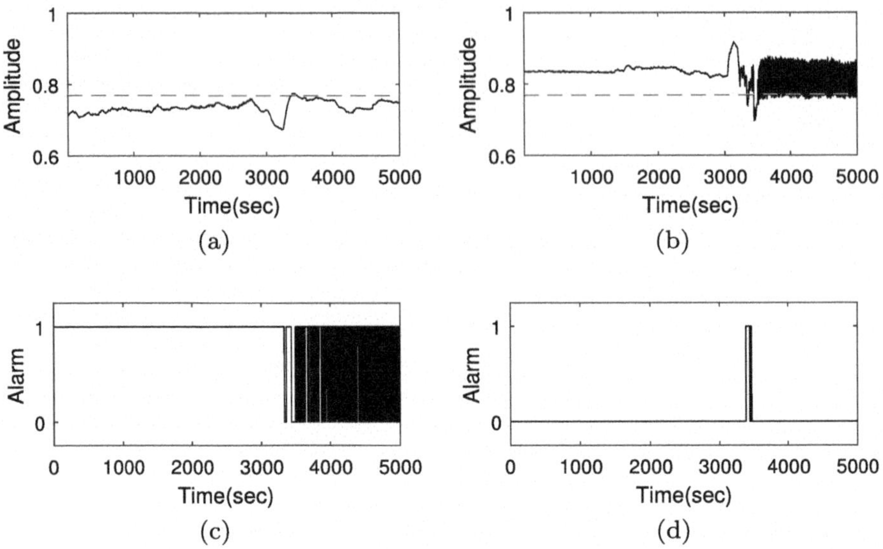

Fig. 3.45 **a** The time sequence plots of $q(t)$ (solid) with a high alarm threshold (dash) and **c** $q_a(t)$ in (3.73): **a**, **c** the third case, **b**, **d** the fourth case, reprinted from Ref. [46], copyright 2018, with permission from IEEE

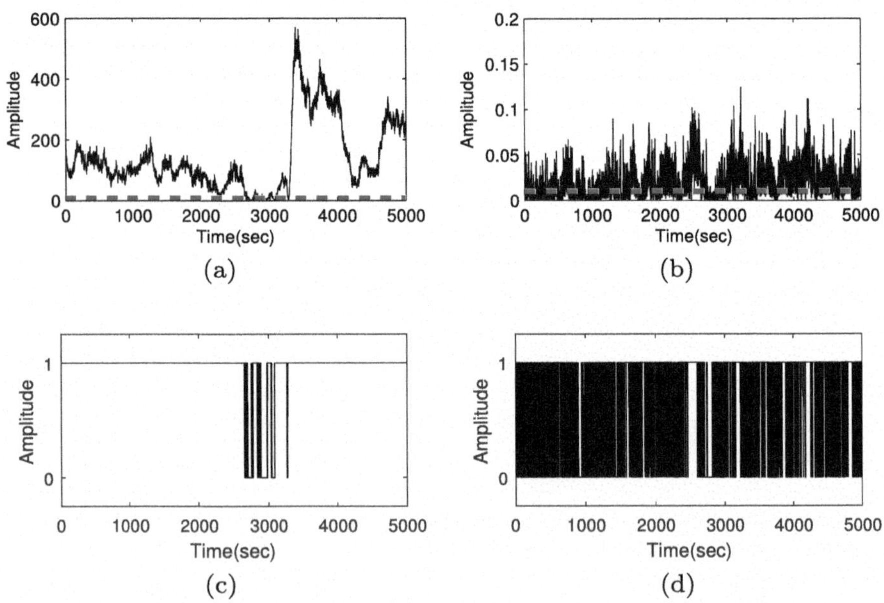

Fig. 3.46 **a** T-squared statistic (solid), **b** SPE statistic (solid) with 99% confidence intervals (dash), and **c**, **d** alarm sequences for the third case, reprinted from Ref. [46], copyright 2018, with permission from IEEE

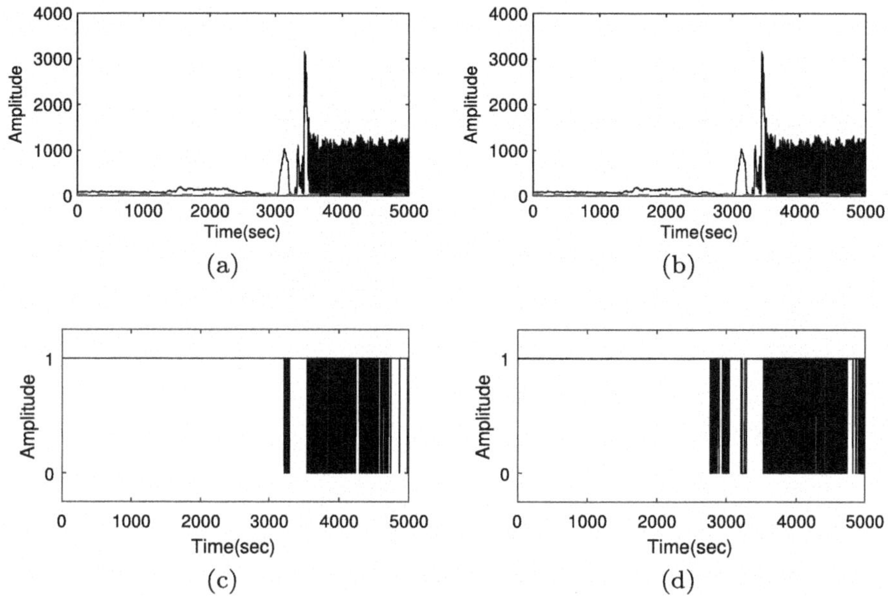

Fig. 3.47 **a** T-squared statistic (solid), **b** SPE statistic (solid) with 99% confidence intervals (dash), and **c**, **d** alarm sequences for the fourth case, reprinted from Ref. [46], copyright 2018, with permission from IEEE

The condenser pressure is the key performance index of condensers. When the condenser pressure is too high or too low, the heat consumption, steam consumption, and power generation of the turbine will be negatively affected. The condenser pressure P_c (in kPa) is mainly affected by several related variables: the cooling water inlet temperature T_1 (in °C), the cooling water outlet temperature T_2 (in °C), and the steam flow rate D_c (in kg/s) entering the condenser. The variation ranges of these variables formulate a high-dimensional geometric space, referred to as the operating zone in this context. The condenser is in abnormal (normal) conditions when a new operating point lies outside (inside) the geometric space or equivalently the operating zone.

Given historical data of the four variables mentioned above, the problem to be solved is to formulate a mathematical model describing the operating zone for operation monitoring of condensers. The operating zone model will be developed based on physical models of condensers and safe operation requirements. An alternative way is to establish the operating zone model based on historical data in normal conditions, but historical data may not be exhaustive enough to cover all operation conditions and are usually incapable of extrapolation for un-experienced zones.

There is a main technical challenge: unknown parameters of physical models are estimated from observed data, so that they generally suffer from estimation uncertainties due to noises and disturbances; as uncertainties of physical model parameters may propagate to the operating zone model and the corresponding operation monitoring, it is necessary to perform a theoretical analysis on false and missed alarms arisen from model uncertainties. For the condenser operation monitoring, most of the existing methods are with two major limitations: first, the univariate-based monitoring method is solely based on the condenser pressure and does not consider relationships among multiple related process variables, which will lead to false and missed alarms; second, the multivariate statistical analysis method for operation monitoring is based on some latent variables that do not have the same physical meanings as condenser process variables, which makes it difficult for power plant operators to understand and accept.

3.4.2 *Steps of the Proposed Method*

This section is on the major steps of the proposed method by building the condenser physical model, formulating the condenser operating zone model and generating alarms if data points are outside of the operating zone.

3.4.2.1 Condenser Physical Modeling

This section presents the condenser physical modeling and the estimation of unknown model parameters, as the condenser physical model is the basis for formulating operating zone models.

The condenser pressure P_c (in kPa) is a critical variable that reflects the operating status of the condenser. It is affected by three measurable variables: the cooling water inlet temperature T_1, the cooling water outlet temperature T_2, and the steam flow rate D_c entering the condenser. P_c generally refers to the saturation pressure corresponding to the steam condensation temperature on the shell side of the condenser. Assuming that the influence of air leakage is not considered, there is a one-to-one correspondence between the saturated steam pressure P_c and its corresponding steam condensation temperature t_s, which can be represented by an empirical formula as [42]

$$P_c = 9.8\left(\frac{t_s + 100}{57.66}\right)^{7.46}. \tag{3.90}$$

As shown in Fig. 3.48, t_s and T_2 are not located in the same position, and they have a distance in space. Thus, there exists a temperature difference δ_t between t_s and T_2,

$$t_s = T_2 + \delta_t, \tag{3.91}$$

where δ_t is defined as the terminal temperature difference [51],

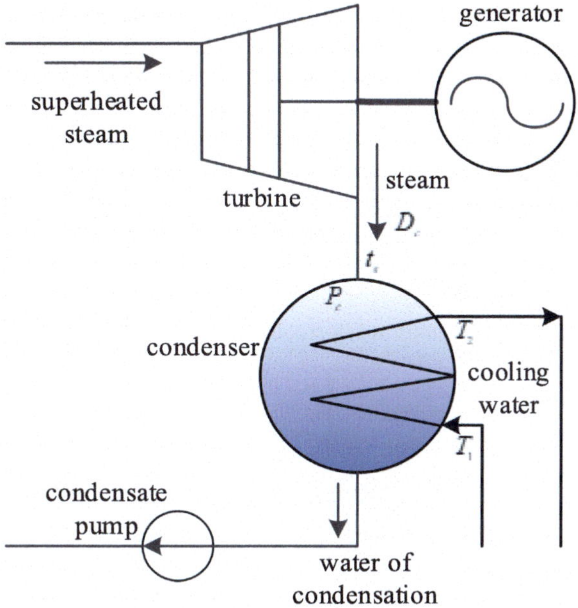

Fig. 3.48 Schematic diagram of a condenser in thermal power plants, reprinted from Ref. [44], copyright 2024, with permission from IEEE

$$\delta_t = \frac{T_2 - T_1}{e^{\frac{\Phi \cdot S}{c_w D_w}} - 1}. \tag{3.92}$$

Here Φ is the overall heat transfer coefficient in the unit kW/m^2°C, S is the heat exchange area of a condenser in the unit m^2, $c_w = 4.1816$ kJ/kg°C is the specific heat capacity of cooling water, and D_w is the cooling water flow rate of a condenser in the unit m^3/s.

The cooling water flow rate D_w in (3.92) is an intermediate unmeasured variable, which can be calculated as [42]

$$D_w = \frac{\Delta_h D_c}{c_w(T_2 - T_1)}, \tag{3.93}$$

where Δ_h means the latent heat of vaporization when 1kg of steam is condensed. S in (3.92) is the heat transfer area

$$S = 2\pi d_1 L n_s, \tag{3.94}$$

where d_1 is the outer diameter of cooling water pipes, L is the length of cooling water pipes, and n_s is the number of cooling water pipes. Φ in (3.92) is calculated by an empirical formula as [22]

$$\Phi = 3.2865 \Phi_w \Phi_t. \tag{3.95}$$

Here Φ_w is a correction factor to account for flow velocity in cooling tubes

$$\Phi_w = \left(\frac{8.8D_w}{\rho \pi n_s d_2^{\frac{9}{4}}} \right)^{\chi}, \tag{3.96}$$

where $\rho = 1000$ kg/m^3 is the cooling water density, d_2 is the inner diameter of cooling water pipes, and χ is

$$\chi = \begin{cases} 0.0969(1 + 0.15T_1), & \text{if } T_1 \le 26.7°\text{C}, \\ 0.4845, & \text{if } T_1 > 26.7°\text{C}. \end{cases} \tag{3.97}$$

Symbol Φ_t is another correction factor considering the cooling water temperature,

$$\Phi_t = \begin{cases} 1 - \frac{0.52-0.0072g_s}{1000}(35 - T_1), & \text{if } T_1 \le 35°\text{C}, \\ 1 + 0.002(T_1 - 35), & \text{if } T_1 > 35°\text{C}, \end{cases} \tag{3.98}$$

where g_s is the steam load per unit area of a condenser.

The physical model of a condenser is developed from (3.90)–(3.98) as

$$P_c = 9.8 \left(\frac{T_2 + \frac{T_2 - T_1}{e^{\frac{2\pi d_1 L n_s (T_2 - T_1)\Phi}{\Delta h D_c}} - 1} + 100}{57.66} \right)^{7.46}, \tag{3.99}$$

where

$$\Phi = 3.2865 \left(\frac{8.8D_c \Delta_h}{\rho c_w \pi n_s d_2^{\frac{9}{4}} (T_2 - T_1)} \right)^{\chi} \Phi_t, \tag{3.100}$$

with χ in (3.97) and Φ_t in (3.98). Thus, given data samples of D_c, T_1 and T_2, the estimated value of P_c at the time instant t in (3.99) is obtained as

$$\hat{P}_c(t; \theta_p) = f(D_c, T_1, T_2; \theta_p), \tag{3.101}$$

where $\theta_p = [\Delta h, n_s, L, g_s, d_2, d_1]'$ is the vector of unknown model parameters. The vector θ_p can be estimated by maximizing the goodness of fit between historical data samples $\{P_c(t)\}_{t=1}^T$ and $\hat{P}_c$ as

$$\hat{\theta}_p = \arg\max_{\theta_p} F(\theta_p). \tag{3.102}$$

Here $F(\theta_p)$ is the goodness of fit,

$$F(\theta_p) = 1 - \frac{\sum_{t=1}^T (P_c(t) - \hat{P}_c(t; \theta_p))^2}{\sum_{t=1}^T (P_c(t) - \bar{P}_c(t; \theta_p))^2}, \tag{3.103}$$

where $\bar{P}_c$ is the mean value of $\{P_c(t)\}_{t=1}^{T}$ as $\bar{P}_c = \frac{\sum_{t=1}^{T}(P_c(t))}{T}$. Based on historical data of P_c, D_c, T_1 and T_2, θ_p is estimated by solving (3.102) through some existing optimization algorithm, e.g., the genetic algorithm.

3.4.2.2 Formulation of Condenser Operating Zones

This section formulates condenser operating zones, establishes a mathematical model for operating zones and generates an alarm variable for condenser operation monitoring.

According to the condenser model in (3.99), there are mainly four variables that reflect the condenser operating conditions: P_c, T_1, T_2, and D_c. A condenser operating zone model is formulated to describe variation spaces in normal conditions based on P_c, T_1, T_2, and D_c. Based on process knowledge and practical experience for safe operation of a condenser, P_c needs to be maintained within a reasonable range as 6 kPa $< P_c <$ 11 kPa. As the heat in the exhaust steam from a turbine will be transferred to the cooling water and there exists a temperature difference δ_t being defined in (3.92), operation requirements for normal conditions are 6 °C $< T_2 - T_1 <$ 12 °C and 1.5 °C $< \delta_t <$ 6 °C, respectively.

Given variation ranges as $P_c = 6{:}\delta_1{:}11$ kPa, $T_2 - T_1 = 6{:}\delta_2{:}12$ °C, and $\delta_t = 1.5{:}\delta_3{:}6$ °C, data samples of P_c, T_1, T_2, and D_c in normal conditions can be generated based on the physical model parameter θ_p estimated in (3.102) as follows. Here the step sizes δ_1, δ_2, and δ_3 are chosen as 0.1 kPa, 0.1 °C, and 0.1 °C according to sensor measuring resolutions, respectively. First, for a value of $P_c = 6{:}\delta_1{:}11$ kPa, the corresponding value of t_s is obtained from (3.90). Second, T_2 is computed from (3.91) with the requirement of $\delta_t = 1.5{:}0.1{:}6$ °C, and T_1 is obtained subject to the requirement $T_2 - T_1 = 6{:}0.1{:}12$ °C. Finally, D_c is calculated based on (3.92) and (3.93) from the above values of P_c, T_1, and T_2. These data samples in normal conditions formulate a four-dimensional geometric space.

As the geometric space may be non-convex, a search-cone-based mathematical model is used to describe the condenser operating zone as [43]

$$M(\theta_p) := \{A_k(\theta_p)S'_{v,\theta_p} - B_k(\theta_p) \leq \mathbf{0}\}_{k=1}^{K}, \tag{3.104}$$

where

$$\begin{cases} A_k = \left[a_{k,1}, \cdots, a_{k,m}, \cdots, a_{k,M_f}\right]', \\ B_k = \left[b_{k,1}, \cdots, b_{k,m}, \cdots, b_{k,M_f}\right]', \\ \mathbf{0} = [0, 0, \cdots, 0]'. \end{cases} \tag{3.105}$$

Here $\{S_{v,\theta_p}(i)\}_{i=1}^{I_v}$ is the set of vertices of K search cones in the operating zone model associated with θ_p in (3.102), where I_v is the number of vertices. Symbol $a_{k,m}$ is the unit normal vector, $b_{k,m}$ is the offset distance to the origin point for the m-th facet in the k-th search cone, and the subscript M_f means the number of facets of a search cone in (3.104). Since the operating zone model in the four-dimensional space cannot be visualized in the Cartesian coordinate, projections on two-dimensional planes of these data samples are shown in Fig. 3.49. Although the projection from four to two dimensions may lose some information, it will not affect the illustration of an operating zone model based on two-dimensional projections.

Each boundary of the operating zone model in Fig. 3.49 has a clear physical meaning. For example, the red lines in Fig. 3.49 are associated with the condition where $P_c \in [6, 11]$ kPa, $T_2 - T_1 = 6\,°\text{C}$, and $\delta_t = 1.5\,°\text{C}$. Its physical meaning is that when the ambient temperature is high $T_2 - T_1$ and δ_t will take their minimum values resulting in T_1 and T_2 being with high values. Thus, in order to keep P_c within normal conditions, D_c must remain at its minimum value. Otherwise, a large value of D_c will lead to the cooling water being unable to cool too much steam, which will cause the increment of P_c for exceeding its normal conditions range.

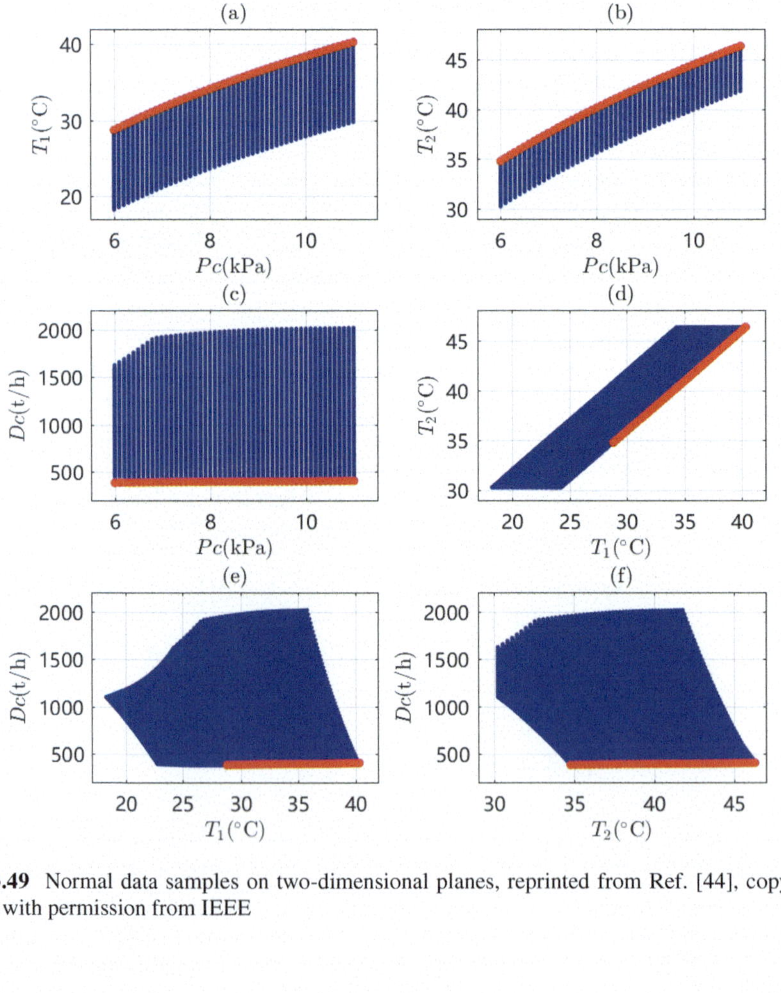

Fig. 3.49 Normal data samples on two-dimensional planes, reprinted from Ref. [44], copyright 2024, with permission from IEEE

An alarm variable $x_a(t)$ is defined after the condenser operating zone model $M(\theta_p)$ in (3.104) has been formulated. In particular, $x_a(t)$ takes the value 1 (0) as the alarm (non-alarm) state, when a new data point $X(t) := [P_c, D_c, T_1, T_2]$ lies inside (outside) one of search cones in the operating zone, i.e.,

$$x_a(t)=\begin{cases}0, & \exists k\in[1, K],\, A_k(\theta_p)X'(t)-B_k(\theta_p)\leq \mathbf{0},\\ 1, & \text{otherwise.}\end{cases} \tag{3.106}$$

3.4.3 Upper Bounds of False and Missed Alarm Rates

This section investigates the performance indices in terms of FAR and MAR for the designed multivariate alarm system. The false and missed alarms are arisen due to the uncertainties of physical model parameters. Upper bounds of the FAR and MAR are established via the Bayesian estimation rule. An optimal operating zone model is obtained by achieving the best trade-off between the FAR and MAR.

The condenser physical model may contain noticeable uncertainties caused by noises and disturbances contaminating measured data samples [37]. An analytical relationship between the uncertainty range of condenser physical model parameters and the counterpart of the condenser operating zone model is difficult to obtain analytically. As an alternative, a numerical relationship can be established as follows. It is a common practice for thermal power plants to collect and store a large amount of historical data. A set of model parameters in θ_p can be estimated in (3.102) from a group of historical data as modeling data. A model validation is adopted to determine whether a condenser model corresponding to θ_p behaves as the condenser does [35]: if $F(\theta_p)$ in (3.103) is larger than a user-selected threshold F_{th}, then θ_p passes the model validation.

Multiple sets of model parameters passing the model validation are used here to obtain the uncertainty range of the operating zone model. The vertices of operating zone models $\{M_j(\theta_{p,j})\}_{j=1}^{I_p}$ being associated with I_p sets of physical model parameters $\{\theta_{p,j}\}_{j=1}^{I_p}$ are saved in dataset $\{S_{\upsilon,\Theta}(i)\}_{i=1}^{I_V}$ to represent the uncertainty range of operating zone models, where $I_V = I_p \cdot I_\upsilon$. For simplicity, it is assumed that $\{M_j(\theta_{p,j})\}_{j=1}^{I_p}$ in Fig. 3.49 are invariant to the number K of search cones in (3.104) and the origin point o of search cones; in other words, $\{M_j(\theta_{p,j})\}_{j=1}^{I_p}$ choose the same values of K and o.

For a set of model parameters in $\theta_{p,j}$, the operating zone model is

$$M_j(\theta_{p,j}) = \{A_{k,j}(\theta_{p,j})S'_{\upsilon,\theta_{p,j}} - B_{k,j}(\theta_{p,j}) \leq \mathbf{0}\}_{k=1}^{K}. \tag{3.107}$$

Thus, $\{S_{\upsilon,\Theta}(i)\}_{i=1}^{I_V}$ corresponds to I_p operating zone models. The real operating zone model is assumed to be one of them, or very close to one of them. The operating zone model to be selected is denoted as the jth one falling inside $\{S_{\upsilon,\Theta}(i)\}_{i=1}^{I_V}$, i.e.,

$$M_{\hat{j}} = \{A_{k,\hat{j}} S'_{v,\Theta} - B_{k,\hat{j}} \leq \mathbf{0}\}_{k=1}^{K}, \tag{3.108}$$

where $A_{k,\hat{j}} = A_{k,j}(\theta_{p,\hat{j}})$, $B_{k,\hat{j}} = B_{k,j}(\theta_{p,\hat{j}})$, and $\hat{j} = 1, 2, 3, \cdots, I_p$.

There are three spaces in $\{S_{v,\Theta}(i)\}_{i=1}^{I_V}$ to be divided by $M_{\hat{j}}$ and $M_j(\theta_{p,j})$, namely, V_1, V_2, and V_0. V_0 stands for the intersection space of $M_{\hat{j}}$ and $M_j(\theta_{p,j})$. V_1 is the space inside $M_j(\theta_{p,j})$ and outside $M_{\hat{j}}$. V_2 is the space inside $M_{\hat{j}}$ and outside $M_j(\theta_{p,j})$. For instance, typical examples of V_0, V_1, V_2 are illustrated in Fig. 3.52c to be given later at Sect. 3.4.4. V_1 is in normal conditions as it locates in $M_j(\theta_{p,j})$, and no alarms should be triggered. Thus, alarms triggered in the space V_1 are false alarms. The FAR of $M_{\hat{j}}$ for $M_j(\theta_{p,j})$ is defined as

$$f_{a,j}(M_{\hat{j}}) = \frac{V(M_j - M_{\hat{j}})}{V(M_{\hat{j}})} = \frac{\sum_{k=1}^{K}\left(b_{k,\hat{j},M_f}^{\frac{N}{2}} \cdot I_{f,k} - b_{k,j,M_f}^{\frac{N}{2}} \cdot I_{f,k}\right)}{\sum_{k=1}^{K}(B'_{k,\hat{j}} \cdot B_{k,\hat{j}})^{\frac{N}{2}}}, \tag{3.109}$$

where $b_{k,\hat{j},M_f}$ is the M_f-th element of $B_{k,\hat{j}}$ in (3.105), and $V(\cdot)$ is the volume of a search cone. Since $\{M_j(\theta_{p,j})\}_{j=1}^{I_p}$ have the same values of K and o, $V(\cdot)$ is only related to $B_{k,\hat{j}}$. $I_{f,k}$ in (3.109) is an indicator scalar,

$$I_{f,k} = \begin{cases} 1, & \text{if } b_{k,\hat{j},M_f} > b_{k,j,M_f}, \\ 0, & \text{otherwise.} \end{cases} \tag{3.110}$$

A value of 1 in $I_{f,k}$ means that the k-th search cone in $M_{\hat{j}}$ is larger than the k-th search cone in $M_j(\theta_{p,j})$.

By contrast, the space V_2 lying outside $M_j(\theta_{p,j})$ should trigger alarms. Thus, V_2 leads to the occurrence of missed alarms. The MAR of $M_{\hat{j}}$ for $M_j(\theta_{p,j})$ is defined as

$$\begin{aligned} m_{a,j}(M_{\hat{j}}) &= \frac{V(M_{\hat{j}} - M_j)}{V(M_{\hat{j}})} \\ &= \frac{\sum_{k=1}^{K}\left(b_{k,j,M_f}^{\frac{N}{2}} \cdot I_{m,k} - b_{k,\hat{j},M_f}^{\frac{N}{2}} \cdot I_{m,k}\right)}{\sum_{k=1}^{K}(B'_{k,\hat{j}} \cdot B_{k,\hat{j}})^{\frac{N}{2}}}, \end{aligned} \tag{3.111}$$

where $I_{m,k}$ is another indicator scalar

$$I_{m,k} = \begin{cases} 1, & \text{if } I_{f,k} = 0, \\ 0, & \text{otherwise.} \end{cases} \tag{3.112}$$

A value of 1 in $I_{f,k}$ means that the k-th search cone in $M_{\hat{j}}$ is smaller than the k-th search cone in $M_j(\theta_{p,j})$.

Each operating zone model $M_j(\theta_{p,j})$ corresponds to a pair of FAR in (3.109) and MAR in (3.111) for $M_{\hat{j}}$. In order to tolerate the worst cases of false and missed alarms, the FAR and MAR of $M_{\hat{j}}$ under $\{S_{v,\Theta}(i)\}_{i=1}^{I_V}$ are defined as

$$f_a(M_{\hat{j}}) = \max\{f_{a,j}(M_{\hat{j}})\}_{j=1}^{I_p}, \tag{3.113}$$

$$m_a(M_{\hat{j}}) = \max\{m_{a,j}(M_{\hat{j}})\}_{j=1}^{I_p}. \tag{3.114}$$

Here, the worst case of false alarms is associated with $M_{\bar{j}_f}$ for

$$\bar{j}_f = \left\{\bar{j}_f \in [1, I_p] | f_{a,\bar{j}_f}(M_{\hat{j}}) = f_a(M_{\hat{j}})\right\}, \tag{3.115}$$

and the worst case of missed alarms is with $M_{\bar{j}_m}$ for

$$\bar{j}_m = \left\{\bar{j}_m \in [1, I_p] | m_{a,\bar{j}_m}(M_{\hat{j}}) = m_a(M_{\hat{j}})\right\}. \tag{3.116}$$

The so-called receiver operating characteristic (ROC) curve is adopted to select $M_{\hat{j}}$. A ROC curve is a plot of $f_a(M_{\hat{j}})$ in (3.113) and $m_a(M_{\hat{j}})$ in (3.114) for different values of $\hat{j} \in [1, I_p]$. Given the weights of FAR and MAR respectively as w and $1-w$, the optimal operating zone model $M_{\hat{j}_{opt}}$ is the one with the corresponding point in the ROC curve having the minimum weighted distance from the origin. Thus, the loss function considering a weighted trade-off between the FAR in (3.113) and MAR in (3.114) is formulated as

$$J(\hat{j}) = w \cdot (f_a(M_{\hat{j}})) + (1 - w) \cdot (m_a(M_{\hat{j}})). \tag{3.117}$$

An optimal parameter $\hat{j}$ in (3.108) is designed as

$$\hat{j}_{opt} = \arg\min_{\hat{j}} J(\hat{j}). \tag{3.118}$$

By using $\hat{j}_{opt}$, $M_{\hat{j}_{opt}}$ in (3.108) is the selected operating zone model.

For an unknown real operating zone model, its FAR and MAR are fixed. However, $f_a(M_{\hat{j}})$ in (3.113) and $m_a(M_{\hat{j}})$ in (3.114) are obtained based on $\{S_{v,\Theta}(i)\}_{i=1}^{I_V}$. With the increment of I_V in $\{S_{v,\Theta}(i)\}_{i=1}^{I_V}$, FAR and MAR being regarded as random variables will gradually approach their real values. Thus, the sample size I_V of $\{S_{v,\Theta}(i)\}_{i=1}^{I_V}$ affects the accuracies of FAR and MAR. By exploiting Bayesian estimation approach [20], confidence intervals of the FAR and MAR are adopted to measure their uncertainties. Theoretical results for estimating the upper bounds of FAR and MAR are provided based on the Bayesian estimation rule in the following propositions.

Proposition 3.1 *For an operating zone model $M_{\hat{j}_{opt}}$ formulated in* (3.108) *and $\hat{j}_{opt}$ in* (3.118)*, the corresponding FAR in* (3.113) *does not exceed an upper bound $\hat{f}_{a,+}$ at a confidence level $1-\alpha$,*

$$\int_{\hat{f}_{a,-}}^{\hat{f}_{a,+}} \frac{\frac{c_1!}{c_1(I_V-c_1)!}(f_a)^{c_1}(1-f_a)^{I_V-c_1}}{\int_0^1 \frac{c_1!}{c_1(I_V-c_1)!}(f_a)^{c_1}(1-f_a)^{I_V-c_1}\,\mathrm{d}f_a}\mathrm{d}f_a = 1-\alpha, \tag{3.119}$$

where f_a is a fictitious element here standing for the FAR of $M_{\hat{j}_{opt}}$, I_V is the number of samples in $\{S_{v,\Theta}(i)\}_{i=1}^{I_V}$, and c_1 is the number of data samples in $\{S_{v,\Theta}(i)\}_{i=1}^{I_V}$ lying inside the space V_1 as

$$c_1 = \left|\left\{X \in S_{v,\Theta} | X \in X_{\bar{j}_f} \& X \notin X_{\hat{j}_{opt}}\right\}\right|_c. \tag{3.120}$$

Symbol $|\cdot|_c$ is the cardinality of a set indicating the number of samples in this set. $X_{\hat{j}_{opt}}$ ($X_{\bar{j}_f}$) is a set containing data samples in $\{S_{v,\Theta}(i)\}_{i=1}^{I_V}$ that lie inside $M_{\hat{j}_{opt}}$ ($M_{\bar{j}_f}$ with $\bar{j}_f$ in (3.115)*) as*

$$X_{\hat{j}_{opt}} = \left\{X \in S_{v,\Theta} | \left\{A_{k,\hat{j}_{opt}} X' - B_{k,\hat{j}_{opt}} \leq \mathbf{0}\right\}_{k=1}^{K}\right\}, \tag{3.121}$$

$$X_{\bar{j}_f} = \left\{X \in S_{v,\Theta} | \left\{A_{k,\bar{j}_f} X' - B_{k,\bar{j}_f} \leq \mathbf{0}\right\}_{k=1}^{K}\right\}. \tag{3.122}$$

Proof of Proposition 3.1 The FAR F_a is regarded as a continuous random variable taking f_a as a realization. The number of data samples in $\{S_{v,\Theta}(i)\}_{i=1}^{I_V}$ lying inside the space V_1 is taken as a realization c_1 of a discrete random variable C_1. Since data samples in $\{S_{v,\Theta}(i)\}_{i=1}^{I_V}$ are mutually independent, C_1 follows the binomial distribution with the probability of false alarm appearances as F_a. Thus, the conditional probability of C_1 on F_a is

$$p_{C_1|F_a}(c_1 \mid f_a) = \frac{c_1!}{c_1(I_V - c_1)!}(f_a)^{c_1}(1-f_a)^{I_V-c_1}. \tag{3.123}$$

Here I_V is the number of samples in $\{S_{v,\Theta}(i)\}_{i=1}^{I_V}$ and c_1 is defined in (3.120). Without any extra information, the prior PDF of F_a is taken as a continuous uniform distribution,

$$p_{F_a}(f_a) = \begin{cases} 1, & \text{if } 0 < f_a < 1, \\ 0, & \text{otherwise.} \end{cases} \tag{3.124}$$

The posterior probability distribution of F_a given C_1 is

$$p_{F_a|C_1}(f_a \mid c_1) = \frac{p_{C_1,f_a}(c_1, f_a)}{p_{C_1}(c_1)}. \tag{3.125}$$

The Bayesian estimate of f_a is often taken as the conditional mean as

$$\begin{aligned}\hat{f}_a &= \int_0^1 f_a p_{F_a|C_1}(f_a \mid c_1)\,\mathrm{d}f_a \\ &= \int_0^1 f_a \cdot \frac{\frac{c_1!}{c_1(I_V-c_1)!}(f_a)^{c_1}(1-f_a)^{I_V-c_1}}{\int_0^1 \frac{c_1!}{c_1(I_V-c_1)!}(f_a)^{c_1}(1-f_a)^{I_V-c_1}\,\mathrm{d}f_a}\,\mathrm{d}f_a.\end{aligned} \tag{3.126}$$

According to the posterior PDF of F_a, the $(1-\alpha)$ confidence interval of $\hat{f}_a$ is obtained as the narrowest interval $\left[\hat{f}_{a,-}, \hat{f}_{a,+}\right]$ satisfying the equality (3.119). The upper bound of FAR for $M_{\hat{j}_{opt}}$ is $\hat{f}_{a,+}$ at a confidence level $1-\alpha$. □

Similar to Proposition 3.1, the upper bound of m_a can also be obtained based on the Bayesian estimation rule in Proposition 3.2.

Proposition 3.2 *For an operating zone model $M_{\hat{j}_{opt}}$ formulated in* (3.108) *and $\hat{j}_{opt}$ in* (3.118)*, the corresponding MAR in* (3.114) *does not exceed an upper bound $\hat{m}_{a,+}$ at a confidence level $1-\alpha$,*

$$\int_{\hat{m}_{a,-}}^{\hat{m}_{a,+}} \frac{\frac{c_2!}{c_2(I_V-c_2)!}(m_a)^{c_2}(1-m_a)^{I_V-c_2}}{\int_0^1 \frac{c_2!}{c_2(I_V-c_2)!}(m_a)^{c_2}(1-m_a)^{I_V-c_2}\,\mathrm{d}m_a}\,\mathrm{d}m_a = 1-\alpha, \tag{3.127}$$

where m_a is a fictitious element here standing for the MAR of $M_{\hat{j}_{opt}}$, and c_2 is the number of data samples in $\{S_{\upsilon,\Theta}(i)\}_{i=1}^{I_V}$ lying inside the space V_2 as

$$c_2 = \left|\left\{X \in S_{\upsilon,\Theta} | X \in X_{\hat{j}_{opt}} \,\&\, X \notin X_{\bar{j}_m}\right\}\right|_c. \tag{3.128}$$

Here $X_{\hat{j}_{opt}}$ is defined in (3.121) *and $X_{\bar{j}_m}$ with $\bar{j}_m$ in* (3.116) *is defined as*

$$X_{\bar{j}_m} = \left\{X \in S_{\upsilon,\Theta} | \left\{A_{k,\bar{j}_m} X' - B_{k,\bar{j}_m} \leq \mathbf{0}\right\}_{k=1}^{K}\right\}. \tag{3.129}$$

Proof of Proposition 3.2 The proof is similar to the counterpart of Proposition 3.1 and is omitted here. □

Given user-selected ratios η_f and η_m, the FAR and MAR upper bounds should satisfy

$$\begin{aligned}\hat{f}_{a,+} &< f_a(M_{\hat{j}_{opt}}) \cdot (1+\eta_f), \\ \hat{m}_{a,+} &< m_a(M_{\hat{j}_{opt}}) \cdot (1+\eta_m).\end{aligned} \tag{3.130}$$

If requirements in (3.130) are not satisfied, it is necessary to use more historical data for estimating model parameters and reducing model uncertainties. The number of samples in $\{S_{\upsilon,\Theta}(i)\}_{i=1}^{I_V}$ is increased by gradually adding 1 to I_p until $\hat{f}_{a,+}$ and $\hat{m}_{a,+}$ satisfy (3.130).

After obtaining $M_{\hat{j}_{opt}}$ in (3.108) with $\hat{j}_{opt}$ in (3.118), as well as (3.130) being satisfied, the alarm variable $X_a(t)$ in (3.131) is updated by replacing $M(\theta_p)$ with $M_{\hat{j}_{opt}}$ in order to achieve the optimal trade-off between the FAR and MAR, i.e.,

$$X_a(t) = \begin{cases} 0, & \exists k \in [1, K],\ A_{k,\hat{j}_{opt}} X'(t) - B_{k,\hat{j}_{opt}} \leq \mathbf{0}, \\ 1, & \text{otherwise}. \end{cases} \quad (3.131)$$

In summary, the proposed method consists of the following steps:

Step 1: For the offline stage, the condenser physical model in (3.99) is built with model parameters to be estimated through the genetic algorithm. If $P_c(t)$ can be well described by its estimate $\hat{P}_c(t;\theta_p)$, then the model quality is satisfactory; otherwise, the model cannot be used.

Step 2: The condenser operating zone model is formulated in (3.108) and (3.118) by considering model parameter uncertainty and computing the upper bounds of its FAR in (3.119) and MAR in (3.127) based on the Bayesian estimation rule.

Step 3: Steps 1 and 2 are repeated by using more historical data for gradually adding 1 to I_p to increase the number of samples in $\{S_{v,\Theta}(i)\}_{i=1}^{I_V}$ until $\hat{f}_{a,+}$ and $\hat{m}_{a,+}$ satisfy the requirements in (3.130).

Step 4: After $M_{\hat{j}_{opt}}$ has been formulated, a corresponding alarm variable $x_a(t)$ in (3.131) is generated from online new data samples for condenser monitoring at the online stage.

3.4.4 Industrial Examples

This section presents industrial examples to demonstrate the effectiveness and practicability of the proposed method, and to compare it with two existing methods for the condenser operation monitoring: the univariate alarm threshold-based method [5] and the PCA-based method [1].

First, one set of condenser physical model parameters θ_p in (3.101) is estimated via (3.102) as listed in Table 3.5 based on industrial historical data in a large-scale 300-MW thermal power plant. The goodness-of-fit in (3.103) is used to determine the fitness between simulated data and industrial historical data of P_c, where the model validation threshold of $F(\theta_p)$ is selected as 0.95. One set of modeling data samples and testing data samples are shown in Figs. 3.50 and 3.51, respectively. As validation, the average goodness-of-fits of this condenser model for different months are shown in Table 3.6. Since the average goodness-of-fits are above 95%, this model can represent the condenser behaviors well.

Table 3.5 Condenser model parameters, reprinted from Ref. [44], copyright 2024, with permission from IEEE

Model parameter	Symbol	Value
Latent heat of vaporization	Δ_h	2152.7 kJ/kg
The number of cooling water pipes	n_s	24613
The length of cooling water pipes	L	14.6996 m
The steam load per unit area	g_s	12.8538 kg/m^{2}s
Inner diameter of cooling water pipes	d_2	0.0182 m
Outer diameter of cooling water pipes	d_1	0.0247 m

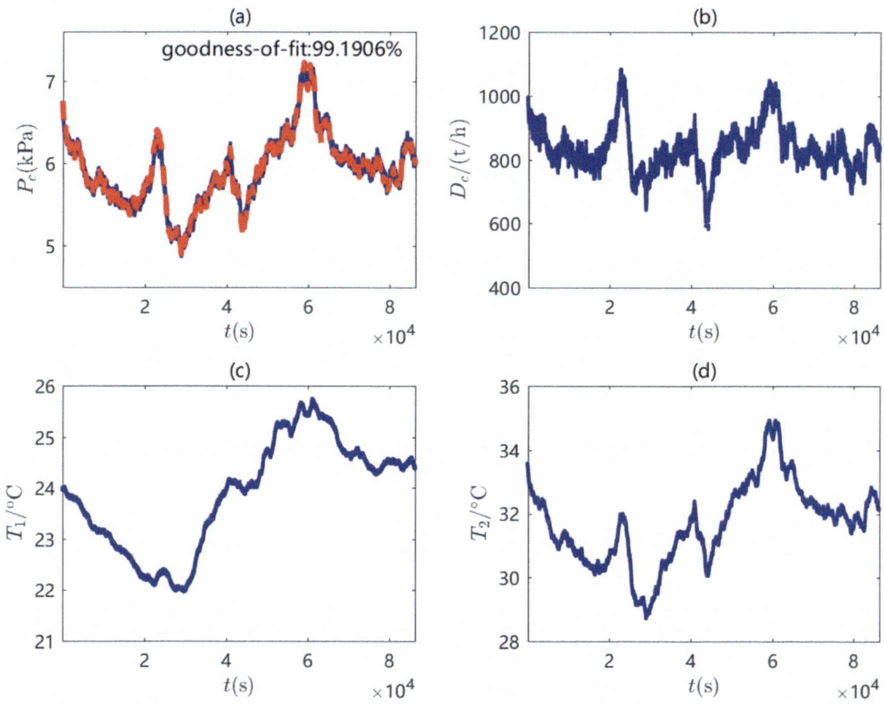

Fig. 3.50 Data samples of four variables for modeling data and simulated data: **a** P_c (blue solid line) and $\hat{P}_c$ (red dashed line), **b** D_c, **c** T_1, **d** T_2, reprinted from Ref. [44], copyright 2024, with permission from IEEE

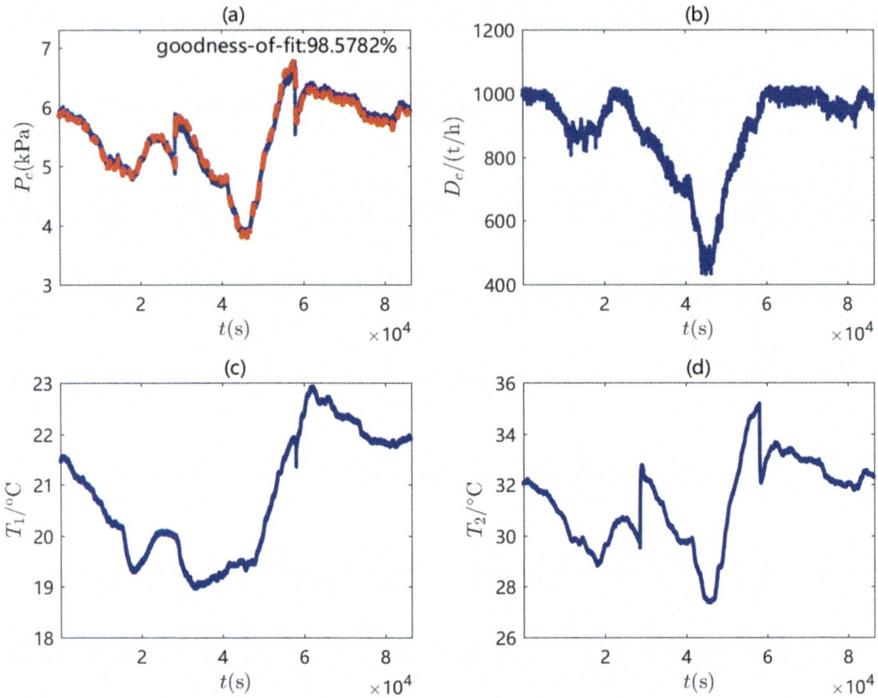

Fig. 3.51 Data samples of four variables for testing data and simulated data: **a** P_c (blue solid line) and $\hat{P}_c$ (red dashed line), **b** D_c, **c** T_1, **d** T_2, reprinted from Ref. [44], copyright 2024, with permission from IEEE

Table 3.6 Averaged goodness-of-fit of the condenser model in different months, reprinted from Ref. [44], copyright 2024, with permission from IEEE

Month	Goodness of fit	Month	Goodness of fit
May	96.3%	August	97.1%
June	98.6%	September	96.5%
July	95.1%	October	98.5%

Second, the uncertainty range $\{S_{v,\Theta}(i)\}_{i=1}^{I_V}$ of the condenser operating zone model in (3.99) is obtained by generating I_p sets of θ_p. By solving the optimization problem in (3.118), the selected operating zone model $M_{\hat{j}_{opt}}$ is associated with $\hat{j}_{opt} = 53$. For illustration, boundaries of the operating zone model on the projection to the plane of T_1 and D_c are shown as the red dashed lines in Fig. 3.52c. $J(\hat{j})$ as a function of $\hat{j}$ is presented in Fig. 3.52a with $w = 0.5$. The ROC curve of FAR and MAR for different values of $\hat{j}$ is given in Fig. 3.52b, where the corresponding FAR is 0.0681 and MAR is 0.0504 for the selected operating zone model as given in Fig. 3.52b.

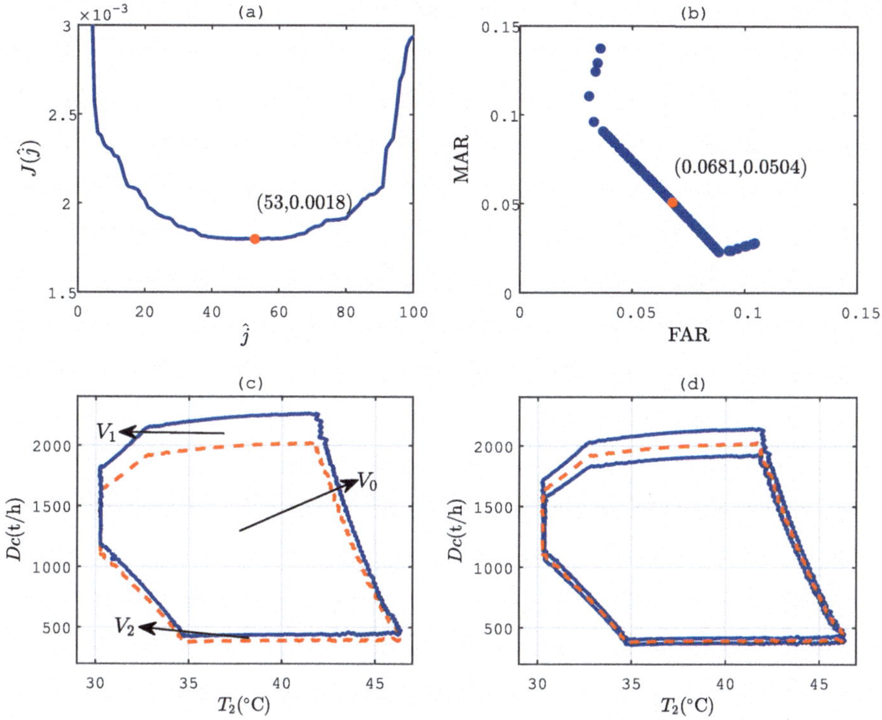

Fig. 3.52 **a** $J(\hat{j})$ as a function of $\hat{j}$, **b** the ROC curve of FAR and MAR for different values of $\hat{j}$, **c** the condenser NOZ $M_{\hat{j}_{opt}}$ (red dash) by the proposed method, $M_j(\theta_{p,j=84})$ (blue solid), and the corresponding V_0, V_1, V_2, **d** $M_{\hat{j}_{opt}}$ (red dash) and another two $M_j(\theta_{p,j})$ (blue solid), reprinted from Ref. [44], copyright 2024, with permission from IEEE

$M_{\hat{j}_{opt}}$ and two other operating zone models of $M_j(\theta_{p,j})$ are shown in Fig. 3.52d to illustrate the model uncertainty range.

Third, the upper bounds of FAR and MAR are obtained as follows. Given $\eta_f = \eta_m = 0.1$ and $\alpha = 0.01$, $f_a(M_{\hat{j}_{opt}}) \cdot (1+\eta_f) \approx 0.075$, and $m_a(M_{\hat{j}_{opt}}) \cdot (1+\eta_m) \approx 0.055$ are obtained from (3.130). By increasing I_V with a step size of 200 samples with $I_v = 360$, the proposed method yields the estimated probability $\hat{f}_a$ with its confidence interval $[\hat{f}_{a,-}, \hat{f}_{a,+}]$ from (3.119) as a function of I_V in Fig. 3.53a. Similarly, the estimated probability $\hat{m}_a$ with its confidence interval $[\hat{m}_{a,-}, \hat{m}_{a,+}]$ from (3.127) are shown in Fig. 3.53b. $\hat{f}_{a,+}$ and $\hat{m}_{a,+}$ gradually stabilize below 0.075 and 0.055 after $I_V > 36000$ and $I_p > 100$. As a validation, $I_p = 100$ sets of new model parameters $\hat{\theta}_p$ are obtained based on historical data from another 100 days. FARs and MARs of these model parameters are shown in Fig. 3.54a, b, with 99 FARs less than 0.075 and 100 MARs less than 0.055. These results validate Propositions 3.1 and 3.2.

Finally, three abnormality detection results from the proposed method and two existing methods are presented, as shown in Table 3.7. Figures 3.55 and 3.56 are

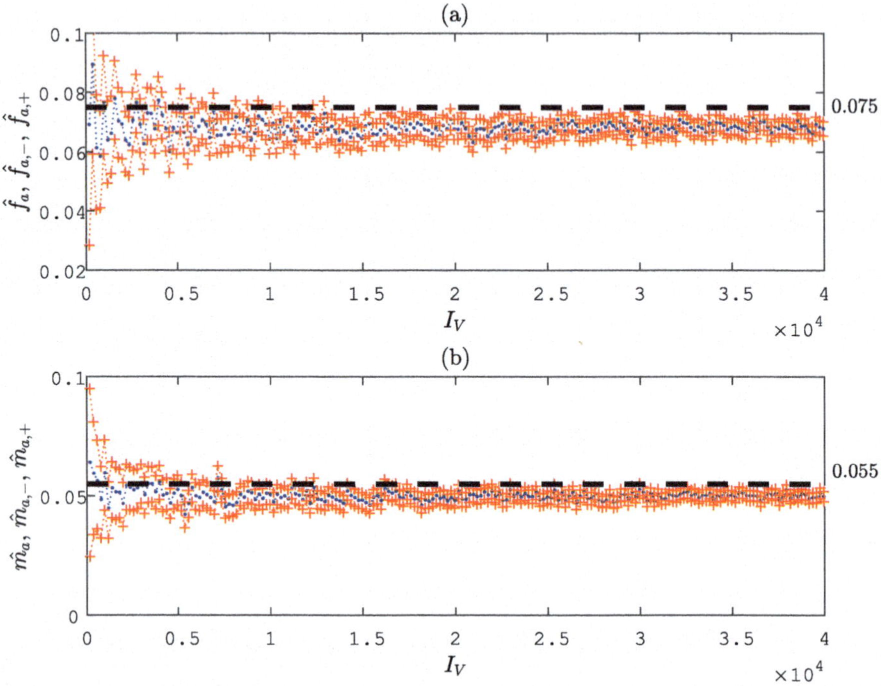

Fig. 3.53 **a** Estimated probability $\hat{f}_a$ (blue dot) and its confidence interval $[\hat{f}_{a,-}, \hat{f}_{a,+}]$ (red plus), **b** estimated probability $\hat{m}_a$ (blue dot) and its confidence interval $[\hat{m}_{a,-}, \hat{m}_{a,+}]$ (red plus), reprinted from Ref. [44], copyright 2024, with permission from IEEE

data samples of four process variables and projections of the condenser operating zone on two-dimensional planes for abnormality #1. Figures 3.57, 3.58, 3.59 and 3.60 are the counterparts for abnormality #2 and abnormality #3, respectively. The univariate alarm threshold-based method detects the abnormalities later than the proposed method. The PCA-based method suffers from nuisance alarms and fails to detect the abnormalities correctly.

A detailed description of abnormality #1 is given below, since the three abnormalities are similar. As shown in Fig. 3.55 with the sampling period 1 sec, 65000 historical data samples on July 18 are used as testing data samples. Projections of the condenser operating zone model on two-dimensional planes are shown in Fig. 3.56, where blue and red dots represent the data points in normal and abnormal conditions, respectively. The proposed method can detect this abnormality at $t = 44424$ sec in the space of original condenser process variables, as revealed in Fig. 3.56f. As shown in Fig. 3.55, the reason for this abnormality is that T_1 and T_2 increase with the increment of D_c while the cooling water cannot cool the steam into the water in time. Then, the condenser operating point crosses the operating zone boundaries at $t = 44424$ sec with $T_1 = 36.08$ °C, $T_2 = 44.95$ °C, $D_c = 906.81$ t/h, and $P_c = 10.79$ kPa. Finally, the condenser pressure gradually crosses the operating zone boundaries, resulting in this abnormality.

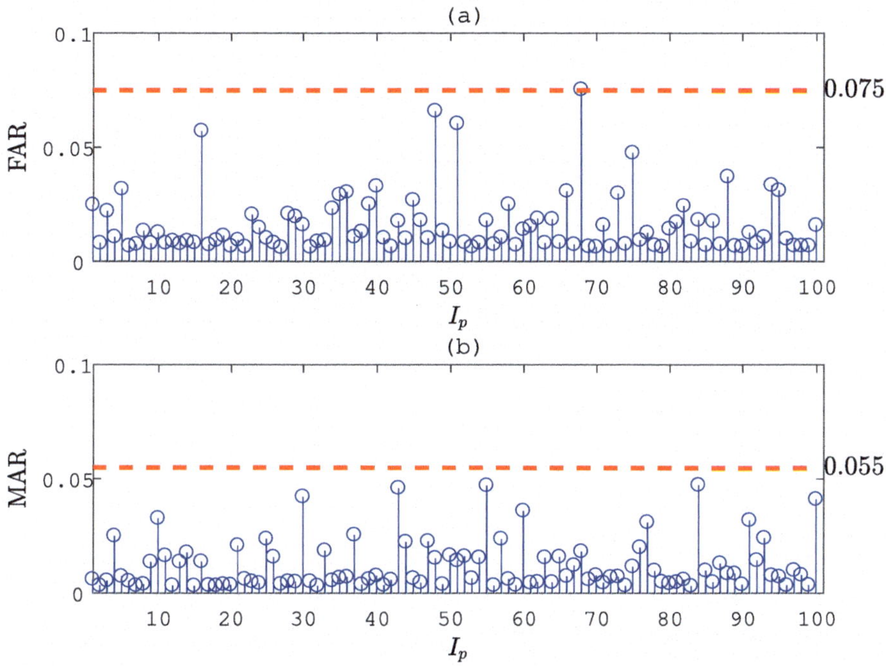

Fig. 3.54 **a** The stem plot of FARs for 100 sets of new model parameters, **b** the stem plot of MARs for 100 sets of new model parameters, reprinted from Ref. [44], copyright 2024, with permission from IEEE

Table 3.7 Results of the proposed method and two existing methods for three abnormalities, reprinted from Ref. [44], copyright 2024, with permission from IEEE

Abnormality	Proposed method	Univariate alarm threshold	PCA-based method
#1	True alarm $t = 44424$ sec	True alarm $t = 45988$ sec	Nuisance alarm $t = 2$ sec
#2	True alarm $t = 51434$ sec	True alarm $t = 53448$ sec	Nuisance alarm $t = 145$ sec
#3	True alarm $t = 54497$ sec	True alarm $t = 55629$ sec	Nuisance alarm $t = 79$ sec

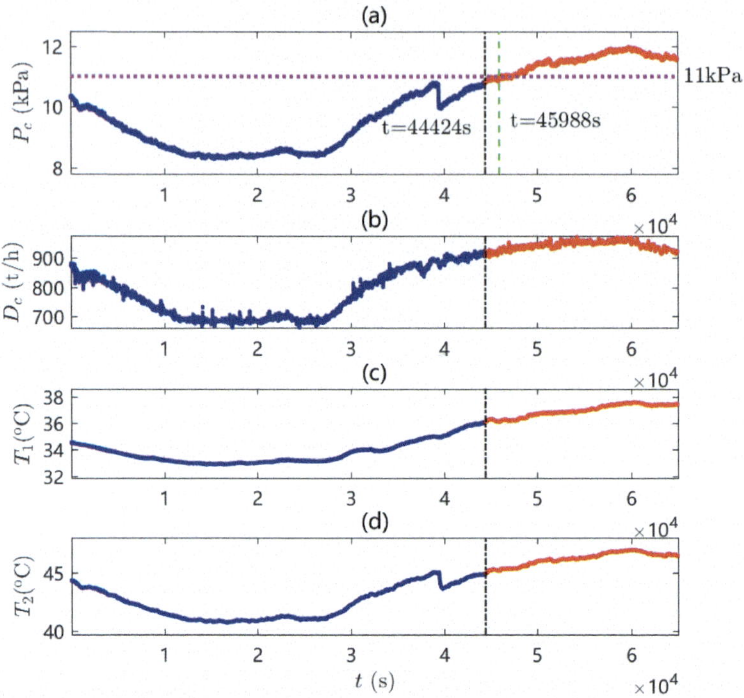

Fig. 3.55 Data samples of four variables for abnormality #1, reprinted from Ref. [44], copyright 2024, with permission from IEEE

The univariate alarm threshold-based method [5] is applied as a comparison; according to the operation requirements, P_c exceeds its high alarm threshold 11kPa for normal conditions at $t = 45988$ sec in Fig. 3.55a. Thus, the proposed method detects this abnormality 1564 sec (about 26 minutes) earlier than the univariate alarm threshold-based method. Industrial plant operators usually look at data samples of multiple variables, instead of the single variable P_c, to confirm the abnormality; such a confirmation can be easily done by looking at the positions of data points with respect to the operating zone boundaries of multiple process variables in Fig. 3.56.

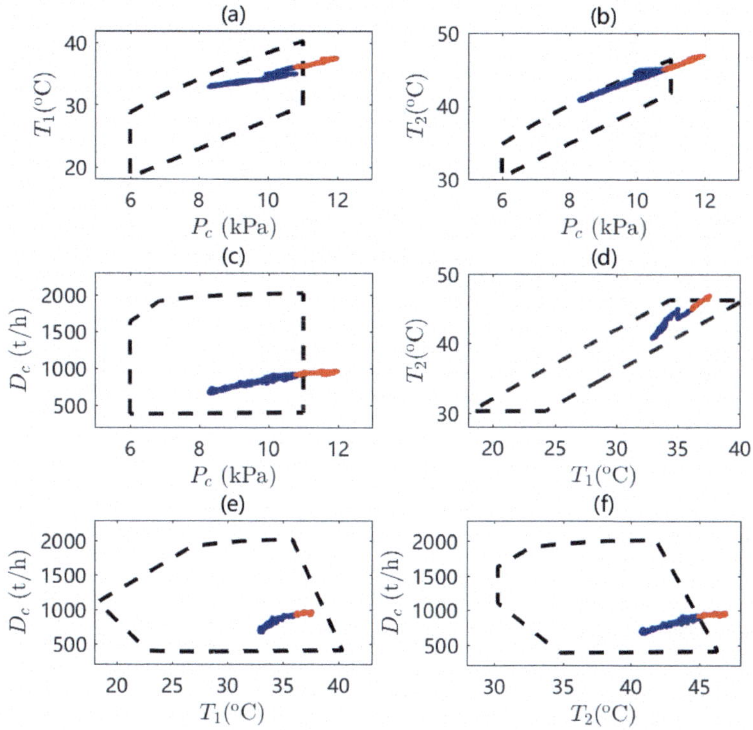

Fig. 3.56 Scatter plots with condenser operating zone model (black dash) on two-dimensional planes for abnormality #1, reprinted from Ref. [44], copyright 2024, with permission from IEEE

As another comparison, the threshold of T^2 statistic for the PCA-based method [1] at 95% confidence level is calculated based on normal data samples of four variables in Fig. 3.50. As shown in Fig. 3.61, it is clear that the T^2 statistic triggers many nuisance alarms along with the abnormality appearing at $t = 44424$ sec. As shown in Fig. 3.62, the 95% confidence ellipse in the PCA-based method does not describe normal data samples well, as they are not with the Gaussian distribution [47]. Thus, the PCA-based method cannot detect the abnormality correctly.

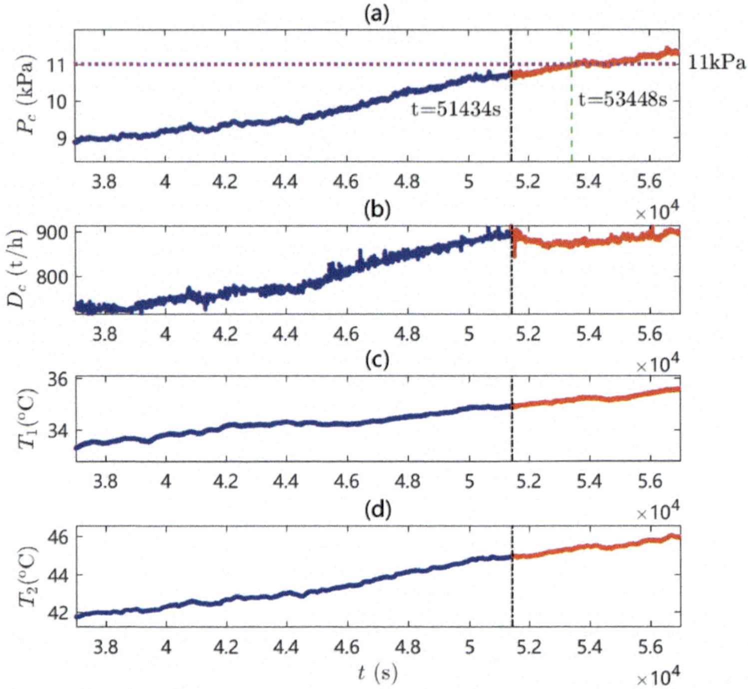

Fig. 3.57 Data samples of four variables for abnormality #2, reprinted from Ref. [44], copyright 2024, with permission from IEEE

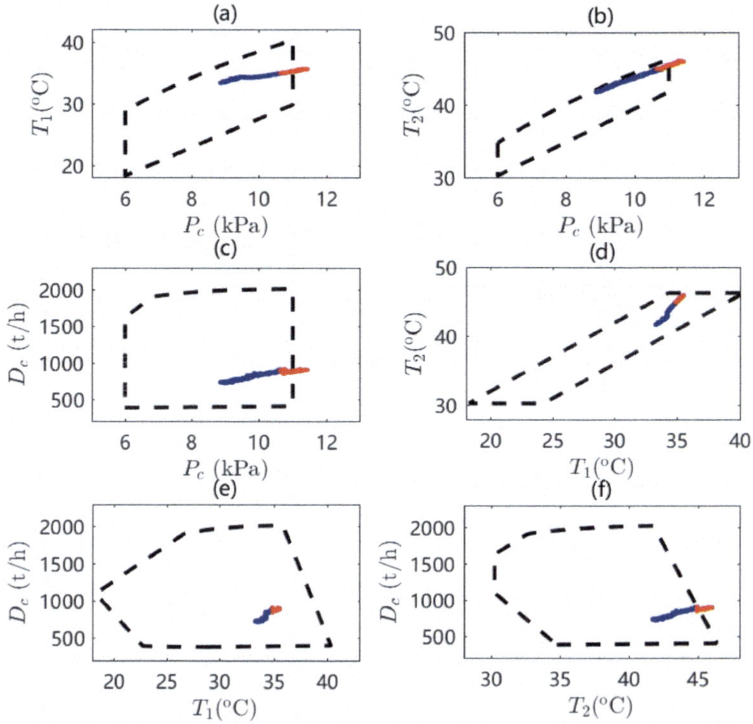

Fig. 3.58 Scatter plots with condenser operating zone model (black dash) on two-dimensional planes for abnormality #2, reprinted from Ref. [44], copyright 2024, with permission from IEEE

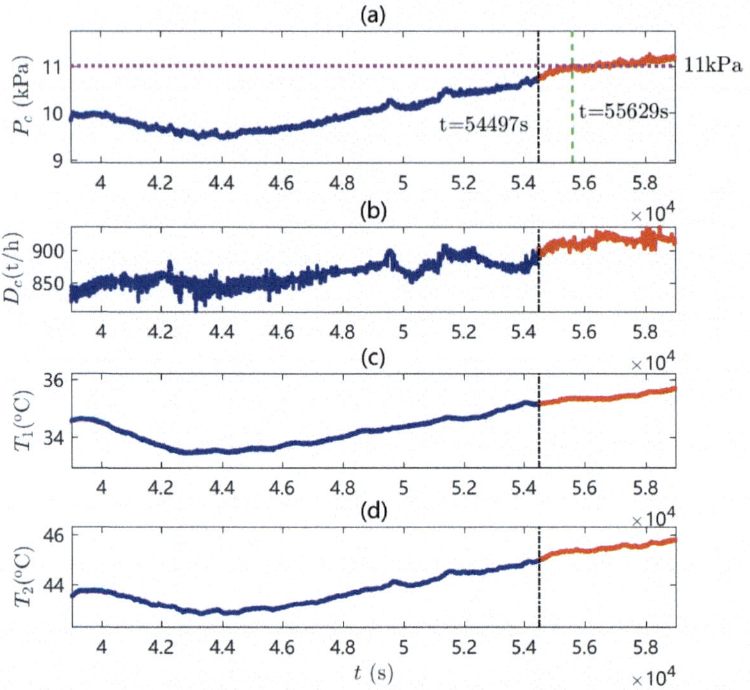

Fig. 3.59 Data samples of four variables for abnormality #3, reprinted from Ref. [44], copyright 2024, with permission from IEEE

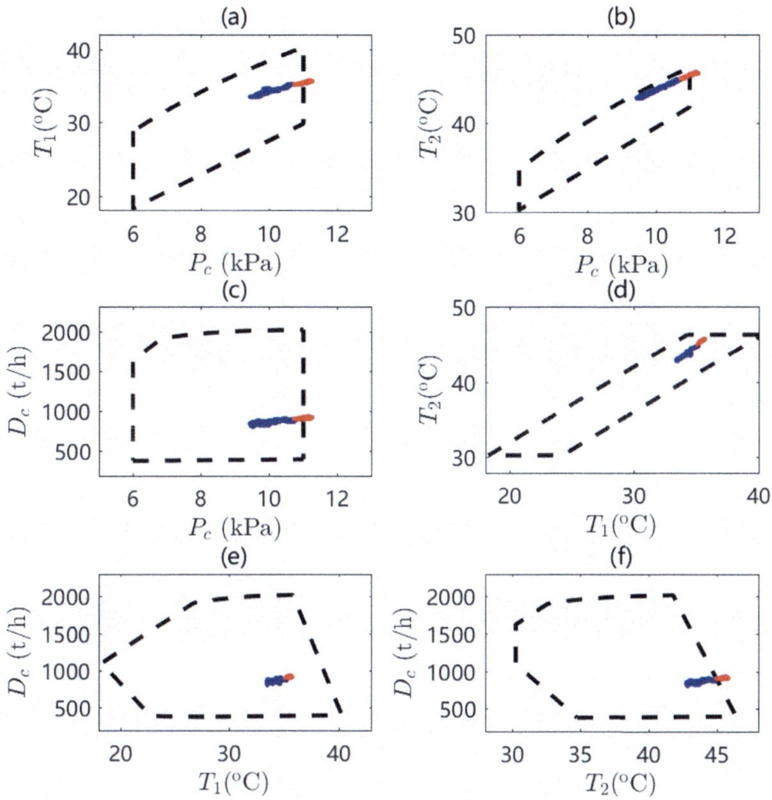

Fig. 3.60 Scatter plots with condenser operating zone model (black dash) on two-dimensional planes for abnormality #3, reprinted from Ref. [44], copyright 2024, with permission from IEEE

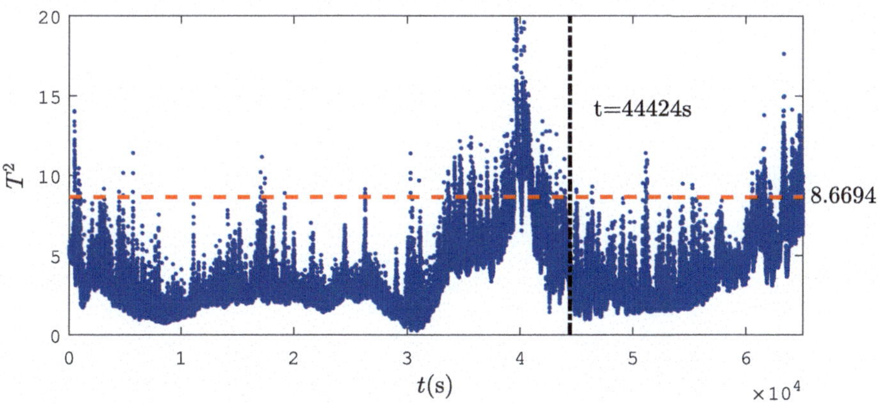

Fig. 3.61 Time sequence plot of T^2 statistic (blue dot) and its control limit (red dash) for abnormality #1, reprinted from Ref. [44], copyright 2024, with permission from IEEE

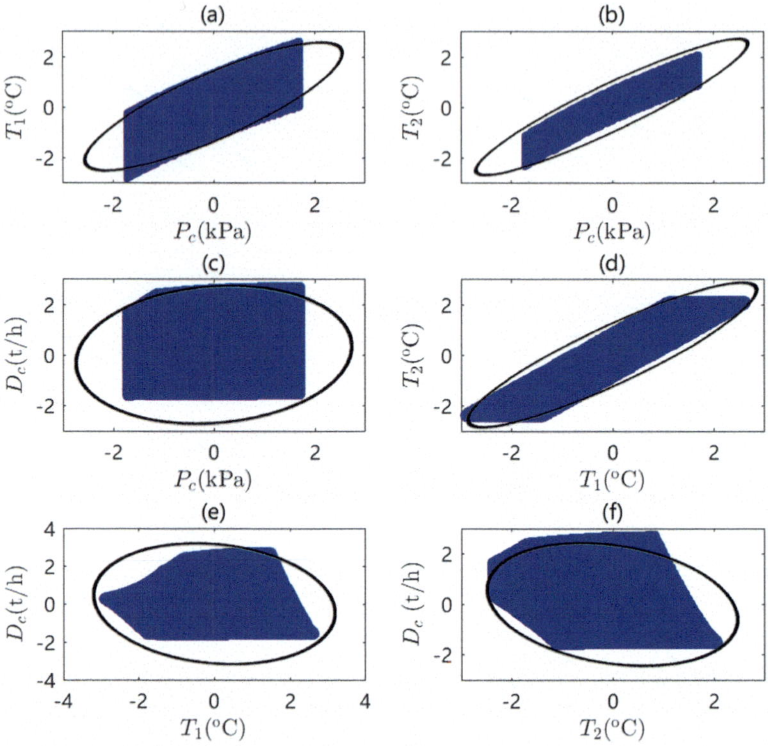

Fig. 3.62 Normalized normal data samples (blue dot) and 95% confidence ellipse (black solid) of PCA method on two-dimensional planes for abnormality #1, reprinted from Ref. [44], copyright 2024, with permission from IEEE

References

1. Ali, A., & Mahdi, D. (2014). Independent component analysis approach for fault diagnosis of condenser system in thermal power plant. *Journal of Central South University, 21*, 242–251.
2. Arastou, A., Ahmadi, P., & Karrari, M. (2022). Modeling and parameter estimation of a steam power plant including condenser back-pressure uncertainty using operational data. *IEEE Systems Journal, 16*, 2979–2988.
3. Bakshi, B. R., & Stephanopoulos, G. (1994). Representation of process trends-III. Multiscale extraction of trends from process data. *Computers & Chemical Engineering, 18*, 267–302.
4. Barber, C. B., Dobkin, D. P., & Huhdanpaa, H. (1996). The quickhull algorithm for convex hulls. *ACM Trans. Mathematical Software, 22*, 469–483.
5. Basu, D., & Pani, A.K. (2020). Back pressure monitoring of power plant condenser using multiple adaptive regression spline. *Emerging Research in Data Engineering Systems and Computer Communications: Proceedings of CCODE 2019*, Singapore: Springer Singapore, 1–10.
6. Blumenson, L. E. (1960). A derivation of n-Dimensional spherical coordinates. *The American Mathematical Monthly, 67*, 63–66.
7. Charbonnier, S., & Portet, F. (2012). A self-tuning adaptive trend extraction method for process monitoring and diagnosis. *Journal of Process Control, 22*, 1127–1138.

8. Chen, H., Kremling, A., & Allgower, F. (1995). Nonlinear predictive control of a benchmark CSTR. *Proceedings of 3rd European Control Conference*, 3247–3252.
9. Chen, K., & Wang, J. (2017). Design of multivariate alarm systems based on online calculation of variational directions. *Chemical Engineering Research and Design, 122*, 11–21.
10. Chen, Z. (2003). Bayesian filtering: From Kalman filters to particle filters, and beyond. *Statistics, 182*, 1–69.
11. EEMUA. (2013). *EEMUA-191: Alarm Systems-A Guide to Design*. Management and Procurement: Engineering Equipment and Materials Users' Association.
12. Flehmig, F., Watzdorf, R. V., & Marquardt, W. (1998). Identification of trends in process measurements using the wavelet transform. *Computers & Chemical Engineering, 22*, S491–S496.
13. Flehmig, F., & Marquardt, W. (2006). Detection of multivariable trends in measured process quantities. *Journal of Process Control, 16*, 947–957.
14. Flynn, D. (2003). *Thermal Power Plant Simulation and Control*. London: Institution of Electrical Engineers.
15. Hodge, V., & Austin, J. (2004). A survey of outlier detection methodologies. *Artificial Intelligence Review, 22*, 85–126.
16. Hsieh, F., Lavori, P., Cohen, H., & Feussner, J. (2003). An overview of variance inflation factors for sample-size calculation. *Evaluation & the Health Professions, 26*, 239–257.
17. ISA. (2009). *ANSI/ISA-18.2: Management of alarm systems for the process industries*. International Society of Automation.
18. Isermann, R. (2011). *Fault Diagnosis Applications: Model-Based Condition Monitoring*. London: Springer.
19. Janusz, M. E., & Venkatasubramanian, V. (1991). Detection of multivariable trends in measured process quantities. *Engineering Applications of Artificial Intelligence, 4*, 329–339.
20. Jean, M., Leroy, T., & Naquet, F. V. (2022). Bayesian estimation and control of engine knocking level for transient operation. *IEEE Transactions on Control Systems Technology, 30*, 944–955.
21. Keogh, E., Chu, S., Hart, D., & Pazzani, M. (2004). Segmenting time series: A survey and novel approach. *Data Mining in Time Series Databases, 57*, 1–22.
22. Kostyuk, A., & Frolov, V. (1998). *Steam and Gas Turbines*. Moscow: Mir Publishers.
23. Kroll, A., & Schulte, H. (2014). Benchmark problems for nonlinear system identification and control using soft computing methods: Need and overview. *Applied Soft Computing, 25*, 496–513.
24. Ljung, L. (1999). *System identification: Theory for the user*. New Jersey: Prentice Hall.
25. Lowry, C. A., Woodall, W. H., Champ, C. W., & Rigdon, S. E. (1992). A multivariate exponentially weighted moving average control chart. *Technometrics, 34*, 46–53.
26. Ma, X., Zhao, X., Zhang, Y., Liu, K., Yang, H., Li, J., Akhlaghi, Y. G., Liu, H., Han, Z., & Liu, Z. (2022). Combined Rankine Cycle and dew point cooler for energy efficient power generation of the power plants-A review and perspective study. *Energy, 238*, 121688.
27. Maestri, M., & Farall, A. (2010). A robust clustering method for detection of abnormal situations in a process with multiple steady-state operation modes. *Computers & Chemical Engineering, 34*, 223–231.
28. Massey, Jr., & Frank, J. E. (1951). The Kolmogorov-Smirnov test for goodness of fit. *Journal of American Statistical Association, 46*, 68–78.
29. Maurya, M. R., Paritosh, P. K., Rengaswamy, R., & Venkatasubramanian, V. (2014). A framework for on-line trend extraction and fault diagnosis. *Computers & Chemical Engineering, 67*, 43–52.
30. Melek, W. W., Lu, Z., Kapps, A., & Fraser, W. D. (2005). A framework for on-line trend extraction and fault diagnosis. *IEEE Transactions on Biomedical Engineering, 52*, 639–651.
31. Pignatiello, J. J., & Runger, G. C. (1990). Comparisons of multivariate CUSUM charts. *Journal of Quality Technology, 22*, 173–186.
32. Pulman, E. (2001). *Steam surface condensers*. New York: ASME Press.
33. Raja, A.K., Srivastava, A.P., & Dwivedi, M. (2006). *Power plant engineering*. New Age International.

34. Rengaswamy, R., & Venkatasubramanian, V. (1995). A syntactic pattern-recognition approach for process monitoring and fault diagnosis. *Engineering Applications of Artificial Intelligence, 8*, 35–51.
35. Sargent, R. G. (2013). Verification and validation of simulation models. *Journal of Simulation, 7*, 12–24.
36. Savitzky, A., & Golay, M. J. (1964). Smoothing and differentiation of data by simplified least squares procedures. *Analytical Chemistry, 36*, 1627–1639.
37. Smith, R.C. (2014). *Uncertainty quantification: Theory, implementation, and applications*, Society for Industrial and Applied Mathematics.
38. Vedam, H., Venkatasubramanian, V., & Bhalodia, M. (1998). A B-spline based method for data compression, process monitoring and diagnosis. *Computers & Chemical Engineering, 22*, S827–S830.
39. Wang, J., & Chen, T. (2014). An online method to remove chattering and repeating alarms based on alarm durations and intervals. *Computers & Chemical Engineering, 67*, 43–52.
40. Wang, J., Wang, Z., Zhou, X., & Yang, F. (2022). Design of delay timers based on estimated probability mass functions of alarm durations. *Journal of Process Control, 110*, 154–165.
41. Wang, J., Sun, S., Wang, Z., Zhou, X., & Gyasi, P. (2023). Alarm deadband design based on maximum amplitude deviations and Bayesian estimation. *IEEE Transactions on Control Systems Technology, 31*, 1941–1948.
42. Wang, W., Zeng, D., Liu, J., Niu, Y., & Cui, C. (2014). Feasibility analysis of changing turbine load in power plants using continuous condenser pressure adjustment. *Energy, 64*, 533–540.
43. Wang, Z., & Wang, J. (2024). Multivariate alarm monitoring for non-convex normal operating zones based on search cones. *IEEE Transactions on Automation Science and Engineering, 21*, 452–462.
44. Wang, Z., Wang, J., & Liu, J. (2024). Multivariate process monitoring for safe operation of condensers in thermal power plants based on normal operating zones. *IEEE Transactions on Control Systems Technology*. https://doi.org/10.1109/TCST.2024.3370036
45. Wise, B. M., & Gallagher, N. B. (1996). The process chemometrics approach to process monitoring and fault detection. *Journal of Process Control, 6*, 329–348.
46. Xiong, W., Wang, J., & Chen, K. (2018). Multivariate alarm systems for time-varying processes using Bayesian filters with applications to electrical pumps. *IEEE Transactions on Industrial Informatics, 14*, 504–513.
47. Xu, Y., Shen, S., He, Y., & Zhu, Q. (2019). A novel hybrid method integrating ICA-PCA with relevant vector machine for multivariate process monitoring. *IEEE Transactions on Control Systems Technology, 27*, 1780–1787.
48. Yadav, R. (2004). *Steam and gas turbines and power plant engineering* (7th ed.). Central Publishing House.
49. Yu, J., Petersen, N., Liu, P., Li, Z., & Wirsum, M. (2022). Hybrid modelling and simulation of thermal systems of in-service power plants for digital twin development. *Energy, 260*, 125088.
50. Yu, Y., & Wang, J. (2020). Alarm monitoring for multivariate processes based on a convex-hull normal operating zone. *IEEE Transactions on Control Systems Technology, 28*, 2649–2656.
51. Zhang, X., Yuan, J., Xu, L., Tian, Z., & Wang, J. (2017). Pseudo-online optimization of condenser pressure for the cold-end system with variable speed pumps. *Applied Thermal Engineering, 126*, 339–349.
52. Zhou, B., Ye, H., Zhang, H., & Li, M. (2017). A new qualitative trend analysis algorithm based on global polynomial fit. *AiChE Journal, 63*, 3374–3383.

Chapter 4
Root-Cause Analysis of Alarm Events

Abstract This chapter is on the root-cause analysis of alarm events based on alarm and process variables. First, two causality inference methods are presented to build connections among alarm and process variables in an off-line manner from historical data. Such connections as information flow paths are helpful to narrow down the search space of alarm and process variables for root causes of abnormalities. Second, root-cause analysis methods are proposed for alarm and process variables to determine the root causes of alarm events in an online manner from recent data. Numerical and industrial examples are given to illustrate these methods.

4.1 Causality Inference for Alarm Variables

This section is on the analysis of cause-effect relationships among binary-valued alarm variables [21]. A normalized transfer entropy (NTE) and a normalized direct transfer entropy (NDTE) are formulated as basic statistical metrics, with consideration of two specific characteristics of alarm variables, namely, random delays and mutual independence among alarm occurrences. A modified statistical test is developed to determine significance thresholds of NTEs and NDTEs, based on which the cause-effect relationships can be inferred.

4.1.1 Problem Description

In this section, alarm variables are denoted as x, y, and z, and their time sequences as x_t, y_t, and z_t, in order to simplify notations. Such simplified notations are taken for definitions of NTE and NDTE in later context, as subscripts and superscripts will be used to represent the extended forms of alarm variables. Two characteristics of alarm variables are assumed:

1. If there is a causal relationship from an alarm variable x to another one y, the time interval λ that an alarm occurrence $y_{t+\lambda} = 1$ follows its counterpart $x_t = 1$

J. Wang et al., *Intelligent Industrial Alarm Systems*,
https://doi.org/10.1007/978-981-97-6516-4_4

in x is a random variable due to disturbances. Such a time interval λ is referred to as the random occurrence delay.

2. Alarm occurrences of each alarm variable are mutually independent,

$$P(x_{t_1} = 1, x_{t_2} = 1) = P(x_{t_1} = 1)P(x_{t_2} = 1), \ \forall t_1 \neq t_2, \tag{4.1}$$

where $P(\cdot)$ indicates the probability of a given event.

The reasons why the aforementioned two assumptions are typically met are as follows: First, disturbances are quite common in industrial processes and often result in the emergence of random occurrence delays [20]. Second, abnormalities that are isolated in time are generally independent of each other, which means that the alarms associated with them are also independent.

With these two assumptions, the goal is to establish the cause-effect connections among alarm variables by examining historical alarm data. Identifying such an information flow path is crucial for reducing the scope of alarm variables when searching for the root causes of abnormalities. It is essential to identify alarms that are directly linked to primary abnormalities and those that are indirectly affected as consequences [27].

The orders of time stamps of alarm occurrences are intended to reveal the cause-effect relationships. However, the orders are often not coherent. In response, a structured approach is introduced here for analyzing the cause-effect relationships among alarm variables. This approach is divided into two main stages: Initially, the NTE and NDTE are established as fundamental statistical measures, taking into account the random occurrence delays and the mutual independence of alarm occurrences; subsequently, a statistical test utilizing surrogate alarm data is employed to ascertain the significance thresholds for the NTE and NDTE, which indicate the causal strengths among the alarm variables.

4.1.2 Modified Transfer Entropies for Alarm Variables

This subsection defines the modified transfer entropies for alarm variables, including the NTE and NDTE.

4.1.2.1 Normalized Transfer Entropy

Given two process variables $\tilde{x}$ and $\tilde{y}$, their time sequences are denoted by $\tilde{X} = \{\tilde{x}_t \in \tilde{\mathcal{X}}; t = 1, 2, \ldots, L\}$ and $\tilde{Y} = \{\tilde{y}_t \in \tilde{\mathcal{Y}}; t = 1, 2, \ldots, L\}$, where L denotes the number of samples. $\tilde{\mathcal{X}}$ and $\tilde{\mathcal{Y}}$ indicate the amplitude ranges of $\tilde{x}_t$ and $\tilde{y}_t$, respectively. The transfer entropy (TE) from $\tilde{X}$ to $\tilde{Y}$ is the amount of uncertainty measured by Shannon entropy reduced in future values of $\tilde{y}_t$ by knowing the past values $\tilde{X}$ given the past values of $\tilde{Y}$ [38]. Given two time sequences $\tilde{X}$ and $\tilde{Y}$, their transfer entropy from $\tilde{X}$ to $\tilde{Y}$ is formulated as

$$T_{\tilde{X}\to\tilde{Y}} = H\left(\tilde{y}_{t+1}|\tilde{\mathbf{y}}_t^{(k)}\right) - H\left(\tilde{y}_{t+1}|\tilde{\mathbf{y}}_t^{(k)}, \tilde{\mathbf{x}}_t^{(l)}\right), \tag{4.2}$$

where

$$H\left(\tilde{y}_{t+1}|\tilde{\mathbf{y}}_t^{(k)}, \tilde{\mathbf{x}}_t^{(l)}\right) = \int_{\tilde{y}_{t+1}\in\tilde{\mathcal{Y}},\tilde{\mathbf{y}}_t^{(k)}\in\tilde{\mathcal{Y}}^k,\tilde{\mathbf{x}}_t^{(l)}\in\tilde{\mathcal{X}}^l} p\left(\tilde{y}_{t+1}, \tilde{\mathbf{y}}_t^{(k)}, \tilde{\mathbf{x}}_t^{(l)}\right) \cdot \log_2 \frac{1}{p\left(\tilde{y}_{t+1}|\tilde{\mathbf{y}}_t^{(k)}, \tilde{\mathbf{x}}_t^{(l)}\right)},$$

$$H\left(\tilde{y}_{t+1}|\tilde{\mathbf{y}}_t^{(k)}\right) = \int_{\tilde{y}_{t+1}\in\tilde{\mathcal{Y}},\tilde{\mathbf{y}}_t^{(k)}\in\tilde{\mathcal{Y}}^k} p\left(\tilde{y}_{t+1}, \tilde{\mathbf{y}}_t^{(k)}\right)\log_2\frac{1}{p\left(\tilde{y}_{t+1}|\tilde{\mathbf{y}}_t^{(k)}\right)}.$$

Here $p(\cdot)$ and $p(\cdot|\cdot)$ are the joint and conditional PDFs, respectively; the symbol $\int$ represents the integration of all possible combinations of probabilities; $\tilde{\mathbf{y}}_t^{(k)} := [\tilde{y}_t, \tilde{y}_{t-\tau}, \ldots, \tilde{y}_{t-(k-1)\tau}]$ and $\tilde{\mathbf{x}}_t^{(l)} := [\tilde{x}_t, \tilde{x}_{t-\tau}, \ldots, \tilde{x}_{t-(l-1)\tau}]$ denote the embedding vectors of past states of $\tilde{Y}$ and $\tilde{X}$, respectively; integers k and l are the embedding dimension of $\tilde{y}_t$ and $\tilde{x}_t$, respectively; the integer τ is the so-called embedding delay. To calculate the TE, a critical step is to estimate the PDFs.

Unlike analog variables, which can take on a range of values, alarm variables are binary-valued, taking the values of 0 or 1. The Transfer Entropy (TE) has been utilized for binary data in fields such as communications and neuroscience [15, 23, 49]. When applying TE to alarm data, it is imperative to take into account the two distinctive characteristics of alarm variables as discussed in Sect. 4.1.1.

First, the TE in (4.2) cannot be directly generalized to alarm variables due to the presence of random occurrence delays stated in Sect. 4.1.1. Given two alarm variables x and y, their time sequences are denoted by $X = \{x_t \in \mathcal{H}; t = 1, 2, \ldots, L\}$ and $Y = \{y_t \in \mathcal{H}; t = 1, 2, \ldots, L\}$, where L denotes the number of samples and $\mathcal{H} = \{0, 1\}$. Figure 4.1 shows a numerical example of random occurrence delays between X and Y. There are four alarm occurrences in X as the 1's in the dashed windows in Fig. 4.1. For each occurrence in X, there is a corresponding alarm occurrence in Y with some occurrence delay λ. The four random occurrence delays are $\lambda(1) = 4$, $\lambda(2) = 3$, $\lambda(3) = 2$, and $\lambda(4) = 5$. Thus, the random occurrence delays lead us to care only about the alarm occurrence in X within a historical window. In view of this, a modified TE is proposed as

$$T_{X\to Y} = H\left(y_{t+1}|\overset{\circ}{y}_t^{(k_1)}\right) - H\left(y_{t+1}|\overset{\circ}{y}_t^{(k_1)}, \overset{\circ}{x}_{t-d_1}^{(l_1)}\right), \tag{4.3}$$

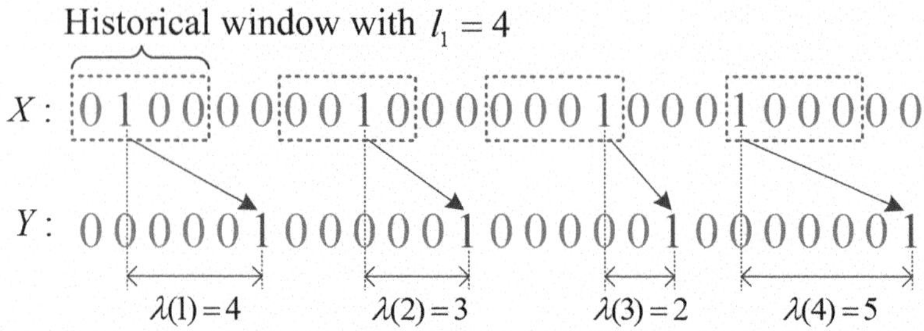

Fig. 4.1 A numerical example for predicting Y based on the history of X under the influence of random occurrence delays, reprinted from Ref. [21], copyright 2017, with permission from Elsevier

where

$$H\left(y_{t+1}|\overset{\circ}{y}_t^{(k_1)}, \overset{\circ}{x}_{t-d_1}^{(l_1)}\right) = \sum_{y_{t+1}, \overset{\circ}{y}_t^{(k_1)}, \overset{\circ}{x}_{t-d_1}^{(l_1)} \in \mathcal{H}^3} p\left(y_{t+1}, \overset{\circ}{y}_t^{(k_1)}, \overset{\circ}{x}_{t-d_1}^{(l_1)}\right) \cdot \log_2 \frac{1}{p\left(y_{t+1}|\overset{\circ}{y}_t^{(k_1)}, \overset{\circ}{x}_{t-d_1}^{(l_1)}\right)},$$

$$H\left(y_{t+1}|\overset{\circ}{y}_t^{(k_1)}\right) = \sum_{y_{t+1}, \overset{\circ}{y}_t^{(k_1)} \in \mathcal{H}^2} p\left(y_{t+1}, \overset{\circ}{y}_t^{(k_1)}\right) \log_2 \frac{1}{p\left(y_{t+1}|\overset{\circ}{y}_t^{(k_1)}\right)}.$$

Here k_1 and l_1 are the time horizons of the effect variable and cause variable, respectively; d_1 denotes the time lag between X and Y; the symbol $\sum$ represents the sum of all possible combinations of probabilities related to y_{t+1}, $\overset{\circ}{y}_t^{(k_1)}$, and $\overset{\circ}{x}_{t-d_1}^{(l_1)}$.

Specifically, $\overset{\circ}{x}_{t-d_1}^{(l_1)}$ and $\overset{\circ}{y}_t^{(k_1)}$ are calculated as

$$\overset{\circ}{x}_{t-d_1}^{(l_1)} = \begin{cases} \bigcup_{u=t-d_1-l_1+1}^{t-d_1} x_u, & \text{if } y_{t+1} = 1, \\ x_{t-d_1}, & \text{otherwise}, \end{cases} \tag{4.4}$$

$$\overset{\circ}{y}_t^{(k_1)} = \begin{cases} \bigcup_{u=t-k_1+1}^{t} y_u, & \text{if } y_{t+1} = 1, \\ y_t, & \text{otherwise}. \end{cases} \tag{4.5}$$

Here x_t and y_t take values from $\{0, 1\}$, $\bigcup_{u=t_1}^{t_2} x_u = x_{t_1} \vee x_{t_1+1} \vee \cdots \vee x_{t_2}$ for $t_1 < t_2$, and the symbol $\vee$ stands for the "or" operation of Boolean algebra. In words, $\overset{\circ}{x}_{t-d_1}^{(l_1)}$ is equal to 1 as long as any samples of x_t in a historical window take the value 1.

Next, a normalized transfer entropy (NTE) is formulated as

$$NT_{X\to Y} = \frac{T_{X\to Y}}{H\left(y_{t+1}|\mathring{y}_t^{(k_1)}\right)} = 1 - \frac{H\left(y_{t+1}|\mathring{y}_t^{(k_1)}, \mathring{x}_{t-d_1}^{(l_1)}\right)}{H\left(y_{t+1}|\mathring{y}_t^{(k_1)}\right)}. \tag{4.6}$$

The conditional entropies $H\left(y_{t+1}|\mathring{y}_t^{(k_1)}\right)$ and $H\left(y_{t+1}|\mathring{y}_t^{(k_1)}, \mathring{x}_{t-d_1}^{(l_1)}\right)$ are positive with $H\left(y_{t+1}|\mathring{y}_t^{(k_1)}\right) \geq H\left(y_{t+1}|\mathring{y}_t^{(k_1)}, \mathring{x}_{t-d_1}^{(l_1)}\right)$. Thus, $NT_{X\to Y}$ is in the range of [0, 1]. The NTE closer to 1 (0) indicates a stronger (weaker) causality.

Equation (4.6) includes three non-negative integer parameters, namely, d_1, k_1, and l_1. The time lag d_1 can be identified as the value that corresponds to the highest $NT_{X\to Y}(d_1)$'s for all nonnegative values of $d_1 \leq D$, where D is a user-defined maximum value of time lags. Since alarm occurrences are mutually independent as outlined in Sect. 4.1.1, there is no necessity to account for the influence of Y's past states. Consequently, the effect variable's time horizon can be fixed at $k_1 = 0$. The parameter l_1 represents the quantity of historical states of X that should be incorporated for predicting Y. The principle for choosing l_1 is that most of the random occurrence delays are smaller than l_1 so that their effects are removed in $\mathring{x}_{t-d_1}^{(l_1)}$ in (4.4). The time stamps for all alarms in X (or Y) are represented by $S^x = [s_1^x, \ldots, s_{M_x}^x]'$ (or $S^y = [s_1^y, \ldots, s_{M_y}^y]'$), where M_x (or M_y) signifies the total count of alarm occurrences for X (or Y). Then, the time intervals between alarm occurrences in Y and the counterparts in X are

$$\Gamma = [\gamma]_{M_y\times M_x} = S^y \cdot (\mathbf{I}_{M_x\times 1})' - \mathbf{I}_{M_y\times 1} \cdot (S^x)', \tag{4.7}$$

where $\mathbf{I}$ is an all-one vector. The minimum positive element in each row is found as

$$\lambda_i = \min\left\{\gamma_{i,j} : \gamma_{i,j} > 0,\, j = 1, 2, \ldots, M_x\right\}, \tag{4.8}$$

where $i = 1, 2, \ldots, M_y$. The majority of random occurrence delays can be found from 1.5×IQR (interquartile range) while the values beyond the range are outliers [34]. Hence, the time horizon of the cause variable can be set as $l_1 = 1.5\times\mathrm{IQR}_\lambda$, where IQR_λ indicates the difference between the third quartile and the first quartile of λ_i, $i = 1, 2, \ldots, M_y$. Further, l_1 should not be set too large so as to avoid the all-one vector of $\mathring{x}_t^{(l_1)}$ in (4.4). An upper bound of l_1 is given as 20, which is the maximum time interval of non-chattering alarms according to the ISA standard [2]. As a result, the parameter l_1 is finally determined as

$$l_1 = \min\{1.5 \times \mathrm{IQR}_\lambda, 20\}. \tag{4.9}$$

4.1.2.2 Normalized Direct Transfer Entropy

The concept of direct transfer entropy (DTE) was introduced in [12] to ascertain if the causal relationship between two continuously-valued process variables is direct or through intermediate variables. This concept is similarly applied to define the DTE for alarm variables. First, similar to (4.6), the NTE from X to the third alarm sequence $Z = \{z_t \in \mathcal{H}; t = 1, 2, \ldots, L\}$ is calculated as

$$NT_{X\to Z} = 1 - \frac{H\left(z_{t+1}|\overset{\circ}{z}{}_t^{(m_1)}, \overset{\circ}{x}{}_{t-d_2}^{(l_2)}\right)}{H\left(z_{t+1}|\overset{\circ}{z}{}_t^{(m_1)}\right)}, \tag{4.10}$$

where $\overset{\circ}{z}{}_t^{(m_1)} = z_t \vee z_{t-1} \vee \ldots \vee z_{t-m_1+1}$; l_2 and m_1 are the time horizons of X and Z, respectively; d_2 indicates the time lag between X and Z. Next, the NTE from Z to Y is

$$NT_{Z\to Y} = 1 - \frac{H\left(y_{t+1}|\overset{\circ}{y}{}_t^{(k_2)}, \overset{\circ}{z}{}_{t-d_3}^{(m_2)}\right)}{H\left(y_{t+1}|\overset{\circ}{y}{}_t^{(k_2)}\right)}, \tag{4.11}$$

where k_2 and m_2 are the time horizons of Y and Z, respectively; d_3 indicates the time lag between Z and Y. Finally, the DTE from X to Y based on Z is formulated as

$$D_{X\to Y|Z} = H\left(y_{t+1}|\overset{\circ}{y}{}_t^{(k')}, \overset{\circ}{z}{}_{t-d_3}^{(m_2)}\right) - H\left(y_{t+1}|\overset{\circ}{y}{}_t^{(k')}, \overset{\circ}{z}{}_{t-d_3}^{(m_2)}, \overset{\circ}{x}{}_{t-d_1}^{(l_1)}\right),$$

where $k' = k_1 = k_2$ and

$$H\left(y_{t+1}|\overset{\circ}{y}{}_t^{(k')}, \overset{\circ}{z}{}_{t-d_3}^{(m_2)}\right) = \sum_{y_{t+1}, \overset{\circ}{y}{}_t^{(k')}, \overset{\circ}{z}{}_{t-d_3}^{(m_2)} \in \mathcal{H}^3} p\left(y_{t+1}, \overset{\circ}{y}{}_t^{(k')}, \overset{\circ}{z}{}_{t-d_3}^{(m_2)}\right) \cdot \log_2 \frac{1}{p\left(y_{t+1}|\overset{\circ}{y}{}_t^{(k')}, \overset{\circ}{z}{}_{t-d_3}^{(m_2)}\right)},$$

$$H\left(y_{t+1}|\overset{\circ}{y}{}_t^{(k')}, \overset{\circ}{z}{}_{t-d_3}^{(m_2)}, \overset{\circ}{x}{}_{t-d_1}^{(l_1)}\right) = \sum_{y_{t+1}, \overset{\circ}{y}{}_t^{(k')}, \overset{\circ}{z}{}_{t-d_3}^{(m_2)}, \overset{\circ}{x}{}_{t-d_1}^{(l_1)} \in \mathcal{H}^4} p\left(y_{t+1}, \overset{\circ}{y}{}_t^{(k')}, \overset{\circ}{z}{}_{t-d_3}^{(m_2)}, \overset{\circ}{x}{}_{t-d_1}^{(l_1)}\right) \cdot \log_2 \frac{1}{p\left(y_{t+1}|\overset{\circ}{y}{}_t^{(k')}, \overset{\circ}{z}{}_{t-d_3}^{(m_2)}, \overset{\circ}{x}{}_{t-d_1}^{(l_1)}\right)}.$$

The normalized direct transfer entropy (NDTE) from X to Z based on Y is then formulated as

$$ND_{X\to Y|Z} = \frac{D_{X\to Y|Z}}{H(y_{t+1}|\overset{\circ}{y}_t^{(k')}, \overset{\circ}{z}_{t-d_3}^{(m_2)})}$$
$$= 1 - \frac{H(y_{t+1}|\overset{\circ}{y}_t^{(k')}, \overset{\circ}{z}_{t-d_3}^{(m_2)}, \overset{\circ}{x}_{t-d_1}^{(l_1)})}{H(y_{t+1}|\overset{\circ}{y}_t^{(k')}, \overset{\circ}{z}_{t-d_3}^{(m_2)})}. \tag{4.12}$$

The range of $ND_{X\to Y|Z}$ is [0, 1]. An NDTE closer to 1 (0) indicates a stronger (weaker) direct causality. The NDTE is able to tell whether the causality from X to Y is direct or through Z, so that the direct information flow path can be inferred.

4.1.3 Causality Inference Method

This section presents a statistical test to determine the significance of the NTE and NDTE, investigates the minimum requirement of alarm occurrences, and presents the steps of the proposed method.

4.1.3.1 Significance Test

The main idea of the significance test for identifying the causality from X to Y is as follows. The null hypothesis posits that $NT_{X\to Y}$ is not significant, implying there is no causality from X to Y. This null hypothesis is tested by contrasting the observed NTE with the counterparts derived from surrogate alarm data without causal relationships.

Given original alarm sequences X and Y, the surrogate alarm sequences X^s and Y^s should have the same length and the same probability of alarm occurrences as X and Y, respectively. Mathematically, X^s is generated with Bernoulli distribution $\mathcal{B}(1, q)$ such that

- the sequence length of X^s is the same as that of X;
- for each sample $x_t^s \in X^s$, the value of x_t^s is either 0 or 1;
- the probability that $x_t^s = 1$ is q, where $q = \frac{M_x}{L}$, M_x is the number of 1's in X, and L is the sequence length of X.

The sequence Y^s is generated analogously. The NTE of X^s and Y^s is calculated as $NT^s_{X^s\to Y^s}$, and the NDTE for surrogate alarm sequences X^s and Y^s based on Z^s is denoted as $ND^s_{X^s\to Y^s|Z^s}$.

The significance test based on a Monte Carlo p-value [14, 19, 36, 38] is used here. The Monte Carlo p-value is formulated as

$$\mathcal{P} = \frac{b+1}{N_s+1}, \tag{4.13}$$

where N_s is the number of simulations, and b indicates the number of NTEs (or NDTEs) from the surrogates larger than the NTE (or NDTE) from the observed data. If $\mathcal{P} < \alpha$, the null hypothesis is rejected at the α significance level. The NTE (or NDTE) of X^s and Y^s is denoted by $NT^s_{X^s \to Y^s}(i)$ (or $ND^s_{X^s \to Y^s|Z^s}(i)$) for the i-th simulation. Here a criterion is chosen that $ND^s_{X^s \to Y^s|Z^s}(i) \geq ND_{X \to Y|Z}$ (or $NT^s_{X^s \to Y^s}(i) \geq NT_{X \to Y}$) is not allowed for the rejection of null hypothesis, i.e., $b = 0$. Given $\alpha = 0.01$, the number of simulations is $N_s = 99$ from (4.13). Thus, the significance threshold γ_T (or γ_D) for NTE (or NDTE) is found as the maximum value of $NT^s_{X^s \to Y^s}(i)$ (or $ND^s_{X^s \to Y^s|Z^s}(i)$), $i = 1, 2, \ldots, N_s$. To make the significance test more robust, the above procedures of Monte Carlo test are repeated for K times. The significance thresholds γ_T and γ_D for NTE and NDTE are calculated respectively as

$$\gamma_T = \frac{1}{K} \sum_{k=1}^{K} \gamma_T(k), \tag{4.14}$$

$$\gamma_D = \frac{1}{K} \sum_{k=1}^{K} \gamma_D(k). \tag{4.15}$$

Here $\gamma_T(k)$ and $\gamma_D(k)$ are respectively the maximum values of $NT^s_{X^s \to Y^s}(i)$ and $ND^s_{X^s \to Y^s|Z^s}(i)$ for $i = 1, 2, \ldots, N_s$ in the k-th Monte Carlo test, i.e.,

$$\gamma_T(k) = \max_{i=1,2,\ldots,N_s} NT^s_{X^s \to Y^s}(i), \tag{4.16}$$

$$\gamma_D(k) = \max_{i=1,2,\ldots,N_s} ND^s_{X^s \to Y^s|Z^s}(i). \tag{4.17}$$

Based on (4.14), the null hypothesis is rejected if $NT_{X \to Y} > \gamma_T$. Otherwise, there is no causality from X to Y. Similarly, based on (4.15), the null hypothesis is rejected if $ND_{X \to Y|Z} > \gamma_D$, so that the direct causality from X to Y based on Z is determined. A numerical example is given next to test the effectiveness of the significance test.

Two alarm sequences X and Y are generated as two independent random binary sequences with data length L that follow the Bernoulli distribution $\mathcal{B}(1, q)$, where q is the probability of alarm occurrences. As a result, X and Y have no causal relationship. First, a case with $q = 0.015$ and $L = 10,000$ is presented in order to show how the significance threshold of NTEs changes with the number K of Monte Carlo simulations. Figure 4.2a shows the significance thresholds of 1000 simulations while Fig. 4.2b gives the trend of variance of γ_T versus K. It can be found that the variations of γ_T become smooth for $K \geq 15$, so that $K = 15$ is a reasonable choice.

Second, more Monte Carlo simulations are implemented with $L = 10,000$ and q varying in each simulation with $q \sim \mathcal{U}(0.005, 0.05)$, where $\mathcal{U}$ is a uniform distribution. There are 1000 pairs of $NT_{X \to Y}$ and γ_T obtained in Fig. 4.3a. Only four pairs out of 1000 are found with $NT_{X \to Y} > \gamma_T$ as indicated in Fig. 4.3b. The detection

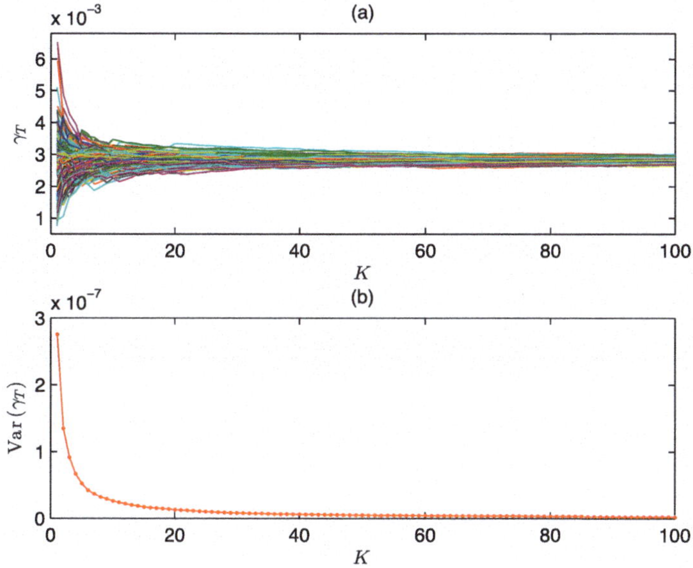

Fig. 4.2 Significance threshold γ_T of NTEs versus K in 1000 Monte Carlo simulations, reprinted from Ref. [21], copyright 2017, with permission from Elsevier

accuracy is 99.6%. Similar results are also observed for NDTEs. Thus, it can be concluded that the significance thresholds γ_T in (4.14) and γ_D in (4.15) are effective.

4.1.3.2 Number of Alarm Occurrences

In the estimation of transfer entropies, a natural question is: What is the minimum number of alarm occurrences required to achieve reliable estimates? This subsection investigates the minimum requirement of alarm occurrences based on Monte Carlo simulations.

A pair of alarm sequences X and Y with a causal relation are created as an example for illustration here. Given the sequence length L and the probability of alarm occurrences q, a binary sequence that follows the Bernoulli distribution $\mathcal{B}(1, q)$ is generated as X. The time stamp of the i-th alarm occurrence in X is denoted as s_i^x. Then the time stamp of the consequential alarm occurrence in Y is

$$s_i^y = s_i^x + \lambda_i, \tag{4.18}$$

where the integer λ_i is the i-th occurrence delay following Gaussian distribution, i.e., $\lambda_i \sim \mathcal{N}(v_m, v_s)$. Figure 4.4 shows an example of X and Y with $L = 20,000$, $q = 0.005$, $v_m = 10$, and $v_s = 1$. The binary sequences of X and Y are shown in Fig. 4.4a and c, respectively. Figure 4.4b and d present the random occurrence delays and their histogram, respectively.

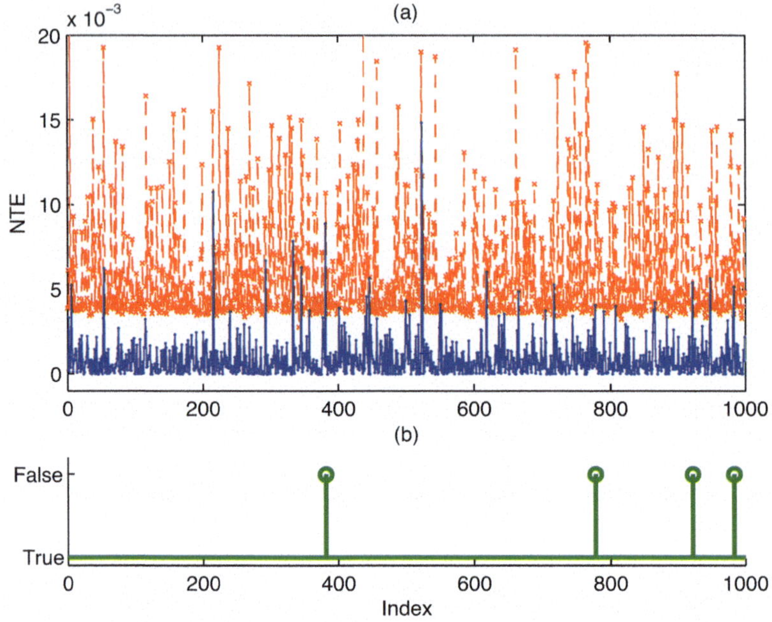

Fig. 4.3 Results of 1000 Monte Carlo simulations for the case without causal relations: **a** NTEs (blue line) and corresponding significance thresholds (red line); **b** correctness of results ("True" for correct estimates while "False" for incorrect estimates), reprinted from Ref. [21], copyright 2017, with permission from Elsevier

More pairs of X and Y are generated in the same manner for different values of q. The mean values of NTEs are computed as $\overline{NT}_{X_s \rightarrow Y_s} = \Sigma_{i=1}^{M_w-N+1} NT_{X_s \rightarrow Y_s}^{(i)}$ for various numbers of alarm occurrences $N \in \{N : N = 5 + 5j,\ j = 1, 2, ..., 39\}$. For $q \in \{q : q = 0.005j,\ j = 1, 2, ..., 10\}$, the trends of $\overline{NT}_{X_s \rightarrow Y_s}$ versus N are shown in Fig. 4.5a. The differences $\Delta\overline{NT}(N) = \overline{NT}_{X_s \rightarrow Y_s}(N) - \overline{NT}_{X_s \rightarrow Y_s}(N-5)$ versus $N = 15, 20, ..., 200$ are shown in Fig. 4.5b. It can be found that the change of $\overline{NT}_{X_s \rightarrow Y_s}$ for $N > 50$ is not significant. Therefore, the minimum number of alarm occurrences $N = 50$ is chosen as a rule of thumb. Note that these alarm occurrences can be distributed in a long time period, and do not have to occur in a short time period.

4.1.3.3 Steps of the Proposed Method

The following conditions are required for the proposed method: (1) At least 50 alarm occurrences should be present in each alarm sequence; (2) the time horizons l_1 and k_1 in (4.4) and (4.5) should be predetermined. The first condition is to ensure reliable estimates of transfer entropies as observed in Sect. 4.1.3.2. For the second condition, the recommended values of l_1 and k_1 have been given at the end of Sect. 4.1.2.1.

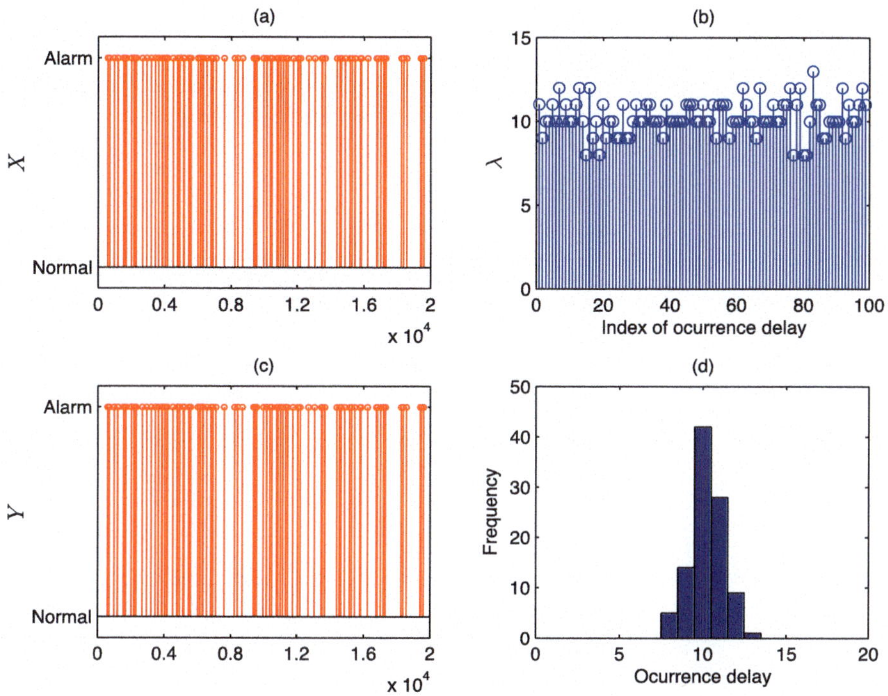

Fig. 4.4 An example of alarm sequences and random occurrence delays: **a** The alarm sequence X, **c** the alarm sequence Y, **b** the random occurrence delays, and **d** the histogram of random occurrence delays, reprinted from Ref. [21], copyright 2017, with permission from Elsevier

The proposed method is composed of two parts. The first part is to obtain a causal map by calculating the NTEs in the following steps:

1. Historical alarm sequences $X_k, k = 1, 2, .., M$, are prepared.
2. By varying the time lag $d \leq D$, $NT_{X_i \to X_j}(d)$, $i \neq j$, is calculated based on (4.6). The NTE from X_i to X_j is computed as $NT_{X_i \to X_j} = \max_d NT_{X_i \to X_j}(d)$ and the corresponding time lag is found as $d_{ij} = \arg\max_d NT_{X_i \to X_j}(d)$.
3. The significance threshold γ_T in (4.14) is calculated using the Monte Carlo test in Sect. 4.1.3.1.
4. If $NT_{X_i \to X_j} > \gamma_T$, $C_{ij} = 1$; otherwise, $C_{ij} = 0$.
5. If NTEs for all pairwise alarm sequences have been calculated, then the proposed method proceeds to the next step; otherwise, it goes back to Step 2 for another pair of alarm sequences by increasing i or j.
6. A causal map is obtained to indicate information flow paths based on the value of C_{ij}. If $C_{ij} = 1$, a line with orientation is drawn to denote the causality from X_i to X_j; otherwise, there is no connection between X_i and X_j.

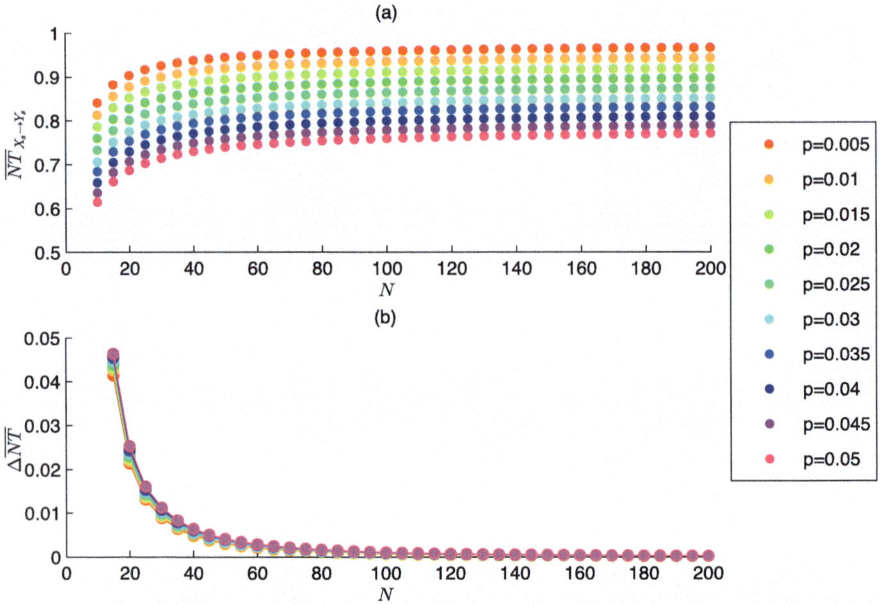

Fig. 4.5 Average NTEs versus numbers of alarm occurrences for different q, reprinted from Ref. [21], copyright 2017, with permission from Elsevier

The second part of the proposed method is to prune superfluous connections of the causal map from the first part based on the NDTEs. This part is composed of the following steps, also depicted as a flowchart in Fig. 4.6:

1. The NDTEs from X_i to X_j based on X_l are calculated by starting from $i = 1$ and $j = 1$.
2. If $C_{ij} = 1$, the computation proceeds to the next step; otherwise, it proceeds to Step 8.
3. The computation starts from X_l with $l = 1$.
4. If $C_{il} = 1$ and $C_{lj} = 1$, the computation proceeds to the next step; otherwise, it proceeds to Step 6.
5. $ND_{X_i \to X_j | X_l}$ is calculated based on (4.12). The time delays d_{ij} and d_{lj} are the ones corresponding to the maximum values of $NT_{X_i \to X_j}(d)$ and $NT_{X_l \to X_j}(d)$, respectively.
6. If NDTEs from X_i to X_j through all intermediate variables have been calculated, then the computation proceeds to the next step; otherwise, it goes back to Step 5 by increasing l by one.
7. Determine the maximum NDTE $ND_{\max}$ from X_i to X_j based on all intermediate alarm variables, and compare $ND_{\max}$ with its corresponding γ_D in (4.15). If $ND_{\max} > \gamma_D$, then the causality from X_i to X_j is direct; otherwise, it is indirect so that the corresponding line from X_i to X_j is excluded from the causal map.

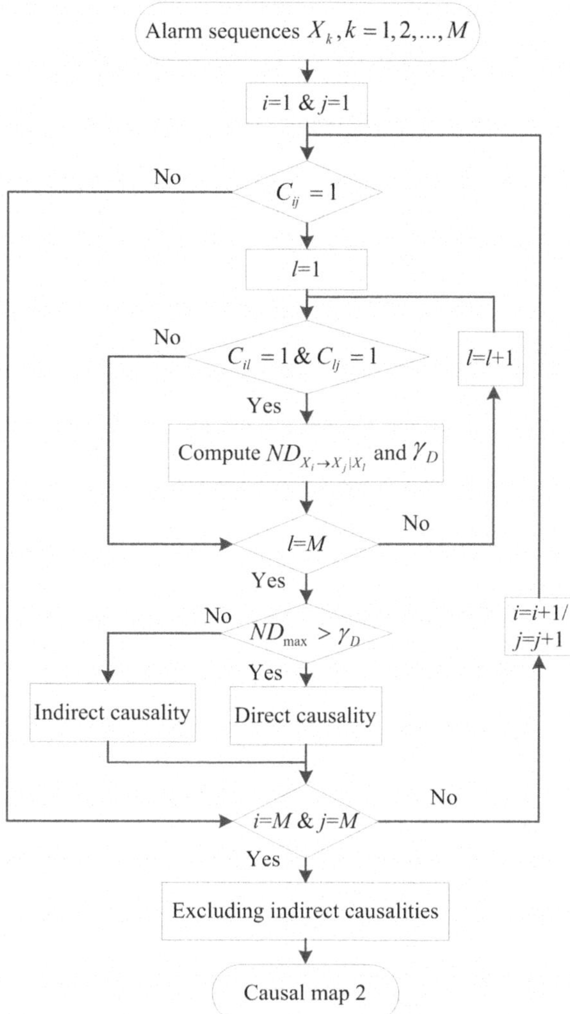

Fig. 4.6 Flowchart for the detection of direct causalities based on the NDTEs, reprinted from Ref. [21], copyright 2017, with permission from Elsevier

8. If NDTEs for all pairwise alarm sequences have been calculated, then the computation stops; otherwise, go back to Step 2 for another pair of alarm sequences by increasing i or j by one.

4.1.4 Numerical and Industrial Examples

This section demonstrates the effectiveness of the proposed method by a numerical example and an industrial example involving real alarm data.

4.1.4.1 Numerical Example

Three alarm sequences with causal relationships as $X_1 \to X_2 \to X_3$ are simulated. First, X_1 is a binary sequence that follows the Bernoulli distribution $\mathcal{B}(1, q)$ with $q = 0.01$. Denote the time stamp of the i-th alarm occurrence in X_1 as s_i^1. The time stamps of the consequential alarm occurrences in X_2 and X_3 are

$$\begin{cases} s_i^2 = s_i^1 + \lambda_i^{(1)}, \\ s_i^3 = s_i^2 + \lambda_i^{(2)}. \end{cases} \tag{4.19}$$

Here $\lambda_i^{(1)}$ and $\lambda_i^{(2)}$ are independent Gaussian random variables, each of which takes the mean value 20 and standard deviation 1. The number of alarm occurrences in each sequence with data length $L = 20,000$ is expected to be about 100, which is sufficient for reliable estimates of transfer entropies as demonstrated in Sect. 4.1.3.2.

The proposed method is applied to the simulated alarm data. First, the NTEs under different time delays ranged from 0 to 100 are presented in Fig. 4.7. The NTEs and their corresponding significance thresholds (in brackets) are given in Table 4.1. The

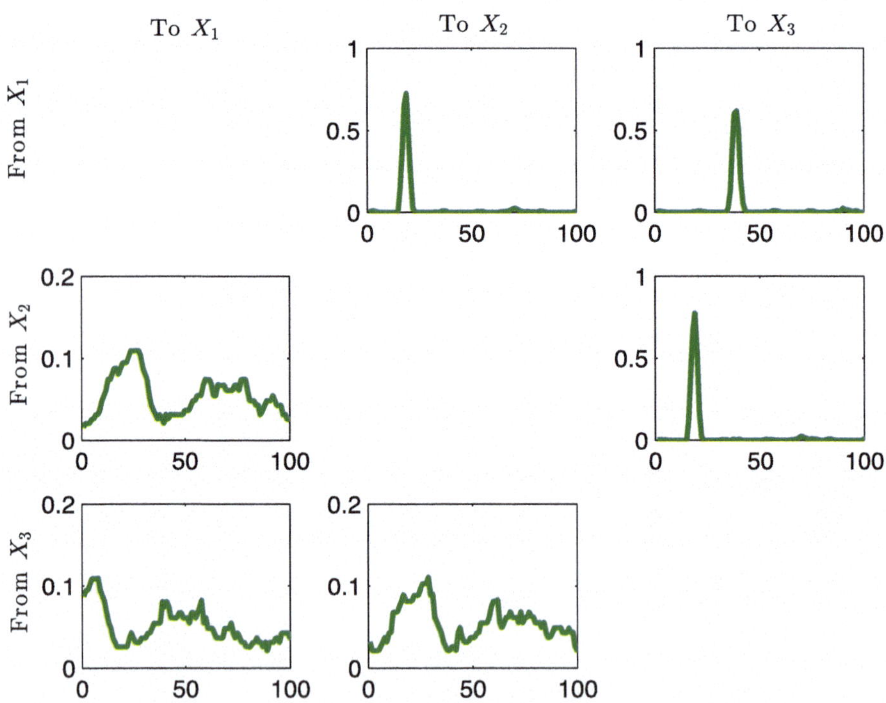

Fig. 4.7 NTEs versus time lags, reprinted from Ref. [21], copyright 2017, with permission from Elsevier

Table 4.1 The NTEs and their significance thresholds (in brackets), reprinted from Ref. [21], copyright 2017, with permission from Elsevier

$T_{X_i \to X_j}$	X_1	X_2	X_3
X_1		0.73(0.022)	0.62(0.023)
X_2	0.109(0.112)		0.78(0.025)
X_3	0.110(0.111)	0.110(0.111)	

cause-effect relationships among the three alarm variables are described by a causal map in Fig. 4.8a.

Second, the NDTE is calculated to be $ND_{X_1 \to X_3|X_2} = 0.0037$ in order to detect whether the causality from X_1 to X_3 is direct or through X_2. The threshold from Monte Carlo tests is obtained as $\gamma_D = 0.025$. Since $ND_{X_1 \to X_3|X_2} < \gamma_D$, the causality from X_1 to X_3 is indirect. Then, the information flow path from X_1 to X_3 should be excluded and a new causal map of direct causality is drawn as Fig. 4.8b, which is consistent with the true relationship.

4.1.4.2 Industrial Example

In an oil refinery, the process encompasses the configuration of over five thousand alarm variables designed to track changes in process variable variations, operational status, and safety conditions. Within a span of 3.5 days, 288 of these alarm variables have recorded alarm events. The introduced method in this section is utilized across the entirety of these alarm variables. Here, a selection of six relevant alarm variables, identified by Tags 1 through 6, are highlighted to illustrate the method's efficacy.

The historical data for these six alarm variables are depicted in Fig. 4.9. Throughout the 3.5-day time period, the alarm occurrences for the six variables amounted to 653, 242, 242, 242, 219, and 106, respectively. By calculating the chattering indices

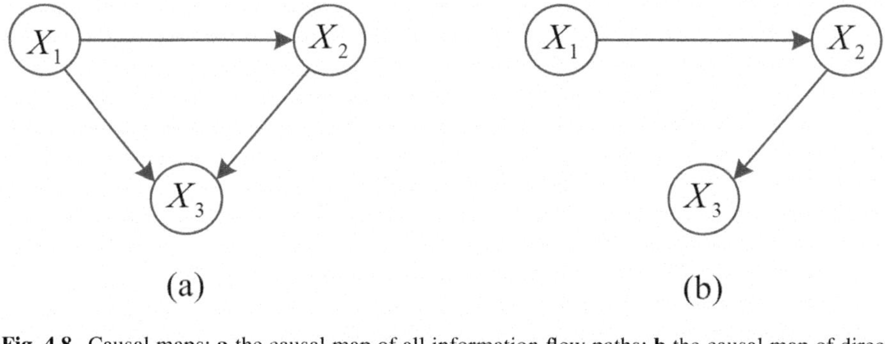

Fig. 4.8 Causal maps: **a** the causal map of all information flow paths; **b** the causal map of direct information flow paths, reprinted from Ref. [21], copyright 2017, with permission from Elsevier

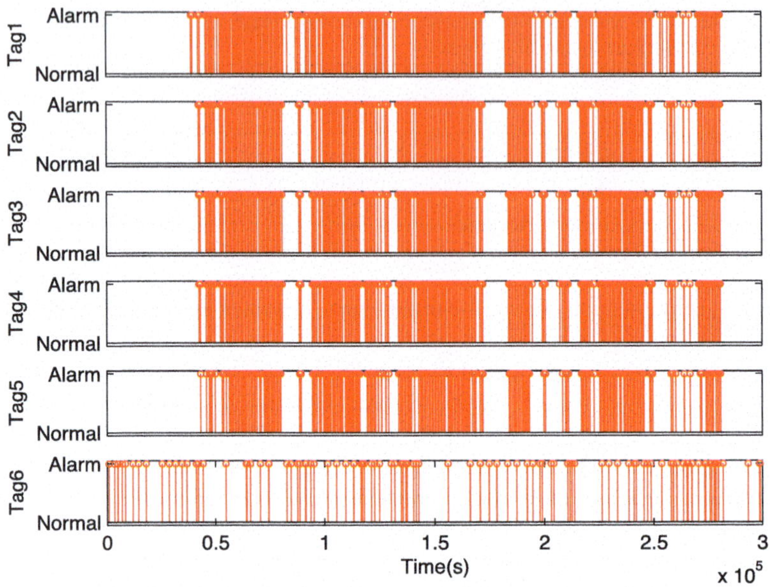

Fig. 4.9 Alarm variables over 3.5 days time period, reprinted from Ref. [21], copyright 2017, with permission from Elsevier

[26, 44], it was determined that the alarm variable with Tag 1 is found to contain chattering alarms. With the implementation of a 30-sample alarm delay timer, the majority of these chattering alarms are eliminated, decreasing the total number of occurrences to 282. According to Sect. 4.1.3.2, the numbers of alarm occurrences for these six variables are sufficient to reach reliable estimates of transfer entropies.

The proposed method is applied to calculate the NTEs among alarm variables in Fig. 4.9. The NTEs under different time delays are shown in Fig. 4.10. The NTEs and their corresponding significance thresholds (in brackets) are given in Table 4.2. As an example, the NTE from Tag 1 to Tag 2 is 0.2577 and the significance threshold is 0.0155, so that there is a causal relationship from Tag 1 to Tag 2. Therefore, a causal map describing the information flow paths is drawn in Fig. 4.11.

Next, the NDTE is calculated for each pair of alarm tags having the NTEs larger than their corresponding thresholds. Taking Tag 1 and Tag 5 as an example, the possible intermediate variables include Tags 2, 3, and 4. Thereby, the NDTEs from Tag 1 to Tag 5 based on Tags 2, 3, and 4 are calculated in Table 4.3. It can be seen that no direct causality is found between the cause variables and effect variables in Table 4.3. By excluding the indirect causalities from the causal map in Fig. 4.11, a causal map describing all direct information flow paths is drawn in Fig. 4.12.

The cause-effect relationships in Fig. 4.12 are consistent with the physical connections of the six alarm variables. The 6th alarm variable is located in a disconnected unit, while the first five alarm variables are located in some interconnected units with the following physical connections: Tag 2 represents the total gas flow in a burner

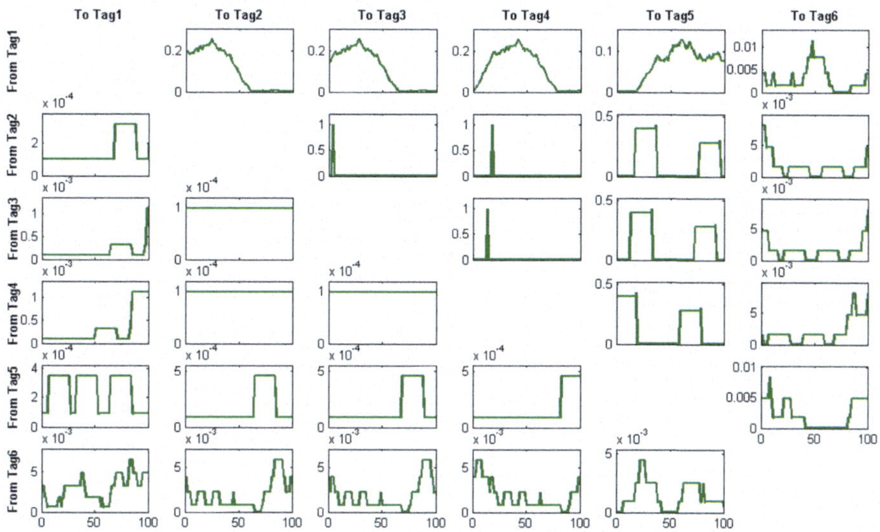

Fig. 4.10 NTEs versus time lags, reprinted from Ref. [21], copyright 2017, with permission from Elsevier

Table 4.2 The NTEs and their significance thresholds (in the brackets), reprinted from Ref. [21], copyright 2017, with permission from Elsevier

	Tag1	Tag2	Tag3
Tag1		0.2577(0.0155)	0.2577(0.0153)
Tag2	0.0003(0.0127)		1(0.0011)
Tag3	0.0011(0.0134)	0.0001(0.0135)	
Tag4	0.0011(0.0130)	0.0001(0.0139)	0.0001(0.0140)
Tag5	0.0004(0.0122)	0.0004(0.0131)	0.0005(0.0131)
Tag6	0.0065(0.0090)	0.0058(0.0089)	0.0058(0.0087)
	Tag4	Tag5	Tag6
Tag1	0.2577(0.0155)	0.1293(0.0151)	0.0115(0.0215)
Tag2	1(0.0013)	0.4336(0.0142)	0.0082(0.0188)
Tag3	1(0.0011)	0.4336(0.0144)	0.0082(0.0183)
Tag4		0.4336(0.0135)	0.0082(0.0204)
Tag5	0.0005(0.0133)		0.0085(0.0167)
Tag6	0.0058(0.0089)	0.0045(0.0102)	

which has two suppliers from cell A and cell B. Tag 1 indicates the gas flow from the feed cell B. Tags 3, 4, and 5 are associated with three products out of the burner. Thus, the information flow paths are Tag 1 → Tag 2 → Tag 3 → Tag 4 → Tag 5. The estimated NTEs in Table 4.2 provides valuable quantitative information. The NTE from Tag 2 to Tag 3, that from Tag 2 to Tag 4, and that from Tag 3 to Tag 4 are

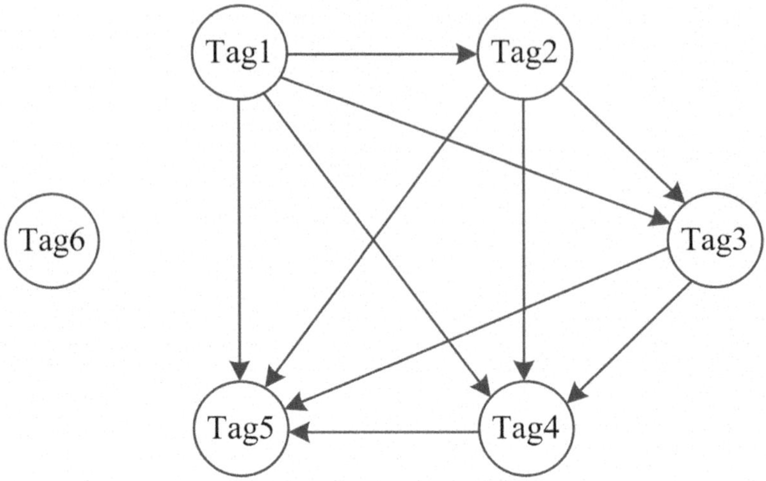

Fig. 4.11 Causal map of information flow paths, reprinted from Ref. [21], copyright 2017, with permission from Elsevier

Table 4.3 Detection of direct causalities based on NDTEs, reprinted from Ref. [21], copyright 2017, with permission from Elsevier

Cause variable	Effect variable	Int. variables	NDTE	Threshold
Tag1	Tag3	Tag2	0	0.0129
Tag1	Tag4	Tag2, Tag3	0	0.0122
Tag1	Tag5	Tag2, Tag3, Tag4	0.00014	0.0118
Tag2	Tag4	Tag3	0	0.0005
Tag2	Tag5	Tag3, Tag4	0	0.0149
Tag3	Tag5	Tag4	0	0.0152

close to one, indicating strong causal relationships among the three alarm variables. By contrast, the NTE from Tag 1 to other alarm tags are relatively small; thus, the abnormalities of Tag 1 for cell B (as one of the two suppliers to the burner) contribute about 25% to the abnormalities in the subsequent alarm variables. From the cause-effect relationships, Tag 1 and Tag 2 are identified as the primary alarm variables, while Tags 3, 4 and 5 are the consequential secondary alarm variables. Thus, the search space of root causes of abnormalities can be quickly narrowed down to those associated with the input gas flow.

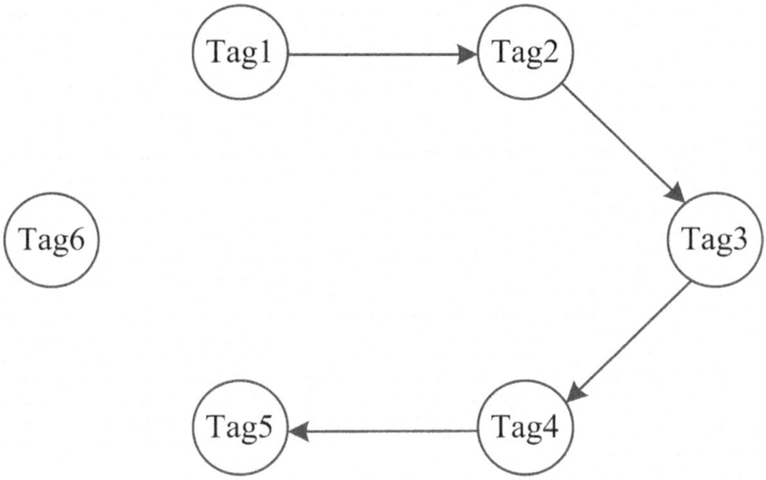

Fig. 4.12 Causal map of direct information flow paths, reprinted from Ref. [21], copyright 2017, with permission from Elsevier

4.2 Causality Inference for Process Variables

This section focuses on inferring causality for process variables by employing a modified version of transfer entropies that leverages information granulation and clustering techniques. The transfer entropy (TE) is a prevalent and potent tool for pinpointing the root causes of disturbances within industrial processes. Nonetheless, TE faces two significant limitations: It requires the industrial processes to be stationary, and it involves substantial computational demands, which restrict its practical use. In light of these issues, a new transfer entropy approach based on the information granulation and clustering, is introduced for causality inference [53]. The information granulation based transfer entropy (IGTE) and information granulation based direct transfer entropy (IGDTE) are proposed to infer causal and direct causal relations. Additionally, a Probability Density Function (PDF) estimator is crafted to gauge joint and conditional probabilities utilizing information granules, complemented by a density-based clustering technique known as Ordering Points To Identify Clustering Structure (OPTICS).

4.2.1 Problem Description

TE serves as a valuable tool for deducing causal relationships among process variables [4, 12, 35, 52]. However, the application of TE is impeded by significant issues, i.e., it is not suitable for nonstationary processes and it entails a high computational load [4, 12]. First, TE requires the process be stationary, whereas the presence of faults may induce nonstationary shifts. Second, the computational demands of TE are substantial, and practical applications might be cost-sensitive in terms of computation. As a result, these limitations present difficulties for TE's practical implementation, especially for real-time root cause analysis. Accurate estimation of probability density is contingent upon stationary data, rendering TE unsuitable for nonstationary data [4]. Moreover, the computation of the Probability Density Function (PDF) and TE within a high-dimensional space are pivotal contributors to the elevated computational complexity of TE [4, 12, 35, 52].

Inspired by the aforementioned problems, this subsection introduces a transfer entropy approach that incorporates information granulation and clustering for the purpose of causality inference. The approach begins with the time sequences of two process variables, represented as $\mathbf{x} = [x_1, x_2, ..., x_N]$ and $\mathbf{y} = [y_1, y_2, ..., y_N]$. The goal is to uncover the causal links between them and to apply these findings to root cause analysis. The framework of the proposed Information Granulation based Transfer Entropy (IGTE) is depicted in Fig. 4.13. Specifically, it includes the following three main stages:

- Information granulation of raw data: This phase involves converting the initial time sequences into a form of abstracted data through the creation of information granules.
- PDF estimation based on the information granule clustering: In this phase, the joint and conditional Probability Density Functions (PDFs) are approximated by clustering the information granules using a density-based clustering technique.
- Causality inference based on IGTE and IGDTE: The final phase constructs a causal map that delineates the relationships between variables, facilitating the identification of root causes by computing both the IGTE and IGDTE.

4.2.2 *Data Abstraction and PDF Estimation*

This section presents the information granulation method for data abstraction and estimated PDFs based on clustering.

4.2.2.1 Data Abstraction via Information Granulation

The traditional TE method has the disadvantages of not being suitable for nonstationary data and high computational complexity. Therefore, how to improve TE becomes a major concern. Symbolic TE and trend TE address these issues by encoding the raw data into symbols or trends, respectively [17, 42]. An information granulation approach is utilized here, which simplifies raw data into granules, making the improved TE suitable for nonstationary data and markedly decreasing the computational complexity. Information granules are essentially collections of data points

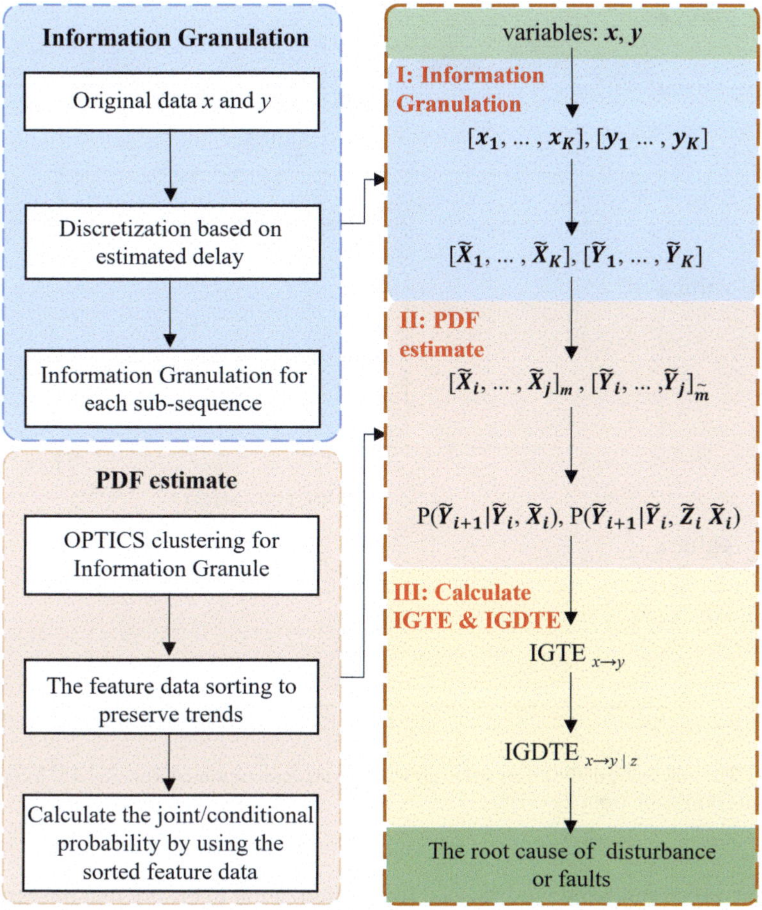

Fig. 4.13 Framework of the proposed approach, reprinted from Ref. [53], copyright 2023, with permission from Elsevier

grouped by similarity or proximity, offering a concise and understandable data representation. Information granulation has gained popularity in time series analysis, providing a basis for data compression and subsequent analysis. The process involves two main stages, as depicted in Fig. 4.14 and outlined below:

- Discretization: For a time sequence $\mathbf{x} = [x_1, x_2, \ldots, x_N]$, K non-overlapping subsequences $\mathbf{x}_1, \mathbf{x}_2, \ldots, \mathbf{x}_K$ are obtained by discretization, where $\mathbf{x}_i = [x_1, x_2, \ldots, x_w]$ $(i = 1, 2, \ldots, K)$, N denotes the length of the time sequence $\mathbf{x}$, and w indicates the discretization window size.
- Information granulation for each subsequence: Each subsequence $\mathbf{x}_i$ $(i = 1, 2, \ldots, K)$ undergoes the information granulation process to create a time-related information granule $\tilde{\mathbf{X}}_i$ that denotes data properties of the subsequence.

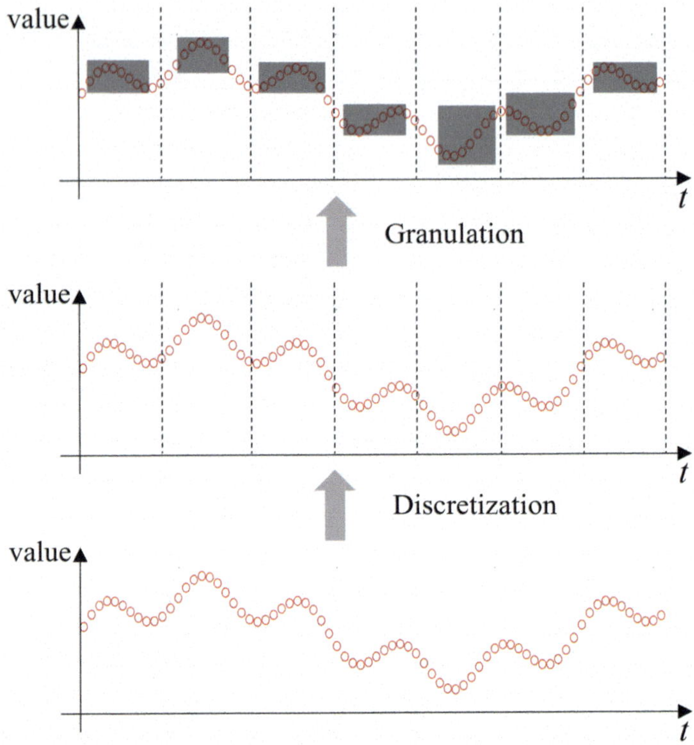

Fig. 4.14 The schematic of the information granulation process, reprinted from Ref. [53], copyright 2023, with permission from Elsevier

Non-overlapping windows are used for discretization to ensure that the resulting information granules adhere to the principle of justifiable granularity, meaning they should extensively cover the data while maintaining specificity and relative independence [11, 18]. It is viable to deduce the discretization window size from the delay between variables, which can be pinpointed by cross-correlation analysis that aligns one time sequence with the other to find the maximum correlation, indicating the delay [5, 52]. The processed raw time sequence $\mathbf{x} = [x_1, x_2, \ldots, x_N]$ can be converted into a granular sequence $\tilde{\mathbf{X}} = [\tilde{\mathbf{X}}_1, \tilde{\mathbf{X}}_2, \ldots, \tilde{\mathbf{X}}_K]$, where $\tilde{\mathbf{X}}_i$ represents the i-th granule, with further details on its definition and calculation provided subsequently.

Various information granulation techniques exist, such as those based on fuzzy sets [11, 33], clustering [11], and intelligent optimization [10, 18]. However, clustering-based methods may result in significant information loss due to their limited data scope, and intelligent optimization-based methods demand excessive computation time, conflicting with the goal of reducing computational complexity in this research. Therefore, the fuzzy set-based granulation is preferred for its effective use of data and computational efficiency. A general definition of fuzzy information granules provided by Zadeh [51] based on fuzzy sets is denoted by

$$g \triangleq (x \text{ is } G) \text{ is } \varrho, \tag{4.20}$$

where x is a variable in the universe U; G is a convex fuzzy set of U and can be described by a membership function μG; ϱ is the probability of x belonging to the set. The main issue of the information granulation using fuzzy sets is to ascertain a membership function $A = \mu G$. The description of information granules generated by the fuzzy set-based information granulation is intimately associated with the form of the membership function. Triangular, trapezoidal, and symmetrical Gaussian functions are the generally used forms [11]. A triangular membership function is given by

$$A(x) = \begin{cases} 0, x < a_1, \\ \frac{x-a_1}{a_2-a_1}, a_1 \leq x \leq a_2, \\ \frac{a_3-x}{a_3-a_2}, a_2 < x \leq a_3, \\ 0, x > a_3, \end{cases} \tag{4.21}$$

where a_1, a_2, and a_3 are the parameters of the membership function. In particular, a_1 and a_3 are called supports, and a_2 is referred to as the core.

For a effective granulation process using the fuzzy set-based information granulation [50], it should make sure two criteria: (i) The information granules adequately describe the original data; (ii) the information granules are sufficiently specialized. The granulation performance is described by a function $Q_A(x)$ with respect to the membership function $A(x)$, i.e.,

$$Q_A = \frac{M_A}{N_A}, \tag{4.22}$$

where $M_A = \sum_{i=1}^{w} A(x_i)$, and maximization of M_A can satisfy the criterion (i); $N_A = \text{measure}(\text{support}(A))$ denotes the length of the interval between the two supports, and minimization of N_A can satisfy the criterion (ii). Here, Q_A should be maximized in light of the aforementioned criteria.

As a result, given each subsequence $\mathbf{x}_i$, its fuzzy information granules are denoted by $\tilde{\mathbf{X}}_i = [\tilde{X}_i^1, \tilde{X}_i^2, \tilde{X}_i^3]$, where $\tilde{X}_i^1$ and $\tilde{X}_i^3$ are the supports, and $\tilde{X}_i^2$ is the core. The parameters a_1, a_2, and a_3 of the triangle membership function in (4.21) are relevant to $\tilde{X}_i^1$, $\tilde{X}_i^2$, and $\tilde{X}_i^3$, respectively. The core of the information granule can be computed as the median of $\mathbf{x}_i$, i.e.,

$$\tilde{X}_i^2 = \text{med}\left(x_1^i, x_2^i, \ldots, x_w^i\right). \tag{4.23}$$

The supports $\tilde{X}^1$ and $\tilde{X}^3$ are determined by the triangular membership function when Q_A is maximal [50] and given by

$$\begin{cases} \tilde{X}_i^1 = \frac{2}{[w/2]} \sum_{j=1}^{[w/2]} x_j^i - \tilde{X}_i^2, \\ \tilde{X}_i^3 = \frac{2}{w-[w/2]-d+1} \sum_{j=[w/2]+d}^{w} x_j^i - \tilde{X}_i^2, \end{cases} \tag{4.24}$$

where $[w/2]$ represents the biggest integer that does not exceed $w/2(w \geq 2)$; x_j^i represents the j-th sample in the i-th subsequence $[x_1^i, x_2^i, \ldots, x_w^i]$. Further, $d = 1$ when w is an even integer; otherwise, $d = 2$. The granular time sequence is acquired for further analysis by the aforementioned procedures.

4.2.2.2 Clustering-Based PDF Estimation

Estimating the PDF is a prerequisite for calculating TE, as it is crucial for both accuracy and computational efficiency. Common PDF estimators such as kernel density, plug-in, and nearest neighbor-based methods have been utilized for TE computations [37]. The kernel density estimator smooths the overall distribution by summing preselected distributions centered on each data point. The plug-in estimator simplifies the process through uniform data binning. The nearest neighbors-based estimator approximates the PDF by calculating local densities around each observation. However, these PDF estimators work for original process signals, but are not directly applicable to the information granule that has a different data structure. Therefore, a clustering based PDF estimator is proposed here for PDF estimation of information granules.

Given $\mathbf{x}$ and $\mathbf{y}$, their granular time sequences are $\tilde{\mathbf{X}} = [\tilde{\mathbf{X}}_1, \tilde{\mathbf{X}}_2, \ldots, \tilde{\mathbf{X}}_K]$ and $\tilde{\mathbf{Y}} = [\tilde{\mathbf{Y}}_1, \tilde{\mathbf{Y}}_2, \ldots, \tilde{\mathbf{Y}}_K]$, respectively, where $\tilde{\mathbf{X}}_i \left(\tilde{\mathbf{Y}}_i\right)$ is the i-th three-dimensional information granule, $i = 1, 2, \ldots, K$. The OPTICS clustering method [1] is then applied to segment these granules. owing to the feature data after OPTICS clustering not retaining the raw data trend altogether, the OPTICS clustering is adapted to get the new features that preserve the trends. Conditional and joint probabilities are estimated from these distinct clusters.

In data mining, density-based clustering algorithms, such as DBSCAN [13], group objects with similar members within the same cluster. A notable drawback of DBSCAN is the substantial impact of input parameters on clustering results. The OPTICS algorithm, an extension of DBSCAN, mitigates this sensitivity and enhances adaptability [1, 54]. The main parameters of the OPTICS are the core distance and the reachability distance. The core distance is the smallest neighborhood radius that makes $\tilde{\mathbf{X}}_i$ a core point. The distance from a sample point $\tilde{\mathbf{X}}_j$ to the core point $\tilde{\mathbf{X}}_i$ is the reachability distance. For given parameters ε and ζ, $N_\varepsilon\left(\tilde{\mathbf{X}}_i\right)$ indicates the nearest neighbor set of $\tilde{\mathbf{X}}_i$ in its ε-neighborhood; if $\left|N_\varepsilon\left(\tilde{\mathbf{X}}_i\right)\right| \geq \zeta$, the core distance D_c and reachability distance D_r are given by

$$D_c\left(\tilde{\mathbf{X}}_i\right) = D\left(\tilde{\mathbf{X}}_i, N_\varepsilon^\zeta\left(\tilde{\mathbf{X}}_i\right)\right), \tag{4.25}$$

$$D_r\left(\tilde{\mathbf{X}}_j, \tilde{\mathbf{X}}_i\right) = \max\left(D_c\left(\tilde{\mathbf{X}}_i\right), D\left(\tilde{\mathbf{X}}_i, \tilde{\mathbf{X}}_j\right)\right), \tag{4.26}$$

where $N_{\varepsilon}^{\zeta}\left(\tilde{\mathbf{X}}_i\right)$ represents the nearest neighbor set of point $\tilde{\mathbf{X}}_i$ with a certain value of ζ. The OPTICS clustering saves the core distance and reachability distance for each sample. It then derives the order of all the items in the data set based on the density clustering structure rather than directly creating clusters.

However, the feature data $\mathbf{c} = [c_1, c_2, \ldots, c_K]$ (where $c_k \in \{1, 2, \ldots, M\}$, M is the cluster number) does not entirely maintain the trend change of the original data $\mathbf{x}$ when the granular time sequence $\tilde{\mathbf{X}}$ is clustered using OPTICS. Changes in trends have an impact on how accurately causal relations between variables may be inferred. The feature data corresponding to the M clusters are sorted according to the magnitude of the core sample $\tilde{\mathbf{X}}_k = [\tilde{X}_k^1, \tilde{X}_k^2, \tilde{X}_k^3]$ of each cluster. Then, the new feature data $\overline{\mathbf{c}}$ can be obtained, and its trend is consistent with the information granule. For $\overline{\mathbf{c}}$, the following condition should be satisfied, i.e.,

$$\overline{c}_{k1} < \overline{c}_{k2}, \text{ if } \tilde{X}_{k1}^2 < \tilde{X}_{k2}^2. \tag{4.27}$$

Accordingly, the trend of the raw data is well kept by the modified OPTICS clustering algorithm. Figure 4.15 presents an example to compare the trends of the data, and it is obvious that the new feature data $\overline{\mathbf{c}}$ can preserve the trends compared to the original feature data.

The modified OPTICS clustering algorithm has obvious advantages that can be seen from the Fig. 4.15. It leads to more accurate PDF estimation by maintaining trends and distributions of the raw data. In addition, the cluster number can be automatically determined in the OPTICS clustering approach; thus, it does not need to choose this parameter beforehand. The distribution of the reachability distances is shown in Fig. 4.16. The cluster number is naturally equal to the peak number plus 1. Hence, there are 9 clusters. The clustering results are shown in Fig. 4.17, and it can be seen that the clustering effect of information granule is obvious. This demonstrates that the improvement is reasonable and effective.

The PDFs of the granular time sequences $\tilde{\mathbf{X}}$ and $\tilde{\mathbf{Y}}$ are indicated by $p\left(\tilde{\mathbf{X}}\right)$ and $p\left(\tilde{\mathbf{Y}}\right)$, respectively, and the joint PDF is $p\left(\tilde{\mathbf{X}}, \tilde{\mathbf{Y}}\right)$. Therefore, the clustering-based PDF estimation method is used to estimate $p\left(\tilde{\mathbf{X}}\right)$. The formula is written as

$$p\left(\tilde{\mathbf{X}}_i\right) = \frac{\Lambda_m\left(\tilde{\mathbf{X}}_i \in m\right)}{\sum_{m=1}^{M} \Lambda_m}, \tag{4.28}$$

where $\Lambda_m(\tilde{\mathbf{X}}_i \in m)$ denotes the sample number belonging to the m-th cluster and M is the cluster number. In addition, the joint PDF $p\left(\tilde{\mathbf{X}}, \tilde{\mathbf{Y}}\right)$ is calculated by

$$p\left(\tilde{\mathbf{X}}_i, \tilde{\mathbf{Y}}_i\right) = \frac{\Lambda_{m,\tilde{m}}\left(\tilde{\mathbf{X}}_i \in m,\ \tilde{\mathbf{Y}}_i \in \tilde{m}\right)}{\sum_{m=1}^{M} \sum_{\tilde{m}=1}^{\tilde{M}} \left(\Lambda_{m,\tilde{m}}\right)}, \tag{4.29}$$

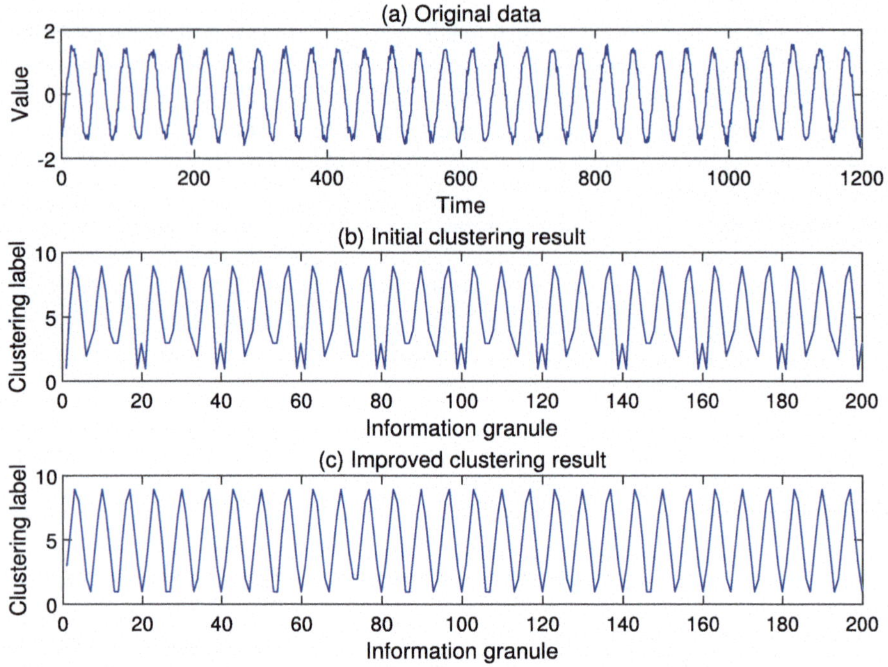

Fig. 4.15 The comparison of the trend (information granulation with $w = 6$), reprinted from Ref. [53], copyright 2023, with permission from Elsevier

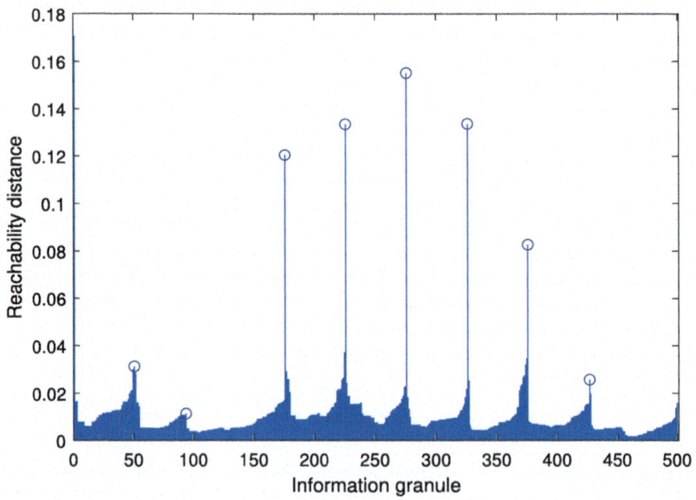

Fig. 4.16 The distribution of the reachability distances, reprinted from Ref. [53], copyright 2023, with permission from Elsevier

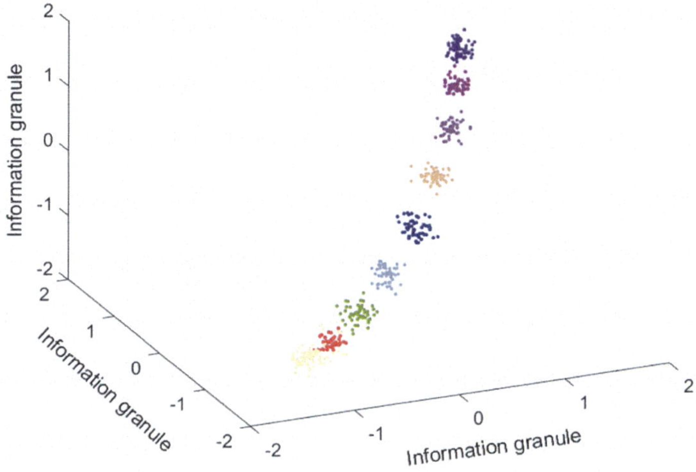

Fig. 4.17 The clustering results of information granule, reprinted from Ref. [53], copyright 2023, with permission from Elsevier

where $\Lambda_{m,\tilde{m}}\left(\tilde{\mathbf{X}}_i \in m, \tilde{\mathbf{Y}}_i \in \tilde{m}\right)$ indicates the sample number in $\tilde{\mathbf{X}}$ belonging to the m-th cluster and the sample number in $\tilde{\mathbf{Y}}$ belonging to the $\tilde{m}$-th cluster; $\tilde{M}$ is the cluster number of $\tilde{\mathbf{Y}}_i$.

4.2.3 *Causality Inference by IGTE and IGDTE*

This section presents how to infer causal relations using IGTE and distinguish direct from indirect causal relations using IGDTE.

4.2.3.1 Causality Inference by IGTE

The original data is replaced by the information granule after clustering, maintaining the dynamic characteristics for variational trends of the raw data. The information granule after clustering, rather than the raw data, is applied to compute TE. Therefore, the proposed IGTE can be applied to nonstationary data. This is similar to the symbolic TE in [42], when calculating TE, not the original data but the symbol is used. Thus, the formula for calculating IGTE is defined as

$$\mathrm{IGTE}_{x\to y} = \sum p\left(\tilde{\mathbf{Y}}_{i+1} \in \tilde{m}, \tilde{\mathbf{Y}}_i^k \in \tilde{m}, \tilde{\mathbf{X}}_i^l \in m\right) \cdot \log \frac{p\left(\tilde{\mathbf{Y}}_{i+1} \in \tilde{m} \mid \tilde{\mathbf{Y}}_i^k \in \tilde{m}, \tilde{\mathbf{X}}_i^l \in m\right)}{p\left(\tilde{\mathbf{Y}}_{i+1} \in \tilde{m} \mid \tilde{\mathbf{Y}}_i^k \in \tilde{m}\right)}, \tag{4.30}$$

where $\tilde{\mathbf{Y}}_i$ ($\tilde{\mathbf{X}}_i$) represents the i-th information granule of the granular time sequence, and is obtained through information granulation and clustering. Further, when calculating IGTE with information granule, both the values in the current window and the historical window are considered. This ensures the rationality of the IGTE calculation results.

There are two key parameters in (4.30), namely, the orders k and l. In the traditional TE, k and l can be selected by the change rate of TE [12]. When the granulation window size equals to the delay between variables, it is rational to select $k = l = 1$ [52]. However, a large delay will cause information granulation to lose more useful information, and may result in erroneous IGTE computation results. Therefore, the 2nd-order is utilized here. This guarantees that useful information is utilized as much as possible from two aspects: (1) The granulation window size only has to be half of the delay, so that the information granulation can retain more information. (2) More historical information is taken into account than in the 1st-order IGTE, leading to more precise computation outputs. The effectiveness of the 2nd-order IGTE is demonstrated in the following example.

Given that x and y are two continuous random variables with nonlinear correlation,

$$y_{i+16} = 2 + 2\,|10 - (0.5x_i + 0.2y_i)| + v_i, \tag{4.31}$$

where $x_i = 10\sin(0.02t) + v$, $v \sim N(0, 0.1)$, $v_i \sim N(0, 0.01)$, and the sampling time t is 1. There are 3000 data samples collected. The delay between x and y is attained as 16, which is the same as the real value. The collected data is abstracted by information granulation with the window size equal to the time delay (1st-order), half of it (2nd-order), and a third of it (3rd-order). Comparison is made in terms of the fluctuation degrees of the calculation results and the computational time. The $\mathrm{IGTE}_{x\to y}$ is calculated by (4.30) under different orders. The results are compared from 10 trials of repeated simulations with random Gaussian noises, as shown in Fig. 4.18.

In order to compare the fluctuation of IGTE under different orders, an indicator λ is defined to measure the fluctuation degree,

$$\lambda = \frac{\varphi}{\mathrm{IGTE}_{\min}}, \tag{4.32}$$

where φ denotes the variance of IGTE calculation results, and $\mathrm{IGTE}_{\min}$ is the minimum value of IGTE. Then, λ is calculated for IGTE of different orders, and the result is shown in Table 4.4. Obviously, the results using the 2nd-order IGTE and the 3rd-order IGTE are relatively more stable, whereas the results based on the 1st-order

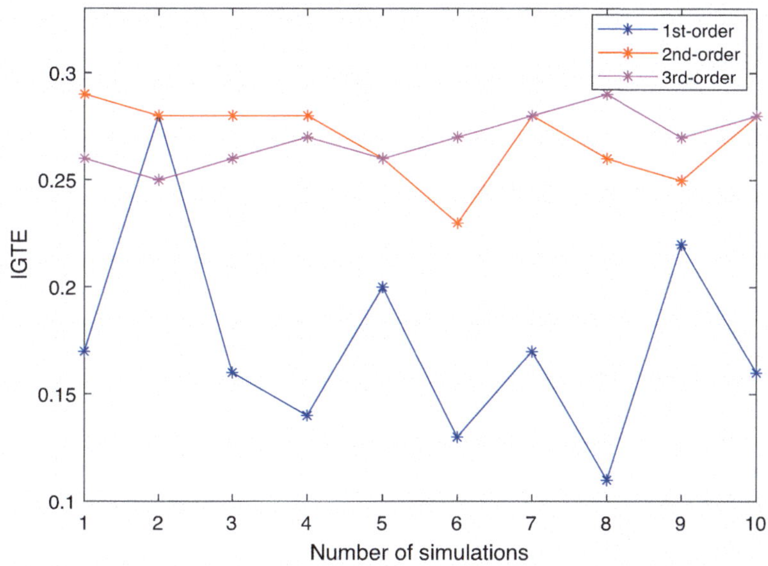

Fig. 4.18 Results of ten trials for IGTE under different orders, reprinted from Ref. [53], copyright 2023, with permission from Elsevier

IGTE fluctuate significantly. The reason for the poor results of the 1st-order IGTE is: The 1st-order IGTE directly using the time delay between variables as the granulation window size may make the granulation process lose more information, while the 2nd-order IGTE considers half of the delay as the granulation window size for information granulation and contains more information. Comparing the 2nd-order and 3rd-order IGTEs, it can also be found that increasing the order does not improve the result in terms of fluctuation degree, but leads to a much increased computational burden. Therefore, the 2nd-order IGTE is more suitable to infer causality and analyze root causes.

In addition, the result of the 2nd-order IGTE with different granulation window sizes is shown in Fig. 4.19. The peak value of the 2nd-order IGTE is achieved at the granulation window size of 8 samples, which is equal to half of the delay between the cause and effect variables in (4.31). The 2nd-order IGTE can find the causal relation from x to y correctly. This proves the effectiveness of the proposed method.

Table 4.4 The fluctuation degree and computational time of IGTE calculation results with different orders, reprinted from Ref. [53], copyright 2023, with permission from Elsevier

	1st-order	2nd-order	3rd-order
λ	0.45	0.08	0.05
Time	3.1 s	4.7 s	26.9 s

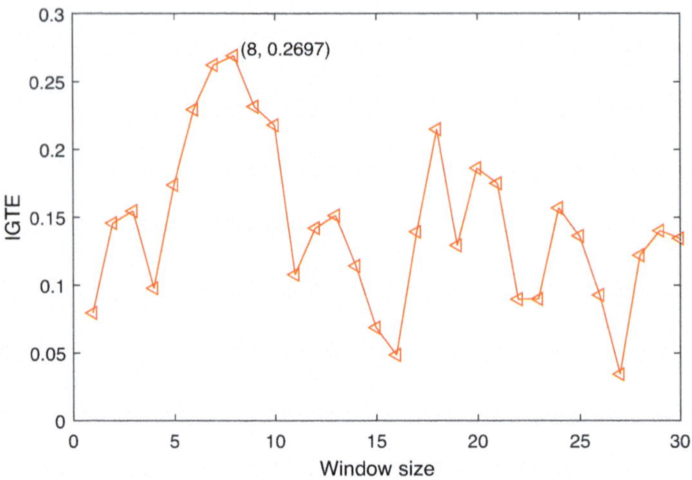

Fig. 4.19 The $\text{IGTE}_{x\to y}$ changes with the window size, reprinted from Ref. [53], copyright 2023, with permission from Elsevier

Comparing the calculated IGTE with a threshold is necessary to determine if a causal relation exists. The threshold can be obtained effectively using the Monte Carlo test based on the surrogate data [4]. The generation of surrogate data must make sure that the statistical features are preserved while entirely destroying the causal relation from x to y. In this context, the PDF estimation is based on the information granule clustering. Therefore, in order to calculate the threshold more reasonably, the surrogate data is generated based on the feature data after clustering instead of the original data. That is, for a surrogate time sequence β_x, it is generated based on the feature data $\bar{\mathbf{c}}$, rather than the raw data $\mathbf{x}$. In addition, it is obvious that using the feature data has less computational complexity.

To determine the threshold $S_{x\to y}$, the IGTEs are calculated from the multiple surrogate time sequences, and the mean and standard deviations of them are then calculated. If $\text{IGTE}_{x\to y} \geq S_{x\to y}$, it suggests that there is a causal link from x to y; if not, it argues that there is no causal link from x to y. The Monte Carlo test based on the surrogate data is also applicable to the online application. However, generating a sequence of surrogate data and calculating their transfer entropies may take some time. As the proposed method has a much lower computational complexity, such a step is more efficient compared to the traditional TE. If there is a high requirement for computational efficiency in the real-time application, it can accelerate computation by generating fewer surrogate data pairs or directly applies a fixed threshold determined offline.

4.2.3.2 Detection of Direct Causality by IGDTE

Some causal relations from the causal network identified by IGTE may be indirect due to the effect of intermediate or confounding variables. In the case of a pair of variables x and y having a causal link, it is important to identify whether the causal link pathway is direct or indirect via a new path from z if the third variable z leads to a triangular network (i.e., z is the intermediate or confounding variable having causal relations with both x and y). Indirect causal pathways can make the causal map more complex and even make it difficult to identify the root cause, especially in a large industrial process. Therefore, the IGDTE can be computed to eliminate indirect causality brought on by intermediate variables or false causality brought on by confounding variables.

Let $\tilde{\mathbf{Z}} = [\tilde{\mathbf{Z}}_1, \tilde{\mathbf{Z}}_2, \ldots, \tilde{\mathbf{Z}}_K]$ represent the granular time sequence of the intermediate variable z. Analogous to DTE [12], the IGDTE is defined as

$$\begin{aligned} \mathrm{IGDTE}_{x\rightarrow y|z} = \sum p\left(\tilde{\mathbf{Y}}_{i+1} \in \tilde{m},\ \tilde{\mathbf{Y}}_i^k \in \tilde{m}, \tilde{\mathbf{Z}}_i^q \in \overline{m}, \tilde{\mathbf{X}}_i^l \in m\right) \\ \cdot \log \frac{p\left(\tilde{\mathbf{Y}}_{i+1} \in \tilde{m} \mid \tilde{\mathbf{Y}}_i^k \in \tilde{m}, \tilde{\mathbf{Z}}_i^q \in \overline{m}, \tilde{\mathbf{X}}_i^l \in m\right)}{p\left(\tilde{\mathbf{Y}}_{i+1} \in \tilde{m} \mid \tilde{\mathbf{Y}}_i^k \in \tilde{m}, \tilde{\mathbf{Z}}_i^q \in \overline{m}\right)}, \end{aligned} \tag{4.33}$$

where $\tilde{\mathbf{Z}}_i$ represents the i-th information granule, q indicates the order of information granule $\tilde{\mathbf{Z}}_i$, and $\overline{m}$ is the m-th cluster of $\tilde{\mathbf{Z}}_i$; in addition, the information granule $\tilde{\mathbf{Z}}_i^q$ is equal to $[\tilde{\mathbf{Z}}_i, \tilde{\mathbf{Z}}_{i-1}, \ldots, \tilde{\mathbf{Z}}_{i-(q-1)}]$. Likewise, the 2nd-order IGDTE is used here to detect direct cause-effect relations.

Comparing the computed IGDTE with a threshold is necessary to evaluate if a direct causal relation exists. The threshold is determined in the same way as described in Sect. 4.2.3.1. The significance level for the IGDTE from x to y can also be calculated using the value of $S_{x\rightarrow y}$. A direct causal link from x to y exists if $\mathrm{IGDTE}_{x\rightarrow y|z} \geq S_{x\rightarrow y}$; otherwise, there is no direct causal link from x to y.

The direct information flow path inferred by the IGDTE can simplify the causal network. Then, it is easy to find the abnormality propagation pathways and identify the root cause. Therefore, detection of direct causality by the IGDTE is essential for inferring the true causal relations. As a further validation, the causal relations are expected to be consistent with the counterparts that are developed from process knowledge.

4.2.4 Numerical Examples

Two numerical examples are provided to demonstrate the effectiveness of the proposed approach based on two models, namely, a two-tank system and the classical Tennessee Eastman Process (TEP).

4.2.4.1 Example 1: Two-Tank System

The causality inference method is utilized to identify the root cause of oscillations in this example. A physical model of the two-tank system is built in Simulink [32]. Many approaches for process monitoring and causality inference have been tested using this model as a benchmark [29–31]. Its process diagram is shown in Fig. 4.20. With

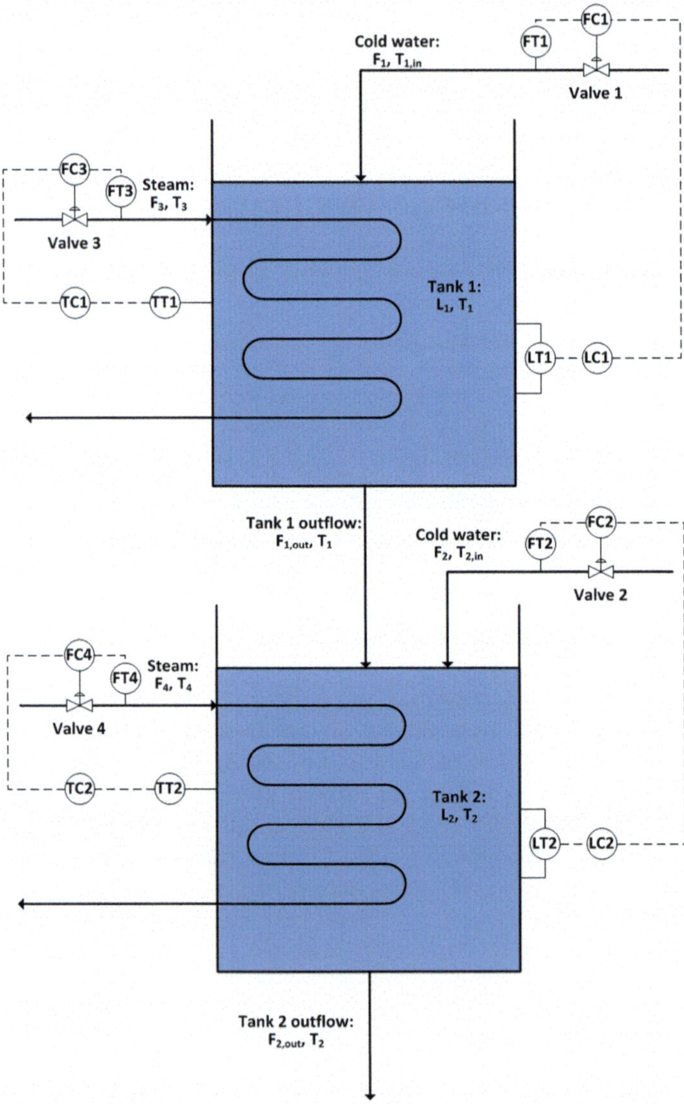

Fig. 4.20 Diagram of the two-tank system, reprinted from Ref. [53], copyright 2023, with permission from Elsevier

heat exchangers, two tanks are connected in series. The cold water flow rates into the tanks determine the tank levels, while steam flow rates through the heating coils determine the tank temperatures. The model is set up with an oscillating disturbance. The parameter T_{1in} is initialized as -50 and its variational amplitude is set to 100. The other parameters are assigned with default values. The time duration is 3000 s and the sampling interval is 0.5 s. Accordingly, a total of 6000 samples are collected. Here, the last 3000 samples under the stationary oscillating state are used for subsequent analysis.

The input temperature $T_{1,in}$ of the cold water experiences an oscillation. This oscillation propagates across the process from $T_{1,in}$ and firstly impacts the temperature T_1 in the first tank. The steam flow rate F_3 is then adjusted by the temperature controller. The second tank's temperature T_2 is impacted as well because the controller is unable to completely reject the input disturbance; the second tank's temperature controller then adjusts the steam flow rate F_4 to make up for this. Data samples of process variables are gathered via simulations, and the process variables also have random noises introduced to them. Figure 4.21 displays the time sequences of original variables. The data obtained from the simulation reveals that the oscillating variables are consistent with the previous analysis.

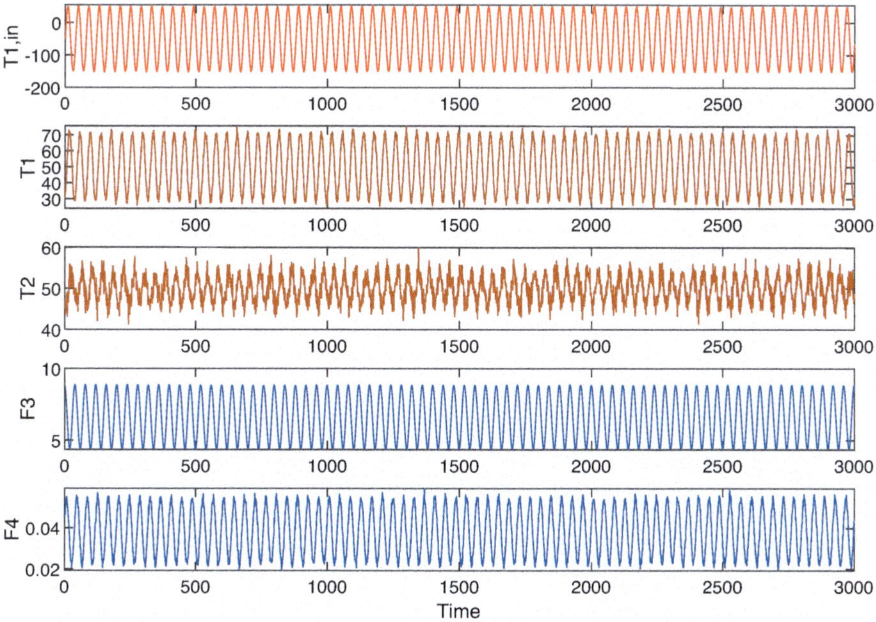

Fig. 4.21 The time sequences of the original variables, reprinted from Ref. [53], copyright 2023, with permission from Elsevier

Table 4.5 The second order IGTEs and thresholds (in the brackets), reprinted from Ref. [53], copyright 2023, with permission from Elsevier

	$T_{1,in}$	T_1	T_2	F_3	F_4
$T_{1,in}$		0.77 (0.08)	1.03 (0.24)	0.01 (0.08)	0
T_1	0		1.17 (0.26)	0.16 (0.07)	0
T_2	0.02 (0.24)	0		0.04 (0.22)	0.42 (0.27)
F_3	0	0.46 (0.10)	1.72 (0.11)		0.02 (0.07)
F_4	0.01 (0.21)	0	1.22 (0.16)	0	

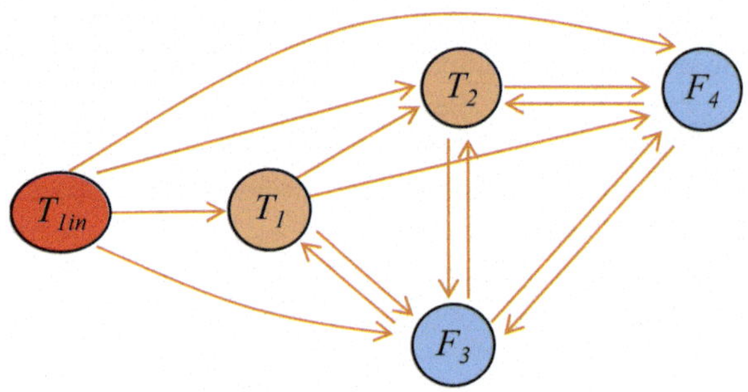

Fig. 4.22 The information flow pathways for the two-tank system by the 1st-order IGTE, reprinted from Ref. [53], copyright 2023, with permission from Elsevier

The information granulation operation is then performed with a window size, determined by the estimated delay between x and y. After information granulation, the OPTICS clustering is performed on the obtained information granule. The 2nd-order IGTEs and thresholds between variables are determined using the proposed approach and are given in Table 4.5. In the IGTE calculation process, the orders of both variables are assigned to be 2 to obtain more information to ensure the validity of the results. In addition, at this time, the window size selection for information granulation only needs to be half of the delay. Thus, less useful information is lost during granulation. This ensures the reliability of the results from another perspective.

Further, to demonstrate the effectiveness and advantages of the proposed 2nd-order IGTE, the causal diagrams based on the calculated IGTEs in different orders are shown in Figs. 4.22 and 4.23, with $k = l = 1$ and $k = l = 2$, respectively.

By comparing the causal diagrams calculated by the 1st-order and 2nd-order IGTEs, it can be seen that the 2nd-order IGTE is more in line with the actual situation. Therefore, the 2nd-order IGTE is more reasonable compared to the 1st-order IGTE. But it can also be observed that the calculated results using the 2nd-order IGTE do not exactly match the actual situation; there are extra causal relations from T_{1in} to T_2 and from F_3 to T_2, which are indirect causal relations. The spurious connection from T_{1in} to T_2 is due to the path via the intermediate variable T_1, and that from F_3 to T_2 is owing to that they have a common confounding variable T_1. In the subsequent analysis, IGDTE is calculated to distinguish direct causal relations from indirect ones, and thus exclude such indirect spurious causal relations.

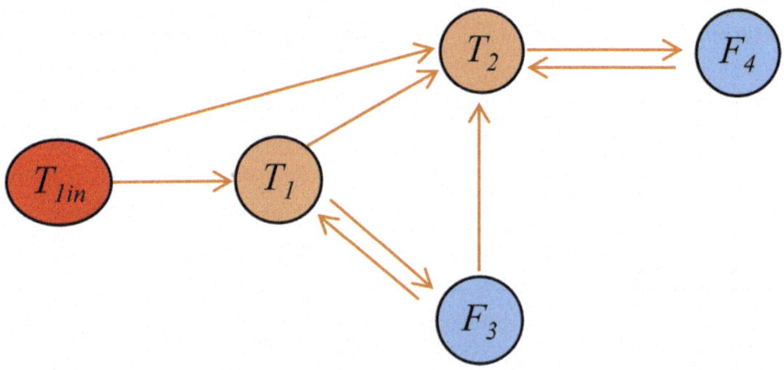

Fig. 4.23 The information flow pathways for the two-tank system by the 2nd-order IGTE, reprinted from Ref. [53], copyright 2023, with permission from Elsevier

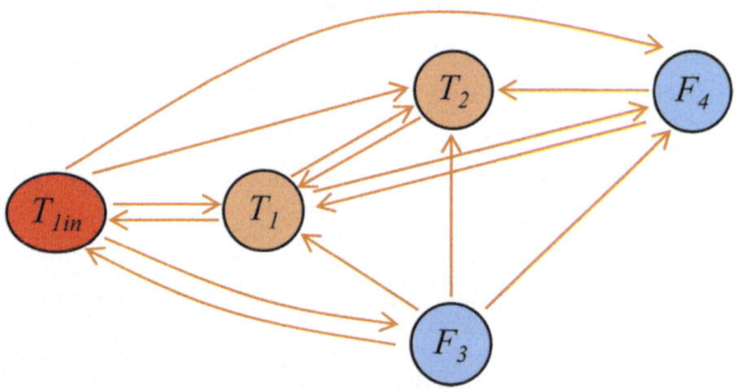

Fig. 4.24 The information flow pathways for the two-tank system by the traditional TE, reprinted from Ref. [53], copyright 2023, with permission from Elsevier

Table 4.6 The accuracy and computational time by calculating IGTE (in different orders) and traditional TE, reprinted from Ref. [53], copyright 2023, with permission from Elsevier

	Accuracy	Computational time
IGTE ($k = l = 1$)	0.6	4.6 s
IGTE ($k = l = 2$)	0.9	6.2 s
Traditional TE ($k = l = 2$)	0.4	20.8 s

Moreover, the traditional TE method based on the kernel estimator is applied to compare with the causal relations obtained using IGTE. The order of the traditional TE is set to 2. The causal relationship calculated by traditional TE is shown in Fig. 4.24. By comparing the causal relations in Figs. 4.23 and 4.24, it can be seen that the IGTE is more precise than the traditional TE in the detection of causal relations.

The accuracy and computational time are compared by calculating IGTE (in different orders) and the traditional TE. The accuracy here is defined as the number of correctly identified relations divided by the total number of relations in two directions for each pair of variables. The comparison results are shown in Table 4.6. It demonstrates that IGTE is more accurate and computationally efficient than the traditional TE. In addition, although adding just a minor amount of computational time compared to the 1st-order IGTE, the 2nd-order IGTE significantly improves the accuracy.

The IGDTEs are computed to distinguish the direct causal links from the indirect ones. The intermediate or confounding variables can be recognized from the triangular connecting pathways of the causal network in Fig. 4.22. The IGDTE should then be computed in order to ascertain whether the causal pathways is direct or indirect. For instance, there are two ways from $T_{1,in}$ to T_2, either directly or indirectly through T_1. Next, the IGDTE from $T_{1,in}$ to T_2 based on T_1 and that from F_3 to T_2 based on T_1 are computed, and found to be less than the thresholds ($\text{IGDTE}_{T_{1,in} \to T_2|T_1} = 0.01 < 0.24$, $\text{IGDTE}_{F_3 \to T_2|T_1} = 0.02 < 11$). Therefore, the causal pathway from $T_{1,in}$ to T_2 and that from F_3 to T_2 are judged to be indirect, and then the indirect connections are eliminated. Similarly, the IGDTE from T_1 to T_2 based on F_3 is found to be greater than the threshold ($\text{IGDTE}_{T_1 \to T_2|F_3} = 1.12 > 0.26$). This implies that the direct path from T_1 to T_2 is true.

As a result, the causal map containing direct causal relations in Fig. 4.25 is constructed. The generated causal map makes it very evident that $F_{1,in}$ is the root cause of the above oscillation. Moreover, the causal relation between the oscillating variables is spelled out in detail. The conclusion is accurate to the situation as it is.

4.2.4.2 Example 2: Tennessee Eastman Process

TEP is a classical chemical process simulator and is extensively utilized for verification of process monitoring and control techniques [7]. The reactor, product condenser, gas-liquid separation tower, product stripper, and recycle compressor are the major units in TEP. In addition, A, C, D, and E are the gaseous reactants; G and H are the products. There are 41 measured and 12 manipulated variables. The measured variables contain 22 process variables sampled every 3 min, and 19 quality variables sampled with dead time and time delays [9]. Due to the existence of dead time, there

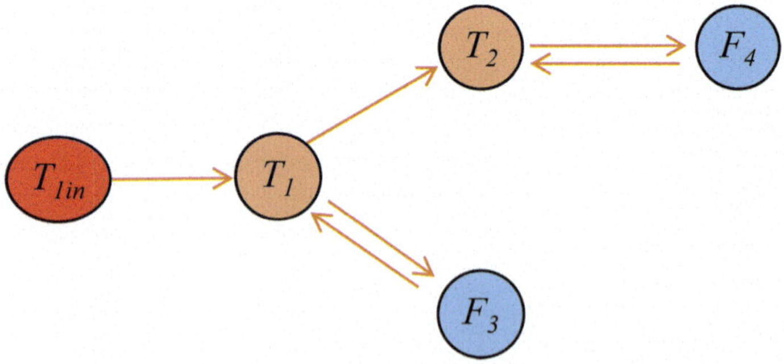

Fig. 4.25 The direct information flow pathways by the IGDTE, reprinted from Ref. [53], copyright 2023, with permission from Elsevier

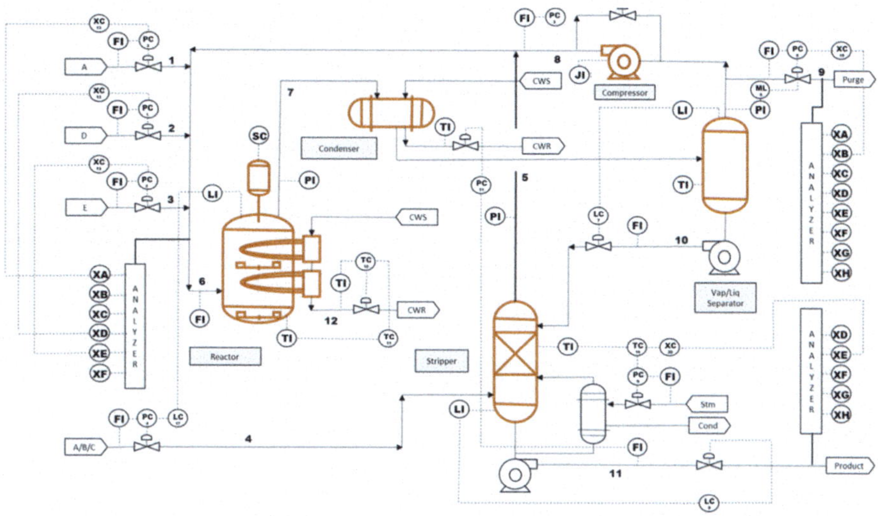

Fig. 4.26 Flowchart of Tennessee Eastman process [48], reprinted from Ref. [53], copyright 2023, with permission from Elsevier

is a time delay in the sampling process. The time delay can be compensated by pre-shifting the data. The flowchart of TEP is shown in Fig. 4.26. Except for the normal situations, there are 21 abnormal conditions that can be simulated in the TEP, indicated as IDV(1), IDV(2), . . ., and IDV(21), involving multiple types of disturbances, valve sticking, and some unknown faults.

The causal relations between variables under IDV(1) are analyzed to find out the root cause. All 52 recorded variables are considered initially. The causal graph obtained with the proposed method is shown in Fig. 4.27. Since IDV(1) shows a step in the A/C feed ratio, only the leftmost among the three groups of causal relations in Fig. 4.27 is relevant. The other two groups of causal relations are not related to IDV(1). It can be clearly seen that many variables do not affect the results of the root cause analysis. This is because these variables are not relevant to IDV(1). Therefore, only the relevant variables of IDV(1) should be considered in the subsequent analysis.

According to the results in Fig. 4.27 and the mechanism of IDV(1), the set of eight relevant variables (given in Table 4.7) are selected as the candidates for causality inference, based on the propagation mechanism of the IDV(1) given as follows [48]:

When IDV(1) occurs in the TEP, the A/C feed ratio changes in stream 4, while the composition of B is held steady. More specifically, the composition of A in the feed stream 4 alters. Meanwhile, the composition of C also changes. However, the compositions in stream 4 are not measured owing to the lack of sensors. Consequently, stream 6 is the one where the composition change of A is detected. The closed loop controller increases the flow rate of stream 1 (V_1) in response to the change. The level controller then adjusts the flow rate of stream 4 (V_4) to boost the reactor level (V_8) as a result of the increased incoming flow rate. Further, the operation of the process is

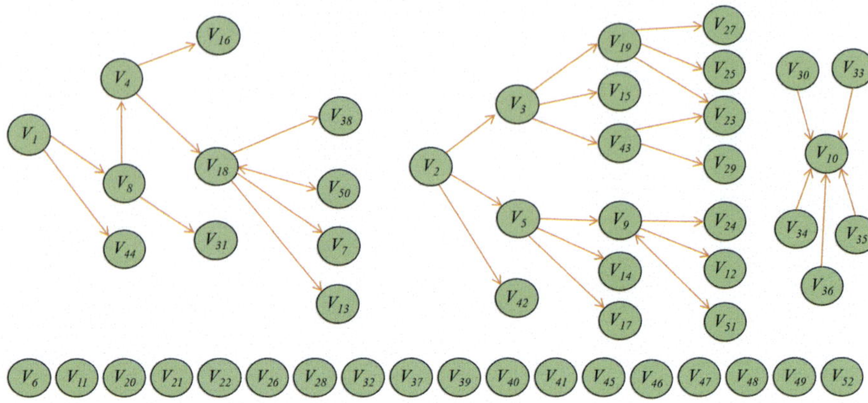

Fig. 4.27 The information flow pathways of all recorded variables for IDV(1) by the proposed IGTE, reprinted from Ref. [53], copyright 2023, with permission from Elsevier

Table 4.7 Variable selection of IDV(1), reprinted from Ref. [53], copyright 2023, with permission from Elsevier

Variable number	Variable meaning
V_1	A feed in stream 1
V_4	A and C feed in stream 4
V_8	Reactor level
V_{16}	Stripper pressure
V_{18}	Stripper temperature
V_{31}	Component C in stream 9
V_{38}	Component E in stream 11
V_{50}	Stripper steam valve

impacted by the increase of the component C. No feedback control system is in place to react to this. As a result, component C in stream 9 changes (V_{31}). The reaction in the reactor between C and E is also isomolar. The consumption of E is impacted by the change of C, which further affects E in the product flow, namely, stream 11 (V_{38}). It should be emphasized that composition of E is under cascade control. The stripper temperature (V_{18}) in this control system is adjusted in order to control component E, while V_{18} is controlled by manipulating the stripper steam valve (V_{50}). Further, the temperature change causes a pressure change in the stripper (V_{16}).

The time sequences of relevant candidate variables under IDV(1) are shown in Fig. 4.28. It is observed that these variables have nonstationary variational trends under IDV(1). The IGTEs and thresholds between these variables are given in Table 4.8. The causal diagram based on the calculated IGTEs is displayed in Fig. 4.29. The root cause of IDV(1) is determined to be the change of V_1, which can be observed in Fig. 4.29, where a causal diagram between the related variables is given. This is generally consistent with the previous propagation path analysis under

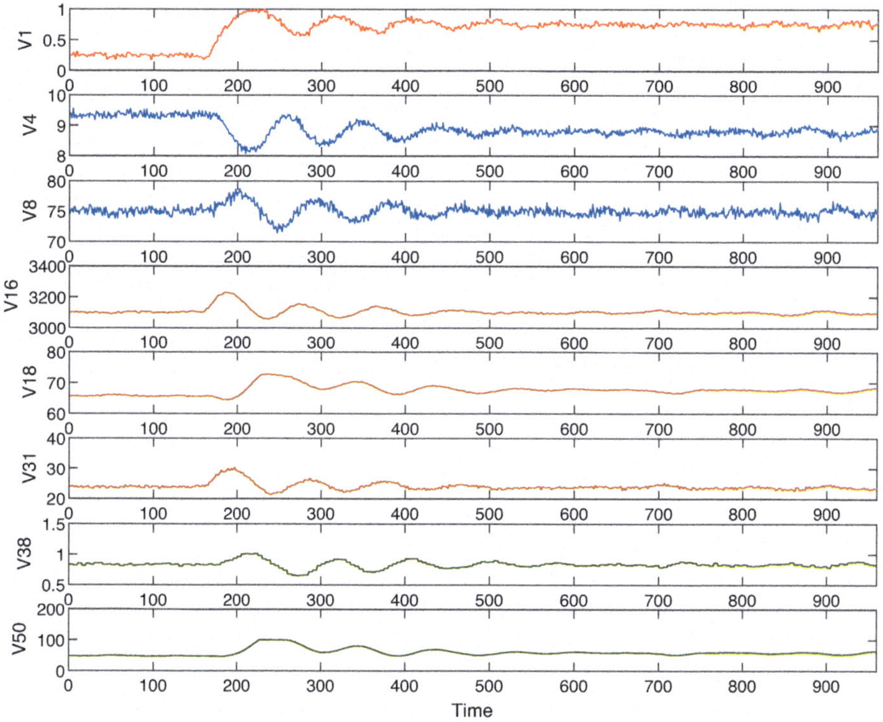

Fig. 4.28 Time sequences of the original variables

IDV(1). The effectiveness of the proposed IGTE for nonstationary processes is also demonstrated.

Moreover, the Symbolic Conditional Transfer Entropy (SCTE) method and the traditional TE method are applied to compare with the causal relations obtained using the proposed IGTE. The traditional TE is based on the kernel estimator, and the order is set to 2. The causal relations calculated by the SCTE and the traditional TE are shown in Figs. 4.30 and 4.31, respectively. Comparing the detected causal relations in Figs. 4.29, 4.30, and 4.31, the results using the IGTE and SCTE are more consistent with the given mechanism of fault propagation. The IGTE identifies the root cause variable of IDV(1) effectively. As for SCTE, most detected causal relations are correct, except for the wrong causal relation from $V_{38} \rightarrow V_1$. In contrast, the causal relations detected by the traditional TE method has too many errors to identify the root cause.

To identify the causal relation is direct or indirect, the IGDTE is calculated. The calculated result of IGDTE from V_1 to V_4 is $\text{IGDTE}_{V_1 \rightarrow V_4 | V_8} = 0.03 < 0.18$. The IGDTE from V_4 to V_{38} is $\text{IGDTE}_{V_4 \rightarrow V_{38} | V_{18}} = 0.05 < 0.09$. The IGDTE from V_4 to V_{50} is $\text{IGDTE}_{V_4 \rightarrow V_{50} | V_{18}} = 0.10 < 0.12$. The IGDTE from V_{16} to V_{31} is $\text{IGDTE}_{V_{16} \rightarrow V_{31} | V_8} = 0.06 < 0.09$. Thus, these causal relations are indirect and the

Table 4.8 The IGTEs and thresholds (in the brackets), reprinted from Ref. [53], copyright 2023, with permission from Elsevier

	V_1	V_4	V_8	V_{16}	V_{18}	V_{31}	V_{38}	V_{50}
V_1		0.33 (0.18)	0.17 (0.09)	0.05 (0.14)	0.06 (0.10)	0.07 (0.15)	0	0
V_4	0		0.11 (0.16)	0.15 (0.11)	0.16 (0.07)	0.14 (0.18)	0.19 (0.09)	0.21 (0.12)
V_8	0.07 (0.16)	0.33 (0.11)		0	0.11 (0.17)	0.25 (0.17)	0	0.09 (0.15)
V_{16}	0	0.12 (0.27)	0.11 (0.21)		0	0.17 (0.09)	0.16 (0.19)	0.18 (0.28)
V_{18}	0	0.09 (0.24)	0.05 (0.14)	0		0	0.22 (0.14)	0.23 (0.12)
V_{31}	0.09 (0.25)	0.17 (0.21)	0	0.04 (0.15)	0.16 (0.22)		0.14 (0.20)	0
V_{38}	0.09 (0.19)	0.12 (0.17)	0.05 (0.13)	0	0.07 (0.18)	0		0.09 (0.22)
V_{50}	0.05 (0.14)	0.13 (0.21)	0	0.11 (0.28)	0.39 (0.20)	0	0	

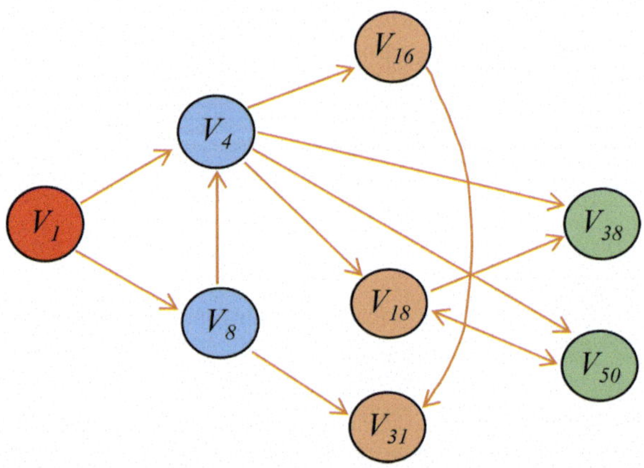

Fig. 4.29 The information flow pathways for IDV(1) by the proposed IGTE, reprinted from Ref. [53], copyright 2023, with permission from Elsevier

pathways are eliminated. Then, the causal map consisting of direct causality is attained as shown in Fig. 4.32. It can be seen that V_1 is the root cause variable of IDV(1) according to the direct causal map in Fig. 4.32 detected by the proposed IGDTE. Moreover, the causal relations among the fault variables are clearly described. The detected result is accurate.

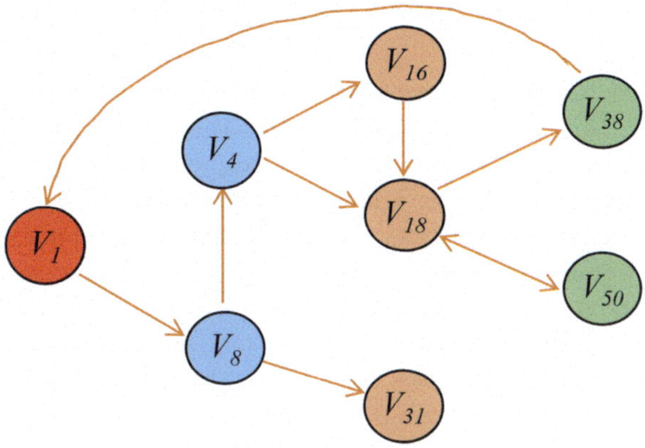

Fig. 4.30 The information flow pathways for IDV(1) by SCTE, reprinted from Ref. [53], copyright 2023, with permission from Elsevier

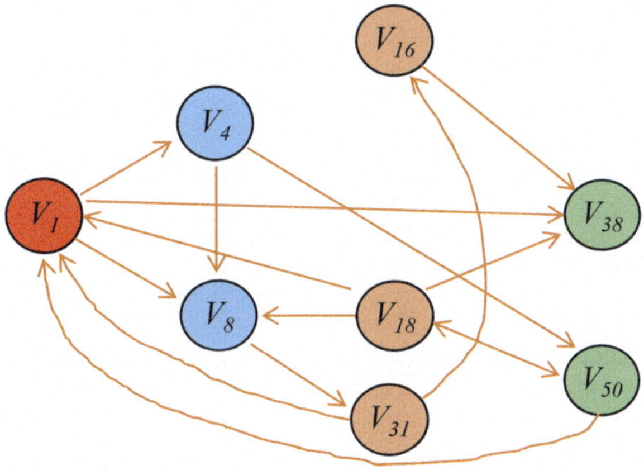

Fig. 4.31 The information flow pathways for IDV(1) by the traditional TE, reprinted from Ref. [53], copyright 2023, with permission from Elsevier

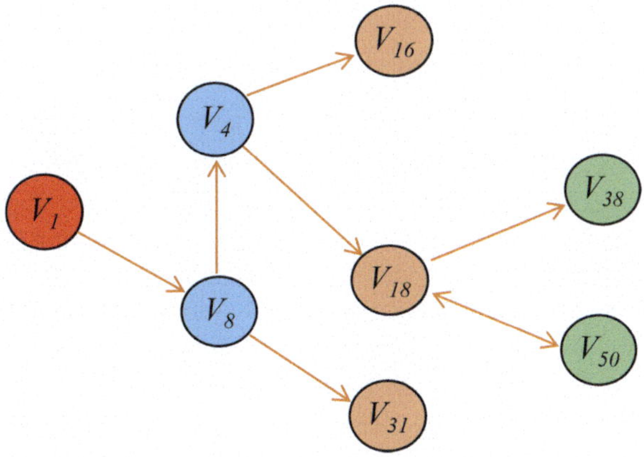

Fig. 4.32 The information flow pathways for IDV(1) by the proposed IGDTE, reprinted from Ref. [53], copyright 2023, with permission from Elsevier

4.3 Root-Cause Analysis for Alarm Variables

The section is on analysis of root causes of occurring alarms, for a special type of industrial alarm variables [46]. This type of alarm variables is known to switch between non-alarm and alarm states, owing to status changes of multiple root-cause variables. The alarm variable and root-cause variables take binary values, namely, 1 for the alarm state and valid status, and 0 for the non-alarm state and invalid status. The relationship between the alarm variable and root-cause variables is described by a Bayesian network with one child node and multiple parent nodes. The probability parameters in the Bayesian network are updated recursively from observed data samples. Root causes of an occurring alarm are determined from the parent node set having the largest posterior conditional possibility. This method can effectively deal with three challenges frequently appeared in industrial environments. That is, it removes negative effects of false and missing alarms, handles the co-existence of multiple root causes, and detects the incompleteness of known root causes. By contrast, a common practice, which finds root causes manually by comparing data samples of the alarm variable and root-cause variables, cannot well deal with these challenges.

4.3.1 Problem Description

Consider an alarm variable X_a taking the value 1 for the alarm state and 0 for the non-alarm state. It is assumed that $X_a = 1$ can be arisen from the status changes of n mutually independent root-cause variables, denoted by $X_1, X_2, \ldots, X_n$. The assumption is usually fulfilled by exploiting some process knowledge. Let the root-cause variables take binary values, i.e., $X_i = 1$ (or $X_i = 0$) for $i \in [1, n]$ represents the status that the i-th root-cause variable X_i is valid (or invalid). If $X_i = 1$, then X_a will run into the alarm state. Given the observed data set $D := \{X(1), \ldots, X(t-1), X(t)\}$ with $X(t) := [X_1(t), X_2(t), \ldots, X_n(t), X_a(t)]'$, the objective is to analyze root causes of an occurring alarm in $X_a(t) = 1$ under the above assumption.

A common practice to reach such an objective is to analyze root causes manually by looking at observed data samples. If $X_a(t) = 1$, then root causes are the ones with $X_i(t) = 1$ for $i \in [1, n]$. However, the common practice often yields inaccurate results due to the following three challenges. First, either X_a or X_i may take the value 1 for a short time duration owing to random noises. Thus, X_a or X_i is contaminated by false alarms, which result in failures of root-cause analysis. Second, multiple root causes may exist simultaneously, instead of the presence of single one root cause. The common practice is manually done by industrial operators. They tend to focus on the first-appeared root cause, and overlook other co-existing root causes. Third, the root causes $X_1, X_2, \ldots, X_n$ may be incomplete. One common reason for the incompleteness is that measurements of some other root-cause variables are unavailable at control centers of industrial plants. The verification of these root causes have to be done on site by industrial engineers. As a result, these root causes are not in the set $\{X_1, X_2, \ldots, X_n\}$, so that they cannot be handled by the common practice. Due to the three challenges, a new method is needed to analyze root causes of occurring alarms in X_a.

4.3.2 Updating Probabilities in a Bayesian Network

A Bayesian network depicted in Fig. 4.33 is used to describe the relationship between the alarm variable X_a and root cause $X_1, X_2, \ldots, X_n$. It is a graphical model that represents a set of random variables and their conditional dependencies via a directed acyclic graph; random variables are represented by nodes in the network and take some discrete values [25]. For the Bayesian network in Fig. 4.33, $X_1, X_2, \ldots, X_n$ are the parent nodes of X_a, and they take binary values 0 and 1 in this context. Denote $\theta_{i,0}$ and $\theta_{i,1}$ as the probabilities of X_i with $i \in [1, n]$ taking the value 0 and 1, respectively, i.e.,

$$\theta_{i,0} := P(X_i = 0),\ \theta_{i,1} := P(X_i = 1),\ i \in [1, n].$$

Let $R := \{X_1, X_2, \ldots, X_n\}$ be the parent node set of X_a, and r_j with $j \in [1, 2^n]$ be the j-th set of possible values taken by the parent nodes of X_a. For instance, if $n = 2$, then there are four sets in R, namely, $r_1 := \{0, 0\}$, $r_2 := \{1, 0\}$, $r_3 := \{0, 1\}$ and $r_4 = \{1, 1\}$. The parameters $\theta_{a,0|j}$ and $\theta_{a,1|j}$ respectively represent the prior conditional probabilities of $X_a = 0$ and $X_a = 1$ given the condition that R takes the value r_j, i.e.,

$$\theta_{a,0|j} := P(X_a = 0|R = r_j),\ \theta_{a,1|j} := P(X_a = 1|R = r_j),\ j \in [1, 2^n].$$

Clearly, these parameters need to satisfy the constraints, $\theta_{i,0} + \theta_{i,1} = 1, \forall i \in [1, n]$ and $\theta_{a,0|j} + \theta_{a,1|j} = 1, \forall j \in [1, 2^n]$.

For the purpose of online root-cause analysis, the probability parameters in the set $\boldsymbol{\theta} := \left\{\theta_{i,0}, \theta_{i,1}, \theta_{a,0|j}, \theta_{a,1|j}\right\}$ for $i = 1, 2, \ldots, n$ and $j = 1, 2, \ldots, 2^n$ in the Bayesian network need to be learned from observed data samples in a recursive manner. More

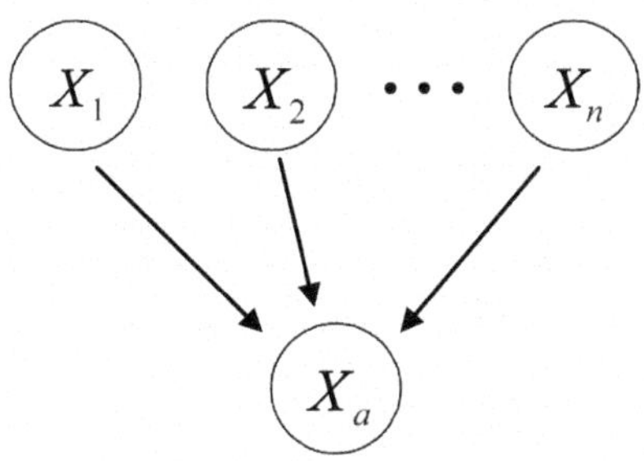

Fig. 4.33 The diagram of a Bayesian network, reprinted from Ref. [46], copyright 2018, with permission from Elsevier

precisely, the parameter set $\boldsymbol{\theta}$ needs to be updated based on a prior parameter set $\bar{\boldsymbol{\theta}}$ and the observed data samples of $X_1, X_2, \ldots, X_n$ and X_a.

Reference [3] proposed a batch learning algorithm being applicable to the historical data set $D := \{X(1), \ldots, X(t-1), X(t)\}$ with $X(t) := [X_1(t), X_2(t), \ldots, X_n(t), X_a(t)]'$. The posterior estimate $\hat{\boldsymbol{\theta}}$ is obtained by maximizing a function $F(\boldsymbol{\theta}) = \lambda L_D(\boldsymbol{\theta}) - d(\boldsymbol{\theta}, \bar{\boldsymbol{\theta}})$, i.e.,

$$\hat{\boldsymbol{\theta}} = \arg\max_{\boldsymbol{\theta}} \left(\lambda L_D(\boldsymbol{\theta}) - d(\boldsymbol{\theta}, \bar{\boldsymbol{\theta}})\right). \tag{4.34}$$

Here $L_D(\boldsymbol{\theta})$ is the averaged logarithm likelihood,

$$L_D(\boldsymbol{\theta}) = \frac{1}{t}\sum_{l=1}^{t} \ln P_{\boldsymbol{\theta}}(X(l)),$$

where $X(l_1)$ and $X(l_2)$ are assumed to be independent with each other for all indices $l_1 \neq l_2$. The penalty term $d(\boldsymbol{\theta}, \bar{\boldsymbol{\theta}})$ is a measure of χ^2 distance between $\boldsymbol{\theta}$ and $\bar{\boldsymbol{\theta}}$. The constant λ is an updating rate parameter. Reference [8] further developed an online updating algorithm based on the batch solution in (4.34). The historical data set D is replaced by the current sample $X(t)$, while $\hat{\boldsymbol{\theta}}(t-1)$ and $\hat{\boldsymbol{\theta}}(t)$ take the place of $\bar{\boldsymbol{\theta}}$ and $\hat{\boldsymbol{\theta}}$, respectively.

The main idea of the online updating algorithm in [8] is adopted here for updating probability parameters of the Bayesian network in Fig. 4.33. Let the prior estimate of $\boldsymbol{\theta}$ be available as $\hat{\boldsymbol{\theta}}(t-1)$ and the current data sample be $X(t) := [X_1(t), X_2(t), \ldots, X_n(t), X_a(t)]'$. Suppose that $X_a(t)$ takes the value 1 and $R(t)$ (the first n elements in $X(t)$) takes the j-th parent node set r_j, i.e., $P(X_a(t) = 1, R(t) = r_j | X(t)) = 1$ and $P(X_a(t) = 1 | X(t)) = 1$. Replacing the historical data set D by the current sample $X(t)$ in the batch solution from [3] (Eq. (13) therein) yields

$$\begin{aligned}\theta_{a,1|j}(t) &= \lambda \frac{P\left(X_a(t) = 1, R(t) = r_j | X(t)\right)}{P\left(X_a(t) = 1 | X(t)\right)} + (1-\lambda)\theta_{a,1|j}(t-1) \\ &= \lambda + (1-\lambda)\,\theta_{a,1|j}(t-1).\end{aligned} \tag{4.35}$$

Meanwhile, the parameter $\theta_{a,0|j}(t)$ has to be updated as

$$\theta_{a,0|j}(t) = (1-\lambda)\,\theta_{a,0|j}(t), \tag{4.36}$$

in order to satisfy the equality $\theta_{a,1|j}(t) + \theta_{a,0|j}(t) = 1$.

For the parent node set r_k with $k \in [1, 2^n]$ and $k \neq j$, the corresponding probability parameters are unchanged, i.e.,

$$\theta_{a,1|k}(t) = \theta_{a,1|k}(t-1), \tag{4.37}$$

$$\theta_{a,0|k}(t) = \theta_{a,0|k}(t-1). \tag{4.38}$$

Since $X_i(t) = 1$ appears in $X(t)$ for $i \in [1, n]$, i.e., $P(X_i(t) = 1|X(t)) = 1$, the corresponding probability parameters are updated as

$$\begin{aligned}\theta_{i,1}(t) &= \lambda P(X_i(t) = 1|X(t)) + (1-\lambda)\theta_{i,1}(t-1) \\ &= \lambda + (1-\lambda)\theta_{i,1}(t),\end{aligned} \tag{4.39}$$

and

$$\theta_{i,0}(t) = (1-\lambda)\theta_{i,0}(t-1). \tag{4.40}$$

Similarly, if $X_a(t) = 0$ and $R(t) = r_j$, the counterparts of updating equations (4.35) and (4.36) are obtained as

$$\theta_{a,0|j}(t) = \lambda + (1-\lambda)\theta_{a,0|j}(t-1), \tag{4.41}$$

$$\theta_{a,1|j}(t) = (1-\lambda)\theta_{a,1|j}(t). \tag{4.42}$$

If $X_i(t) = 0$, the counterparts of (4.39) and (4.40) are

$$\theta_{i,0}(t) = \lambda + (1-\lambda)\theta_{i,0}(t-1), \tag{4.43}$$

$$\theta_{i,1}(t) = (1-\lambda)\theta_{i,1}(t-1). \tag{4.44}$$

Thus, the probability parameter set $\boldsymbol{\theta}$ is updated via (4.35) and (4.36) (or (4.41) and (4.42)), (4.37), (4.38) (4.39) and (4.40) (or (4.43) and (4.44)) in an online manner, i.e.,

$$\boldsymbol{\theta}(t) = \left\{\theta_{i,0}(t), \theta_{i,1}(t), \theta_{a,0|j}(t), \theta_{a,1|j}(t)\right\}, \text{ for } i = 1, 2, \ldots, n,\ j = 1, 2, \ldots, 2^n. \tag{4.45}$$

4.3.3 Root Cause Determination for Occurring Alarms

First, denote x_a as the value of X_a, and $x_{i,j}$ as the value of X_i in the j-th parent node set r_j. From the Bayesian rule, the posterior conditional probability of r_j based on x_a is obtained as

$$\begin{aligned}P\left(R = r_j|X_a = x_a\right) &= \frac{P(R = r_j, X_a = x_a)}{P(X_a = x_a)} \\ &= \frac{P(X_1 = x_{1,j}, \ldots, X_n = x_{n,j}, X_a = x_a)}{\sum_j P(X_1 = x_{1,j}, \ldots, X_n = x_{n,j}, X_a = x_a)}.\end{aligned} \tag{4.46}$$

According to the chain rule and owing to the mutual independence among $X_1, \ldots, X_n$, the joint probability can be rewritten as

$$\begin{aligned}
&P(X_1 = x_{1,j}, \ldots, X_n = x_{n,j}, X_a = x_a) \\
&= P\left(X_a = x_a | X_1 = x_{1,j}, \ldots, X_n = x_{n,j}\right) \\
&\cdot P\left(X_1 = x_{1,j} | X_2 = x_{2,j}, \ldots, X_n = x_{n,j}\right) \\
&\cdots P\left(X_{n-1} = x_{n-1,j} | X_n = x_{n,j}\right) \cdot P\left(X_n = x_{n,j}\right) \\
&= \prod_{i=1}^{n} P\left(X_i = x_{i,j}\right) \cdot P(X_a = x_a | R = r_j).
\end{aligned} \tag{4.47}$$

Using the total probability in (4.47) and the updated probability parameters in $\boldsymbol{\theta}(t)$ from (4.45), the posterior probability $P\left(R = r_j | X_a = x_a\right)$ in (4.46), denoted by $\theta_{j|x_a}$, is updated in an online manner as

$$\theta_{j|x_a}(t) = \frac{\prod_{i=1}^{n} \theta_{i,x_{i,j}}(t) \cdot \theta_{a,x_a|j}(t)}{\sum_j \prod_{i=1}^{n} \theta_{i,x_{i,j}}(t) \cdot \theta_{a,x_a|j}(t)}.$$

Without loss of generality, let r_{2^n} represent the parent node set $\{X_1(t) = 0, \ldots, X_n(t) = 0\}$. When X_a is in the non-alarm state, there is no need of analyzing alarm root causes. Since the denominators of $\theta_{j|x_a}(t)$'s for $X_a = 1$ are the same, it is sufficient to compare the numerators. Denote $\theta_{j|x_a}^{(N)}(t)$ as the numerator of $\theta_{j|x_a}(t)$. A vector is defined for the posterior probabilities with $X_a = 1$,

$$\boldsymbol{\theta}_p := \left[\theta_{j=1|x_a=1}^{(N)}(t), \theta_{j=2|x_a=1}^{(N)}(t), \ldots, \theta_{j=2^n|x_a=1}^{(N)}(t)\right].$$

Sorting the elements of $\boldsymbol{\theta}_p$ in the descending order yields a sorted vector $\boldsymbol{\theta}_{p,s}$,

$$\left(\boldsymbol{\theta}_{p,s}, \boldsymbol{\eta}\right) = \max\left(\boldsymbol{\theta}_p\right), \tag{4.48}$$

where the vector $\boldsymbol{\eta}$ contains the indices of the sorted elements. The root cause is determined as

$$\hat{j}(t) = \boldsymbol{\eta}(1), \tag{4.49}$$

where $\boldsymbol{\eta}(1)$ denotes the first element of $\boldsymbol{\eta}$. In words, if $\theta_{j|1}(t)$ for $j \in [1, 2^n - 1]$ is the largest posterior probability, then X_a is in the alarm state, and the root causes are the parent nodes $X_1, \ldots, X_n$ who take the value 1 in the set r_j; if the last element $\theta_{2^n|1}(t)$ is the largest, then it is concluded that $X_a = 1$ is caused by some root causes other than the known ones $X_1, X_2, \ldots, X_n$.

Denote $\theta_{a,0}$ and $\theta_{a,1}$ as the probabilities of X_a taking the value 0 and 1, respectively. If $X_a(t) = 1$, then online updates of $\theta_{a,0}(t)$ and $\theta_{a,1}(t)$ are obtained analogously to

$\theta_{i,0}(t)$ and $\theta_{i,1}(t)$ in (4.39) and (4.40) as

$$\theta_{a,1}(t) = \lambda + (1-\lambda)\,\theta_{a,1}(t-1)\,,$$
$$\theta_{a,0}(t) = (1-\lambda)\,\theta_{a,0}(t-1)\,.$$

If $X_a(t) = 0$, then the counterparts of (4.43) and (4.44) are

$$\theta_{a,0}(t) = \lambda + (1-\lambda)\,\theta_{a,0}(t-1)\,,$$
$$\theta_{a,1}(t) = (1-\lambda)\,\theta_{a,1}(t-1)\,.$$

If $\theta_{a,0}(t) > \theta_{a,1}(t)$, then the root cause analysis result $\hat{j}(t)$ is set to zero. Therefore, together with (4.49), the root cause of an occurring alarm is

$$\hat{j}(t) = \begin{cases} \eta(1)\,, & \text{if } \theta_{a,0}(t) \le \theta_{a,1}(t), \\ 0, & \text{if } \theta_{a,0}(t) > \theta_{a,1}(t). \end{cases} \quad (4.50)$$

Next, let us investigate the accuracies of root cause analysis results. First, it is followed directly from Theorem 1 in [8] that if $0 < \lambda < 1$, each estimate in $\boldsymbol{\theta}(t)$ from (4.45), denoted by $\theta(t)$, is unbiased with a finite variance, i.e.,

$$E(\theta(t)) = \theta^*,$$
$$Var(\theta(t)) = \frac{\lambda}{2-\lambda}\theta^*(1-\theta^*),$$

where θ^* is the true value of $\theta(t)$. Second, all nodes of the Bayesian network in Fig. 4.33 take binary values 0 and 1. Due to effects of random noises, $X_1, \ldots, X_n$ and X_a may take the value 1 or 0 even if they are actually in the opposite state (status) 1 or 0. The erroneous values are referred to as false alarms for the former case and missing alarms for the later case. If the probabilities of false and missing alarms are not larger than 0.5, then the next proposition proves that the proposed method gives the true root causes of alarms.

Proposition 4.1 *For the Bayesian network in Fig. 4.33, assume that the true root cause of $X_a = 1$ is associated with $R = r_{j^*}$. If the probabilities of false and missing alarms in $X_1, \ldots, X_n$ and X_a are less than 0.5, then (4.50) gives the true root cause r_{j^*} for a sufficient large value of t.*

Proof of Proposition 4.1 It is possible to have $\theta_{i,x_{i,j}}(t) \neq 0$ and $\theta_{a,x_a|j}(t) \neq 0$ for $j \neq j^*$ and $i \in [1, n]$, owing to the presence of false and missing alarms. However, if the probabilities of false and missing alarms are less than 0.5 and $X_a = 1$ is caused by $R = r_{j^*}$, then $\theta_{i,x_{i,j^*}}(t) \geq 0.5$, $\theta_{i,x_{i,j}}(t) < 0.5$, $\theta_{a,1}(t) \geq 0.5$, $\theta_{a,0}(t) < 0.5$, $\theta_{a,1|j^*}(t) \geq 0.5$ and $\theta_{a,1|j}(t) < 0.5$ for $j \neq j^*$ when the updated probability parameters are in the steady-state conditions. Thus, the following inequality holds,

Table 4.9 Recommended values of m from the industrial standards [2]

Variable type	m (s)
Flow rate	15
Level	60
Pressure	15
Temperature	60

$$\prod_{i=1}^{n} \theta_{i,x_{i,j}}(t) \cdot \theta_{a,1|j}(t) < \prod_{i=1}^{n} \theta_{i,x_{i,j^*}}(t) \cdot \theta_{a,j^*|1}(t)\,.$$

Because the denominator of $\theta_{1|1}(t), \theta_{2|1}(t), \ldots, \theta_{2^n|1}(t)$ are the same, the online alarm root analysis in (4.50) gives the true root cause $\hat{j}(t) = j^*$. This completes the proof. □

Finally, the updating rate parameter λ in (4.34) is designed. The valid range of λ is [0, 1], in order to achieve unbiased estimate of $\boldsymbol{\theta}(t)$. A physical rule is adopted for designing λ, namely, if X_i for $i \in [1, n]$ (or X_a) switches its value from 0 to 1, and m subsequent consecutive samples of X_i (or X_a) take the value 1, then the actual state of X_i (or X_a) is switched from 0 to 1; otherwise, the short-duration of changes between 0 and 1 are regarded as false or missing alarms. The physical rule has been widely used in practice to remove nuisance alarms [2]. The industrial standard [2] recommends the values of m in Table 4.9 based on variation rates of different types of process variables. The updating rate of $\theta_{i,j}$ needs to be consistent with the value of m when X_i is changing between the states 0 and 1. The updating equations in Sect. 4.3.2 can be represented as

$$\theta(t) = \lambda + (1 - \lambda)\theta(t - 1).$$

It is a first-order difference equation. If $\theta(0)$ denotes the initial state, then

$$\theta(t) = 1 - (1 - \lambda)^t(1 - \theta(0)). \tag{4.51}$$

Proposition 4.1 says that it is necessary for $\theta(t)$ reaching $\theta(\infty) \geq 0.5$, in order to ensure that the true root cause is obtained. The initial value $\theta(0)$ is usually set to 0, and it should take at least m samples from $\theta(0)$ to be close to the final state $\theta(\infty)$. The value of λ is developed from (4.51),

$$\lambda = 1 - (0.5)^{\frac{1}{m}}. \tag{4.52}$$

4.3.4 Industrial Examples

This section presents industrial examples to illustrate the effectiveness of the proposed method.

The examples are from a large-scale thermal power generation unit. The special type of alarm variables being considered in this context is mainly formulated by two ways. First, the alarm variable and root-cause variables are connected via physical laws such as mass and energy balance laws. The root causes are usually defined from process knowledge. For instance, the water level in a boiler drum is a critical process variable configured with alarms; the alarm state can be caused by abnormalities in the feed-water flow, main steam flow and boiler drum pressure as well as faults in water level sensors. Second, an alarm variable is formulated via some logic mechanisms composed by AND, OR and NOT modules in the DCS. Such alarm variables can be found by reading configuration diagrams in the DCS. Up to now, the proposed method has been applied to 211 out of 2226 alarm variables in total. Hence, it is practically useful to have the proposed method for analyzing root causes of occurring alarms for such alarm variables. Two industrial alarm variables are presented next as representatives of the alarm variables formulated in the two ways.

4.3.4.1 Example 1: Coal Feeders

Coal feeders are critical devices providing raw coals into coal grinding mills to produce pulverized coal powders for burning. One important operational requirement is to maintain sufficient feeding coal flows from coal feeders. The feeding coal flow X_c is configured with a low-alarm threshold 10 t/h. It is known from process knowledge that the alarm state in X_a can be arisen from multiple causes. For the ease of presentations, two main causes are considered, namely, the shortage of inletting coal as X_1 and electrical motor fault as X_2. If the inletting coal as the input to the coal feeder is reduced to a certain level, a digital sensor will send the value 1 to the control center of the power generation unit, and the coal feeder is either shut down or switched to an idle state to prevent device damages. The electrical motor provides the driving force for the coal feeder. If there is a fault in the electrical motor, the coal feeder has to be shut down until the fault is removed. Thus, X_1 and X_2 are digital variables in nature. Once X_a runs into the alarm state, the root cause needs to be analyzed accurately. Corresponding corrective actions are taken to get the coal feeder back into normal operations. Otherwise, generated electrical powers have to be reduced soon due to the shortage of coal powders, with a negative consequence of economic losses.

Figure 4.34 presents some collected data samples of $X_1(t)$, $X_2(t)$, $X_c(t)$ and $X_a(t)$. The sampling period is 1 s. For some cases, e.g., $X_a(t) = 1$ for $t \in [662, 1001]$, a common practice via visual investigations can easily tell that X_2 is the alarm root cause. However, all the three challenges discussed in Sect. 4.3.1 are presented in the

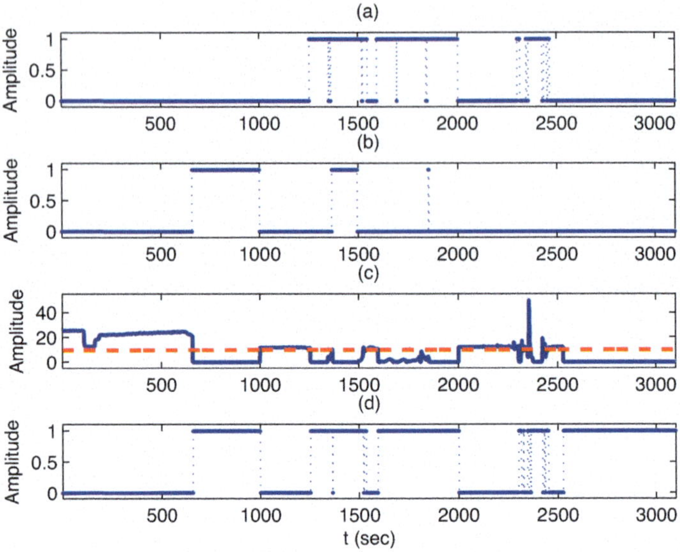

Fig. 4.34 The time sequence plots: **a** $X_1(t)$, **b** $X_2(t)$, **c** $X_c(t)$ (blue solid) and its low-alarm threshold (red dash), and **d** $X_a(t)$, reprinted from Ref. [46], copyright 2018, with permission from Elsevier

rest time instants, hindering the accuracies of analyzing alarm root causes by the common practice via visual investigations.

The proposed method is applied to $X_1(t)$, $X_2(t)$ and $X_a(t)$. The updating rate parameter $\lambda = 0.0341$ is determined from (4.52) by selecting $m = 20$ samples. As the sampling period is 1 s, the time period 20 s is selected based on a rule of thumb that alarms repeating more than three times per minute are considered as nuisance alarms [2]. The updated prior and posterior probabilities are given in Figs. 4.35 and 4.36, respectively. All initial values of probability parameters are set to zero. The root-cause analysis results are given in Fig. 4.37, where $X_1(t)$, $X_2(t)$ and $X_a(t)$ are respectively shifted in amplitude for -1.2, 1.2 and 2.4 for a better visualization. The indices $j = 0, 1, 2, 3, 4$ respectively represent the cases $\{X_a = 0\}$, $\{X_1 = 1, X_2 = 0, X_a = 1\}$, $\{X_1 = 0, X_2 = 1, X_a = 1\}$, $\{X_1 = 1, X_2 = 1, X_a = 1\}$ and $\{X_1 = 0, X_2 = 0, X_a = 1\}$.

The three challenges discussed in Sect. 4.3.1 have been successfully dealt with by the proposed method. First, the proposed method removes the negative effects of false and missing alarms in X_1, X_2 and X_a with time durations less than $m = 20$ samples. Here X_1 has false alarms ($X_1(t) = 1$) for the time intervals $t \in [2303, 2316]$ and $t \in [2348, 2351]$, and missing alarms ($X_1(t) = 0$) for the time intervals $t \in [1354, 1355]$, $t \in [1362, 1364]$, $t \in [1522, 1525]$, $t = 1698$, $t \in [1844, 1849]$, $t \in [2429, 2436]$ and $t \in [2453, 2457]$; X_2 has false alarms ($X_2(t) = 1$) for the time interval $t \in [1854, 1856]$; X_a has false alarms ($X_a(t) = 1$) for the time intervals $t \in [2305, 2317]$, $t \in [2328, 2330]$, $t \in [2349, 2355]$ and $t \in [2363, 2364]$, and missing alarms ($X_a(t) = 0$) for the time intervals $t \in [1367, 1368]$, $t \in [1524, 1527]$, $t \in$

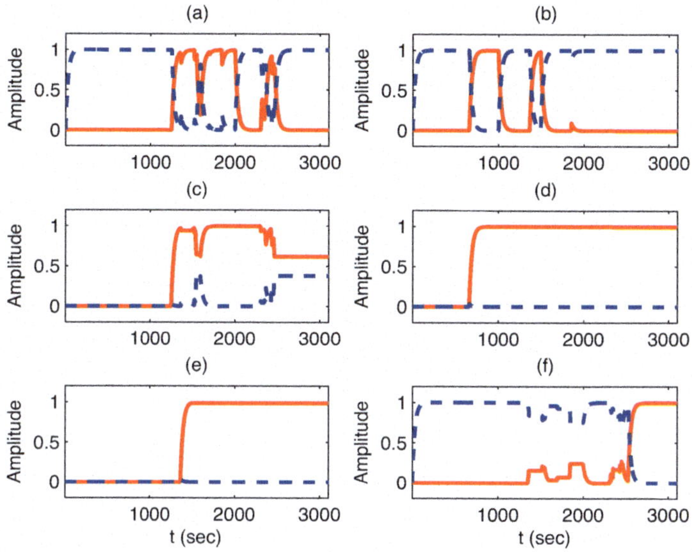

Fig. 4.35 The time sequence plots of prior probability parameters: **a** $\theta_{1,0}(t)$ (blue dash), $\theta_{1,1}(t)$ (red solid), **b** $\theta_{2,0}(t)$ (blue dash), $\theta_{2,1}(t)$ (red solid), **c** $\theta_{a,0|j=1}(t)$ (blue dash), $\theta_{a,1|j=1}(t)$ (red solid), **d** $\theta_{a,0|j=2}(t)$ (blue dash), $\theta_{a,1|j=2}(t)$ (red solid), **e** $\theta_{a,0|j=3}(t)$ (blue dash), $\theta_{a,1|j=3}(t)$ (red solid), **f** $\theta_{a,0|j=4}(t)$ (blue dash), $\theta_{a,1|j=4}(t)$ (red solid), reprinted from Ref. [46], copyright 2018, with permission from Elsevier

$[2425, 2432]$ and $t \in [2436, 2439]$. By contrast, $\hat{j}(t)$ is not affected by these false and missing alarms. Second, the proposed method can handle the co-existence of multiple root causes. Around the time interval $t \in [1250, 1600]$, X_1 is the first root cause leading to the alarm state of X_a, and X_2 becomes valid after a while. The proposed method yields the correct result $\hat{j}(t) = 3$ for the co-existence of two root causes. Third, the proposed method can detect the incompleteness of known root causes. Around the time interval $t \in [2530, 3000]$, X_a is in the alarm state, but both X_1 and X_2 take the value 0. Thus, the occurring alarm is caused by root causes other than X_1 and X_2. This situation is successfully detected by the proposed method with $\hat{j} = 4$.

If the common practice is used here by looking at observed data samples, then root causes of $X_a(t) = 1$ are the ones with $X_i(t) = 1$ for $i \in [1, n]$. In this context, the root-cause result is denoted by $\tilde{j}(t)$, i.e.

$$\tilde{j}(t) = \begin{cases} 0, & \text{if } X_a(t) = 0, \\ 1, & \text{if } X_1(t) = 1, X_2(t) = 0, X_a(t) = 1, \\ 2, & \text{if } X_1(t) = 0, X_2(t) = 1, X_a(t) = 1, \\ 3, & \text{if } X_1(t) = 1, X_2(t) = 1, X_a(t) = 1. \end{cases}$$

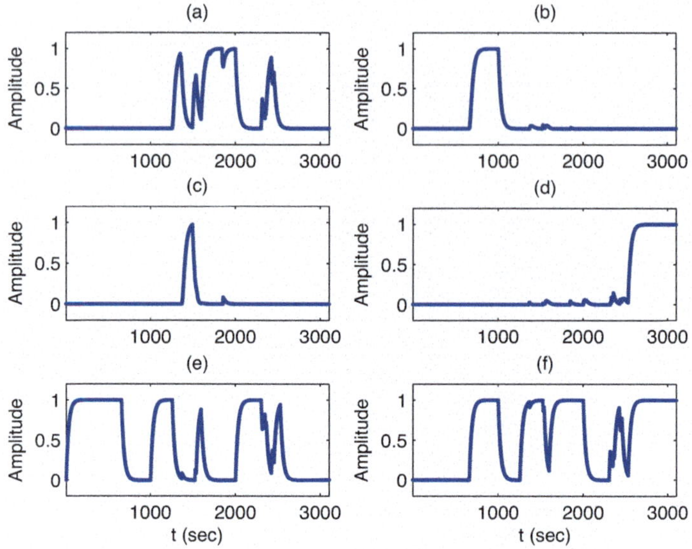

Fig. 4.36 The time sequence plots of the numerators of posterior probability parameters: **a** $\theta_{1|1}^{(N)}(t)$, **b** $\theta_{2|1}^{(N)}(t)$, **c** $\theta_{3|1}^{(N)}(t)$, **d** $\theta_{4|1}^{(N)}(t)$, and probabilities of X_a: **e** $\theta_{a,0}(t)$, **f** $\theta_{a,1}(t)$, reprinted from Ref. [46], copyright 2018, with permission from Elsevier

Figure 4.38 is the counterpart of Fig. 4.37, where $\tilde{j}(t)$ is severely affected by false alarms in $X_1(t)$ and $X_2(t)$, and cannot handle the third challenge. Thus, $\hat{j}(t)$ has a much better performance than $\tilde{j}(t)$, with some sacrifices in detection time delays.

A detection time delay is the time difference between the occurrence of an actual alarm in X_a and the detection of corresponding root causes from $\hat{j}(t)$. As discussed in Sect. 4.3.3, it takes m samples for a probability parameter $\theta(t)$ changing from $\theta(0)$ to $\theta(\infty)$. Since $\hat{j}(t)$ is obtained from the products of probability parameters. The detection time delay would be equal to or larger than m, depending on the number of probability parameters to be updated. For instance, $X_a(t)$ changes its value from 0 to 1 at $t = 662$, and $\hat{j}(t)$ does so from 0 to '2' at $t = 690$, so that the detection time delay is 28 s; Fig. 4.35 reveals that $\theta_{2,1}(t)$ and $\theta_{a,1|j=2}(t)$ are updating for this change. As another example, $X_a(t)$ changes from 0 to 1 at $t = 2530$, and $\hat{j}(t)$ does so from 0 to '4' at $t = 2550$, so that the detection time delay is 20 s; only $\theta_{a,1|j=4}(t)$ and $\theta_{a,0|j=4}(t)$ are updating for this change as shown in Fig. 4.35. The detection time delays may be reduced by exploiting different updating parameters for probabilities in the Bayesian network. Thus, one future work is to design these updating parameters in a systematic manner.

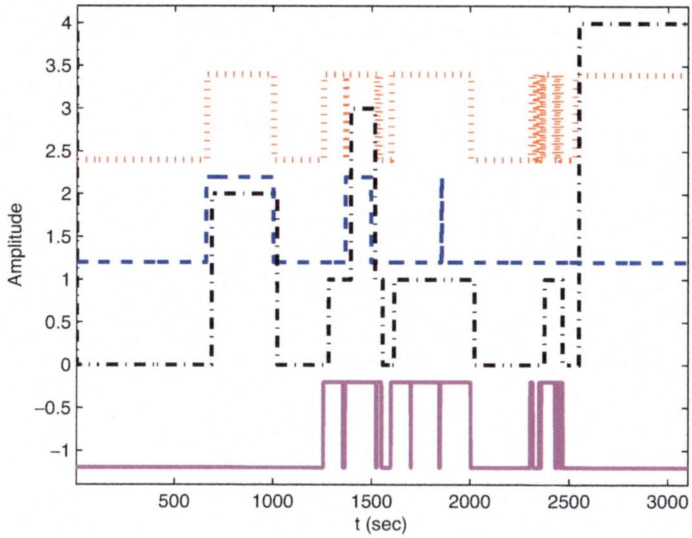

Fig. 4.37 Time sequence plots of $X_1(t)$ (magenta solid, shifted downwards for -1.2), $X_2(t)$ (blue dash, shifted upwards for 1.2), $X_a(t)$ (red dotted, shifted upwards for 2.4) and $\hat{j}(t)$ from the proposed method (black dash-dotted), reprinted from Ref. [46], copyright 2018, with permission from Elsevier

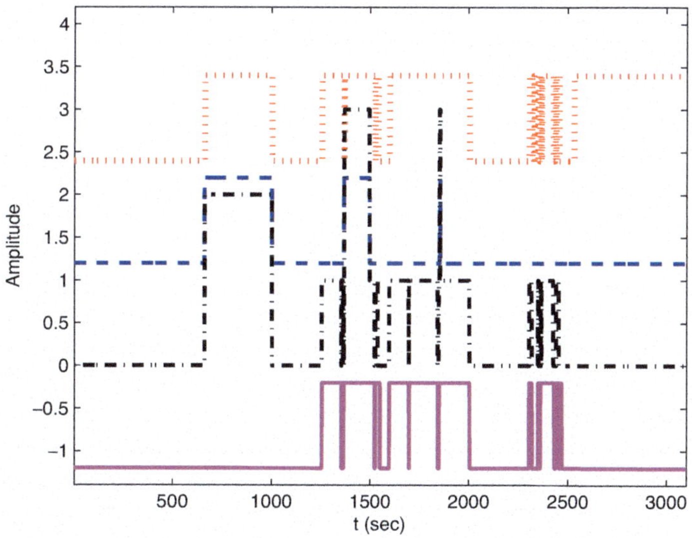

Fig. 4.38 Time sequence plots of $X_1(t)$ (magenta solid), $X_2(t)$ (blue dash), $X_a(t)$ (red dotted) and $\tilde{j}(t)$ from the common practice (black dash-dotted), reprinted from Ref. [46], copyright 2018, with permission from Elsevier

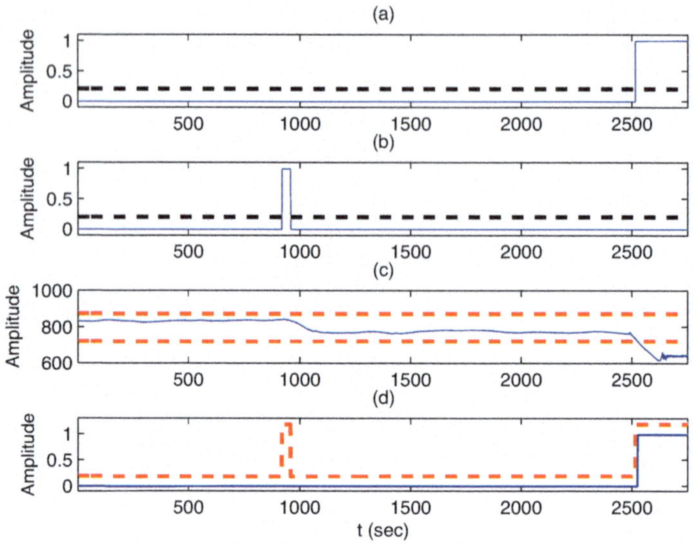

Fig. 4.39 Collected data samples: **a** $X_1(t)$ (blue solid), $X_2(t)$ (black dash, shifted upwards for 0.1), **b** $X_3(t)$ (blue solid), $X_4(t)$ (black dash, shifted upwards for 0.1), **c** $X_c(t)$ (blue solid) with high and low thresholds (red dash), and **d** $X_5(t)$ (blue solid) and $X_a(t)$ (red dash, shifted upwards for 0.1), reprinted from Ref. [46], copyright 2018, with permission from Elsevier

4.3.4.2 Example 2: Condensers

The water level of a condenser hot well is one important process variable. If the water level is too high, the heat exchange space and the degree of condenser vacuum would be reduced so that power generation efficiency could be deteriorated; if the water level is too low, then outlet water pumps would suffer from cavitation phenomena and the deaerator water level could be negatively affected. Thus, an alarm variable X_a is configured to indicate the occurrence of abnormal situations in the water level of the condenser hot well. It is generated via a logic mechanism where 5 root causes can result in alarms in X_a. The root causes are given as follows: X_1 and X_2 represent the faults in outlet condenser water pumps A and B, respectively; X_3 (X_4) is a hardware indicator saying that the water level of the condenser hot well is higher (lower) than 876 mm (300 mm); X_5 is the output of an univariate alarm system taking the value 1 when an analog measurement X_c of the water level of the condenser hot well is either larger than 876 mm or lower than 724 mm. In other words, X_a is in the alarm state when any one of $X_1, X_2, \ldots, X_5$ takes the value 1.

Figure 4.39 gives some data samples of $X_a(t)$, $X_c(t)$, $X_1(t)$, $X_2(t)$, ..., $X_5(t)$. There are two time intervals that X_a runs into the alarm state. First, it is clear by visualization that $X_3(t)$ and $X_a(t)$ take the value 1 for $t \in [922, 960]$; however, X_c is obviously inside the normal operating range [724, 876] mm formulated by the low and high alarm thresholds. Thus, these alarms are the false ones. As a matter of facts, X_3 often gives false alarms, even though X_c is inside the normal operating range. As an illustration, Fig. 4.40 presents four sets of historical data samples of X_3 and X_c, where the first three sets in Fig. 4.40a–c yield false alarms, and the last set in Fig. 4.40d gives true alarms. Second, $X_a(t)$, $X_1(t)$ and $X_5(t)$ switches from 0 to 1 at the time instances 2518, 2518 and 2528 s, respectively. Thus, $X_a(t)$ runs into the alarm state firstly due to X_1, as verified by the abrupt change of the current of outlet water pump A going from non-zero values to zero in Fig. 4.41a. Owing to an inter-lock mechanism, the outlet condenser water pump B in Fig. 4.41b is switched on automatically and the condenser outlet water flow rate in Fig. 4.41c is increased quickly to an excessively large value. As a result, X_c is decreased to be smaller than the lower alarm threshold 247 mm, so that X_5 switches its value from 0 to 1. It is important for operators to notice the co-existence of two abnormal situations in X_3 and X_5 so that a corrective action can be taken accordingly.

The proposed method is applied here with the updating rate parameter $\lambda = 0.0115$ from (4.52) by selecting $m = 60$ samples in Table 4.9. The root-cause analysis results are given in Fig. 4.42, where $X_3(t)$, $X_5(t)$ and $X_a(t)$ are respectively shifted in amplitude for -0.4, -0.2 and 0.2 for a better visualization. The indices $j = 0, 1, 2, 3, 4$ respectively represent the cases $X_a = 0$, $\{X_3 = 1, X_5 = 0, X_a = 1\}$, $\{X_3 = 0, X_5 = 1, X_a = 1\}$, $\{X_3 = 1, X_5 = 1, X_a = 1\}$ and $\{X_3 = 0, X_5 = 0, X_a = 1\}$. The result $\hat{j}(t)$ is not affected by the false alarms in $X_3(t)$ for $t \in [922, 960]$, and successfully detects the co-existence of abnormalities in X_3 and X_5 at the time instant $t = 2598$.

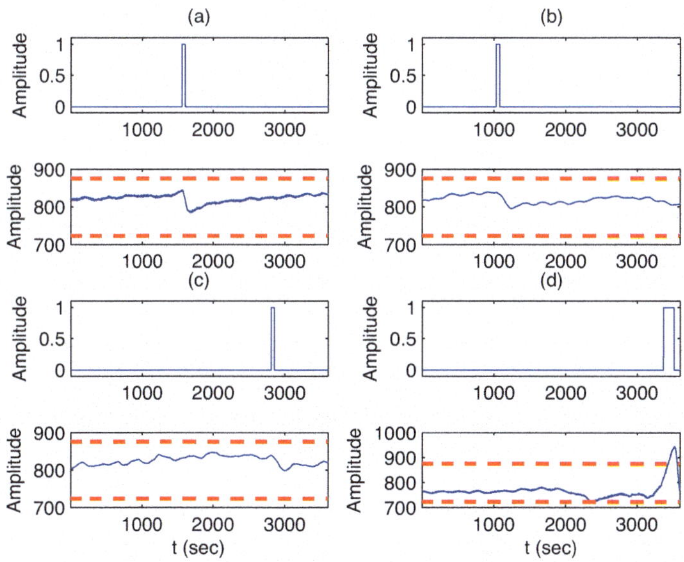

Fig. 4.40 Four sets of historical data samples of X_3 and X_c: X_3 (the upper subplot) and $X_c(t)$ with high and low thresholds (red dash in the lower subplot), reprinted from Ref. [46], copyright 2018, with permission from Elsevier

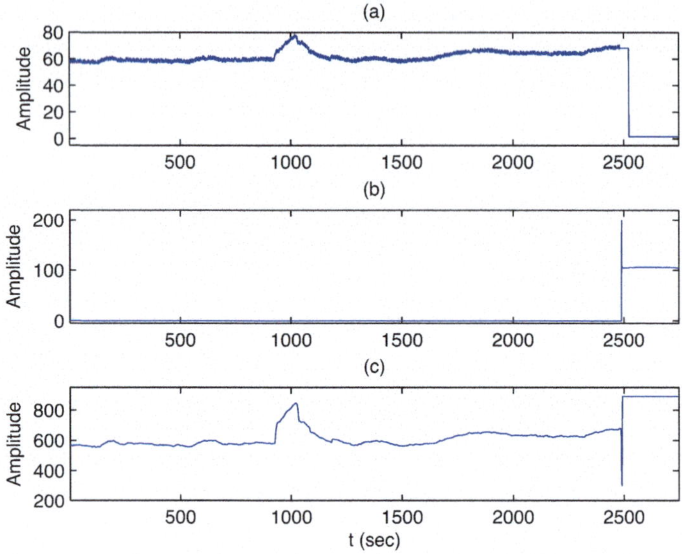

Fig. 4.41 Data samples of related variables: **a** the current of the outlet water pump A, **b** the current of the condenser outlet water pump B, **c** the condenser outlet water flow rate, reprinted from Ref. [46], copyright 2018, with permission from Elsevier

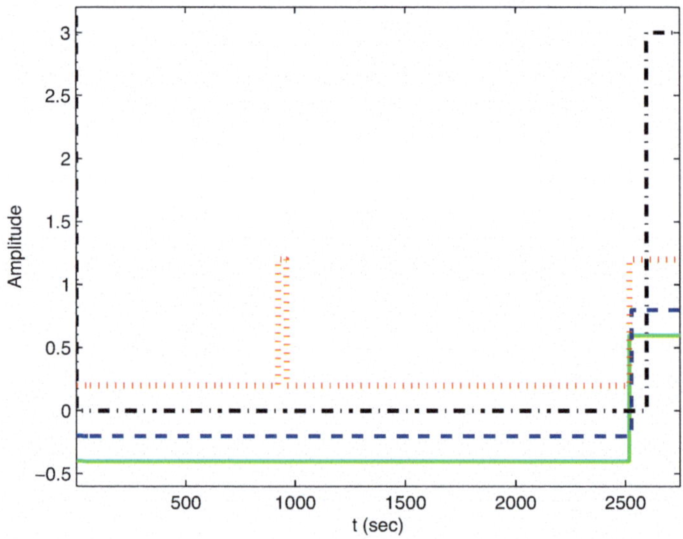

Fig. 4.42 Results of root cause analysis: $X_3(t)$ (green solid, shifted downwards for -0.4), $X_5(t)$ (blue dash, shifted downwards for -0.2), $X_a(t)$ (red dotted, shifted upwards for 0.2) and $\hat{j}(t)$ (black dash-dotted), reprinted from Ref. [46], copyright 2018, with permission from Elsevier

4.4 Root-Cause Analysis for Process Variables

This section is on root-cause analysis based on process variables. Determination of root causes for occurring abnormalities is a sophisticated work in process monitoring of complex process industries. Many existing methods are based on an assumption that causal-effect relations are time invariant, whereas industrial processes may have complex dynamic characteristics, leading to time-varying cause-effect relations. Accordingly, a new cause-effect detection method is proposed to analyze time-varying cause-effect relations between the key variable and influence variables [22]. This method is motivated by a common observation in practice that if the key variable takes an increasing/decreasing amplitude change, then corresponding influence variables are with amplitude changes in the same/opposite qualitative trends, and such relations between trend changes may change with time.

4.4.1 Problem Description

In complex processes, there usually exist a few variables pivotal to the production quality and process safety, e.g., the condenser vacuum in thermal power plants is such a key variable that affects the power generation cost and process safety. Operators make empirical evaluations to the plant operating status by monitoring these key variables. However, the problem plaguing operators is how to find out the root cause when such key variables get into abnormal situations. In general, there may exist multiple variables that affect the key one owing to the complex interconnections. For instance, the condenser vacuum is mainly affected by the inlet steam temperature, the condensate flow rate, and the inlet circulating water temperature. The three variables do not contribute to an increment/decrement of the condenser vacuum all the time. The cause-effect relations between the influence variables and the vacuum are time varying, and thus the root cause variables can be different, i.e., the change in the condenser vacuum could be caused by one, two or all of the three variables over different time periods. Therefore, it needs to judge among these influence variables which one is the root cause variable by detecting the time-varying cause-effect relations when the monitored key variable is found in an abnormal condition.

In practice, a causal map can be obtained with all information prorogation paths among process variables based on process knowledge. However, in real-time operation, not all paths will be executed; the cause-effect relations between process variables may change with time, causing difficulty in correctly identifying the root causes. For example, in the illustrative example in Fig. 4.43a, the causal map obtained based on process knowledge shows that there exist multiple paths representing how influence variables affect the key variable y. However, in Fig. 4.43b, the paths show that x_1 is the only root cause variable significantly influencing the key variable y over a certain time period during process operation, whereas other paths do not hold.

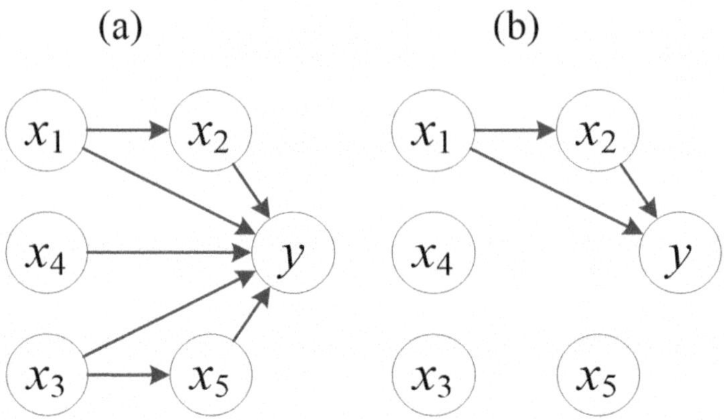

Fig. 4.43 Illustrative example for time-varying cause-effect relations: **a** Causal map with all information prorogation paths obtained based on process knowledge; **b** Causal map with executed paths over a certain time period during process operation, reprinted from Ref. [22], copyright 2022, with permission from Elsevier

In view of the above practical problem, this section proposes a new method to detect the root cause, which is based on the contributions of the influence variables to the amplitude change of the key variable. In practice, all influence variables x_i's for $i = 1, 2, \ldots, M$ that have physical connections with a key variable y can be found from process knowledge. Given data samples of y and x_i's, the problem to be solved is: When the key variable y takes an increasing/decreasing amplitude change, which one of x_i's is the root-cause variable (experiencing amplitude changes in the same/opposite qualitative trend) that leads to the amplitude change in y. An assumption required by the proposed method is that the key variable and the influence variables are related in terms of the increasing/decreasing of their amplitude changes. This assumption usually holds in practice, owing to physical principles among these process variables.

The proposed method consists of three main steps:

- The time sequence of the key variable y is separated into several short straight lines to formulate piecewise linear representation (PLR) segments.
- Lagged correlation analysis is conducted in a piecewise manner to estimate the time delay between each influence variable x_i and y. Given the estimated time delay λ_i between x_i and y, the time sequence of x_i moves forward by λ_i samples, so as to realign with y. Then, the qualitative trends in PLR segments are captured from the time sequence of y and the realigned time sequence of x_i.
- For each PLR segment, contribution factors of x_i's to changes of y are calculated by building a multiple linear regression model for normalized amplitude changes of all involved variables.

These steps are elaborated in the next two subsections.

4.4.2 Quantitative Trends Extraction

This subsection presents the representation of piecewise linear segments and the extraction of qualitative trends.

4.4.2.1 Piecewise Linear Representations

To learn the relations between influence variables x_i's and the key variable y over different time periods, the foremost step is to split the time sequences into shorter segments and capture the corresponding time intervals, such that during each time interval these variables should hold constant relations. In this context, the data segmentation is conducted based on the variational trends of the key variable y. Specifically, the time sequence $y(1:N) = [y(1), y(2), \ldots, y(N)]$ is partitioned into multiple short straight lines via the PLR technique, which is a commonly used technique for extraction of intrinsic information [6]; it can partition a time sequence into short segments and represent them by straight lines that present variational trends.

Assume that $y(1:N)$ is segmented into K segments connected in an end-to-end manner. The k-th segment is denoted by $y(t_k : t_{k+1} - 1)$, where t_k and $t_{k+1} - 1$ are the first and last sample instants of the segment, respectively. Samples in each data segment $y(t_k : t_{k+1} - 1)$ are assumed with a linear regression model, i.e.,

$$y(t) = a_k + b_k t + e_k(t), \tag{4.53}$$

where a_k and b_k are the intercept and slope parameters in the k-th segment, respectively; $e_k(t)$ is assumed to be the Gaussian white noise with zero mean and variance σ_k^2. The model in (4.53) is approximated by

$$\hat{y}(t) = \hat{a}_k + \hat{b}_k t, \tag{4.54}$$

where $\hat{a}_k$ and $\hat{b}_k$ are respectively the estimates of a_k and b_k, and can be easily obtained using the least squares method [39]. By minimizing the fitting error $\sum_{t=t_k}^{t_{k+1}-1} \left(y(t) - \hat{y}(t)\right)^2$, the two parameters $\hat{a}_k$ and $\hat{b}_k$ are calculated as

$$\begin{cases} \hat{a}_k = \bar{y} - \hat{b}_k \bar{t}, \\ \hat{b}_k = \dfrac{\sum_{t=t_k}^{t_{k+1}-1} (y(t)-\bar{y})(t-\bar{t})}{\sum_{t=t_k}^{t_{k+1}-1} (t-\bar{t})^2}, \end{cases} \tag{4.55}$$

where $\bar{y} = \frac{1}{t_{k+1}-t_k} \sum_{t=t_k}^{t_{k+1}-1} y(t)$ and $\bar{t} = \frac{1}{t_{k+1}-t_k} \sum_{t=t_k}^{t_{k+1}-1} t$.

An unbiased estimate $\tilde{\sigma}_k^2$ of the variance σ_k^2 of the Gaussian white noise $e_k(t)$ in (4.53) is given by [39]

$$\tilde{\sigma}_k^2 = \frac{\sum_{t=t_k}^{t_{k+1}-1} \left[y(t) - \hat{y}(t)\right]^2}{t_{k+1} - 1 - t_k}. \tag{4.56}$$

The estimated model error $\hat{e}_k(t) = \hat{y}(t) - y(t)$ at a particular time instant t' is then calculated as [39]

$$\hat{\sigma}_k^2(t') = \tilde{\sigma}_k^2 \left[1 + \frac{1}{t_{k+1} - t_k} + \frac{(t' - \bar{t})^2}{\sum_{t=t_k}^{t_{k+1}-1} (t - \bar{t})^2} \right]. \tag{4.57}$$

The confidence interval $[y_L(t'), y_H(t')]$ of $y(t')$ at a significance level α is

$$\begin{cases} y_L(t') = \hat{y}(t') - \iota_{\alpha/2,m} \hat{\sigma}_k^2(t'), \\ y_H(t') = \hat{y}(t') + \iota_{\alpha/2,m} \hat{\sigma}_k^2(t'), \end{cases} \tag{4.58}$$

where $\iota_{\alpha/2,m}$ represents the $\alpha/2$ critical value of Students's t-distribution with $m = t_{k+1} - 2 - t_k$ degrees of freedom.

In the above segmentation, it must decide how many segments a time sequence should be partitioned into, i.e., the value of K needs to be determined prior to segmentation. Here K is determined as follows [47]. Denote the parallelogram-shaped space between confidence intervals of the estimates $\hat{y}(t_k : t_{k+1} - 1)$ by A_k. Such a space A_k is determined by four vertices, namely, $[t_k, y_L(t_k)]$, $[t_{k+1} - 1, y_L(t_{k+1} - 1)]$, $[t_{k+1} - 1, y_H(t_{k+1} - 1)]$, and $[t_k, y_H(t_k)]$, which are obtained by (4.58). In addition, given a sequence of original data samples $y(t_k : t_{k+1} - 1)$ in the time period $[t_k, t_{k+1} - 1]$, the convex hull composed by these samples is formed as B_k. Then, an index η_k representing the percentage of overlapped area in A_k is given by

$$\eta_k = \frac{|A_k \cap B_k|}{|A_k|}. \tag{4.59}$$

An index function of K is formulated as

$$\eta(K) = \sum_{k=1}^{K} \frac{t_{k+1} - t_k}{N} \eta_k, \tag{4.60}$$

which is essentially the weighted average of the indices $\eta_k, k = 1, 2, \ldots, K$. The number of segments is eventually obtained as

$$\hat{K} = \arg\max_K \eta(K), \tag{4.61}$$

such that the confidence intervals include as many data points as possible while minimizing the fitting errors. Then, each segment of y is denoted by $y(t_k : t_{k+1} - 1)$, $k = 1, 2, \ldots, K$.

4.4.2.2 Extraction of Qualitative Trends

For the key variable y, its amplitude change A_k^y in the k-th data segment is calculated as

$$A_k^y = \hat{y}(t_{k+1} - 1) - \hat{y}(t_k). \tag{4.62}$$

Then, the qualitative trend ψ_k^y is extracted by comparing A_k^y with a significance threshold A_0^y,

$$\psi_k^y = \begin{cases} 1, & \text{if } A_k^y \geq A_0^y, \\ 0, & \text{if } -A_0^y < A_K^y < A_0^y, \\ -1, & \text{if } A_k^y \leq -A_0^y. \end{cases} \tag{4.63}$$

Here 1 and -1 denote significant increasing and decreasing variational trends, respectively; 0 means non-significant variational change.

The significance threshold A_0^y is given by [45]

$$A_0^y = \sqrt{\frac{12R_0^2\hat{\sigma}_v^2}{1 - R_0^2}}, \tag{4.64}$$

where σ_v^2 is calculated as

$$\sigma_v^2 = \frac{1}{K}\sum_{k=1}^{K}\left[\frac{1}{t_{k+1} - t_k}\sum_{t=t_k}^{t_{k+1}-1}\left(y(t) - \hat{y}(t)\right)\right], \tag{4.65}$$

and R_0^2 is the user-selected lower limit of R_k^2, which represents the R-squared statistic that measures the goodness of fit of the linear regression in (4.54) for the k-th data segment. The formula to calculate R_k^2 is given by

$$R_k^2 = 1 - \frac{\sum_{t=t_k}^{t_{k+1}-1}\left(y(t) - \hat{y}(t)\right)}{\sum_{t=t_k}^{t_{k+1}-1}\left(y(t) - \bar{y}\right)}, \tag{4.66}$$

where $\bar{y}$ represents the sample mean of y in the k-th data segment.

The time sequence of y may be misaligned with counterparts of influence variables due to the presence of time delays. Thus, the time delay between x_i and y is crucial to realign the time sequence. Considering the time varying relation between x_i and y, the method in [5] is extended here for the estimation of time delays in a piecewise manner. The method in [5] detects the correlation between two variables and estimates their time delay by comparing the maximum lagged correlation with an empirical threshold. The major advantages of this method are the ease of implementation and the simple interpretation of obtained results. However, the limitation is that this method is only suitable for time-invariant linear processes, whereas in this context the process can be time-varying and thus the correlation may change with time. To handle this issue, a modified method is proposed by incorporating PLR segments. The method is presented as follows:

Based on the definitions in [5], the cross-correlation function (CCF), as well as its maximum and minimum values, and maximum time lagged correlation for piecewise estimation are calculated as follows: Given the k-th segment of y, the CCF $\phi^k(\tau)$ for x_i and y is given by

$$\phi_i^k(\tau_i^k) = \frac{1}{N_k} \sum_{t=t_k}^{t_{k+1}-1} \tilde{x}_i(t + \tau_i^k)\tilde{y}(t), \tag{4.67}$$

where $N_k = t_{k+1} - t_k$ denotes the length of the k-th data segment; $\tilde{x}_i(t)$ and $\tilde{y}(t)$ represent the normalized variables with zero means and unit standard deviations. Since x_i is an influence variable of y, the time delay τ_i^k between x_i and y should be a non-negative integer. The maximum and minimum values of the CCF for different τ_i^k's are obtained as

$$\begin{cases} \phi_{i,\max}^k &= \max_{\tau_i^k} \phi_i^k(\tau_i^k), \\ \phi_{i,\min}^k &= \min_{\tau_i^k} \phi_i^k(\tau_i^k). \end{cases} \tag{4.68}$$

The maximum time lagged correlation ρ_i^k for the k-th segment pair is

$$\rho_i^k = \max\left\{\phi_{i,\max}^k, \left|\phi_{i,\min}^k\right|\right\}. \tag{4.69}$$

The correlation between x_i and y in the k-th segment is determined with a one-sided test, i.e., x_i and y are correlated in the k-th segment if $\rho_i^k \geq \rho_{th}^k$, where ρ_{th}^k represents the threshold of the correlation index; otherwise, x_i and y are uncorrelated in the k-th segment. The value of ρ_{th}^k is determined by massive simulations with uncorrelated surrogate data. An empirical estimate of the threshold is given as [5]

$$\rho_{th}^k = 1.85N_k^{-0.41} + 2.37N_k^{-0.53}. \tag{4.70}$$

If $\rho_i^k \geq \rho_{th}^k$, the correlation between x_i and y holds and the time delay from the k-th data segment is

$$\lambda_i^k = \arg\max_{\tau_i^k} \left|\phi_i^k(\tau_i^k)\right|. \tag{4.71}$$

Accordingly, in each data segment with $\rho_i^k \geq \rho_{th}^k$, there is one time delay λ_i^k calculated. Eventually, a weighted value of the time delay between x_i and y is determined as

$$\lambda_i = \frac{\sum_{k=1}^K \lambda_i^k \times v_i^k \times \rho_i^k}{\sum_{k=1}^K v_i^k \times \rho_i^k}, \tag{4.72}$$

where $v_i^k = 1$ if $\rho_i^k \geq \rho_{th}^k$. It is noteworthy that the estimation of the time delay using (4.72) assigns a higher weight to λ_i^k for the segment pair with a higher correlation and excludes λ_i^k for the uncorrelated segment pair by assigning a zero weight.

Eventually, the time delays for all influence variables $x_i, i = 1, 2, \ldots, M$ are obtained as $\lambda_i, i = 1, 2, \ldots, M$. Then, the time sequence of each influence variable x_i moves forward by λ_i samples, so as to realign with the time sequence of the key variable y. In other words, the k-th segment $y(t_k : t_{k+1} - 1)$ is aligned with $x_i(t_k + \lambda_i : t_{k+1} - 1 + \lambda_i)$.

After obtaining the time delays, the qualitative trends for each influence variable x_i can be obtained in the same way as that for y. Given the shifted time sequence $x_i(t_k + \lambda_i : t_{k+1} - 1 + \lambda_i)$ in the k-th segment, a linear regression model is obtained via (4.54). Then, the amplitude change $A_k^{x_i}$ for x_i in the k-th segment is derived by substituting the estimated samples of shifted time sequence of x_i into (4.62). Denote the significance threshold of x_i by $A_0^{x_i}$. Then, the qualitative trend $\psi_k^{x_i}$ for the k-th segment of x_i is obtained via (4.63) by comparing $A_k^{x_i}$ with $A_0^{x_i}$,

$$\psi_k^{x_i} = \begin{cases} 1, & \text{if } A_k^{x_i} \geq A_0^{x_i}, \\ 0, & \text{if } -A_0^{x_i} < A_k^{x_i} < A_0^{x_i}, \\ -1, & \text{if } A_k^{x_i} \leq -A_0^{x_i}. \end{cases} \tag{4.73}$$

Eventually, a combination of the qualitative trends of is

$$\Psi_k = (\Psi_k^x; \psi_k^y) = (\psi_k^{x_1}, \psi_k^{x_2}, \ldots, \psi_k^{x_M}; \psi_k^y). \tag{4.74}$$

For instance, $\Psi_k = (1, -1, 0; 1)$ indicates that the key variable shows a significant increasing trend, and the three influence variables show the increasing trend, decreasing trend, and non-significant change, respectively.

4.4.3 Determination of Root Cause Variables

In different scenarios, the contributions of influence variables to the amplitude change of the key variable y could be very different. Based on such contributions, it can tell which influence variables have significant impacts to the change of the key variable. To quantify such contributions, a contribution factor is formulated; it is defined as the proportion of x_i contributing to the amplitude change in the time sequence of the key variable y in a certain time period.

Given that the variational trends of x_i and y are monotonous, the amplitude changes of x_i and y in the k-th segment can be represented by $\Delta x_i = x_i^e - x_i^s$ and $\Delta y = y^e - y^s$, respectively. Here x_i^s and x_i^e indicate the first and last elements of x_i in the k-th segment, namely, $x_i^s = x_i(t_k + \lambda_i)$ and $x_i^e = x_i(t_{k+1} - 1 + \lambda_i)$; similarly, $y^s = y(t_k)$ and $y^e = y(t_{k+1} - 1)$. A contribution factor λ_i^k in the k-th data segment is formulated as

$$\lambda_i^k = \frac{\Delta y_i}{\Delta y}, \text{ s.t. } \lambda_i^k \in [0, 1], \sum_{i=1}^{M} \lambda_i^k = 1, \tag{4.75}$$

where Δy_i is the i-th component of Δy contributed by Δx_i; $\Delta y = \Delta y_1 + \Delta y_2 + \cdots, \Delta y_M$. To ensure $\lambda_i^k \in [0, 1]$, the signs of Δy_i and Δy should be the same. Thus, to measure the contribution factor, it should estimate the i-th component Δy_i of Δy contributed by Δx_i.

Considering that the data segments are represented by piecewise straight lines, it is reasonable to assume linear relations between the key variable y and the influence variables x_i's. As a result, y^s (y^e) holds linear relations with x_i^s's (x_i^e's). Then, estimating Δy_i based on Δy and Δx_i can be achieved if the model between y and x_i's is obtained. This study calculates contribution factors of influence variables to the change of the key variable in each data segment using the multiple linear regression (MLR). The relation between y and x_i's is represented by an MLR model as

$$y = \beta_0 + \beta_1 x_1 + \beta_2 x_2 + \cdots + \beta_M x_M + \epsilon, \tag{4.76}$$

where β_0 is a constant variable, ϵ denotes the residual, and β_i represents the regression coefficient for x_i.

The calculation of contribution factors is based on the regression coefficients, which are estimated by taking into account the following aspects:

- The root cause variables are determined for segments with either $\psi_k^y = 1$ or $\psi_k^y = -1$, whereas $\psi_k^y = 0$ indicates a non-significant change, so that there is no need to detect root cause variables for data segments with $\psi_k^y = 0$.
- Since process variables usually have different measurement scales, they should be normalized to the same scale, so as to make the obtained regression coefficients comparable.

In the k-th segment, if it is detected that the amplitude change is significant, namely, $\psi_k^y = 1$ or $\psi_k^y = -1$, an MLR model is constructed based on $x_i(t_k + \lambda_i : t_{k+1} - 1 + \lambda_i)$, $i = 1, 2, \ldots, M$, and $y(t_k : t_{k+1} - 1)$. By normalizing the data and incorporating the time delays, the regression model in (4.76) becomes

$$\check{y}(t) = \beta_0^k + \beta_1^k \check{x}_1(t + \lambda_1) + \beta_2^k \check{x}_2(t + \lambda_2) + \cdots + \beta_M^k \check{x}_M(t + \lambda_M), \tag{4.77}$$

where $t = t_k, t_k + 1, \ldots, t_{k+1} - 1$; $\boldsymbol{\beta} = [\beta_0^k, \beta_1^k, \ldots, \beta_M^k]'$ presents the vector of regression coefficients; $\check{y}(t)$ and $\check{x}_i(t)$ denote the normalized time sequences of y and x_i in the k-th segment, respectively. The normalization of time sequence is conducted based on the significance thresholds of amplitude changes, namely, $A_0^{x_i}$ and A_0^y obtained via (4.64). Given the original time sequence $x_i(t_k + \lambda_i : t_{k+1} - 1 + \lambda_i)$ of x_i in the k-th segment, the normalized time sequence is $\check{x}_i(t_k + \lambda_i : t_{k+1} - 1 + \lambda_i)$ with each element $\check{x}_i(j)$ given by

$$\check{x}_i(t) = \frac{x_i(t) - \bar{x}_i}{A_0^{x_i}}. \tag{4.78}$$

Here $t = t_k + \lambda_i, \ldots, t_{k+1} - 1 + \lambda_i$; $\bar{x}_i$ represents the mean value of $x_i(t_k + \lambda_i : t_{k+1} - 1 + \lambda_i)$. For the key variable y, its normalized time sequence $\check{y}(t_k : t_{k+1} - 1)$ in each segment can be obtained in the same manner.

Considering the directions of variational trends, the above regression model in (4.77) can be further modified as

$$\begin{aligned}\check{y}(t) = \beta_0^k + \beta_1^k \varsigma_1^k \check{x}_1(t + \lambda_1) + \beta_2^k \varsigma_1^k \check{x}_2(t + \lambda_2) + \\ \cdots + \beta_M^k \varsigma_1^k \check{x}_M(t + \lambda_M) + \epsilon(t),\end{aligned} \tag{4.79}$$

where ς_i^k is calculated for x_i and y as

$$\varsigma_i^k = \psi_k^{x_i} \psi_k^y, \tag{4.80}$$

such that $\varsigma_i^k = -1$ if x_i and y have different variational change directions in the k-th segment, and $\varsigma_i^k = 1$ if x_i and y have the same change direction. If the amplitude change of x_i is non-significant in the k-th segment, namely, $\psi_k^{x_i} = 0$, it is obtained that $\varsigma_i^k = 0$, which gets x_i excluded from the regression model in (4.79). Thus, $\check{y}(t)$ always has the same direction of the variational change as that of $\varsigma_i^k \check{x}_i(t + \lambda_i)$ with $\psi_k^{x_i} \neq 0$. Accordingly, it can assume that the regression coefficients are non-negative, namely, $\beta_i^k \geq 0, i = 0, 2, \ldots, M$.

The least square estimate $\hat{\boldsymbol{\beta}} = [\hat{\beta}_0^k, \hat{\beta}_1^k, \ldots, \hat{\beta}_M^k]'$ of $\boldsymbol{\beta}$ is obtained as

$$\hat{\boldsymbol{\beta}} = \arg\min_{\boldsymbol{\beta}} ||\mathbf{X}_k \boldsymbol{\beta} - \mathbf{y}_k||_2, \text{ s.t. } \boldsymbol{\beta} \geq 0, \tag{4.81}$$

where $\mathbf{y}_k = [\check{y}(t_k), \check{y}(t_k + 1), \ldots, \check{y}(t_{k+1} - 1)]'$ denotes the vector of the normalized values of y in the k-th segment; $\mathbf{X}_k$ represents the matrix of the explanatory variables,

$$\mathbf{X}_k = \begin{pmatrix} 1 & \check{x}_1(t_k + \lambda_1) & \cdots & \check{x}_M(t_k + \lambda_M) \\ 1 & \check{x}_1(t_k + 1 + \lambda_1) & \cdots & \check{x}_M(t_k + 1 + \lambda_M) \\ \vdots & \vdots & \ddots & \vdots \\ 1 & \check{x}_1(t_{k+1} - 1 + \lambda_1) & \cdots & \check{x}_M(t_{k+1} - 1 + \lambda_M) \end{pmatrix}. \tag{4.82}$$

Equation (4.81) is a typical non-negative least squares problem, which can be solved using an active-set method in [28].

To get the amplitude changes of x_i and y in the k-th data segment (namely, $\Delta x_i = x_i^e - x_i^s$ and $\Delta y = y^e - y^s$, respectively), substituting x_i^s, x_i^e, y^s and y^e into (4.79) yields

$$\frac{y^e - \bar{y}}{A_0^y} = \hat{\beta}_0^k + \hat{\beta}_1^k \varsigma_1^k \frac{x_1^e - \bar{x}_1}{A_0^{x_1}} + \hat{\beta}_2^k \varsigma_2^k \frac{x_2^e - \bar{x}_2}{A_0^{x_2}} + \cdots + \hat{\beta}_M^k \varsigma_M^k \frac{x_M^e - \bar{x}_M}{A_0^{x_M}}, \tag{4.83}$$

$$\frac{y^s - \bar{y}}{A_0^y} = \hat{\beta}_0^k + \hat{\beta}_1^k \varsigma_1^k \frac{x_1^s - \bar{x}_1}{A_0^{x_1}} + \hat{\beta}_2^k \varsigma_2^k \frac{x_2^s - \bar{x}_2}{A_0^{x_2}} + \cdots + \hat{\beta}_M^k \varsigma_M^k \frac{x_M^s - \bar{x}_M}{A_0^{x_M}}. \tag{4.84}$$

Subtracting (4.83) from (4.84) gives

$$\Delta y = \frac{\hat{\beta}_1^k A_0^y}{A_0^{x_1}} \varsigma_1^k \Delta x_1 + \frac{\hat{\beta}_2^k A_0^y}{A_0^{x_2}} \varsigma_2^k \Delta x_2 + \cdots + \frac{\hat{\beta}_M^k A_0^y}{A_0^{x_M}} \varsigma_M^k \Delta x_M. \tag{4.85}$$

Based on (4.85), it can tell how much the amplitude change of each influence variable x_i contributes to the amplitude change of the key variable y. Here, a contribution coefficient for each influence variable x_i is defined as

$$\varphi_i^k = \frac{\hat{\beta}_i^k A_0^y}{A_0^{x_i}}. \tag{4.86}$$

A contribution factor for each influence variable x_i is defined as the proportion of the contribution of Δx_i in Δy,

$$\gamma_i^k = \left| \frac{\varphi_i^k \Delta x_i}{\Delta y} \right|. \tag{4.87}$$

To alleviate noise effects, Δx_i and Δy are replaced by the estimated amplitude changes $A_k^{x_i}$ and A_k^y in Sect. 4.4.2.2. Accordingly, the contribution factor γ_i for x_i becomes

$$\gamma_i^k = \left| \frac{\varphi_i^k A_k^{x_i}}{\sum_{i=1}^{M} \varphi_i A_k^{x_i}} \right|, \tag{4.88}$$

such that $\gamma_i^k \in [0, 1]$ and $\sum_{k=1}^{K} \gamma_i^k = 1$. In essence, γ_i^k represents the percentage contribution of the amplitude change of x_i to that of y in the k-th data segment. Overall, the contribution factors in the k-th data segment are

$$\Gamma_k = [\gamma_1^k, \gamma_2^k, \ldots, \gamma_M^k]. \tag{4.89}$$

As a result, it not only tells how the variational trend of each influence variable x_i affects the change direction of y based on Ψ_k in (4.74), but also indicates how much x_i contributes to the amplitude change of y based on Γ_k in (4.89). For instance, given $\Psi_k = (1, -1, 0; 1)$ and $\Gamma_k = [0.9, 0.1, 0]$, it can conclude that the increasing of y is mainly caused by the increasing of x_1 with a contribution of 90% and partially caused by the decreasing of x_2 with a contribution of 10%, and x_3 has no significant influence to y.

To facilitate a better understanding of the proposed method, the major steps in the above subsections are integrated as an systematic algorithm, namely, the root cause detection based on qualitative trends and change amplitudes in Algorithm 4.3. The input is the historical data $\mathbb{D}$ consisting of time sequence of the key variable and M influence variables. The final outputs are the combination of qualitative trends

Algorithm 4.3 Root cause detection based on qualitative trends and change amplitudes.

1: Input Argument: historical data $\mathbb{D}$
2: Output Argument: Ψ_k and Γ_k
3: Given $y(1:N)$ from $\mathbb{D}$, calculate K via Eq. (4.61)
4: **for** $i = 1 : M$ **do**
5:     **for** $k = 1 : K$ **do**
6:         Calculate ρ^k for x_i in $\mathbb{D}$ via Eq. (4.69)
7:         Calculate ρ_{th}^k based on $N_k = t_{k+1} - t_k$
8:         **if** $\rho^k \geq \rho_{th}^k$ **then**
9:             $\nu^k = 1$
10:             Calculate λ^k via Eq. (4.71)
11:         **else**
12:             $\nu^k = 0$
13:         **end if**
14:     **end for**
15:     Calculate λ_i for x_i via Eq. (4.72)
16: **end for**
17: **for** $k = 1 : K$ **do**
18:     Calculate A_k^y and A_0^y via Eqs. (4.62) and (4.64)
19:     Calculate ψ_k^y via Eq. (4.63) by comparing A_k^y with A_0^y
20:     **for** $i = 1 : M$ **do**
21:         Calculate $A_k^{x_i}$ and $A_0^{x_i}$ based on $x_i(t_k + \lambda_i : t_{k+1} - 1 + \lambda_i)$
22:         Calculate $\psi_k^{x_i}$ by comparing $A_k^{x_i}$ with $A_0^{x_i}$
23:     **end for**
24:     $\Psi_k = (\psi_k^{x_1}, \psi_k^{x_2}, \ldots, \psi_k^{x_M}; \psi_k^y)$
25: **end for**
26: **for** $k = 1 : K$ **do**
27:     **if** $\psi_k^y \neq 0$ **then**
28:         $\mathbf{y}_k = [\check{y}(t_k), \check{y}(t_k + 1), \ldots, \check{y}(t_{k+1} - 1)]'$
29:         $\mathbf{X}_k = \mathbf{1}_{N_k \times 1}$
30:         **for** $i = 1 : M$ **do**
31:             $\varsigma_i^k = \psi_k^{x_i} \psi_k^y$
32:             **if** $\varsigma_i^k \neq 0$ **then**
33:                 $\mathbf{x}_i = [\check{x}_i(t_k + \lambda_i), \ldots, \check{x}_i(t_{k+1} - 1 + \lambda_i)]'$
34:                 $\mathbf{X}_k = [\mathbf{X}_k, \mathbf{x}_i]$
35:             **end if**
36:         **end for**
37:         Obtain $\hat{\boldsymbol{\beta}}$ by solving Eq. (4.81)
38:         Calculate γ_i^k for $x_i, i = 1, 2, \ldots, M$ via Eq. (4.88)
39:         $\Gamma_k = [\gamma_1^k, \gamma_2^k, \ldots, \gamma_M^k]$
40:         Judge the root cause based on Ψ_k and Γ_k
41:     **end if**
42: **end for**

Ψ_k and the vector of the contribution factor Γ_k in each data segment. The first for-loop from line 4 to line 15 conducts lagged correlation analysis to estimate the time delays. The second for-loop from line 16 to line 24 extracts qualitative trends in each segment. The last for-loop from line 25 to line 41 conducts multiple linear regression to calculate the contribution factors.

To help industrial plant operators with the judgment of root causes in real-time applications, the proposed method can be applied in the following way: (i) A qualitative trend database $\mathbb{Q}$ can be established by grouping trend combinations; (ii) for online monitoring, industrial plant operators judge variational trends of the influence and key variables for recent data samples, and determine root causes by querying the database $\mathbb{Q}$. In other words, the proposed method extracts the qualitative trends and identifies the root cause variable over different time periods, and then the results will be used online to determine root causes by matching with the trend combinations in the historical database.

4.4.4 Numerical and Industrial Examples

This section presents numerical and industrial examples to demonstrate the effectiveness of the proposed method.

4.4.4.1 Numerical Example

In this numerical example, the key variable y is affected by x_1 and x_2. The time delays for them followed by y are set to 10 and 8, respectively. Their time sequences are comprised by 4 parts. In each part, the time sequences of x_1 and/or x_2 are produced in sinusoidal signal forms. The relations between y and the independent variables are time varying, such that the relations are different in the four parts. The first part from $t = 1$ to $t = 800$ is given by

$$\begin{cases} x_1(t) = 5\sin(\frac{\pi}{100}t - \frac{\pi}{2}) + 5 + e_1(t), \\ x_2(t) = 2.5\sin(\frac{\pi}{100}t - \frac{\pi}{2}) + 2.5 + 1.5e_2(t), \\ y(t) = 1.5x_1(t-10) + x_2(t-8) + e_3(t), \end{cases} \tag{4.90}$$

such that y holds correlations with both x_1 and x_2, and is mainly affected by x_1; $e_1(t)$, $e_2(t)$, and $e_3(t)$ are Gaussian noises with zero means and unit standard deviations. The second part from $t = 801$ to $t = 1600$ is

$$\begin{cases} x_1(t) = 5\sin(\frac{\pi}{100}t - \frac{\pi}{2}) + 5 + e_1(t), \\ x_2(t) = 1.5e_2(t), \\ y(t) = 1.5x_1(t-10) + e_3(t), \end{cases} \tag{4.91}$$

such that y is correlated with x_1 only. The third part from $t = 1601$ to $t = 2400$ is

$$\begin{cases} x_1(t) = 2.5\sin(\frac{\pi}{100}t - \frac{\pi}{2}) + 2.5 + 1.5e_1(t), \\ x_2(t) = 5\sin(\frac{\pi}{100}t - \frac{\pi}{2}) + 5 + e_2(t), \\ y(t) = x_1(t-10) + 1.5x_2(t-8) + e_3(t), \end{cases} \tag{4.92}$$

such that y is correlated with both x_1 and x_2, and is mainly affected by x_2. The fourth part from $t = 2401$ to $t = 3200$ is

$$\begin{cases} x_1(t) = 1.5e_1(t), \\ x_2(t) = 5\sin(\frac{\pi}{100}t - \frac{\pi}{2}) + 5 + e_2(t), \\ y(t) = 1.5x_2(t-8) + e_3(t), \end{cases} \tag{4.93}$$

such that y is correlated with x_2 only. Then, the time sequence of y consists of 32 segments with increasing and decreasing trends. The time sequences of y, x_1, and x_2 are shown as the blue lines in Fig. 4.44.

First, the time sequence of y is partitioned into $K = 32$ PLR segments. The results are shown in Fig. 4.44, where the original time sequence is segmented by dashed vertical lines that represent the split time points. It can be seen that the linear model fits each segment well. Second, the correlation coefficients and time delays between x_1 and y in 32 PLR segments are shown in Fig. 4.45a and c. It can be seen that x_1 and y have very strong correlations in the first 16 segments and moderate correlations from the 17th segment to the 24th segment; in the last 8 segments, x_1 and y are detected to be not correlated given their correlations smaller than the thresholds. Based on (4.72), the time delay between x_1 and y is calculated as $\lambda_1 = 10$. In the

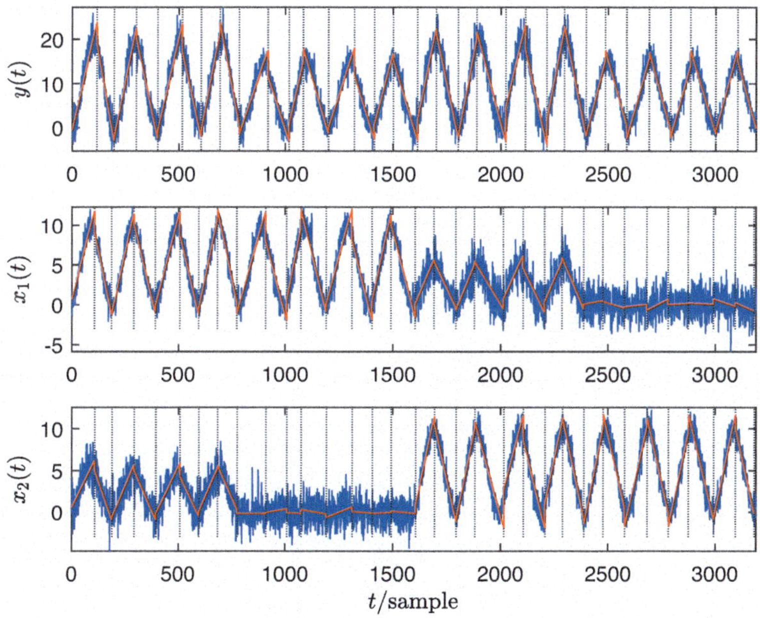

Fig. 4.44 Piecewise linear representations for the time sequences of y, x_1, and x_2. The solid blue lines, solid red lines, and dashed black lines, indicate the original time sequences, straight lines, and split time points, respectively, reprinted from Ref. [22], copyright 2022, with permission from Elsevier

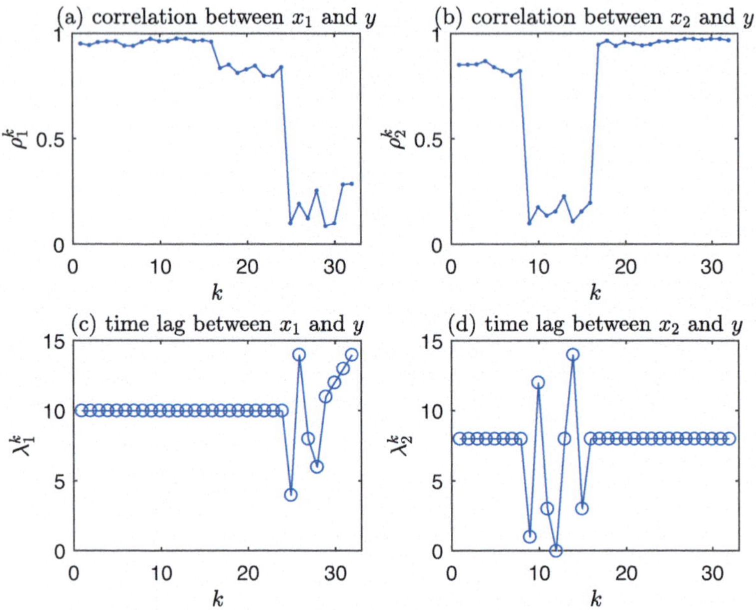

Fig. 4.45 Detected correlations and time delays for the two influence variables and the key variable in each segment, reprinted from Ref. [22], copyright 2022, with permission from Elsevier

same way, the time delay between x_2 and y is calculated as $\lambda_2 = 8$. Therefore, time delays are correctly estimated.

Next, qualitative trends are captured and shown in Fig. 4.46. It can be seen that the detected qualitative trends are in accordance with trends in the actual sinusoidal waves given by (4.90)–(4.93). In total, there are 8 segments with $\Psi_k = (1, 1; 1)$, 8 segments with $\Psi_k = (-1, -1; -1)$, 4 segments with $\Psi_k = (1, 0; 1)$, 4 segments with $\Psi_k = (-1, 0; -1)$, 4 segments with $\Psi_k = (0, 1; 1)$, and 4 segments with $\Psi_k = (0, -1; -1)$. Finally, contribution factors for x_1 and x_2 are calculated, as given in Fig. 4.47. It is found that y is mainly affected by x_1 in the first 16 segments and x_2 in the last 16 segments. In the first 8 segment, x_1 and x_2 contribute around 80% and 20% of the amplitude change of y, respectively. From the 17th segment to the 24th segment, x_2 contributes more to the amplitude change of y compared to x_1. Thus, the conclusion is in line with the actual relations in (4.90)–(4.93).

For a comparison, two typical data-driven causality inference methods based on the Granger causality and the transfer entropy are applied to the same data:

- A tool for Granger causal connectivity analysis in [40] is used to construct autoregressive (AR) models of the three variables and determine the cause-effect relations based on the model residual variances. For all time sequences in Fig. 4.44, the residual variance of the AR model for y alone is $\varepsilon_y = 5.3449$, while the residual variance of the AR model for x_1 and y is $\varepsilon_{x_1 \to y} = 5.2191$, and the counterpart for x_2 and y is $\varepsilon_{x_2 \to y} = 5.2582$. Given a fact that $\varepsilon_{x_1 \to y} < \varepsilon_y$ and $\varepsilon_{x_2 \to y} < \varepsilon_y$, it is

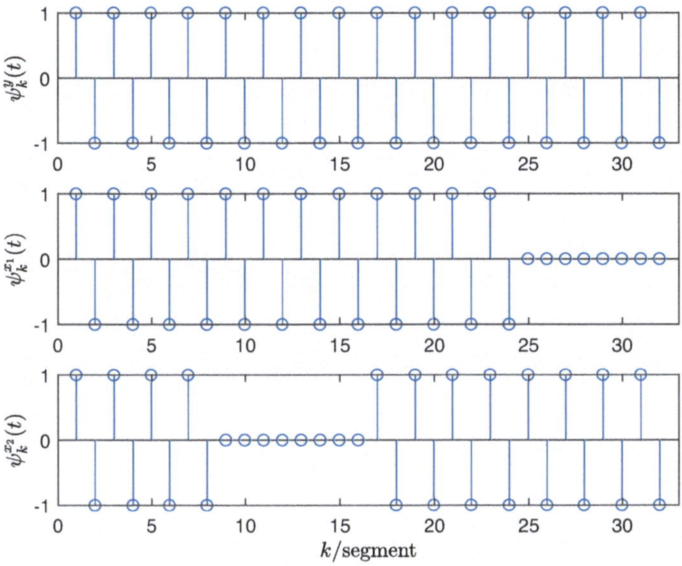

Fig. 4.46 Extracted qualitative trends of the influence and key variables in each segment, reprinted from Ref. [22], copyright 2022, with permission from Elsevier

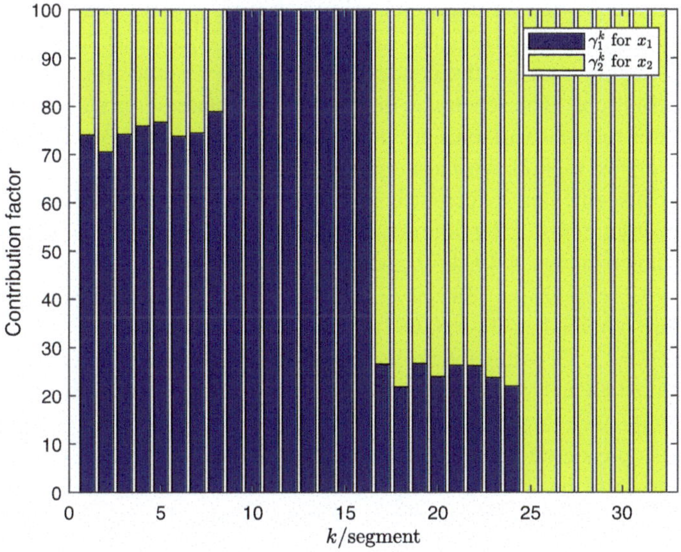

Fig. 4.47 Contribution factors of the two influence variables to the amplitude change in the time sequence of the key variable in each data segment, reprinted from Ref. [22], copyright 2022, with permission from Elsevier

concluded that the past information of either x_1 or x_2 is helpful in predicting y. Thus, both x_1 and x_2 are the causes of y. Clearly, the Granger causality captures the overall cause-effect relations for all time sequences, but cannot indicate actual cause-effect relations in different short data segments.

- A non-parametric causality inference method based on the transfer entropy [4, 32] is exploited to determine the cause-effect relations. The transfer entropy value between x_1 and y is calculated to be $T_{x_1 \to y} = 0.7491$, and the threshold obtained from Monte Carlo simulations is $T^{th}_{x_1 \to y} = 0.5829$. For x_2 and y, the transfer entropy value and threshold are calculated as $T_{x_2 \to y} = 0.7718$ and $T^{th}_{x_2 \to y} = 0.5835$, respectively. Given that $T_{x_1 \to y} > T^{th}_{x_1 \to y}$ and $T_{x_2 \to y} > T^{th}_{x_2 \to y}$, it can be concluded that both x_1 and x_2 are the causes of y. The conclusion is consistent with that based on the Granger causality, and only the overall cause-effect relations are captured.

4.4.4.2 Industrial Example

The industrial process in this example is a condenser of a real thermal power plant. The function of the condenser is to condense the low pressure steam from the turbine exhaust such that the steam is converted to liquid. The condenser vacuum is a key process variable playing an important role for operating safety and efficiency of a thermal power plant. The influence variables of the condenser vacuum y are the inlet steam temperature x_1, the condensate flow rate x_2, and the inlet circulating water temperature x_3. The schematic diagram of the condenser is shown in Fig. 4.48, where the four relevant variables are presented. Historical process data samples are collected over two months with a sampling period 1 s.

The data is segmented by piecewise linear representation for the condenser vacuum y. Three segments of the time sequence and the piecewise linear representation results are shown in Figs. 4.49, 4.50, and 4.51. The time delays between x_i's and y are estimated as $\lambda_1 = 2$ s, $\lambda_2 = 23$ s and $\lambda_3 = 27$ s. The significance thresholds for y, x_1, x_2, and x_3 are obtained as $A_0^y = 0.0918$ KPa, $A_0^{x_1} = 0.1764$ °C, $A_0^{x_2} = 19.3767$ t/h, and $A_0^{x_3} = 0.0623$ °C, respectively; the standard deviations of the errors are calculated as 0.04284 KPa, 0.08289 °C, 8.5017 t/h, and 0.0284 °C, respectively. Compared with the significance thresholds, the standard deviations of errors are much smaller. A variable is considered in the calculation of contribution factors if only its change amplitude exceeds the corresponding significance threshold; in such scenarios, the variation of noise is minor compared to the change amplitude of the variable, and thus has little influence to the change of the condenser vacuum. In each segment, qualitative trends of the influence and key variables are captured. To ease the presentation, let us only consider decrements of the condenser vacuum.

Taking the data segment highlighted by yellow areas in Fig. 4.49 as an example, the qualitative trends in this segment are $\Psi_k = (1, 1, 1; -1)$. It can be observed that the key variable y shows a significant decreasing trend while all other influence variables have apparent increasing trends. The contribution factors are calculated to

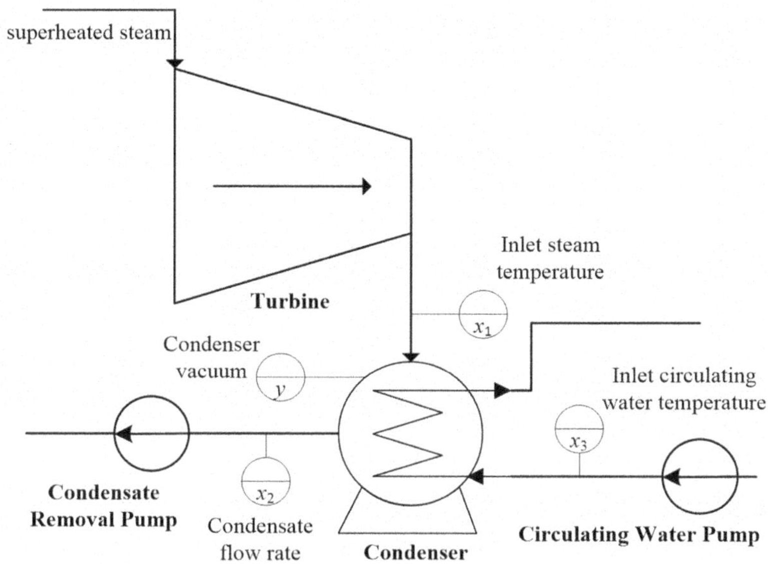

Fig. 4.48 The schematic diagram of the condenser, reprinted from Ref. [22], copyright 2022, with permission from Elsevier

be 0.8269, 0.1306, and 0.0425 for x_1, x_2, and x_3, respectively. Thus, it is concluded that the decrement of y is mainly caused by the increment of x_1 with a contribution factor 82.69%, and partially caused by the increment of x_2 and x_3 with contribution factors 13.06% and 4.25%, respectively. In fact, as the inlet steam temperature x_1 rises, the effect of condensing the inlet steam into water becomes worse and thus the steam volume in the condenser grows, leading to decreased condenser vacuum y. During the highlighted period, the temperature x_1 climbs up by around 0.4 °C, which is much larger than the amplitude change threshold $A_0^{x_1} = 0.1764$ °C. The increasing of the condensate flow rate x_2 also contributes to the vacuum decreasing, but is not the major cause; x_2 rises up by 27.1 t/h, which is greater than the threshold $A_0^{x_2} = 19.3767$ t/h. In addition, the circulating water exchanges heat with the steam entering the condenser and thus may also affect the steam condensation. Here, the inlet circulating water temperature x_3 climbs up by around 0.12 °C ($> A_0^{x_3} = 0.0623$ °C) and also contributes to the vacuum decreasing in some extent. Since the quality of the MLR model affects the calculation of contribution factors and poor model quality may lead to false conclusions, it is necessary to evaluate the model quality. Here, the R^2 value of the MLR model is calculated to be 0.9111, which shows that the model in this example is satisfactory and implies that the variance of y can be well explained by the independent variables in the MLR model. Thus, the conclusion is reliable.

In Fig. 4.50, the qualitative trends in this segment are $\Psi_k = (1, 1, 0; -1)$. It can be observed that the key variable y shows a significant decreasing trend, x_1 and x_2 have apparent increasing trends, and x_3 displays a slowly descending yet non-significant trend in the highlighted area. The contribution factors for x_1, x_2, and x_3 are 0.1519, 0.8481, and 0, respectively. Thus, it is concluded that the decrement of y is mainly caused by the increment of x_2 with a contribution factor 84.81%, and partially caused by the increment of x_1 with a contribution factors 15.19%, and x_3 has no significant influence to y. In this example, the inlet steam temperature x_1 rises up by around 0.4 °C ($> A_0^{x_1} = 0.1764$ °C); the increment is almost the same as that in Fig. 4.49. The condensate flow rate x_2 grows up by around 59 t/h, which is significantly larger than the threshold $A_0^{x_2} = 19.3767$ t/h and also bigger than the increment in Fig. 4.49. Thus, it is reasonable that the contribution of x_2 is larger than that in Fig. 4.49 and becomes greater than x_1 in this example. As for the inlet circulating water temperature

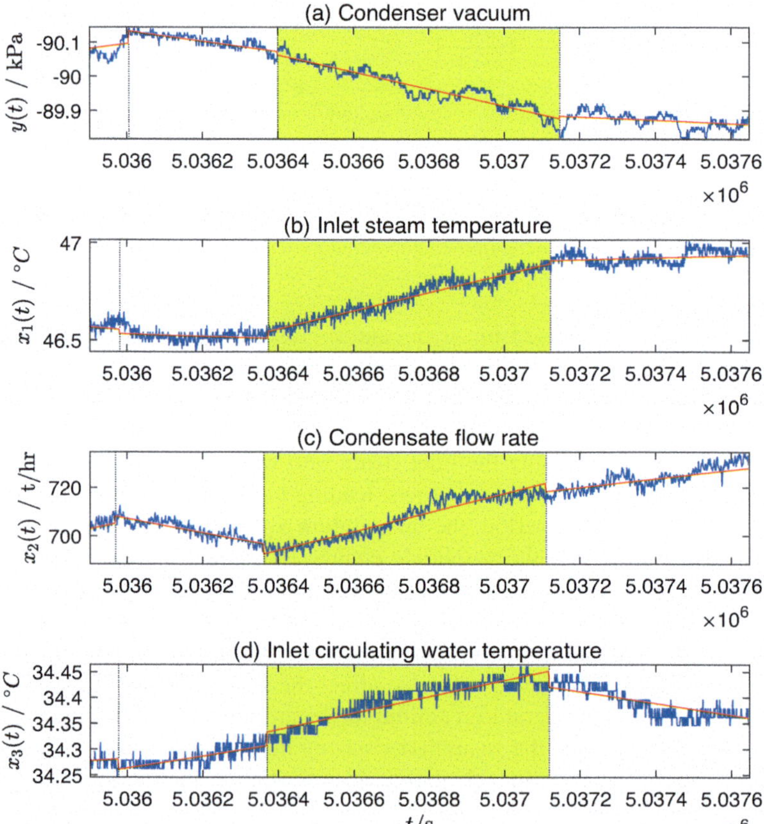

Fig. 4.49 Piecewise linear representations for the time sequences of y, x_1, x_2, and x_3. Note that the tick labels in the vertical axis of the top subplot is arranged in a reverse order that the largest absolute value (smallest original value) of the condenser vacuum appears at the top. The yellow areas highlight the data segment where the condenser vacuum y has a decreasing trend. The combination of qualitative trends for x_1, x_2, x_3, and y in the highlighted data segment is $\Psi_k = (1, 1, 1; -1)$, reprinted from Ref. [22], copyright 2022, with permission from Elsevier

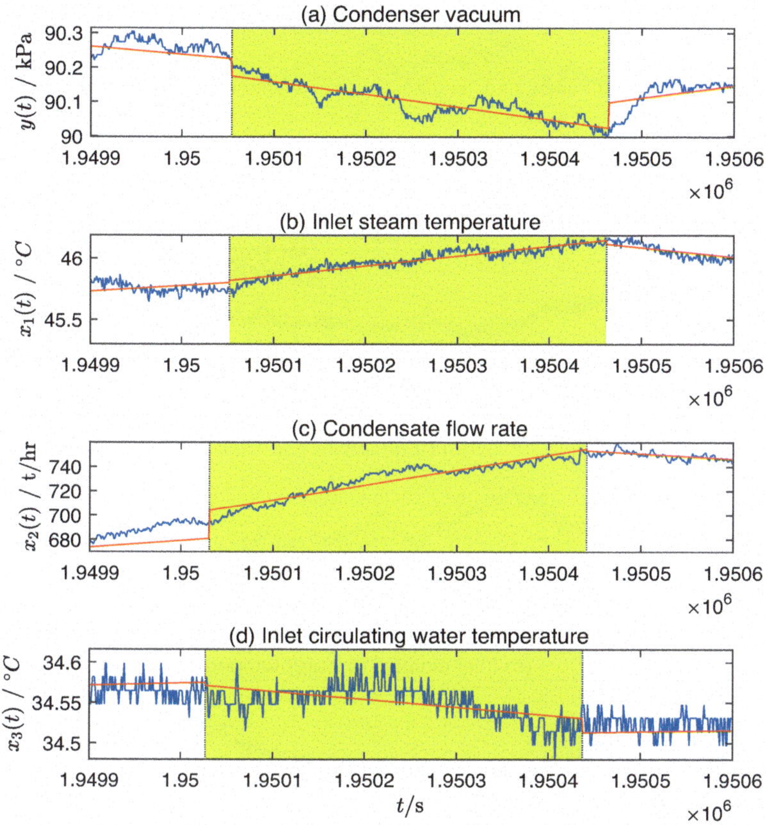

Fig. 4.50 Piecewise linear representations for the time sequences of y, x_1, x_2, and x_3. The combination of qualitative trends for x_1, x_2, x_3, and y in the highlighted data segment is $\Psi_k = (1, 1, 0; -1)$, reprinted from Ref. [22], copyright 2022, with permission from Elsevier

x_3, it does not change significantly (x_3 drops by around 0.05 °C, which is smaller than the threshold $A_0^{x_3} = 0.0623$ °C) and thus has no significant influence to the change of the condenser vacuum. Here, the R^2 value of the MLR model is calculated to be 0.8299, so that the above conclusion is reliable.

In Fig. 4.51, the qualitative trends in this segment are $\Psi_k = (0, 0, 1; -1)$, indicating that x_3 is the only root cause variable that has a significant contribution to the decreasing of y. In other words, x_3 has a contribution factor of 100% while the contribution factors for x_1 and x_2 are 0's. In this example, the inlet steam temperature x_1 and the condensate flow rate x_2 rise up by around 0.13 °C and 14 t/h, respectively. Both are smaller than the corresponding thresholds $A_0^{x_1} = 0.1764$ °C and $A_0^{x_2} = 19.3767$ t/h, implying that the two variables do not change significantly. By contrast, the increment of the inlet circulating water temperature x_3 is around

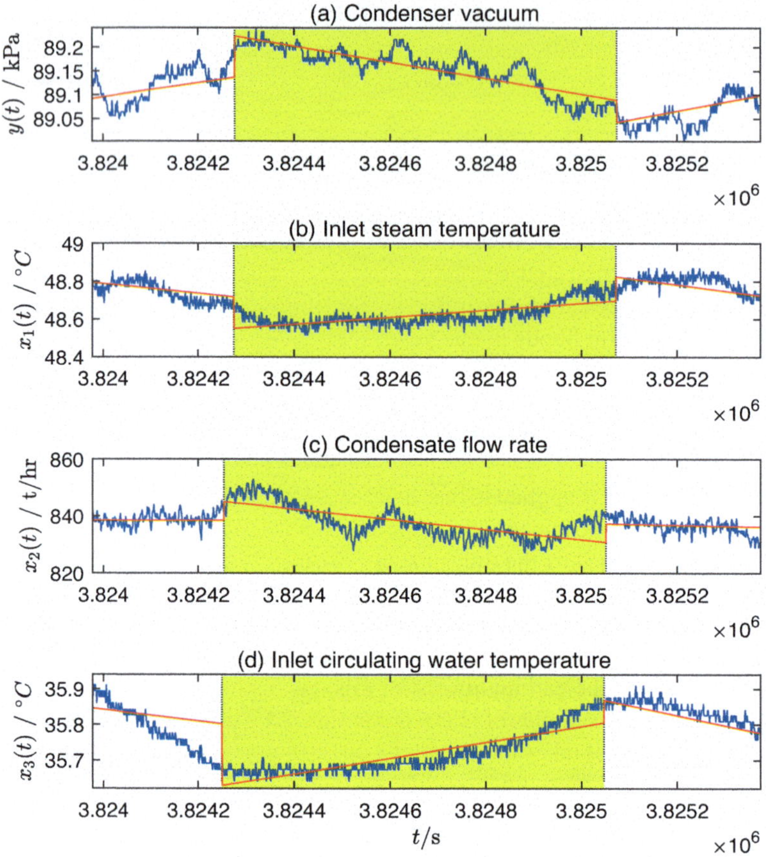

Fig. 4.51 Piecewise linear representations for the time sequences of y, x_1, x_2, and x_3. The combination of qualitative trends for x_1, x_2, x_3, and y in the highlighted data segment is $\Psi_k = (0, 0, 1; -1)$, reprinted from Ref. [22], copyright 2022, with permission from Elsevier

0.18 °C, which is much larger than the threshold $A_0^{x_3} = 0.0623$ °C. Thus, it is reasonable that x_3 is identified as the root cause for the decreasing of the condenser vacuum. Here, the R^2 value of the MLR model is calculated to be 0.8211. Thus, the above conclusion is reliable.

For a long-term application, Table 4.10 presents different qualitative trends of the influence and key variables in the first four columns, and their counts in the last column. For instance, there are 253 segments with $\Psi_k = (1, 1, 1; -1)$, indicating that y drops with the increments of all three influence variables. For the segments with $\Psi_k = (1, 0, 0; -1)$ and $\Psi_k = (0, 0, 1; -1)$, one single influence variable makes a contribution. For the other segments, the contribution factors of x_1, x_2, and x_3 are shown in Fig. 4.52. Based on the results, it is observed that the decrements of

Table 4.10 Different combinations of qualitative trends of the influence and key variables, and their counts in the data, reprinted from Ref. [22], copyright 2022, with permission from Elsevier

$\psi_k^{x_1}$	$\psi_k^{x_2}$	$\psi_k^{x_3}$	ψ_k^{y}	Count
1	1	1	−1	253
1	1	0	−1	187
1	1	−1	−1	111
1	0	1	−1	348
1	0	0	−1	176
0	0	1	−1	31

y are mainly caused by the increments of x_1 in most segments, and are also significantly affected by the increments of x_3 in a few segments with $\Psi_k = (1, 0, 1; -1)$ as indicated by the long yellow bars in Fig. 4.52d.

For a comparison, two typical data-driven causality inference methods based on the Granger causality [16] and the transfer entropy are applied to the same industrial data:

- The tool for Granger causal connectivity analysis in [40] is used to determine the cause-effect relations. Given the entire data, the residual variance of the AR model for y alone is $\varepsilon_y = 7.0381 \times 10^{-5}$, while the residual variance of the AR model for x_1 and y is $\varepsilon_{x_1 \to y} = 7.0131 \times 10^{-5}$, the counterpart for x_2 and y is $\varepsilon_{x_2 \to y} = 7.0352 \times 10^{-5}$, and the counterpart for x_3 and y is $\varepsilon_{x_3 \to y} = 7.0379 \times 10^{-5}$. Given a fact that $\varepsilon_{x_1 \to y} < \varepsilon_y$, $\varepsilon_{x_2 \to y} < \varepsilon_y$, and $\varepsilon_{x_2 \to y} < \varepsilon_y$, it is concluded that the past information of either x_1, x_2, or x_3 is helpful in predicting y. Thus, x_1, x_2, and x_3 are the causes of y. It can be seen that the Granger causality captures the overall cause-effect relations for all time sequences, but cannot indicate actual cause-effect relations in different short data segments.
- The transfer entropy [4, 32] is exploited to determine the cause-effect relations. The transfer entropy value between x_1 and y is calculated to be $T_{x_1 \to y} = 0.3248$, and the threshold obtained from Monte Carlo simulations is $T^{th}_{x_1 \to y} = 0.1816$. For x_2 and y, the transfer entropy value and threshold are calculated as $T_{x_2 \to y} = 0.3222$ and $T^{th}_{x_2 \to y} = 0.2002$, respectively. For x_3 and y, the transfer entropy value and threshold are calculated as $T_{x_3 \to y} = 0.2950$ and $T^{th}_{x_3 \to y} = 0.1152$, respectively. Given that $T_{x_1 \to y} > T^{th}_{x_1 \to y}$, $T_{x_2 \to y} > T^{th}_{x_2 \to y}$, and $T_{x_3 \to y} > T^{th}_{x_3 \to y}$, it can be concluded that x_1, x_2, and x_3 are the causes of y. The conclusion is consistent with that based on the Granger causality, and only the overall cause-effect relations are captured.

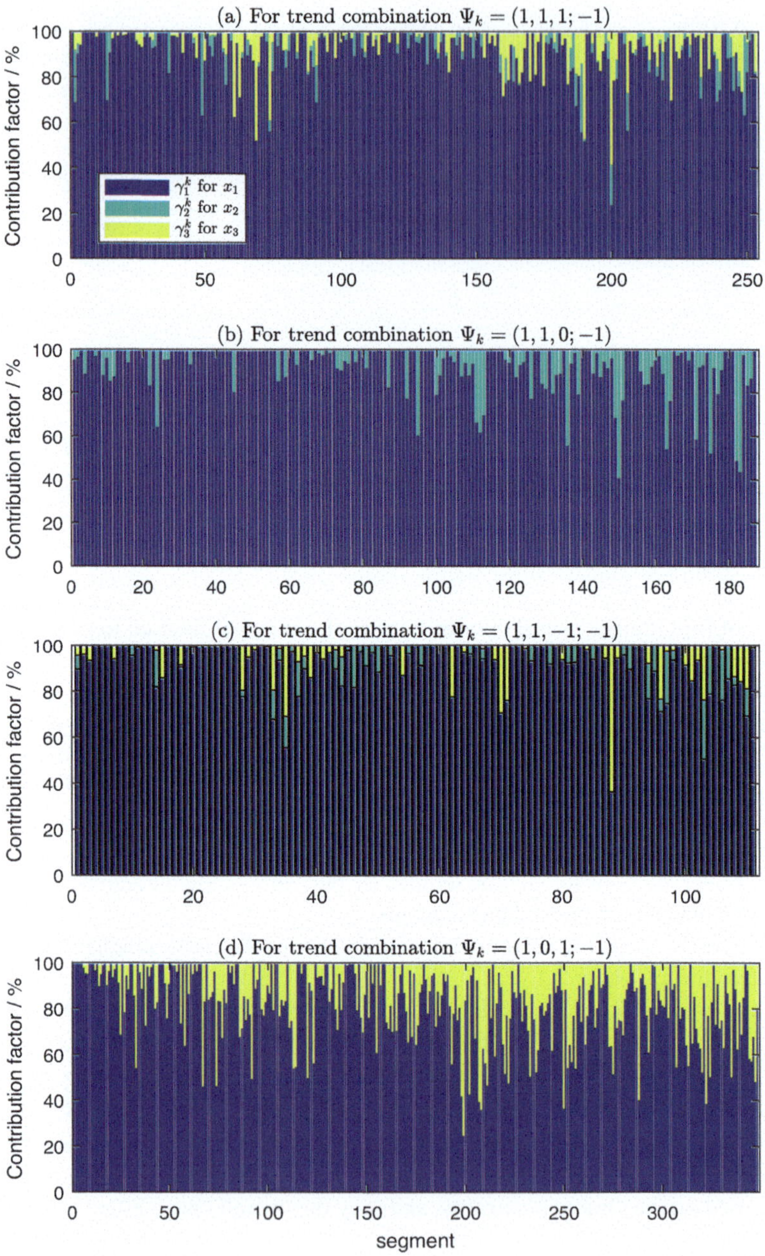

Fig. 4.52 Contribution factors of the three influence variables to the amplitude change in the condenser vacuum for trend combinations, **a** $\Psi_k = (1, 1, 1; -1)$, **b** $\Psi_k = (1, 1, 0; -1)$, **c** $\Psi_k = (1, 1, -1; -1)$, and **d** $\Psi_k = (1, 0, 1; -1)$, reprinted from Ref. [22], copyright 2022, with permission from Elsevier

References

1. Ankerst, M., Breunig, M., Kriegel, H., & Sander, J. (1999). OPTICS: Ordering points to identify the clustering structure. *ACM Sigmod Record, 28*, 49–60.
2. ANSI/ISA-18.2. (2016). *ANSI/ISA-18.2: Management of Alarm Systems for the Process Industries*. Durham, NC USA: ISA (International Society of Automation).
3. Bauer, E., Koller, D., & Singer, Y. (1997). Update rules for parameter estimation in Bayesian networks. In *The 13th Annual Conference on Uncertainty in Artificial Intelligence* (pp. 3–13).
4. Bauer, M., Cox, J., Caveness, M., Downs, J., & Thornhill, N. (2007). Finding the direction of disturbance propagation in a chemical process using transfer entropy. *IEEE Transactions on Control Systems Technology, 15*, 12–21.
5. Bauer, M., & Thornhill, N. (2008). A practical method for identifying the propagation path of plant-wide disturbances. *Journal of Process Control, 18*, 707–719.
6. Cheung, J., & Stephanopoulos, G. (1990). Representation of process trends-Part I. A formal representation framework. *Computers & Chemical Engineering, 14*, 495–510.
7. Chiang, L., Russell, E., & Braatz, R. (2000). Fault diagnosis in chemical processes using Fisher discriminant analysis, discriminant partial least squares, and principal component analysis. *Chemometrics and Intelligent Laboratory Systems, 50*, 243–252.
8. Cohen, I., Bronstein, A., & Cozman, F. G. (2001). Online learning of Bayesian network parameters. *Technical Report No. HPL-2001-55R1*.
9. Dong, Y., & Qin, S. (2018). A novel dynamic PCA algorithm for dynamic data modeling and process monitoring. *Journal of Process Control, 67*, 1–11.
10. Du, S., Wu, M., Chen, L., Cao, W., & Pedrycz, W. (2022). Information granulation with rectangular information granules and its application in time series similarity measurement. *IEEE Transactions on Fuzzy Systems, 30*, 4069–4081.
11. Du, S., Wu, M., Chen, L., Hu, J., Jin, L., Cao, W., & Pedrycz, W. (2020). Operating mode recognition based on fluctuation interval prediction for iron ore sintering process. *IEEE/ASME Transactions on Mechatronics, 25*, 2297–2308.
12. Duan, P., Yang, F., Chen, T., & Shah, S. (2013). Direct causality detection via the transfer entropy approach. *IEEE Transactions on Control Systems Technology, 21*, 2052–2066.
13. Ester, M., Kriegel, H. P., Sander, J., & Xu, X. (1996). A density-based algorithm for discovering clusters in large spatial databases with noise. *KDD, 96*, 226–231.
14. Gandy, A. (2009). Sequential implementation of Monte Carlo tests with uniformly bounded resampling risk. *Journal of American Statistical Association, 104*(488), 1504–1511.
15. Gourévitch, B., & Eggermont, J. J. (2007). Evaluating information transfer between auditory cortical neurons. *Journal of Neurophysiology, 97*(3), 2533–2543.
16. Granger, C. (1969). Investigating causal relations by econometric models and cross-spectral methods. *Econometrica, 37*, 424–438.
17. Guo, C., Yang, F., & Yu, W. (2015). A causality capturing method for diagnosis based on transfer entropy by analyzing trends of time series. *IFAC-PapersOnLine, 48*, 778–783.
18. Guo, H., Wang, L., Liu, X., & Pedrycz, W. (2021). Information granulation-based fuzzy clustering of time series. *IEEE Transactions on Cybernetics, 51*, 6253–6261.
19. Hope, A. C. (1968). A simplified Monte Carlo significance test procedure. *Journal of the Royal Statistical Society. Series B (Methodological)*, 582–598.
20. Hu, W., Wang, J., & Chen, T. (2015). A new method to detect and quantify correlated alarms with occurrence delays. *Computers & Chemical Engineering, 80*, 189–198.
21. Hu, W., Wang, J., Chen, T., & Shah, S. (2017). Cause-effect analysis of industrial alarm variables using transfer entropies. *Control Engineering Practice, 64*, 205–214.
22. Hu, W., Wang, J., Yang, F., Han, B., & Wang, Z. (2022). Analysis of time-varying cause-effect relations based on qualitative trends and change amplitudes. *Computers & Chemical Engineering, 162*, 107813.
23. Ito, S., Hansen, M. E., Heiland, R., Lumsdaine, A., Litke, A. M., & Beggs, J. M. (2011). Extending transfer entropy improves identification of effective connectivity in a spiking cortical network model. *PloS One, 6*(11), e27431.

24. Jiang, H., Patwardhan, R., & Shah, S. L. (2009). Root cause diagnosis of plant-wide oscillations using the concept of adjacency matrix. *Journal of Process Control, 19*(8), 1347–1354.
25. Koller, D., & Friedman, N. (2009). *Probabilistic Graphical Models: Principles and Techniques*. The MIT Press.
26. Kondaveeti, S. R., Izadi, I., Shah, S. L., Shook, D. S., Kadali, R., & Chen, T. (2013). Quantification of alarm chatter based on run length distributions. *Chemical Engineering Research and Design, 91*, 2550–2558.
27. Larsson, J. E. (1994). Diagnostic reasoning strategies for means-end models. *Automatica, 30*(5), 775–787.
28. Lawson, C., & Hanson, R. (1995). *Solving Least Squares Problems*. Society for Industrial and Applied Mathematics.
29. Lindner, B., Auret, L., & Bauer, M. (2017). Investigating the impact of perturbations in chemical processes on data-based causality analysis. Part 1: Defining desired performance of causality analysis techniques. *IFAC-PapersOnLine, 50*, 3269–3274.
30. Lindner, B., Auret, L., & Bauer, M. (2017). Investigating the impact of perturbations in chemical processes on data-based causality analysis. Part 2: Testing Granger causality and transfer entropy. *IFAC-PapersOnLine, 50*, 3275–3280.
31. Lindner, B., Auret, L., & Bauer, M. (2020). A systematic workflow for oscillation diagnosis using transfer entropy. *IEEE Transactions on Control Systems Technology, 28*, 908–919.
32. Lindner, B., Auret, L., Bauer, M., & Groenewald, J. (2019). Comparative analysis of Granger causality and transfer entropy to present a decision flow for the application of oscillation diagnosis. *Journal of Process Control, 79*, 72–84.
33. Lu, W., Pedrycz, W., Liu, X., Yang, J., & Li, P. (2014). The modeling of time series based on fuzzy information granules. *Expert Systems with Applications, 41*, 3799–3808.
34. Montgomery, D. C., & Runger, G. C. (2010). *Applied Statistics and Probability for Engineers*. Wiley.
35. Naghoosi, E., Huang, B., Domlan, E., & Kadali, R. (2013). Information transfer methods in causality analysis of process variables with an industrial application. *Journal of Process Control, 23*, 1296–1305.
36. Phipson, B., & Smyth, G. K. (2010). Permutation P-values should never be zero: Calculating exact P-values when permutations are randomly drawn. *Statistical Applications in Genetics and Molecular Biology, 9*(1).
37. Rozo, A., Morales, J., Moeyersons, J., Joshi, R., Caiani, E., Borzée, P., Buyse, B., Testelmans, D., Van Huffel, S., & Varon, C. (2021). Benchmarking transfer entropy methods for the study of linear and nonlinear cardio-respiratory interactions. *Entropy, 23*, 939.
38. Schreiber, T., & Schmitz, A. (2000). Surrogate time series. *Physica D: Nonlinear Phenomena, 142*, 346–382.
39. Seber, G., & Lee, A. (2012). *Linear Regression Analysis*. Wiley.
40. Seth, A. (2010). A MATLAB toolbox for Granger causal connectivity analysis. *Journal of Neuroscience Methods, 186*, 262–273.
41. Shamos, M., & Hoey, D. (1976). Geometric intersection problems. In *Proceedings of the 17th Annual Symposium on Foundations of Computer Science* (pp. 208–215).
42. Staniek, M., & Lehnertz, K. (2008). Symbolic transfer entropy. *Physical Review Letters, 100*, 158101.
43. Thambirajah, J., Benabbas, L., Bauer, M., & Thornhill, N. (2009). Cause-and-effect analysis in chemical processes utilizing XML, plant connectivity and quantitative process history. *Computers & Chemical Engineering, 33*, 503–512.
44. Wang, J., & Chen, T. (2013). An online method for detection and reduction of chattering alarms due to oscillation. *Computers & Chemical Engineering, 54*, 140–150.
45. Wang, J., Pang, X., Gao, S., Zhao, Y., & Cui, S. (2019). Assessment of automatic generation control performance of power generation units based on amplitude changes. *International Journal of Electrical Power & Energy Systems, 108*, 19–30.
46. Wang, J., Yang, Z., Su, J., Zhao, Y., Gao, S., & Pang, X. (2018). Root-cause analysis of occurring alarms in thermal power plants based on Bayesian networks. *Electrical Power and Energy Systems, 103*, 67–74.

47. Wang, J., Yu, Y., & Chen, K. (2019). Determining the number of segments for piece-wise linear representation of discrete-time signals. *Computers & Chemical Engineering, 120*, 46–53.
48. Wen, C., Yan, Z., Liang, Y., Wu, H., Zhou, L., & Yao, Y. (2022). A control chart-based symbolic conditional transfer entropy method for root cause analysis of process disturbances. *Computers & Chemical Engineering, 164*, 107902.
49. Wyner, A. D., & Ziv, J. (1973). A theorem on the entropy of certain binary sequences and applications-I. *IEEE Transactions on Information Theory, 19*(6), 769–772.
50. Yu, F., & Pedrycz, W. (2009). The design of fuzzy information granules: Tradeoffs between specificity and experimental evidence. *Applied Soft Computing, 9*, 264–273.
51. Zadeh, L. (1979). Fuzzy sets and information granularity. *Advances in Fuzzy Set Theory and Applications, 11*, 3–18.
52. Zhang, X., Hu, W., & Yang, F. (2022). Detection of cause-effect relations based on information granulation and transfer entropy. *Entropy, 24*, 212.
53. Zhang, X., Hu, W., Yang, F., Cao, W., & Wu, M. (2023). A new transfer entropy approach based on information granulation and clustering for root cause analysis in complex industrial processes. *Control Engineering Practice, 140*, 105669.
54. Zhao, Y., Li, H., Yu, X., Ma, N., Yang, T., & Zhou, J. (2022). An independent central point OPTICS clustering algorithm for semi-supervised outlier detection of continuous glucose measurements. *Biomedical Signal Processing and Control, 71*, 103196.

Chapter 5
Analysis of Industrial Alarm Floods

Abstract This chapter is on the analysis of industrial alarm floods, mainly involving the detection, similarity analysis, pattern mining, and prediction of alarm floods. New criteria are designed for detecting the presence of alarm floods in both online and offline manners. For offline analysis of industrial alarm floods, a local alignment algorithm for efficient similarity analysis is proposed to find clusters of similar alarm flood sequences, and a pattern mining approach is developed to find interesting alarm patterns in alarm floods from historical alarm data. For online analysis, a maximum-entropy-based method is presented to predict upcoming alarms for an occurring alarm flood. These methods provide effective solutions to improve operators' awareness to critical alarms in alarm flood periods, reduce irrelevant alarm flood sequences, and achieve early warning and prediction of alarm floods.

5.1 Detection of Alarm Floods

Detection of alarm floods is the foremost step in alarm flood analysis. This section is on the criteria and methods to detect the presence of industrial alarm floods in both online and offline manners [39]. Two basic criteria in practice are analyzed for detecting alarm floods, based on the number of alarm occurrences and the number of alarm variables in the alarm state. Due to the omnipresence of nuisance alarms, the two basic criteria have a drawback of resulting in false detected alarm floods. Distinguishing nuisance alarms from true alarms is a difficult problem in general. As a result, nuisance alarms cannot be completely avoided. Hence, it is desirable to detect alarm floods in a way being less affected by nuisance alarms. In order to alleviate the drawback, a new criterion is formulated based on the number of alarm variables newly appeared in the alarm state. A new method is proposed based on the new criterion to detect an occurring alarm flood or the presence of alarm floods in historical data sets.

J. Wang et al., *Intelligent Industrial Alarm Systems*,
https://doi.org/10.1007/978-981-97-6516-4_5

5.1.1 Problem Description

This subsection describes the problem to be solved and analyzes two basic criteria to solve the problem of detecting alarm floods. In particular, two criteria are shown to have the drawback of resulting in false detected alarm floods, due to chattering alarms and long-standing alarms.

An alarm data set of alarm variables $x_1, x_2, \ldots, x_M$ can be mathematically represented as

$$X(t-N+1:t) \triangleq [x_1(t-N+1:t), \ldots, x_M(t-N+1:t)]. \tag{5.1}$$

Here $x_i(t-N+1:t) \triangleq [x_i((t-N+1)T), \ldots, x_i(tT)]'$ for $i \in [1, M]$ is the data set of the i-th alarm variable x_i, where T is a real-valued sampling period, t is a positive integer-valued sample index, and N is another positive integer as the number of samples. In order to ease the notations, the sampling period T is assumed to be 1 s and is omitted in the sequel. Thus, the sample index t has the physical interpretation as the sampling time instant in second. Without loss of generality, the alarm variable x_i takes the value of 1 (0) when x_i is in the alarm (non-alarm) state. Our objective is to determine from the alarm data set $X(t-N+1:t)$ in (5.1) whether the alarm sequences in the time period $[t-N+1, t]$ of a subset of alarm variables $x_1, x_2, \ldots, x_M$ satisfy a certain criterion for alarm floods, so that the presence of alarm floods can be detected.

5.1.1.1 Criterion Based on Number of Alarm Occurrences

Choosing a proper metric is the first step in designing a criterion for detecting alarm floods. A widely-accepted definition of alarm floods is that 10 or more alarms occur within a 10-min time period [5, 22]. However, the definition is vague in terms of two different understandings on the word 'alarms', namely, alarm occurrences and alarm variables. The first understanding leads to taking the number of alarm occurrences as the metric.

The alarm occurrence (alarm clearance) is defined as the event that the i-th alarm variable x_i switches from 0 to 1 (from 1 to 0). To capture the alarm occurrence, another related form of x_i is formulated by taking the value of 1 only at the time instants of alarm occurrences and the value of 0 at other time instants, i.e.,

$$\hat{x}_i(t) = \begin{cases} 1, & \text{if } x_i(t-1) = 0 \text{ and } x_i(t) = 1, \\ 0, & \text{otherwise.} \end{cases} \tag{5.2}$$

If the number of alarm occurrences is the metric to be used, then the alarm flood is present if there are 10 or more alarm occurrences within a 10-min time period, i.e.,

Criterion A: An alarm flood is present in the 10-min

data set $X(t-599:t)$, if $\left(\sum_{i=1}^{M}\sum_{l=t-599}^{t}\hat{x}_i(l)\right) \geq 10.$

This criterion has been widely applied in recent studies on alarm floods, e.g., [1, 11, 12, 14, 24, 27]. However, the presence of nuisance alarms in practice severely deteriorates the performance of this criterion.

Nuisance alarms are the ones that have no effects even if the occurring alarms are ignored by plant operators. By contrast, true alarms require operators to pay attentions or take actions in a prompt manner; otherwise, the associated abnormal conditions would have negative effects on the safety and/or efficiency of industrial plants. Chattering alarms make quick transitions between alarm and non-alarm states without requiring attentions or actions from operators, and are the most common nuisance alarms [19]. The presence of nuisance alarms, especially the chattering alarms, implies a large number of alarm occurrences and thus is very likely to result in false detected alarm floods if criterion A is exploited.

Distinguishing nuisance alarms from true ones is based on whether an occurring alarm is associated with an abnormal condition. The removal of chattering alarms is feasible for some special situations, e.g., alarm durations or intervals are short or periodic. The alarm duration T_1 is the duration of consecutive 1's in x_i from an alarm occurrence to the next clearance, i.e.,

$$T_1 \triangleq t_2 - t_1 + 1, \tag{5.3}$$

where

$$x_i(t_1-1)=0,\ x_i(t_2+1)=0,$$
$$\sum_{t=t_1}^{t_2} x_i(t) = t_2 - t_1 + 1,\ \text{for } t_2 > t_1.$$

The alarm interval, denoted as T_0, is the time interval from an alarm clearance to the next alarm occurrence, i.e.,

$$T_0 \triangleq t_2 - t_1 + 1, \tag{5.4}$$

where

$$x_i(t_1-1)=1,\ x_i(t_2+1)=1,$$
$$\sum_{t=t_1}^{t_2}(1-x_i(t)) = t_2 - t_1 + 1,\ \text{for } t_2 > t_1.$$

A rule of thumb is that alarms repeating more than three times per minute are considered as chattering alarms [5]. Hence, a rule to detect chattering alarms is formulated as "if the alarm duration T_1 or the alarm interval T_0 is less than 20 s, then the chattering alarm is present" [36]. This rule will be used later to remove chattering alarms by the proposed method.

To illustrate that the performance of criterion A is severely affected by chattering alarms, an example with an alarm system for a thermal power plant is provided here. The DCS records the values of 24079 process variables every second. There are 2226 process variables being configured with alarm variables. All the subsequent examples are from the same alarm system.

Criterion A is applied to the one-hour data set of the 2226 alarm variables from 7 to 8 am on June 1st, and 16 alarm variables are found to experience alarm occurrences in the hour. The time sequences of the alarm variables are given in Fig. 5.1. The numbers of alarm occurrences in the 6 non-overlapping consecutive 10-min time periods are respectively 245, 260, 195, 225, 198 and 212, given at the vertical dashed lines separating the 10-min periods. The numbers of alarm occurrences are much larger than the threshold 10 in criterion A, indicating the presence of alarm floods. However, the time sequences in Fig. 5.1 reveal that chattering alarms are obviously present, because the alarm durations T_1 in (5.3) or the alarm intervals T_0 in (5.4) are very short for most alarm occurrences or clearances of the 16 alarm variables. Therefore, the chattering alarms lead to the false detected alarm floods.

In practice, it is impossible to remove all chattering alarms. Thus, it is difficult to avoid false detected alarm floods by using criterion A. Figure 5.2 illustrates this difficulty by showing the time sequences of 9 alarm variables experiencing alarm

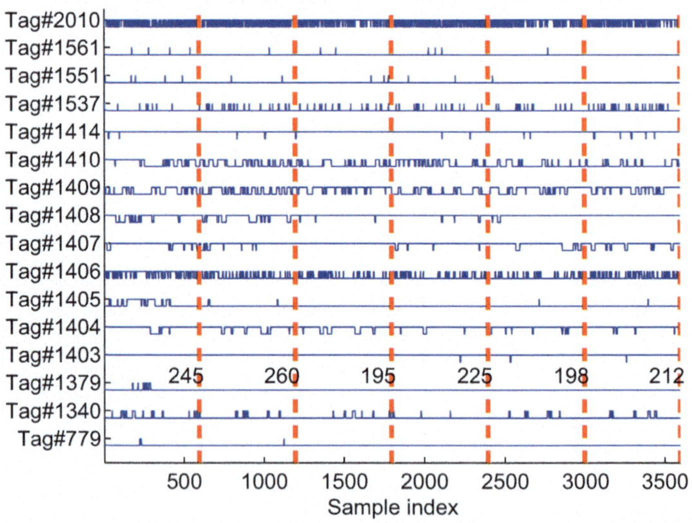

Fig. 5.1 Time sequences of alarm variables experiencing alarm occurrences from 7 to 8 am on June 1st (the lower level for each alarm variable represents the value 0, and the higher level represents the value 1), reprinted from Ref. [39], copyright 2018, with permission from IEEE

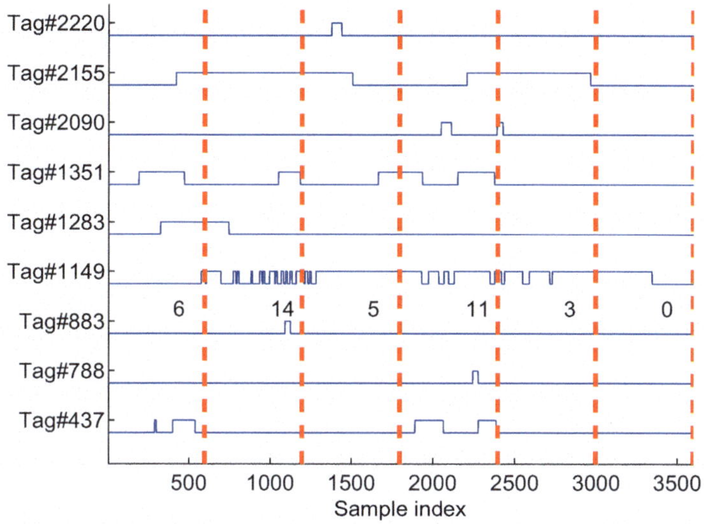

Fig. 5.2 Time sequences of alarm variables experiencing alarm occurrences from 10 to 11 am on June 8th, reprinted from Ref. [39], copyright 2018, with permission from IEEE

occurrences in one hour from 10 to 11 am on June 8th. The numbers of alarm occurrences in the 6 non-overlapping consecutive 10-min time periods are 6, 14, 5, 11, 3 and 0, respectively. Criterion A says that alarm floods are present for two time periods, namely, one from 10:10 to 10:20 am and another from 10:30 to 10:40 am. However, the time sequences of the involved alarm variables reveal that the alarm variable with Tag #1149 is a major culprit for the false detection of alarm floods. This alarm variable experienced several alarm occurrences with short alarm durations, which result in the numbers of alarm occurrences exceeding the detection threshold 10. Thus, no actual alarm floods are present.

5.1.1.2 Criterion Based on Number of Variables in Alarm State

Another metric for alarm flood detection is the number of alarm variables in the alarm state, based on the second understanding of the word "alarms" as alarm variables. Industrial plant operators are concerned about whether an alarm variable is in the alarm state, instead of the repeated alarm occurrences of the same alarm variable in 10 min.

The criterion based on the number of alarm variables in the alarm state is described as follows. First, define the total alarm duration $T_{1,i}$ for the alarm variable x_i in the 10-min time period $[t - 599, t]$ as

$$T_{1,i} := \sum_{l=t-599}^{t} x_i(l).$$

Second, an indicator variable is defined to indicate whether x_i has been in the alarm state in the 10 min, i.e.,

$$\tilde{x}_i = \begin{cases} 1, \text{ if } T_{1,i} > 0, \\ 0, \text{ otherwise.} \end{cases}$$

Next, if there are more than 10 alarm variables in the alarm state within a 10-min time period, then an alarm flood is present, i.e.,

$$\text{Criterion B: An alarm flood is present in the 10-min}$$
$$\text{data set } X(t-599:t), \text{ if } \left(\sum_{i=1}^{M} \tilde{x}_i\right) \geq 10.$$

Clearly, this criterion is much less affected by chattering alarms. However, it is severely affected by the so-called long-standing alarms. Long-standing alarms are the ones continuously remaining in the alarm state for a long time period, e.g., 8 or 24 h [5, 19, 22]. Long-standing alarms could be either true alarms or nuisance ones: for the former type, a long-standing alarm is an indication of an abnormal condition that cannot be resolved in a short term [9, 19], and for the latter type, the main causes include the incorrectly configured alarm variables and the design of alarm variables being inconsistent with operating states. For the second cause, nuisance long-standing alarms could be removed by introducing state-based alarm generation mechanisms [37]. The presence of nuisance long-standing alarms certainly does not indicate any actual abnormal conditions, and may result in false detected alarm floods. For the true long-standing alarms, their alarm durations are so long such that operators have already known the corresponding abnormal conditions. Hence, there is no need to keep counting these long-standing alarms as the ones to which operators should pay attentions every 10 min.

To illustrate that criterion B suffers from the negative effects of long-standing alarms, an example is provided here. For the first alarm data set in Fig. 5.1, namely, the one from 7 to 8 am on June 1st, the numbers of alarm variables in the alarm state in the 6 non-overlapping consecutive 10-min time periods are 227, 225, 223, 223, 224 and 222, respectively. If criterion B is used, then alarm floods are present in the entire one hour. However, there are 210 alarm variables in the alarm state throughout the one hour, among which there are nuisance alarms and true alarms.

As an example of nuisance long-standing alarms, an alarm variable of the manual mode of a feedback control loop for an electrical feedwater pump has been in the alarm state, as shown in Fig. 5.3a. However, the alarm state does not indicate any abnormal conditions. A common practice for power plants is to use steam-driven feedwater pumps in the normal operation, and to exploit the electrical feedwater pump only at the startup stage or emergency situations. Thus, the electrical feedwater

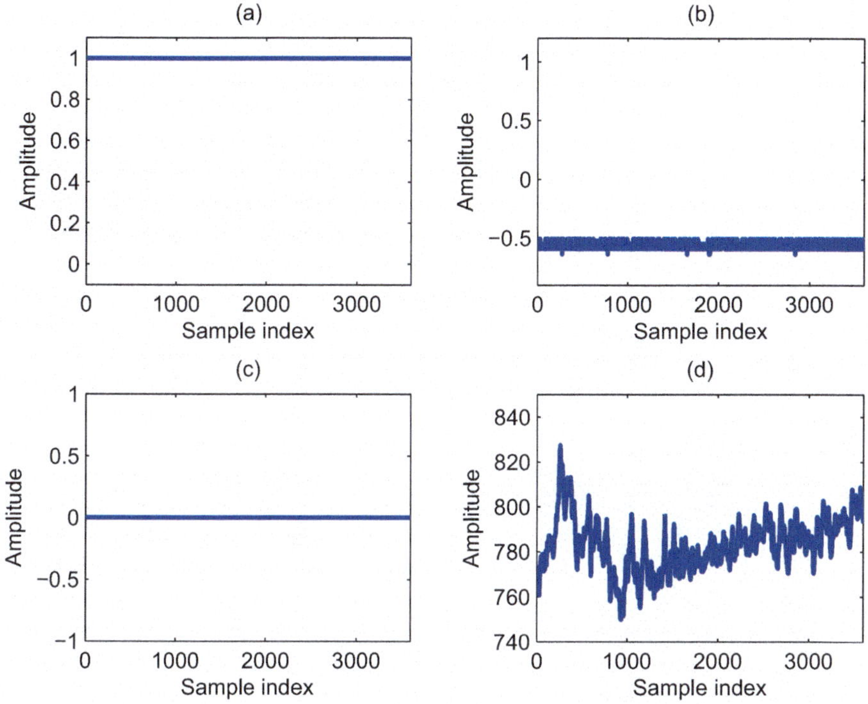

Fig. 5.3 Time sequences of an alarm variable representing the manual mode of a feedback control loop (**a**), the electrical current (**b**), and the inlet flow (**c**) for an electrical feedwater pump, and that of the main feedwater flow (**d**) from 7 to 8 am on June 1st, reprinted from Ref. [39], copyright 2018, with permission from IEEE

pump has been in the shutdown status, as indicated by its electrical current and inlet flow in Fig. 5.3b and c respectively. As a comparison, the main feedwater flow is in the normal condition, as shown in Fig. 5.3d. Thus, it is normal to see that the feedback control loop for the electrical feedwater pump is in the manual mode. The presence of such a nuisance long-standing alarm obviously may result in false detected alarm floods.

Figure 5.4a presents an example of true long-standing alarms. An alarm variable, being in the alarm state for the entire one hour, represents the pressure difference over 140 pa between two layers of a selective catalytic reduction of NOx using NH3. The calculated pressure difference and the measurements of pressures at the two layers are shown in Fig. 5.4b–d, respectively. The high pressure difference usually indicates an ill condition of fly ashes blocking the two layers. However, the soot blowing action to remove the ill condition could only be implemented at the next round of equipment maintenance, usually in the period of 8 h. Consequently, such an alarm variable is expected to be in the alarm state for a long time period. Hence, it

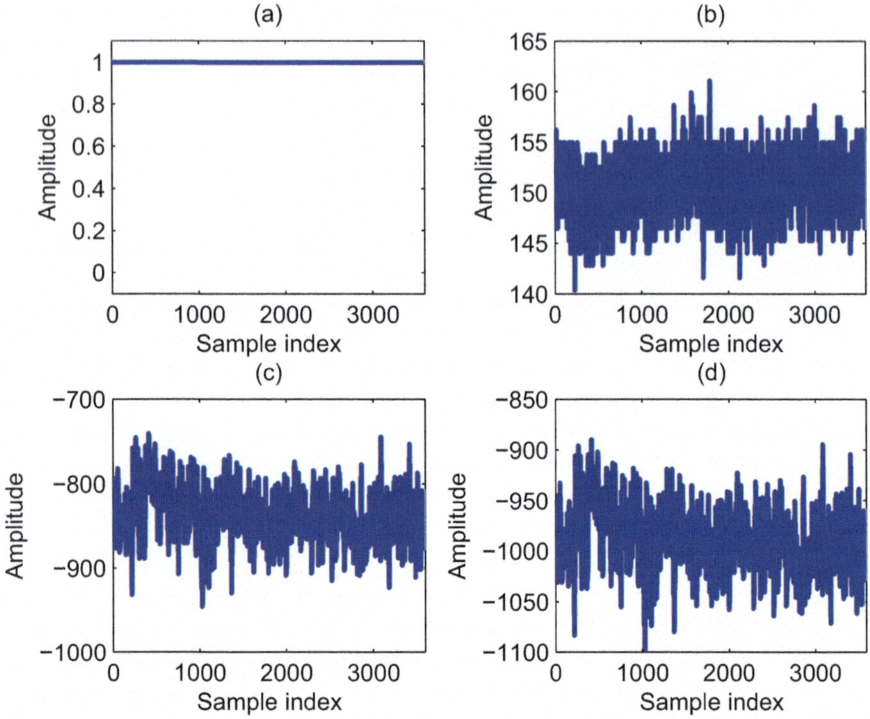

Fig. 5.4 Time sequences of an alarm variable representing the pressure difference over 140pa between two layers (**a**), the pressure difference between two layers (**b**), the pressure at the upper layer (**c**), and the pressure at the lower layer (**d**) of a selective catalytic reduction of NOx with NH3 from 7 to 8 am on June 1st, reprinted from Ref. [39], copyright 2018, with permission from IEEE

would be meaningless to keep counting it as the one to which operators should pay attentions every 10 min.

5.1.2 New Criterion for Alarm Flood Detection

Even though there are some techniques to deal with nuisance alarms in the literature, chattering alarms and long-standing alarms cannot be completely removed in practice. As a result, the two basic criteria in Sect. 5.1.1 often lead to false detected alarm floods. This section formulates a new criterion for the detection of alarm floods based on the number of alarm variables newly appeared in the alarm state, which is a metric suffering much less from the negative effects of chattering alarms and long-standing alarms.

This metric is defined as follows. First, let $I_1(t)$ be the index set of alarm variables $x_1, x_2, \ldots, x_M$ in the time period $[t - 599, t]$ that experience alarm occurrences, i.e.,

$$I_1(t) \triangleq \{i\} \text{ such that } \left(\sum_{l=t-599}^{t} x_i'(l) \right) > 0,$$

where $x_i'(t)$ is defined in (5.2). A second set $I_2(t)$ is defined as the index set of alarm variables that are in the alarm state for the entire 10 min, i.e.,

$$I_2(t) \triangleq \{i\} \text{ such that } \left(\sum_{l=t-599}^{t} x_i(l) \right) = 600.$$

A third set $I_3(t)$ is the index set of alarm variables that are in the alarm state for recent 30 min, i.e.,

$$I_3(t) \triangleq \{i\} \text{ such that } \left(\sum_{l=t-1799}^{t} x_i(l) \right) = 1800.$$

Second, the index set of alarm variables newly appeared in the alarm state, denoted as $I(t)$, is determined as

$$I(t) \triangleq I_1(t) \cup \left(I(t-600) \cap I_2(t) \cap \overline{I_3(t)} \right), \tag{5.5}$$

where $\overline{I_3(t)}$ is the complement set of $I_3(t)$, and $I(t-600)$ is the counterpart of $I(t)$ in the previous 10 min. Finally, the new criterion is formulated,

Criterion C: An alarm flood is present in the 10-min data set $X(t-599:t)$, if $|I(t)| \geq 10$,

where $|I(t)|$ is the cardinality of $I(t)$.

The set $I(t)$ in (5.5) is composed of the indices of alarm variables from two groups. The first group includes the alarm variables in $I_1(t)$, running into the alarm state in the recent 10-min time period $[t-599, t]$. The designing rationale for the first group is transparent, because industrial plant operators need to analyze the alarm variables having alarm occurrences in the recent 10 min. The chattering alarms affect the set $I_1(t)$: If an alarm variable is solely experiencing chattering alarms, then the index of this alarm variable will be one element of $I_1(t)$, which may result in a false detection of alarm floods. The negative effect of chattering alarms on criterion C is much less than criterion A, because the number of alarm occurrences is the metric used in criterion A, while the concern of the set $I_1(t)$ is whether an alarm variable has non-zero alarm occurrences. Nevertheless, the effect of chattering alarms on criterion C will be partially removed by introducing alarm delay timers in Sect. 5.1.3. An important observation for designing the set $I_1(t)$ is that the alarm variables in the first group could be in the non-alarm state at the current time instant t. This observation is illustrated by a concrete example here. For the industrial alarm system in the previous example, the alarm variable with Tag #2090 is generated from a complex logic with 81 root causes, indicating some abnormal conditions in the coal system of the thermal power plant. The alarm state is set to be cleared automatically in 60 s after each alarm occurrence in order to raise an alarm when another root cause becomes valid afterwards. Hence, once the alarm variable has been in the alarm state recently, operators have to find the corresponding root cause even if the alarm variable is in the non-alarm state at the time instant t.

The designing principle for the second group of alarm variables in the set $\left(I(t-600) \cap I_2(t) \cap \overline{I_3(t)}\right)$ is to inherit some alarm variables from the previous index set $I(t-600)$. The newly-appeared alarm variables in the previous 10 min, as elements in $I(t-600)$, could stay in the alarm state in the recent 10 min, and operators may take more than 10 min to complete the analysis on them. The indices of such alarm variables are captured by the intersection of $I(t-600)$ and the second set $I_2(t)$ that contains the indices of alarm variables staying in the alarm state in the recent 10 min. However, if some alarm variables have been in the alarm state for more than 30 min, i.e., those in the third set $I_3(t)$, then they should have been well understood by operators, and deserve no further attentions. Thus, excluding the alarm variables in $I_3(t)$ is to avoid the negative effects of long-standing alarms that have been discussed in Sect. 5.1.1.2. The sets $I_2(t)$ and $I_3(t)$ respectively find the alarm variables staying in the alarm state in the recent 10 and 30 min, and obviously, $I_3(t)$ is a subset of $I_2(t)$. The time window of 30 min for $I_3(t)$ is up to the choice of users, and should be selected as the required time period for operators to complete the analysis of an occurring alarm. The change of the time window to other values is straightforward, e.g., if one hour is used, then $I_3(t)$ becomes

$$I_3(t) \triangleq \{i\} \text{ such that } \left(\sum_{l=t-3599}^{t} x_i(l) \right) = 3600.$$

If the time window is larger, then $I(t)$ would be more likely to be negatively affected by long-standing alarms and would result in more false detected alarm floods. For instance, if a 8-hour time window is used, then $I(t)$ will erroneously enclose the indices of alarm variables that ran into the alarm state a long time ago and have been in the alarm state for nearly 8 h.

An example is provided here to illustrate the calculation of the index set $I(t)$ and the rationale of the proposed criterion C. Figure 5.5 presents the time sequences of alarm variables in the index set $I(t)$ for one hour from 12 am to 13 pm on July 1st. The cardinality of $I(t)$ in the 6 non-overlapping consecutive 10-min time periods are 2, 0, 5, 5, 4 and 3, respectively. The corresponding tag numbers of alarm variables in each 10-min time period are also given in Fig. 5.5. The elements in the sets $I_1(t)$, $\left(I(t-600) \cap I_2(t) \cap \overline{I_3(t)}\right)$ and $I(t)$ for each 10-min time period are presented in Table 5.1. In the first 10-min time period, the sets $I(t-600)$ and $I_3(t)$ for $t = 600$ are initialized as empty sets. Thus, the index set $I(t = 600)$ contains two alarm variables experiencing alarm occurrences, namely, the alarm variables

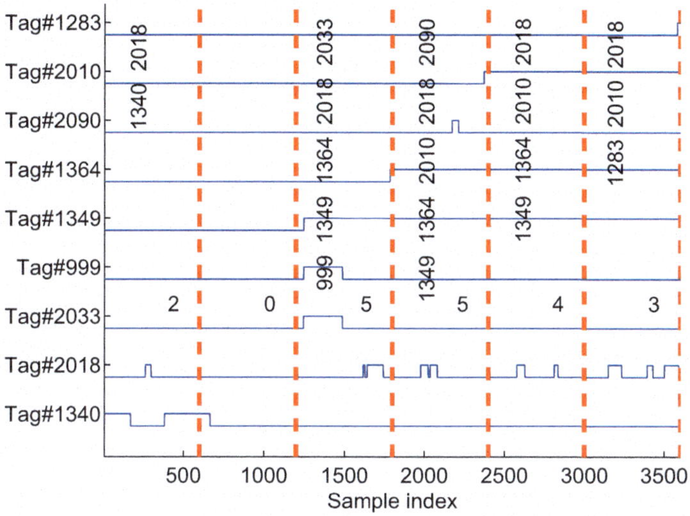

Fig. 5.5 Time sequences of alarm variables in the index set $I(t)$ from 12 am to 13 pm on July 1st, reprinted from Ref. [39], copyright 2018, with permission from IEEE

Table 5.1 Elements in the index sets, reprinted from Ref. [39], copyright 2018, with permission from IEEE

t	$I_1(t)$	$I(t-600) \cap I_2(t) \cap \overline{I_3(t)}$	$I(t)$
600	{#1340, #2018}	{}	{#1340, #2018}
1200	{}	{}	{}
1800	{#999, #1349, #1364, #2018, #2033}	{}	{#999, #1349, #1364, #2018, #2033}
2400	{#2010, #2018, #2090}	{#1349, #1364}	{#1349, #1364, #2010, #2018, #2090}
3000	{#2018}	{#1349, #1364, #2010}	{#1349, #1364, #2010, #2018}
3600	{#1283, #2018}	{#2010}	{#1283, #2010, #2018}

with Tags #1340 and #2018. In the second 10-min time period, both Tag #1340 and Tag #2018 return to the non-alarm state, and no other alarm variables are in the alarm state, so that $I(t=1200)$ is empty. In the third 10-min time period, $I(t=1800)$ contains five alarm variables experiencing alarm occurrences, namely, Tags #999, #1349, #1364, #2018 and #2033. In the fourth 10-min time period, three alarm variables with Tags #2010, #2018 and #2090 experience alarm occurrences, so that they are the elements in the set $I_1(t=2400)$, while Tags #1349 and #1364 are in the alarm state throughout the 10 min and are inherited from the last 10 min as the elements in the second group $\left(I(t-600) \cap I_2(t) \cap \overline{I_3(t)}\right)$ for $t = 2400$; thus, there are 5 elements in the index set $I(t = 2400)$. In the fifth 10-min time period, the set $I_1(t = 3000)$ has one element Tag #2018 experiencing alarm occurrences, and Tags #1349, #1364 and #2010 stay in the alarm state and are inherited from the set $I(t = 2400)$; thus, the set $I(t = 3000)$ is composed by these four alarm variables. In the last 10-min time period, two alarm variables with Tags #1349 and #1364 have been in the alarm state for 30 min, so that they do not deserve attentions from operators, and are not included in the index set $I(t = 3600)$.

5.1.3 Steps of the Proposed Method

This section proposes a new method based on criterion C to detect the presence of alarm floods.

Given the alarm data set $X(t - N + 1 : t)$ in (5.1), the proposed method consists of the following steps:

1. Remove chattering alarms with alarm durations and alarm intervals less than 20 s via alarm delay timers.
2. Implement the four sub-steps in the k-th iteration:
   a. Prepare the data set $X(kT - 599 : kT))$, where the updating window size $1 \leq T \leq 600$ is a user-defined positive integer;
   b. Obtain the index set $I(kT)$ as that in (5.5) based on the index set $I((k-1)T)$ in the previous $(k-1)$-th iteration, i.e.,

$$I(kT) \triangleq I_1(kT) \cup \left(I((k-1)T) \cap I_2(kT) \cap \overline{I_3(kT)}\right);$$

   c. Calculate the cardinality $|I(kT)|$ as the number of elements in $I(kT)$;
   d. Detect the presence of alarm floods by setting a detection variable equal to 1 only if criterion C is satisfied, i.e.,

$$x_{AF}(kT) = \begin{cases} 1, & \text{if } |I(kT)| \geq 10, \\ 0, & \text{otherwise.} \end{cases} \tag{5.6}$$

3. Verify the detection result by manually analyzing the time sequences of the alarm variables in $I(kT)$.

The proposed method takes the alarm data set $X(t - N + 1 : t)$ as the input, and yields the alarm-flood detection variable $x_{AF}(kT)$ and the index set $I(kT)$ as the outputs. By using the updating window size T, the proposed method unifies both the online and offline detections of alarm floods. If $T = 1$ s, then the index kT is the same as the sample index t, and the proposed method detects the alarm floods in an online manner.

Criterion C is insensitive to long-standing alarms, owing to the exclusion of alarm variables with alarm durations more than 30 min in the set I_3. However, the chattering alarms affect the set I_1, and it is impossible to remove all chattering alarms, as discussed in Sect. 5.1.1. In order to reduce the negative effects of chattering alarms, the m-sample alarm delay timer can be introduced in the step 2-(d) as

$$x_{AF}^{(m)}(kT) = \begin{cases} 1, \text{ if } x_{AF}((k-m+1)T : kT) = 1 \text{ and } x_{AF}^{(m)}((k-1)T) = 0, \\ 0, \text{ if } x_{AF}((k-m+1)T : kT) = 0 \text{ and } x_{AF}^{(m)}((k-1)T) = 1, \\ x_{AF}^{(m)}((k-1)T), \text{ otherwise.} \end{cases} \tag{5.7}$$

Thus, alarm floods are detected if m consecutive samples of $x_{AF}(kT)$ take the value of 1, or equivalently, Criterion C has been valid for a time period mT seconds.

Chattering alarms may cause $x_{AF}(kT)$ take the value of 1 in a short period, which can be tolerated by the alarm delay timer so that $x_{AF}^{(m)}(kT)$ is less affected by chattering alarms. As a sacrifice, the delay timer induces a time delay of m samples to detect alarm floods. This sacrifice is not an issue for offline detection of alarm floods. For online detection ($T = 1$ s), the determination of m is subject to the maximum allowable induced detection delay. As an example, for thermal power plants, one minute time delay is usually acceptable. Thus, a default value $m = 60$ is used here.

5.1.4 *Industrial Examples*

This section illustrates the effectiveness of the proposed method via three industrial examples. The first and second examples evaluate the proposed method for the absence and presence of alarm floods. The third example applies the proposed method to detect alarm floods in one year.

5.1.4.1 Example 1: Absence of Alarm Floods

The proposed method is applied to one-hour data set of 2226 alarm variables from 10 to 11 am on June 25. The updating window size T is set to 1 s, so that the index kT can be equivalently replaced by the time index t. The factor $m = 60$ is used for the alarm delay timer in (5.7). In order to have fair comparisons with Criteria A and B, the first step in the proposed method is not implemented, i.e., all criteria are applied to the original alarm data set without removing chattering alarms via alarm delay timers. Figure 5.6a presents the cardinality $|I(t)|$ together with the detection threshold 10 (note that the first 599 samples of $|I(t)|$ are initialized as zeros). The variable $x_{AF}(t)$ in Fig. 5.6b (solid line) takes the value of 1 for 2 time periods $t \in [1985, 2037]$ and $t \in [2228, 2244]$. The two time periods are less than $m = 60$ s, so that the alarm delay timer in (5.7) yields $x_{AF}^{(m)}(t)$ in Fig. 5.6b (dashed line) saying the absence of alarm floods.

This result is validated by visualizing the time sequences of alarm variables in the index set $I(t)$ in Fig. 5.7. The red dashed vertical lines therein denote the staring and ending positions of the 10-min time period $t \in [1368, 1985]$. There are 10 alarm variables in the index set $I(t = 1985)$, namely, Tags #833, #844, #848, #1149, #1353, #1359, #1406, #1407, #1537 and #1790. The green dot-dashed vertical lines stand for the staring and ending positions of the second time period $t \in [1629, 2228]$. The elements in $I(t = 2228)$ are Tags #833, #844, #848, #1147, #1149, #1353, #1406, #1407, #1537 and #1790. However, Tag #1537 is clearly associated with chattering alarms, because it quickly switches between the alarm and non-alarm states. Tags #1406 and #1407 are in the alarm state for most of time, with occasional small alarm intervals (defined in (5.4)), so that they are indeed long-standing alarms. Therefore, the total number of alarm variables having true alarms is less than 10, so that alarm floods are not present. These nuisance alarms cause $x_{AF}(t)$ in (5.6) equal to 1 for a short time period, but $x_{AF}^{(m)}(t)$ in (5.7) is not affected by the nuisance alarms.

As a comparison, the two basic criteria in Sect. 5.1.1 are exploited. Figure 5.8a, b respectively present the number of alarm occurrences and the number of alarm variables in the alarm state as the counterparts of $|I(t)|$ in Fig. 5.6a. Note that the

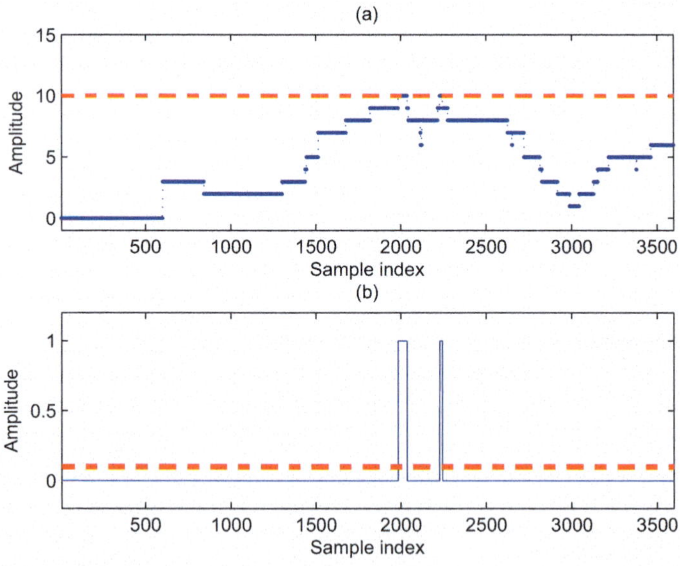

Fig. 5.6 Detection of alarm floods from 10 to 11 am on June 25: **a** the cardinality $|I(t)|$ (solid) and the detection threshold 10 (dash), **b** $x_{AF}(t)$ (solid) and $x_{AF}^{(m)}(t)$ (dash, move upwards for 0.1 for a better visualization), reprinted from Ref. [39], copyright 2018, with permission from IEEE

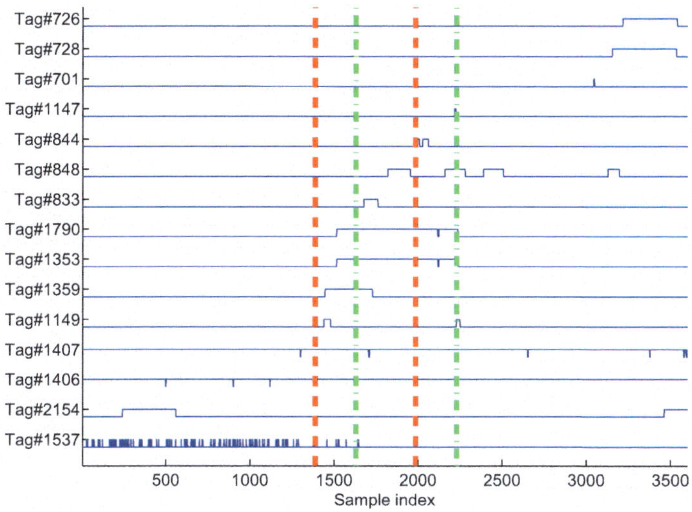

Fig. 5.7 Time sequences of alarm variables in $I\ (t)$ from 10 to 11 am on June 25, reprinted from Ref. [39], copyright 2018, with permission from IEEE

first 599 samples of the two metrics in Fig. 5.8a, b are initialized as zeros. The criterion A says that alarm floods are present. As revealed in Fig. 5.7, such a false detection is mainly due to the presence of chattering alarms in Tag #1537. The criterion B tells that alarm floods are present, too. The false detection is due to a fact that there are 10 alarm variables being in the alarm state in the entire one hour (note that the time sequences of the 10 long-standing alarms are not shown here). Even though chattering alarms and long-standing alarms are present, the proposed method does not lead to false detection of alarm floods, but the two basic criteria do.

5.1.4.2 Example 2: Presence of Alarm Floods

This example presents the detection of alarm floods from one-hour data set of 2226 alarm variables from 10 to 11 am on June 15, as shown in Fig. 5.9a. The configuration is the same as the first example, with the updating window size $T = 1$ s and the factor $m = 60$ s. The detection variable $x_{AF}^{(m)}\ (t)$ takes the value of 1 for one time period $t \in [2530, 3023]$. The detection of alarm floods is validated by the time sequences of all alarm variables in the index set $I\ (t = 2530)$ for all time instants in Fig. 5.10. The physical descriptions of these alarm variables are given in Table 5.2. The vertical dashed lines denote the starting and ending positions of the 10-min time period $t \in [1931, 2530]$. In this time period, 13 alarm variables are in the set $I\ (t = 2530)$, namely, Tags #369, #372, #1484, #1505, #1634, #1812, #1843, #2016, #2018, #2090, #2128, #2130 and #2131.

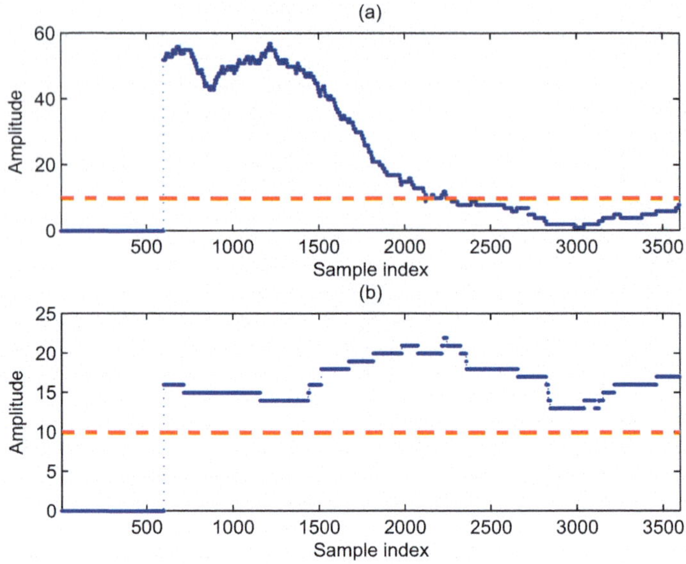

Fig. 5.8 Detection of alarm floods from 10 to 11 am on June 25: **a** the number of alarm occurrences (solid) and the threshold (dash), **b** the number of alarm variables in the alarm state (solid) and the threshold (dash), reprinted from Ref. [39], copyright 2018, with permission from IEEE

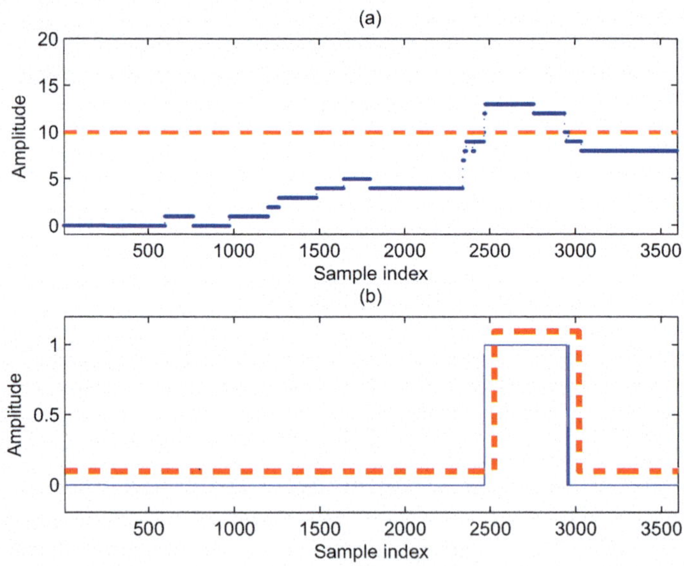

Fig. 5.9 Detection of alarm floods from 10 to 11 am on June 15: **a** the cardinality $|I(t)|$ (solid) and the detection threshold 10 (dash), **b** $x_{AF}(t)$ (solid) and $x_{AF}^{(m)}(t)$ (dash, move upwards for 0.1 for a better visualization), reprinted from Ref. [39], copyright 2018, with permission from IEEE

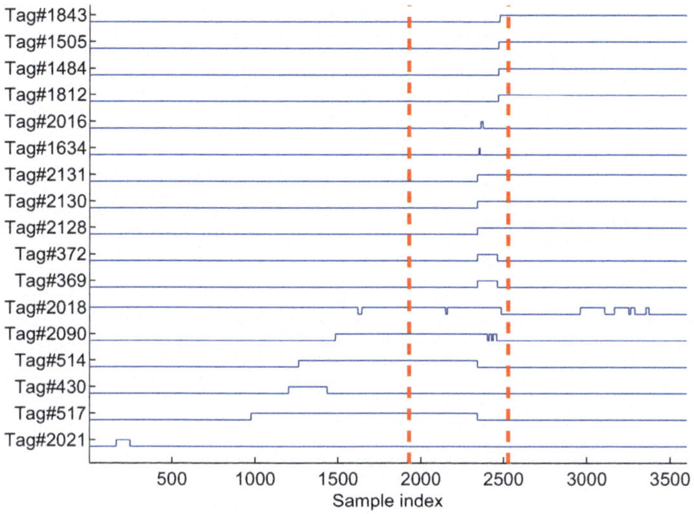

Fig. 5.10 Time sequences of alarm variables in $I\ (t)$ from 10 to 11 am on June 15, reprinted from Ref. [39], copyright 2018, with permission from IEEE

Table 5.2 Alarm variables during an alarm flood from 10 to 11 am on June 15, reprinted from Ref. [39], copyright 2018, with permission from IEEE

Tag #	Description
2021	Reactor A outlet ash NOx density high
517	Mill C drive-end bearing temperature #2 high
430	Mill A coal bunkering capacity low
514	Mill C drive-end bearing temperature #1 high
2090	Coal system in abnormal
2018	Reactor A inlet ash NOx density high
369	Mill C1 primary air fan in manual
372	Mill C2 primary air fan in manual
2128	Mill C emergency shutdown
2130	Mill C1 emergency shutdown
2131	Mill C2 emergency shutdown
1634	Furnace air pressure in abnormal
2016	Reactor B layer pressure difference high
1812	Automatic generation control in manual
1484	Boiler main control in manual
1505	Primary air fan control in manual
1843	O2 control in manual

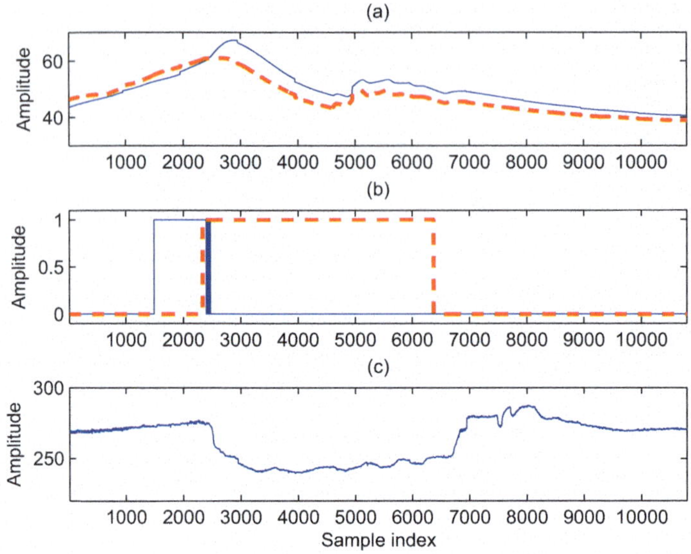

Fig. 5.11 Time sequences of related variables from 10 am to 13 pm on June 15: **a** the mill C drive-end bearing temperature #1 (solid) and mill C drive-end bearing temperature #2 (dash), **b** Tag #2090 (solid) and Tag #2128 (dash), and **c** actual generated power, reprinted from Ref. [39], copyright 2018, with permission from IEEE

The detected alarm flood was mainly caused by an abnormality in the mill C due to the high temperature of the mill drive-end bearing, as indicated by two alarm variables with Tags #517 and #514. The measurements of the two temperature sensors are shown in Fig. 5.11a. Tags #517 and #514 are in the alarm state if the temperatures are larger than 51, while another related alarm variable with Tag #2090 is in the alarm state if one temperature is larger than 55 and another is over 60. This abnormality was ignored by operators. After a while about 22 min, the mill C was forced into an emergency shutdown (Tag #2128) by a safety protection logic driven by an event that the two temperatures were larger than 60, as shown by the three alarm variables with Tags #2128, #2130 and #2131. As consequences, a series of alarm variables related to the mill C ran into the alarm state, namely, Tags #369, #372, #1484, #1505, #1634, #1812, #1843. The emergency shutdown of the mill C led to an abrupt reduction of actual generated power from 275 to 240 MW, as shown in Fig. 5.11c. The abnormal situation caused a significant economic loss to the thermal power plant.

5.1.4.3 Example 3: Long-term Application

The proposed method is applied to one-year data of 2226 alarm variables. The updating window size T is set to 600 s. The first step is implemented to remove chattering alarms in the collected alarm data set via m-sample alarm delay timers using $m = 20$.

The three criteria are applied to the same alarm data set preprocessed by the first step. Figure 5.12a presents the cardinality $|I\ (kT)|$ for criterion C. Figure 5.12b, c respectively give the number of alarm occurrences for criterion A and the number of alarm variables in the alarm state for criterion B as the counterparts of $|I\ (kT)|$. To avoid the domination of large values, all values in Fig. 5.12 are in the logarithm to the base 10, e.g., $\log_{10}(|I\ (kT)|)$ in Fig. 5.12a.

The metrics for Criteria A, B and C in Fig. 5.12 overpass the threshold 1 (after taking the logarithm operation) for 1790, 52047 and 396 times, respectively. The presence of long-standing alarms result in a complete failure of criterion B. It is impossible to remove all chattering alarms, even though the alarm data set has been preprocessed by m-sample alarm delay timers. As a result, the remaining chattering alarms severely affect the number of alarm occurrences as the metric of criterion A, and lead to many false detected alarm floods. By contrast, the set $I_1(kT)$ for criterion C only considers the presence of alarm occurrences and is less affected by chattering alarms. Thus, criterion C performs much better than criterion A. Note that the results from criterion C should be further refined by using $T = 1$ s for small-sized dat sets around the time period for each detected alarm flood in Fig. 5.12a, as discussed in Sect. 5.1.3.

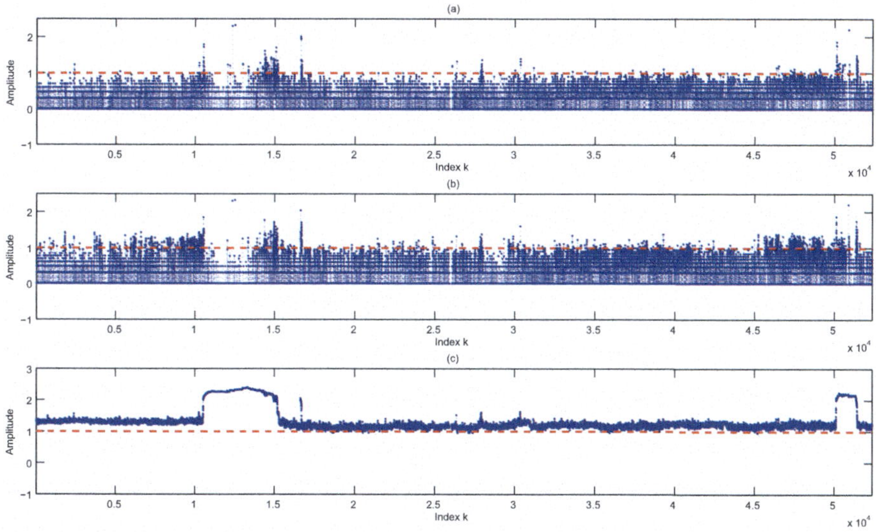

Fig. 5.12 The metrics of Criteria A, B and C with the detection threshold (dash) for the alarm data set in one year: **a** the cardinality $|I\ (kT)|$ (solid), **b** the number of alarm occurrences (solid), **c** the number of alarm variables in the alarm state. All values are presented in the logarithm to the base 10, reprinted from Ref. [39], copyright 2018, with permission from IEEE

5.2 Similarity Analysis of Alarm Floods

This section is on similarity analysis of industrial alarm flood sequences through sequence alignment, which compares alarm sequences by matching the common subsequences and determines the groups of similar alarm floods that likely have the same root causes. A new local alignment algorithm is proposed for efficient similarity analysis of industrial alarm floods based on the basic local alignment search tool (BLAST) [24]. To accelerate the computation, a set-based pre-matching mechanism is designed to avoid unnecessary computations by excluding all irrelevant alarm floods and alarm tags. In addition, to improve the accuracy while reducing the computation, a priority-based similarity scoring strategy is designed to make the sequence alignment more sensitive to alarms having higher alarm priorities.

5.2.1 Problem Description

An alarm flood sequence is composed of a series of chronologically sorted alarms and can be described as

$$X =< x_1, x_2, ..., x_M >, \tag{5.8}$$

where the symbol $< \cdot >$ indicates a sequence, the length M is the total number of occurred alarms in X, and the element x_i indicates the i-th alarm occurred in the chronological order. Each alarm occurrence event x_i is denoted by a tuple with three attributes, i.e.,

$$x_i = (e_i, t_i, p_i), \tag{5.9}$$

where e_i is the alarm tag of x_i; t_i and p_i are the corresponding time stamp and alarm priority, respectively. For ease of computation, the alarm tag e_i is in a numerical form so that all distinct alarm tags in words are mapped to numerical symbols. Thus, a numerical alphabet can be constructed as $\Sigma = \{1, 2, ..., V\}$, where V represents the size of the alphabet, equal to the total number of distinct alarm tags. Clearly, the functional relationship between Σ and all distinct alarm tags is bijective. Table 5.3 shows an industrial example of an alarm flood: The first column lists the alarm tags in the chronological order, the second column gives the numerical symbols of alarm tags, the third column indicates the time stamps, and the last column presents the alarm priorities.

The reoccurrences of a fault usually cause similar alarm floods consisting of the same series of alarms in a certain chronological order. Based on this observation, the similarity analysis can be conducted to detect similar alarm floods, so as to facilitate the early warning of alarm floods and the root cause analysis. That is, by comparing an incoming alarm sequence with potentially similar alarm floods in the historical database, it is possible to achieve an early warning of abnormalities, predict the cause

Table 5.3 An industrial example of an alarm flood sequence, reprinted from Ref. [24], copyright 2016, with permission from Elsevier

Alarm tag	Symbol (e_i)	Time stamp (t_i)	Priority (p_i)
T09.PVLO	9	10/5 02:20:01	Low
T05.PVHI	5	10/5 02:20:55	Low
T06.PVLO	6	10/5 02:22:02	Low
T02.PVHI	2	10/5 02:22:42	High
T01.LOLO	1	10/5 02:23:35	High
T03.PVHI	3	10/5 02:24:22	Low
T04.PVLO	4	10/5 02:25:11	Low
T03.PVHI	3	10/5 02:27:02	Low
T04.PVLO	4	10/5 02:27:03	Low
T07.PVLO	7	10/5 02:27:03	Emergency
T04.PVLO	4	10/5 02:29:05	Low
T05.LOLO	5	10/5 02:30:05	Low

of the incoming alarm flood, and take proactive operational actions to prevent the occurrence of an alarm flood and its negative consequences.

Sequence alignment is a way commonly used in bioinformatics to compare biological sequences to identify similar regions. There exist two categories of computational approaches for sequence alignment, namely, the global alignment and local alignment. The SW algorithm is a basic local sequence alignment method that was initially proposed by T. F. Smith and M. S. Waterman [34]. It identifies similar regions by comparing segments instead of the entire sequences. In the algorithm, the alignment of two symbolic sequences is represented by matches, mismatches, insertions, and deletions. Among them, both the insertions and deletions insert gaps in the sequences; such gaps are usually represented by dashes in illustrative examples. For instance, given two symbolic sequences [A,D,E,G] and [A,B,G], an alignment result of them is that [A,D,E,G] and [A,B,-,G], where there are two pairs of matches for A and G, the alignment between D and B is a mismatch, and E is aligned with a gap "-" such that the two sequences are of equal length.

Given two alarm food sequences $X =< x_1, x_2, ..., x_M >$ and $Y =< y_1, y_2, ..., y_N >$, a scoring matrix H of size $(M + 1) \times (N + 1)$ is calculated. The elements in the first row and column are initialized as 0's, i.e.,

$$\begin{cases} H_{i,1} = 0, i = 1, 2, \ldots, M + 1, \\ H_{1,j} = 0, j = 1, 2, \ldots, N + 1. \end{cases} \tag{5.10}$$

Then, the other entries of H are calculated from upper left to lower right. The entry on the i-th row and j-th column of H is obtained as

$$H_{i,j} = \max\{H_{i-1,j-1} + s(x_i, y_j), H_{i,j-1} + \delta, H_{i-1,j} + \delta, 0\}, \tag{5.11}$$

where $s(x_i, y_j)$ denotes the basic match score between two elements x_i and y_j; δ indicates the gap penalty; $i = 2, 3, ..., M + 1$ and $j = 2, 3, ..., N + 1$. The details about $s(x_i, y_j)$ and δ will be elaborated in the next subsection. After the full matrix H is obtained, the best local alignment can be found by a backtrack step, which starts from the entry with the highest score and goes backward until reaching an entry with 0.

However, the above local alignment algorithm has several limitations in alarm flood similarity analysis: First, it aligns all alarms even if two alarm floods share almost no common alarm tags, while such a computation should be avoided. Second, the computation complexity of the SW algorithm is too high, which may prevent it from online prediction of upcoming alarm floods. Last, as important attributes of alarm variables, alarm priorities have not been considered by the SW algorithm yet, while it is an intuitively reasonable choice to weight more on alarms with higher priorities in the similar alarm sequence alignment.

Motivated by addressing the above limitations, a new accelerated local alignment algorithm is presented here to find similar alarm flood sequences based on the BLAST formulated in [3, 4]. Comparing with the SW algorithm, this proposed algorithm is much faster in computation and provides a higher alignment accuracy for similar alarm sequences. These improvements are owing to the following three novelties:

- A set-based pre-matching mechanism is introduced to exclude the comparison between alarm floods with few common alarm tags, and to exclude irrelevant alarm tags in order to avoid their distractions on the subsequent alarm sequence alignment.
- The seeding and extending steps of the conventional BLAST are adapted for alarm floods, where only regions of high similarities are preserved, so that the searching space is reduced significantly.
- A priority-based similarity scoring strategy is developed so that the proposed algorithm is more sensitive to alarms having higher alarm priorities.

5.2.2 *Similarity Scores for Alarm Floods*

This section defines a priority-based similarity scoring strategy for comparing two alarm flood sequences. To measure the similarity in terms of either alarm sets or alarm sequences, it should firstly gives the basic scoring strategy for alarm pairs. In the sequence alignment, there are three basic scores, including the match score ϕ, mismatch score μ, and gap penalty δ.

Given two time-stamped alarm events $x_i = (e_i^x, t_i^x, p_i^x)$ and $y_j = (e_j^y, t_j^y, p_j^y)$ in two alarm floods X and Y, respectively, the similarity score $s(x_i, y_j)$ of two elements x_i and y_j is formulated as

$$s(x_i, y_j) = \begin{cases} \phi & \text{if } e_i^x = e_j^y, \\ \mu & \text{if } e_i^x \neq e_j^y, \end{cases} \tag{5.12}$$

where ϕ is the match score that takes a positive value and μ is the mismatch score that takes a negative value. For the gapped alignment [14, 34], x_i may be aligned with a gap "-" rather than a mismatched alarm in Y; in this case, the similarity score is equal to a constant gap penalty $\delta < 0$, i.e.,

$$s(x_i, -) = \delta. \tag{5.13}$$

In order to prefer a gapped alignment instead of the alignment of mismatched alarm tags, the mismatch score and the gap penalty should satisfy the following condition that $\mu < 2\delta < 0$.

The above similarity measurement is based on the matches of alarms and treats each alarm equally, whereas the alarms in an alarm flood are usually not identically important. Thus, reference [24] improved the similarity scoring strategy by incorporating the alarm priority information. Alarm priorities indicate the relative importance of alarm tags, based on the seriousness of consequences of ignoring alarms, and the allowable response time to handle abnormalities associated with alarms [5]. Three or four priorities are often adopted, possibly labeled by different names as shown in Table 5.4. As a good practice guide, the numbers of alarm tags assigned with three priorities from lower to higher are recommended to have the 80%/15%/5% distribution [5]. Higher priorities are configured to a smaller number of alarm tags, and they are expected to occur rarely; however, their occurrences indicate the presence of severe abnormalities. By contrast, lower priorities are assigned to most alarms associated with less severe abnormalities. Thus, it is reasonable that the similarity analysis of alarm floods is more sensitive to the occurrences of alarms with higher priorities.

Accordingly, a match score with respect to the alarm priority can be used, i.e., the match score ϕ in (5.12) is replaced by $\phi(p_l)$, where $\phi(\cdot)$ is a function of alarm priority. For a list of priorities p_1, p_2, ..., p_L in a descending order of importance, the match scores $\{\phi(p_1), \phi(p_2), \ldots, \phi(p_L)\}$ should satisfy the following inequalities,

$$\phi(p_l) > \phi(p_{l+1}) > 0, \text{ for } l = 1, 2, \ldots, L-1. \tag{5.14}$$

Table 5.4 Alarm priorities in alarm systems, reprinted from Ref. [24], copyright 2016, with permission from Elsevier

Alarm priority	List 1	List 2	List 3
Priority 1 (p_1)	Emergency	Emergency	Critical
Priority 2 (p_2)	High	High	Warning
Priority 3 (p_3)	Medium	Low	Advisory
Priority 4 (p_4)	Low		

Table 5.5 Basic similarity scores, reprinted from Ref. [24], copyright 2016, with permission from Elsevier

Item	Symbol	Score
Match for p_l	$\phi(p_l)$	$3 + 1.5(L - l)$
Mismatch	μ	-2.5
Gap penalty	δ	-1

As a smaller index l indicates a higher alarm priority, the match score $\phi(p_l)$ should be monotonically deceasing with respect to l, e.g.,

$$\phi(p_l) = \alpha + \beta(L - l), \tag{5.15}$$

where α and β are two positive constants.

Table 5.5 presents an example of basic similarity scores, including the linear match score function in (5.15) with parameters $\alpha = 3$ and $\beta = 1.5$, the mismatch score $\mu = -2.5$, and the gap penalty $\delta = -1$. The choices of the match score, mismatch score, and gap penalty are not unique. The distance between $\phi(p_l), l = 1, 2, ..., L$ is based on the importance of alarm priorities. For instance, given three priorities "Emergency", "High", and "Low", the scores based on Table 5.5 are 6, 4.5, and 3, which indicate that the importance of an "Emergency" alarm is equal to that of two "Low" alarms, and the importance of two "High" alarm is equal to that of three "Low" alarms. The distance between $\phi(p_l), l = 1, 2, ..., L$, can be changed based on the experts' experience on their assessment of relative importance of different alarm priorities in alarm management. Generally, if $\mu < 2\delta < 0$ and $\phi(p_l) > |\mu|, l = 1, 2, ..., L$, are satisfied, the choices of $\phi(p_l)$, μ, and σ have no significant effects on the final results.

5.2.3 *Sequence Alignment of Alarm Floods*

This section firstly revisits the main idea of the basic local alignment search tool (BLAST), and then presents the detailed steps of the new local alignment algorithm for alarm flood similarity analysis.

5.2.3.1 Main Steps of Accelerated Sequence Alignment

The sequence alignment is a typical dynamic programming process that has a high computational complexity. In view of the requirement for the computational time in practice, it is necessary to improve the efficiency in calculation. Reference [24] proposed an accelerated method for alarm floods based on the BLAST formulated in [3, 4]. The BLAST locates similar segments between a query sequence and object sequences [3, 4]. The BLAST is composed of two main steps, namely, the seeding and extending steps. In the seeding step, both the query and object sequences are broken into short words having a fixed size. A lookup table is built for all possible assemblies. By indexing short words in the lookup table, similar segments of the query and object sequences are quickly located. The pairs of words with similarity scores exceeding a threshold are kept as seeds while others are discarded. Next, in the extending step, all seeds are extended in two directions until the similarity score falls below a cutoff threshold. The alignments with similarity scores larger than a certain threshold are called the High Scoring segment Pairs (HSPs). The HSPs with the highest score are treated as the most similar segments between the query and object sequences.

With four unique alarms $\Sigma = \{1, 2, 3, 4\}$, a lookup table is built for all $k = 3$ short words so that the table has $4^3 = 64$ assemblies. Two alarm sequences <3, 2, 1, 4, 3, 2, 2> and <3, 4, 2, 1, 4, 2> are used for comparison. The first alarm sequence can be broken into <3, 2, 1>, <2, 1, 4>, <1, 4, 3>, <4, 3, 2>, and <3, 2, 2>. The second alarm sequence can also be broken into 4 short words in the same manner. By indexing the lookup table and determining which words from the two sequences have the same index, the matched pairs can be quickly found. As a result, <2, 1, 4> is a completely matched pair and is treated as a seed. In the extending step, the seed <2, 1, 4> is extended in two directions using a gapped extension scheme. By allowing one gap alignment, the best alignment for the two alarm sequences is achieved as

$$\begin{array}{ccccccc} 3 & - & 2 & 1 & 4 & 3 & 2 & 2 \\ | & & | & | & | & & | & \\ 3 & 4 & 2 & 1 & 4 & - & 2 & - \end{array}, \tag{5.16}$$

where '$-$' denotes a gap insertion, and '|' indicates a matched pair.

Comparing with the Smith-Waterman algorithm [34], the BLAST does not align all elements in sequences. Its main merit is to achieve fast sequence alignments by locating regions having high similarities and pruning most search space. Such a merit is important, especially for the pairwise alignment of a long sequence with the other sequences in a large query database.

To make the BLAST more suitable for alarm floods, some significant modifications are made as follows to achieve high computational efficiency and better alignment results. First, considering a large variety of alarm tags, a set based matching mechanism is designed as a pre-step to exclude irrelevant alarm floods and alarm tags. Second, the conventional BLAST requires a predefined look up table comprised of all possible short words of a fixed size. The construction and query of such a lookup table become infeasible if there are too many unique components. To solve this problem, the seeding stage is modified to search all matched pairs of different sizes. Third, specialized for alarm floods, a priority based scoring scheme is used to make the algorithm more sensitive to alarms of higher priorities. Finally, the conventional BLAST does not take the time information into account while the alarms are time stamped. Thus, in the extending stage, a time ambiguity tolerance strategy is proposed to make the algorithm less sensitive to orders of alarms occurring almost simultaneously. The details of these modifications are described in the following subsections.

5.2.3.2 Set-Based Pre-Matching Mechanism

A set-based pre-matching mechanism is proposed here, in order to exclude irrelevant alarm floods and alarm tags. First, to find common alarm tags between alarm flood sequences $X =< x_1, x_2, ..., x_M >$ and $Y =< y_1, y_2, ..., y_N >$, two binary-valued indexing vectors are formulated as

$$\begin{cases} A_X = [a_1^x \ a_2^x \ \cdots \ a_M^x], \\ A_Y = [a_1^y \ a_2^y \ \cdots \ a_N^y], \end{cases} \tag{5.17}$$

where

$$a_i^x = \begin{cases} 1 & \text{if } \exists j \in \{1, 2, ..., N\} \text{ s.t. } e_j^y = e_i^x, \\ 0 & \text{otherwise}, \end{cases} \tag{5.18}$$

$$a_i^y = \begin{cases} 1 & \text{if } \exists j \in \{1, 2, ..., M\} \text{ s.t. } e_j^x = e_i^y, \\ 0 & \text{otherwise}. \end{cases} \tag{5.19}$$

Here e_i^x and e_j^y are the alarm tags of x_i in X and y_j in Y, respectively. Based on the indexing vectors A_X and A_Y, a set-based similarity index is formulated as

$$S_{set}(X, Y) = \frac{\sum_{i=1}^{M}(a_i^x \times \phi(p_i^x)) \sum_{i=1}^{N}(a_i^y \times \phi(p_i^y))}{\sum_{i=1}^{M}(\phi(p_i^x)) \sum_{i=1}^{N}(\phi(p_i^y))}. \tag{5.20}$$

The set-based similarity index $0 \leq S_{set}(X, Y) \leq 1$ is used to measure whether two alarm flood sequences are alike without considering the chronological orders. In other words, $S_{set}(X, Y) = 1$ indicates alarm floods X and Y have the same alarm set, while $S_{set}(X, Y) = 0$ means that X and Y share no common alarm. If more alarm tags are shared by two alarm floods, the set-based similarity index is closer to 1.

The calculation of $S_{set}(X, Y)$ is based on alarm sets rather than alarm sequences, and is a quick preliminary step to select suitable alarm floods for subsequent similarity analysis. That is, the set-based similarity index between X and Y is said to be too low, if

$$S_{set}(X, Y) \leq \gamma, \tag{5.21}$$

where γ is a user-selected threshold. In this case, there is no need to proceed to the next step of sequence alignment for alarm floods. The simplest choice is $\gamma = 0$, indicating that the computation will not proceed if two alarm floods share no common alarms. The value of γ can be set to be larger than 0 if there are a large number of alarm floods, so as to avoid time consumptions in comparing alarm floods with few common alarm tags.

Second, even if two alarm floods X and Y have a set-based similarity index $S_{set}(X, Y)$ larger than γ, they may still contain many uncommon alarm tags. It is meaningless to align these uncommon alarm tags, and is even distractive to have them in the subsequent sequence alignment. Hence, all irrelevant alarms are removed from the original alarm flood sequences X and Y, and only the common alarm tags are preserved. Based on the indexing vectors A_X and A_Y in (5.17), the alarms x_i with $a_i^x = 1$ and y_j with $a_j^y = 1$ are extracted from X and Y to construct two shorter sequences $\tilde{X} =< \tilde{x}_1, \tilde{x}_2, ..., \tilde{x}_{\tilde{M}} >$ and $\tilde{Y} =< \tilde{y}_1, \tilde{y}_2, ..., \tilde{y}_{\tilde{N}} >$. Here $\tilde{M} = \sum_{i=1}^{M} a_i^x$ and $\tilde{N} = \sum_{i=1}^{N} a_i^y$ are the numbers of common alarm tags in X and Y, respectively. Clearly, $S_{set}(\tilde{X}, \tilde{Y}) = 1$, i.e., all alarm tags in $\tilde{X}$ can also be found in $\tilde{Y}$, and vice versa.

To illustrate the above calculation, a numerical example is presented. In Table 5.6, two alarm sequences X and Y of different lengths ($M = 12$ and $N = 14$) are given. By comparing the alarm symbols, two indexing vectors A_X and A_Y are calculated as

$$\begin{cases} A_X = [1\ 1\ 1\ 1\ 1\ 1\ 1\ 1\ 1\ 1\ 1\ 1], \\ A_Y = [1\ 1\ 1\ 1\ 1\ 1\ 1\ 1\ 1\ 0\ 1\ \ 0\ \ 1\ 0]. \end{cases} \tag{5.22}$$

The indexing vector A_X is an all-one vector, indicating that all alarm tags in X can be found in Y. The 10th, 12th, and 14th elements in A_Y are zeros, owing to the alarm tag with symbol 8 that is not present in X. There are three ($L = 3$) alarm priorities in Table 5.6, namely, "Emergency" (p_1), "High" (p_2), and "Low" (p_3). Based on

Table 5.6 Alarm floods X and Y with time and priority information, reprinted from Ref. [24], copyright 2016, with permission from Elsevier

Sequence X			Sequence Y		
Symbol	Time stamp	Priority	Symbol	Time stamp	Priority
9	00:00:01	Low	9	11:00:01	Low
5	00:00:55	Low	5	11:01:11	Low
6	00:02:02	Low	2	11:01:59	High
2	00:02:42	High	1	11:02:18	High
1	00:03:35	High	3	11:03:11	Low
3	00:04:22	Low	4	11:04:01	Low
4	00:05:11	Low	7	11:06:32	Emergency
3	00:07:02	Low	3	11:06:33	Low
4	00:07:03	Low	4	11:06:33	Low
7	00:07:03	Emergency	8	11:07:16	Low
4	00:09:05	Low	9	11:08:45	Low
5	00:10:05	Low	8	11:09:22	Low
			6	11:09:56	Low
			8	11:09:57	Low

the priority-based similarity score in Table 5.5, the scores assigned to alarms of the three priorities are $\phi(p_1) = 3 + 1.5 \times (3 - 1) = 6$, $\phi(p_2) = 3 + 1.5 \times (3 - 2) = 4.5$, and $\phi(p_1) = 3 + 1.5 \times (3 - 3) = 3$, respectively. Using (5.20), the set-based similarity index is calculated as $S_{set}(X, Y) = \frac{42\times39}{42\times48} = 0.8125$. Thus, the set-based pre-matching mechanism says that X and Y are very similar alarm sequences. Next, two subsequences $\tilde{X}$ and $\tilde{Y}$ are obtained as $\tilde{X} =< 9, 5, 6, 2, 1, 3, 4, 3, 4, 7, 4, 5 >$ and $\tilde{Y} =< 9, 5, 2, 1, 3, 4, 7, 3, 4, 9, 6 >$ by removing the alarm tag 8 from Y. The two subsequences will be used in the similar alarm sequence alignment in the following subsections.

5.2.3.3 The Seeding Step

The seeding step in the conventional BLAST establishes a lookup table for all possible assemblies of words having a fixed size. Such a lookup table is feasible for biological sequences and text strings, because the numbers of basic components are limited as shown in Table 5.7. However, for industrial alarm systems, the number of distinct alarm tags is quite large, usually larger than 1000 [29]. Thus, it is infeasible to build a lookup table for alarm sequences. For example, if a 3-word table is built for 1000 alarm tags, then the table size would be $1000^3 = 10^9$ that is too large for computation. Hence, the seeding step of the conventional BLAST is not directly applicable to alarm sequences.

To solve this problem, the seeding step needs to be modified. The main idea is to find all matched pairs rather than short words of a fixed size. A matched pair $Z(i), i = 1, 2, \ldots, K_b$, is indexed by its positions in two alarm sequences $\tilde{X}$ and $\tilde{Y}$ as

$$\mathbf{z}_i = (z_{i,1}, z_{i,2}, z_{i,3}), \tag{5.23}$$

where $z_{i,1}$ and $z_{i,2}$ are the start positions of the matched pair $Z(i)$ in $\tilde{X}$ and $\tilde{Y}$, respectively; $z_{i,3}$ represents the length of $Z(i)$; K_b indicates the total number of matched pairs. The matched pair $Z(i)$ satisfies two properties as follows. First, all alarm tags of $Z(i)$ in two alarm sequences $\tilde{X}$ and $\tilde{Y}$ should be identical, i.e. $e^x_{z_{i,1}+k} = e^y_{z_{i,2}+k}$ for $k = 0, 1, \ldots, z_{i,3} - 1$, where $e^x_{z_{i,1}+k}$ and $e^y_{z_{i,2}+k}$ are the alarm tags of the $(z_{i,1} + k)$th and $(z_{i,2} + k)$th alarms in $\tilde{X}$ and $\tilde{Y}$, respectively. Second, any matched pair should not be overlapping with other matched pairs. Given one matched pair $Z(i)$, two

Table 5.7 Basic components of different sequences, reprinted from Ref. [24], copyright 2016, with permission from Elsevier

	Genome	Protein	Text string	Alarm flood
Component	Nucleotide	Amino acid	Alphabet	Alarm tag
Number of types	5	23	26	>1000

conditions $z_{i,1} \in [z_{j,1}, z_{j,1} + z_{j,3} - 1]$ and $z_{i,2} \in [z_{j,2}, z_{j,2} + z_{j,3} - 1]$ cannot be true simultaneously for any $Z(j)$, $j \neq i$.

To select seeds from the matched pairs, the seeding score, namely, the similarity score of each matched pair $Z(i)$ is calculated as

$$h(Z(i)) = \sum_{k=0}^{z_{i,3}-1} s(\tilde{x}_{z_{i,1}+k}, \tilde{y}_{z_{i,2}+k}), \tag{5.24}$$

where $s(\tilde{x}_{z_{i,1}+k}, \tilde{y}_{z_{i,2}+k})$ is obtained from (5.12). The matched pairs having large seeding scores are preserved as seeds, while other pairs are discarded. The seeds and their seeding scores are denoted as $\tilde{Z}(k)$ and $h(\tilde{Z}(k))$ for $k = 1, ..., K$, respectively, where K is the number of seeds selected as the matched pairs of top K seeding scores. An example is presented here to illustrate the modified seeding step.

Based on the two shorter sequences $\tilde{X} =< 9, 5, 6, 2, 1, 3, 4, 3, 4, 7, 4, 5 >$ and $\tilde{Y} =< 9, 5, 2, 1, 3, 4, 7, 3, 4, 9, 6 >$ obtained in Example 2, the matched pairs shown in Table 5.8 are found in the modified seeding step. Taking $\mathbf{z}_4 = (4, 3, 4)$ as an example, the matched pair is $Z(4) =< 2, 1, 3, 4 >$ with 4 alarms, and it starts at the 4th position of $\tilde{X}$ and the 3rd position of $\tilde{Y}$. Using the priority-based similarity score in Table 5.5, the similarity scores of all matched pairs are calculated as shown in the fifth column of Table 5.8. The seeding score of $Z(4)$ is $h(Z(4)) = 4.5 + 4.5 + 3 + 3 = 15$. By choosing $K = 1$, the matched pairs with the largest seeding scores is preserved as the seed, namely, $\tilde{Z}(1) =< 2, 1, 3, 4 >$ with $h(\tilde{Z}(1)) = 15$.

Table 5.8 The matched alarm pairs, reprinted from Ref. [24], copyright 2016, with permission from Elsevier

Matched pair	$z_{i,1}$	$z_{i,2}$	$z_{i,3}$	$h(Z(i))$	Sequences
$Z(1)$	1	1	2	6	$< 9, 5 >$
$Z(2)$	1	10	1	3	$< 9 >$
$Z(3)$	3	11	1	3	$< 6 >$
$Z(4)$	4	3	4	15	$< 2, 1, 3, 4 >$
$Z(5)$	6	8	2	6	$< 3, 4 >$
$Z(6)$	8	5	3	12	$< 3, 4, 7 >$
$Z(7)$	8	8	2	6	$< 3, 4 >$
$Z(8)$	11	6	1	3	$< 4 >$
$Z(9)$	11	9	1	3	$< 4 >$
$Z(10)$	12	2	1	3	$< 5 >$

5.2.3.4 The Extending and Backtracking Steps

Once seeds are obtained, the conventional BLAST extends the seeds in two directions to find the HSPs, as stated in Sect. 5.2.3.1. In this subsection, the extending step is adapted to alarm sequences based on the priority-based similarity score in Sect. 5.2.2. There are two types of extensions, namely, the ungapped extension and the gapped one [3, 4, 26]. In order to have more flexibility in aligning alarm sequences, the gapped extension, which aligns irrelevant components with gaps, is used in this context.

Denote the subsequences of $\tilde{X}$ and $\tilde{Y}$ based on the k-th seed $\tilde{Z}(k)$ for extension in one direction as X_s of length m and Y_s of length n. An extending score matrix H is calculated in three ways as shown in Fig. 5.13. Initially, the first element of H is defined as

$$H_{1,1} = h(\tilde{Z}(k)) + U, \tag{5.25}$$

where $h(\tilde{Z}(k))$ is the seeding score of the seed $\tilde{Z}(k)$ and U is the cutoff threshold used to stop the extension. In principle, the extension grows until the extending score $H_{i,j}$ falls more than a threshold below the maximum extending score, namely, $H_{i,j} < H_{max} - U$, where $H_{\max}$ denotes the highest extending score achieved before the extension proceeds to $H_{i,j}$. Initially, $H_{\max} = H_{1,1}$. Thus, the incorporation of U in $H_{1,1}$ avoids the degenerated case that $H_{1,1} < U$.

For all $i = 2, 3, ..., m + 1$ and $j = 2, 3, ..., n + 1$, the element $H_{i,j}$ on the i-th row and j-th column of the matrix H is obtained as the first way in Fig. 5.13 and is calculated as

$$H_{i,j} = \max\{H_{i-1,j-1} + s(x_i, y_j), H_{i,j-1} + \delta, H_{i-1,j} + \delta, 0\}, \tag{5.26}$$

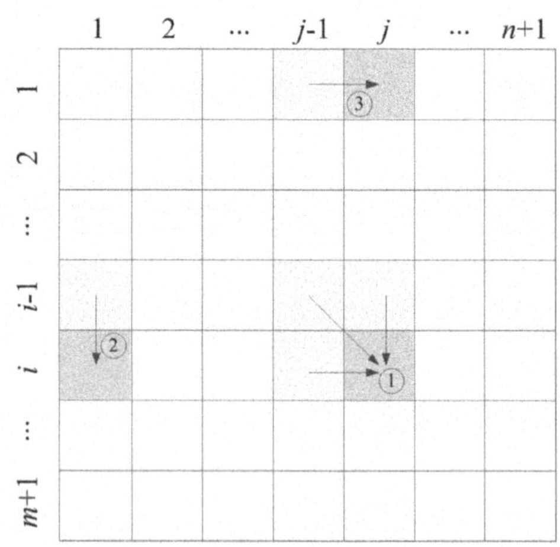

Fig. 5.13 The calculation of the extending score matrix H in the gapped extension. The circled numbers 1, 2, and 3 indicate the first, second, and third ways to calculate the element $H_{i,j}$ associated with (5.26), (5.27), and (5.28), reprinted from Ref. [24], copyright 2016, with permission from Elsevier

where $s(x_i, y_j)$ is the basic similarity score between x_i and y_j with $x_i \in X_s$ and $y_j \in Y_s$. For all $i = 2, 3, ..., m + 1$ and $j = 1$, $H_{i,1}$ is calculated as the second way in Fig. 5.13, i.e.,

$$H_{i,1} = \max\{H_{i-1,1} + \delta, 0\}. \tag{5.27}$$

For all $i = 1$ and $j = 2, 3, ..., n + 1$, $H_{1,j}$ is calculated as the third way in Fig. 5.13, i.e.,

$$H_{1,j} = \max\{H_{1,j-1} + \delta, 0\}. \tag{5.28}$$

Algorithm 5.4 Gapped extension

Input Argument : Subsequences X_s and Y_s; seed score $H_s(\tilde{Z}(k))$;
Output Argument: Score matrix H; Parameter: Cutoff threshold U;
$H = \mathbf{0}_{m+1,n+1}$; $Q = \mathbf{0}_{m+1,n+1}$; $H_{1,1} = h(\tilde{Z}(k)) + U$; $H_{\max} = H_{1,1}$;
for $i = 2$ to $m + 1$ **do**
  if $H_{i-1,1} \geq H_{\max} - U$ **then**
    $H_{i,1} = \max\{H_{i-1,1} + \delta, 0\}$; $Q_{i,1} = 3$;
  end if
end for
for $j = 2$ to $n + 1$ **do**
  if $H_{1,j-1} \geq H_{\max} - U$ **then**
    $H_{1,j} = \max\{H_{1,j-1} + \delta, 0\}$; $Q_{1,j} = 2$;
  end if
  for $i = 2$ to $m + 1$ **do**
    if $\max\{H_{i-1,j-1}, H_{i,j-1}, H_{i-1,j}\} \geq H_{\max} - U$ **then**
      $H_{i,j} = \max\{H_{i-1,j-1} + s(x_i, y_j), H_{i,j-1} + \delta, H_{i-1,j} + \delta, 0\}$;
      if $H_{i,j} = H_{i-1,j-1} + s(x_i, y_j)$ **then** $Q_{i,j} = 1$;
      else if $H_{i,j} = H_{i,j-1} + \delta$ **then** $Q_{i,j} = 2$;
      else if $H_{i,j} = H_{i-1,j} + \delta$ **then** $Q_{i,j} = 3$;
      end if
      if $H_{i,j} > H_{\max}$ **then** $H_{\max} = H_{i,j}$;
      end if
    end if
  end for
  if $\max_{1 \leq i \leq m+1} H_{i,j} < H_{\max} - U$ **then** break;
  end if
end for

According to Fig. 5.13, there are three stop conditions respectively corresponding to the calculations of $H_{i,j}$ in (5.26), (5.27), and (5.28). First, for all $i = 2, 3, ..., m + 1$ and $j = 2, 3, ..., n + 1$, the extension to $H_{i,j}$ stops at $H_{i,j}$ if

$$\max\{H_{i-1,j-1}, H_{i,j-1}, H_{i-1,j}\} < H_{\max} - U, \tag{5.29}$$

Second, for all $i = 2, 3, ..., m + 1$ and $j = 1$, the extension to $H_{i,1}$ stops at $H_{i,1}$ if

$$H_{i-1,1} < H_{\max} - U. \tag{5.30}$$

Third, for all $i = 1$ and $j = 2, 3, ..., n + 1$, the extension to $H_{1,j}$ stops at $H_{1,j}$ if

$$H_{1,j-1} < H_{\max} - U. \tag{5.31}$$

The gapped extension in one direction can be described via the pseudo codes in Algorithm 5.4. The inputs of Algorithm 5.4 are the subsequences X_s, Y_s and the seeding score $h(\tilde{Z}(k))$, while the output is the extending score matrix H. Once the extension is stopped, the best alignment can be obtained by a backtracking step, which is based on the indicator matrix Q obtained in Algorithm 5.4. It starts at the element having the maximum score. If $Q_{i,j}$ is equal to 1, 2 or 3, the backtracking proceeds to $Q_{i-1,j-1}$, $Q_{i,j-1}$, or $Q_{i-1,j}$, respectively.

Denote the maximum extending scores in the backward and forward extensions as H_b and H_f, respectively. The final similarity score of the best alignment based on the seed $\tilde{Z}(k)$ is calculated as

$$S(\tilde{Z}(k)) = H_b + H_f - h(\tilde{Z}(k)) - 2U. \tag{5.32}$$

When several seeds are found, the best alignments for these seeds may be different. The subsequences with the highest score is treated as the best alignment. The corresponding score is regarded as the final best score between $\tilde{X}$ and $\tilde{Y}$, denoted by $S(\tilde{X}, \tilde{Y})$.

To illustrate the gapped extension, the seed $\tilde{Z}(1) =< 2, 1, 3, 4 >$ with $h(\tilde{Z}(1)) = 15$ is investigated in details here. The basic similarity score in Table 5.5 is used. The extension threshold is set to $U = 2|\delta| = 2$, i.e., two gaps in the alignment are allowed. Thus, $H_{1,1} = h(\tilde{Z}(1)) + U = 15 + 2 = 17$. The seed $\tilde{Z}(1) =< 2, 1, 3, 4 >$ is firstly extended to the forward direction subsequences $X_f =< 3, 4, 7, 4, 5 >$ and $Y_f =< 7, 3, 4, 9, 6 >$. The extending score matrix H for X_f and Y_f is generated as Table 5.9. The backtracking in the forward extension yields the best alignment, as shown by the alarm tags with scores marked by underlines in Table 5.9. The seed $\tilde{Z}(1) =< 2, 1, 3, 4 >$ is secondly extended to the backward direction subsequences $X_b =< 6, 5, 9 >$ and $Y_b =< 5, 9 >$. The extending score matrix H for X_b and Y_b is obtained as Table 5.10. Figure 5.14 presents the extensions in two directions based on the seed $\tilde{Z}(1)$. The green squares indicate the matched alarm pairs and the grey squares represent the entries where the extension proceeds. It is obvious that most search space marked by the white squares is pruned in the extensions, so that the computation cost is greatly reduced.

Combining the extension in the two directions, the best alignment is finally achieved as

$$\begin{array}{l} \tilde{X}: 9\ 5\ 6\ 2\ 1\ 3\ 4\ 3\ 4\ 7\ -\ 4 \\ \quad\ \ |\ |\quad\ |\ |\ |\ |\quad\quad\ |\quad | \\ \tilde{Y}: 9\ 5\ -\ 2\ 1\ 3\ 4\ -\ -\ 7\ 3\ 4 \end{array}. \tag{5.33}$$

Table 5.9 Extending score matrix H in the forward extension based on the seed $\tilde{Z}(1) =< 2, 1, 3, 4 >$, reprinted from Ref. [24], copyright 2016, with permission from Elsevier

		7	3	4	9	6
	17	16	0	0	0	0
3	16	15	0	0	0	0
4	15	14	0	0	0	0
7	14	21	20	19	0	0
4	0	20	19	23	22	21
5	0	19	18	22	21	20

Table 5.10 Extending score matrix H in the backward extension based on the seed $\tilde{Z}(1) =< 2, 1, 3, 4 >$, reprinted from Ref. [24], copyright 2016, with permission from Elsevier

		5	9
	17	16	0
6	16	15	0
5	15	19	18
9	14	18	22

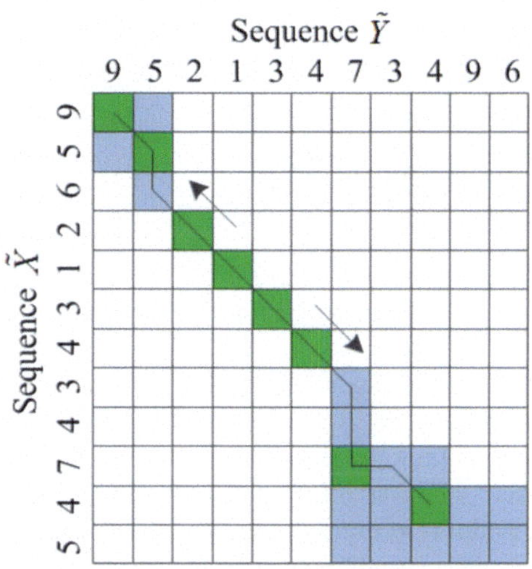

Fig. 5.14 Graphic illustration of the gapped extension, reprinted from Ref. [24], copyright 2016, with permission from Elsevier

In particular, the alarm tag 7 having a higher alarm priority rather than the alarm tag 3 is found as the matched alarm tag. The priority-based basic similarity score in Sect. 5.2.2 makes the Algorithm 5.4 more sensitive to alarms with higher priorities. Based on $\tilde{Z}(1)$, the maximum scores of the backward and forward extensions are $H_b = 22$ and $H_f = 23$, respectively. Then, the final score is $S(\tilde{Z}(1)) = 22 + 23 - 15 - 2 \times 2 = 26$.

5.2.3.5 Time Ambiguity Tolerance Strategy

Physically connected alarm tags often arise almost simultaneously, but they may have different chronological orders in different alarm floods. Thus, a strategy of incorporating time information of alarms was proposed to tolerate certain time ambiguity of these alarms occurring almost simultaneously [14]. This strategy is adopted here in the proposed algorithm as follows. To redefine the basic similarity score $s(x_i, y_j)$ in (5.26), two vectors are defined, namely, the time distance vector $\mathbf{d}_i$ and the time weight vector $\mathbf{w}_i$. The distance vector $\mathbf{d}_i$ for the ith alarm x_i in $\tilde{X}$ is formulated as

$$\mathbf{d}_i = [d_{i1}\ d_{i2}\ \cdots\ d_{iV}]^T, \tag{5.34}$$

where

$$d_{iv} = \min_{1\le k\le m} \left\{ |t_i - t_k| : e_k^x = v \right\}. \tag{5.35}$$

Here e_k^x is the alarm tag of the k-th alarm x_k in $\tilde{X}$, and v is the numerical symbol in the alarm tag alphabet Σ of size V. Each entry d_{iv} represents the time gap between the ith alarm and the nearest alarm with alarm tag v. The time weigh vector $\mathbf{w}_i$ for x_i in $\tilde{X}$ is

$$\mathbf{w}_i = [w_{i1}\ w_{i2}\ \cdots\ w_{iV}]^T, \tag{5.36}$$

where $w_{iv} = f(d_{iv})$. The weighting function $f(\cdot)$ for $\tilde{X}$ can be chosen as the scaled Gaussian function [14],

$$f(d_{iv}) = e^{-\frac{d_{iv}^2}{2\sigma^2}}, \tag{5.37}$$

while the counterpart for $\tilde{Y}$ is determined as

$$f(d_{iv}) = \begin{cases} 1 \text{ if } d_{iv} = 0, \\ 0 \text{ if } d_{iv} \neq 0. \end{cases} \tag{5.38}$$

Here the standard deviation σ in (5.37) is a user selected parameter. The value of σ decides how much the time difference can be tolerated. In general, the time difference between simultaneously occurred alarms is in the level of a few seconds. If a large value of σ is used, then the orders of two alarms with a large time difference are regarded as being not important. Eventually, the basic similarity score $s(x_i, y_j)$ in (5.12) is reformulated as

$$s(x_i, y_j) = \mu + (\phi(p_j^y) - \mu) \max_{1\le v\le V} w_{iv}^x w_{jv}^y, \tag{5.39}$$

where w_{iv}^x and w_{jv}^y are the time weights for the ith alarm x_i in $\tilde{X}$ and the jth alarm y_j in $\tilde{Y}$, respectively. The gap penalty is preserved the same as (5.13).

From Table 5.6, it is obvious that the alarm tags at the 8th, 9th and 10th positions in X have very close time stamps. The same case can be found at the 7th, 8th, and 9th

Table 5.11 Extending score matrix H of $\tilde{Z}(1)$ in the forward extension with time stamp information, reprinted from Ref. [24], copyright 2016, with permission from Elsevier

		7	3	4	9	6
	$\underline{17}$	16	0	0	0	0
3	16	$\underline{19.3537}$	19	0	0	0
4	15	18.3537	$\underline{22.3537}$	22	0	0
7	14	21	23.3549	$\underline{27.3549}$	26.3549	25.3549
4	0	20	24	26.3549	25.3549	24.3549
5	0	19	23	25.3549	24.3549	23.3549

positions in Y; the chronological orders for these alarms should be not important in the sequence alignment. Thus, the time ambiguity tolerance strategy is adopted. The standard deviation of the Gaussian function is set as $\sigma = 2$. By using the reformulated similarity score $s(x_i, y_j)$ in (5.39), the score matrix H for the gapped extension of $\tilde{Z}(1) =< 2, 1, 3, 4 >$ in the forward direction sequences X_f and Y_f is obtained in Table 5.11. In the backward extension, the best alignment is the same as that in Table 5.10. Combining the extensions in two directions, the final best alignment is obtained as

$$\begin{array}{l} \tilde{X}: 9\ 5\ 6\ 2\ 1\ 3\ 4\ \underline{3}\ \underline{4}\ \underline{7} \\ \quad\ \ |\ |\ \ \ \ |\ |\ |\ |\ |\ |\ | \\ \tilde{Y}: 9\ 5\ -\ 2\ 1\ 3\ 4\ \underline{7}\ \underline{3}\ \underline{4} \end{array}. \tag{5.40}$$

Here the alarm tags marked with underlines are not exactly matched, but are treated as matched pairs owing to their close time stamps. For the gapped extension based on $\tilde{Z}(1)$, the maximum scores for the backward and forward extensions are $H_b = 22$ and $H_f = 27.3549$, respectively, so that the best score is $S(\tilde{Z}(1)) = 22 + 27.3549 - 15 - 2 \times 2 = 30.3549$.

5.2.3.6 Main Procedures

The proposed local alignment algorithm is composed of the following steps, which are also depicted as the flowchart in Fig. 5.15.

1. Two alarm floods X and Y are prepared with chattering alarms removed.
2. The set-based pre-matching mechanism is exploited for X and Y, where $S_{set}(X, Y)$ in (5.20) is calculated. If $S_{set}(X, Y) > \gamma$, then the subsequences $\tilde{X}$ and $\tilde{Y}$ are obtained from X and Y by preserving common alarm tags in both X and Y; otherwise, the computation is terminated.
3. In the seeding step, the matched segments between $\tilde{X}$ and $\tilde{Y}$ are indexed by $Z = (z_1, z_2, z_3)$. Their seeding scores $h(Z)$ in (5.24) are calculated. The segments having larger seeding scores are preserved as seeds $\tilde{Z}(k), k = 1, ...K$, while others are discarded. The seeding score of each seed is denoted as $h(\tilde{Z}(k))$.

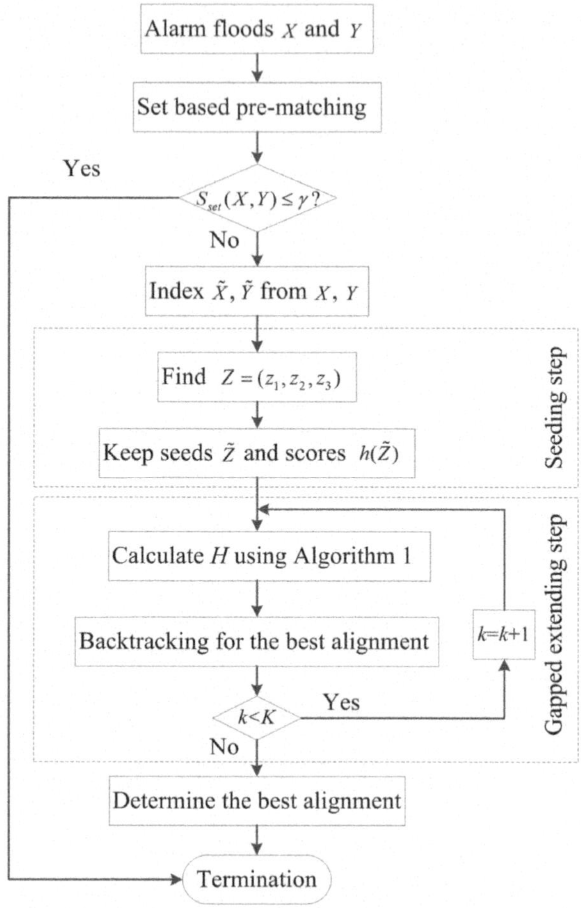

Fig. 5.15 Flowchart of the proposed local alignment algorithm, reprinted from Ref. [24], copyright 2016, with permission from Elsevier

4. The gapped extension is implemented to each seed $\tilde{Z}(k)$, $k = 1, \dots K$ using Algorithm 5.4. In each iteration, $s(x_i, y_j)$ is computed based on (5.34)–(5.39), $H_{i,j}$ is calculated based on (5.26)–(5.28), and $Q_{i,j}$ is recorded. The extension is terminated if the stop conditions (5.29)–(5.31) are satisfied. The extending score for the best alignment of $\tilde{Z}(k)$ is calculated using (5.32).
5. Find the aligned subsequence having the highest score as the final best sequence alignment of X and Y. To measure the computational accuracy and efficiency for comparing X and Y, two quantities are defined as: the number of matched pairs between X and Y, denoted as $C(X, Y)$, is counted to measure the alignment accuracy, and the computation time T is recorded for applying the proposed algorithm to X and Y.

The proposed algorithm has a much lower computational cost than the modified SW algorithm. Given two alarm floods X and Y containing M and N alarms respectively, the implementation of the modified SW algorithm consists of two steps: (a) the dynamic programming step calculates a score matrix H for all $M \times N$ elements; (b) the backtracking step finds the best alignment based on an indicator matrix associated with H. Thus, the computational complexity of the modified SW algorithm is $O(MN)$.

In the proposed algorithm, the set-based pre-matching mechanism decreases the number of computations by excluding irrelevant alarm floods and alarm tags. The computation of this step is a constant time operation $O(1)$. If the set-based similarity index is lower than the threshold γ, the computation is terminated. If the set-based similarity index is larger than the threshold, two shorter alarm sequences $\tilde{X}$ and $\tilde{Y}$ are obtained (denote the length difference between X and $\tilde{X}$ as Δ_1, and the counterpart between Y and $\tilde{Y}$ as Δ_2.). The computation proceeds to the sequence alignment, which consists of a seeding step and an extending step: (a) The seeding step includes the indexing of all matched pairs $Z(i), i = 1, 2, \ldots, K_b$, and the calculation of each similarity score $h(Z(i))$. The indexing of all matched pairs is a constant time operation $O(1)$. Since the matched pairs can be found from either $\tilde{X}$ or $\tilde{Y}$, the maximum computational complexity should be $O(M - \Delta_1)$ or $O(N - \Delta_2)$. Thus the total computational complexity of the seeding step is $O(1 + M - \Delta_1)$ or $O(1 + N - \Delta_2)$. (b) In the extending step, the computational complexity for the dynamic programming and the backtracking with stopping strategy is $O(2(M - \Delta_1)(N - \Delta_2)\eta)$. Here $\eta \in (0, 1]$ indicates the ratio of cells to be calculated, and the value of η depends on the number of seeds K and the extension threshold U. Thus, the total computational complexity of the proposed method is $O(2 + M - \Delta_1 + 2(M - \Delta_1)(N - \Delta_2)\eta)$ or $O(2 + N - \Delta_2 + 2(M - \Delta_1)(N - \Delta_2)\eta)$.

5.2.3.7 Parameter Determination

The seeding step of the proposed local alignment algorithm finds all the matched alarm segments, among which the segments having the K largest seeding scores are preserved as seeds while others are discarded. In the extending step, the gapped extension stops till the extending score falls more than the threshold U below the maximum score. Thus, K and U are two important parameters influencing the computational cost and the alignment accuracy. To investigate the effects of K and U on the proposed algorithm, Monte Carlo simulations are implemented in the following example.

Here, 1000 Monte Carlo simulations are implemented. In each simulation, a pair of symbolic sequences X_l and $Y_l, l = 1, 2, ..., 1000$ are randomly generated. The symbolic tags come from a numeric alphabet $\Sigma = \{1, 2, ..., V\}$, where three priorities $\{p_1, p_2, p_3\}$ with a distribution 5%/15%/80% are assigned to V symbolic tags in Σ. The basic similarity scores in Table 5.5 are used. The sequence lengths L_x and L_y are uniform random integers in the range $[L_1, L_2]$. The time stamps $t_i^x, i = 1, 2, ..., L_x$ and $t_j^y, j = 1, 2, ..., L_y$ of symbolic tags in X and Y are Gaussian random variables in

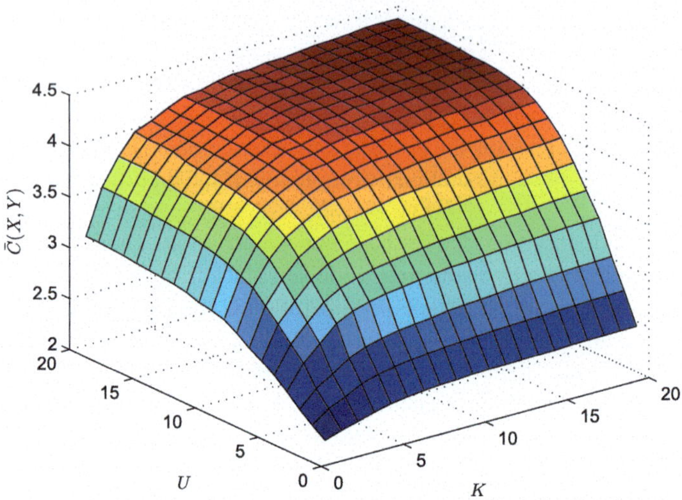

Fig. 5.16 Average number of matched pairs $\bar{C}(X, Y)$ versus K and U, reprinted from Ref. [24], copyright 2016, with permission from Elsevier

the range $[1, t_{max}]$, where t_{max} is a uniform variable in $[t_{max,1}, t_{max,2}]$. The parameters are set as $V = 100$, $L_1 = 50$, $L_2 = 500$, $t_{max,1} = 4320\,\text{s}$ (0.05 day), $t_{max,2} = 86300\,\text{s}$ (1 day). The algorithm is carried out at a personal computer with 3.3 GHz CPU, 4G RAM and 64bit operating system. The number $C(X, Y)$ of matched pairs between X and Y is counted to measure the alignment accuracy, and the computation time T in applying the proposed algorithm to X and Y is recorded. Figure 5.16 presents the average numbers of matched pairs for 1000 pairs of X_l and Y_l, i.e., $\bar{C}(X_l, Y_l) := \frac{1}{1000}\sum_{l=1}^{1000} C(X_l, Y_l)$, for different values of K and U. Figure 5.17 is the counterpart of Fig. 5.16 for the average computation time $\bar{T} := \frac{1}{1000}\sum_{l=1}^{1000} T_l$.

The following observations can be obtained from Figs. 5.16 and 5.17. Given a fixed number K of seeds $\bar{C}(X, Y)$ drastically increases with the increment of U for $U \leq 10$, and does not significantly improve with the increment of U for $U > 10$. Given a fixed cutoff threshold U, similar observation can be drawn with $K \leq 7$ and $K > 7$. It is obvious that the increment of the two parameters increases the computation time. To achieve an accurate alignment result within a short time period, U and K are recommended as $U = 10$ and $K = 7$.

5.2.4 Industrial Examples

This section provides industrial examples to validate the performance of the proposed method. In the first example, multiple alarm floods are studied for multiple queries

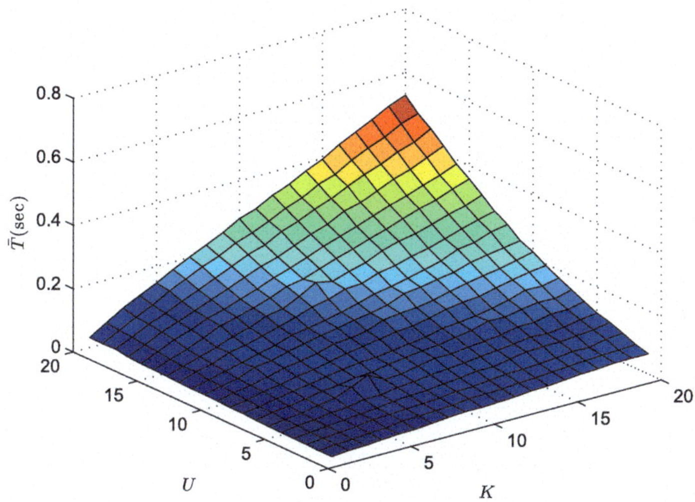

Fig. 5.17 Average computation time $\bar{T}$ versus K and U, reprinted from Ref. [24], copyright 2016, with permission from Elsevier

to a historical sequence database for one year and half. In the second example, a detailed analysis for the query of one particular alarm flood is carried out.

5.2.4.1 Example 1: Long-Term Application

Consider an oil conversion industrial process, being quite common in the petrochemical industry. It transforms the crude oil into valuable petroleum products, such as naphtha, gasoline, and diesel fuel [6]. The oil conversion plant is comprised of a number of interconnected pipelines, heaters, reactors, separators and vessels, as shown in Fig. 5.18 [16]. The crude oil and the hydrogen gas are heated to certain temperatures in the feed heater and hydrogen heater, and are contacted with a catalyst bed in the two reactors. The gas and liquid effluent overflows from the reactors to a common outlet line directed to the separators, where the vapor, liquid hydrocarbon, and water phases are separated. The liquid hydrocarbon streams are carried to the atmospheric tower and fractionated into desired products, which are stripped by gas oil strippers. The refining processes are operated by a distributed control system. To ensure the safe operation of the oil conversion plant, a large number of alarm variables are configured in different sections.

Some historical alarm data with 1547 unique alarms and 3 alarm priority levels was collected from the oil conversion plant over the time period from May 12th to November 12th in the next year. 389 alarm floods excluding chattering alarms were detected and saved as an database of alarm floods $\mathbb{B} = \{Y_l : l = 1, 2, ..., 389\}$. The longest alarm flood contains 1840 alarms while the shortest alarm flood includes 10 alarms. The average length of alarm floods is 69.2.

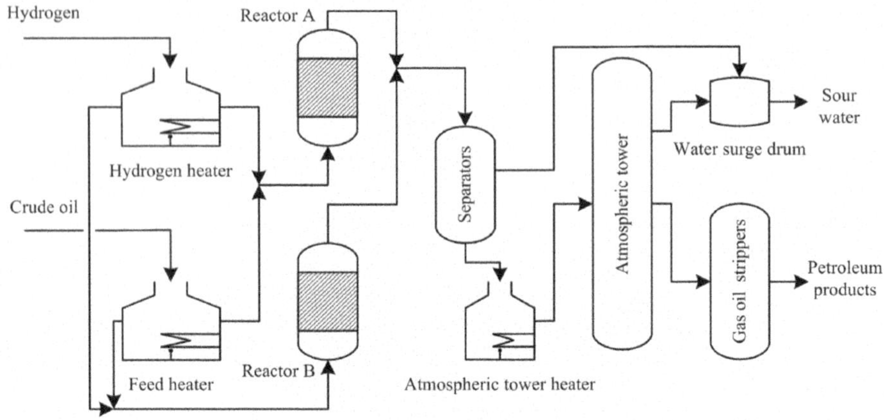

Fig. 5.18 Simplified schematic diagram of the oil conversion plant, reprinted from Ref. [24], copyright 2016, with permission from Elsevier

Table 5.12 The alignment accuracy measured by the average number of the matched pairs $\bar{C}(X, Y)$ and the total computation time T in querying $\mathbb{B}$ for X_i ($i = 1, ..., 10$). $L(X_i)$ indicates the number of alarms contained in X_i. $N(Y)$ denotes the number of relevant alarm floods Y found in $\mathbb{B}$, reprinted from Ref. [24], copyright 2016, with permission from Elsevier

Index	$L(X_i)$	$N(Y)$	Proposed algorithm		Modified SW algorithm	
			$\bar{C}(X, Y)$	T (s)	$\bar{C}(X, Y)$	T (s)
1	1010	130	6.65	8.32	3.48	924.62
2	848	341	7.07	11.52	4.73	644.69
3	661	382	8.21	10.02	4.60	419.06
4	420	291	6.91	5.53	3.73	239.02
5	347	327	7.65	7.63	3.76	203.51
6	290	305	5.91	4.83	3.29	162.16
7	249	320	5.71	5.66	3.20	176.37
8	196	223	6.75	3.53	4.77	123.86
9	186	276	6.19	4.42	3.42	99.12
10	114	125	5.49	1.86	3.14	50.37

To illustrate the efficiency of the proposed local alignment algorithm, 10 alarm floods X_i for $i = 1, ..., 10$ occurred at different time periods are studied for the queries to the database $\mathbb{B}$. The number of alarms in X_i is given in the second column of Table 5.12. First, using the set-based pre-matching mechanisms with $\gamma = 0$, irrelevant object alarm floods are excluded from the database $\mathbb{B}$, while others are preserved for sequence alignment. The number of relevant object alarm floods for each query alarm flood X_i is listed in the third column of Table 5.12. Second, both the proposed algorithm and the modified SW algorithm [14] are applied to the query of X_i from the preserved object alarm floods. The same parameters $U = 10$ and $K = 7$

are used to implement the proposed algorithm. The alignment accuracy measured by the average number of matched pairs $\bar{C}(X, Y) := \frac{1}{389}\sum_{l=1}^{389} C(X_i, Y_l)$ and the total computation time T for querying X_i are presented in the last four columns of Table 5.12. Since alarm priorities are not considered in the modified SW algorithm, a unified match score rather than the priority based scoring strategy is utilized. That is, the match score 4.5, the average value of the match scores in Table 5.5, is used for all matched alarm tags.

Comparing the average numbers of the matched pairs in the 4th and 6th columns of Table 5.12, it is obvious that the proposed algorithm achieves more accurate alarm sequence alignments than the modified SW algorithm. That is, more matched pairs are found using the proposed algorithm. This improvement is owing to the utilization of the set-based pre-matching mechanism that removes irrelevant alarm tags. Another comparison of the computation times in the 5th and 7th columns of Table 5.12 clearly says that the proposed algorithm is much more efficient in computation. In this example, the proposed algorithm is about $20 \sim 111$ times faster than the modified SW algorithm.

5.2.4.2 Example 2: Analysis for One Alarm Flood

A detailed analysis of the sequence alignment for the 10th query alarm flood in the first example is presented here. As shown in Table 5.12, this alarm flood X contains 114 alarms. 125 object alarm floods from $\mathbb{B}$ are found containing common alarms in X. For the proposed algorithm and the modified SW algorithm, the number of matched pairs $C(X, Y_i)$ and computation time T_i for X and $Y_i, i = 1, 2, .., 125$ are shown as the dotted lines in Fig. 5.19. In particular, for the queries to the 3rd, 4th, 36th, 38th, and 55th object alarm floods, the computation times of the proposed algorithm are much smaller than the modified SW algorithm; in addition, the alignment accuracies of the proposed algorithm are higher.

As detected by the proposed algorithm, the 117th object alarm flood Y as shown in Fig. 5.19 was found to be the most similar to X. The best alignment between X and Y was determined by the backtracking step. Excluding irrelevant alarms, the alignment of matched alarms is shown in Fig. 5.20. The alarms in the two alarm floods X and Y were chronologically sorted while the occurrence dates of X and Y were different. Among the 40 matches, 30 pairs of alarms were perfectly matched with the same chronological order while other 10 pairs were matched in disorder. Observing the time stamps of these disordered matches, it can be found that these alarms occurred almost simultaneously. Thus, their orders in the alignment should be not important.

The alarm flood Y appeared due to an inadequate feed of the purified recycle gas for one of the two hydrogen supplies in Fig. 5.18. For a clearer illustration, the simplified schematic diagram of the hydrogen heater is presented in Fig 5.21. The inadequate feed of supply led to the overheat of the convection skin of outlets. To maintain the outlet temperatures within the designed ranges, a series of automatic regulations and protective strategies were activated. Based on the knowledge about

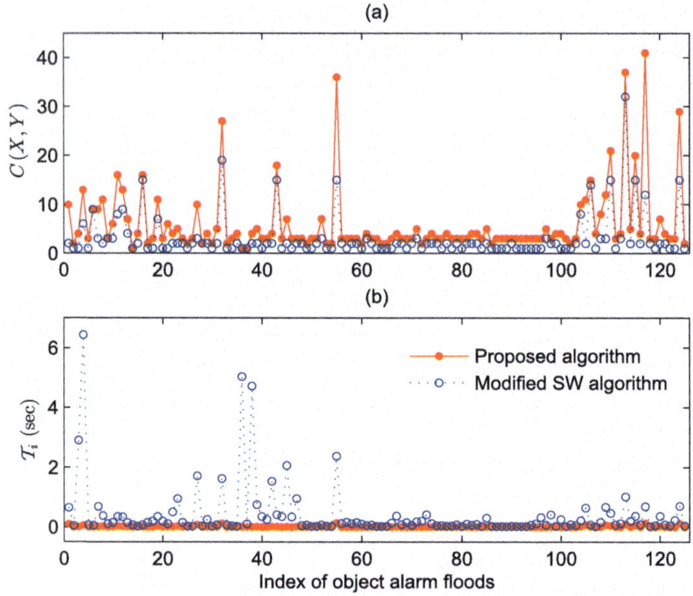

Fig. 5.19 Performance statistics of the proposed local alignment algorithm (dot) and the modified SW algorithm (circle-dash) for the query of X to 125 object alarm floods: **a** the number of matched pairs; **b** the computation time, reprinted from Ref. [24], copyright 2016, with permission from Elsevier

the cause of Y from the alarm flood database $\mathbb{B}$, one might predict that the alarm flood X was caused by the inadequate feed of hydrogen supply. By checking the historical records, it was found that the supply of the makeup hydrogen (one of the two hydrogen supplies) was decreased, which supported the similarity between X and Y.

5.3 Pattern Mining of Alarm Floods

This section is on detection of frequent alarm patterns from historical alarm flood sequences. In practice, it is common that some abnormal events may happen frequently, causing a series of alarms; if such repeated series of alarms can be detected from historical data, it will be helpful in configuring dynamic alarm suppression modules. Accordingly, a data driven method is proposed here to find interesting alarm patterns in alarm floods from historical alarm data based on itemset mining approaches [23]. The identification and extraction of alarm floods are formulated according to industrial benchmarks. Then, frequent alarm patterns are defined and itemset mining approaches are adapted to discover meaningful patterns in alarm floods.

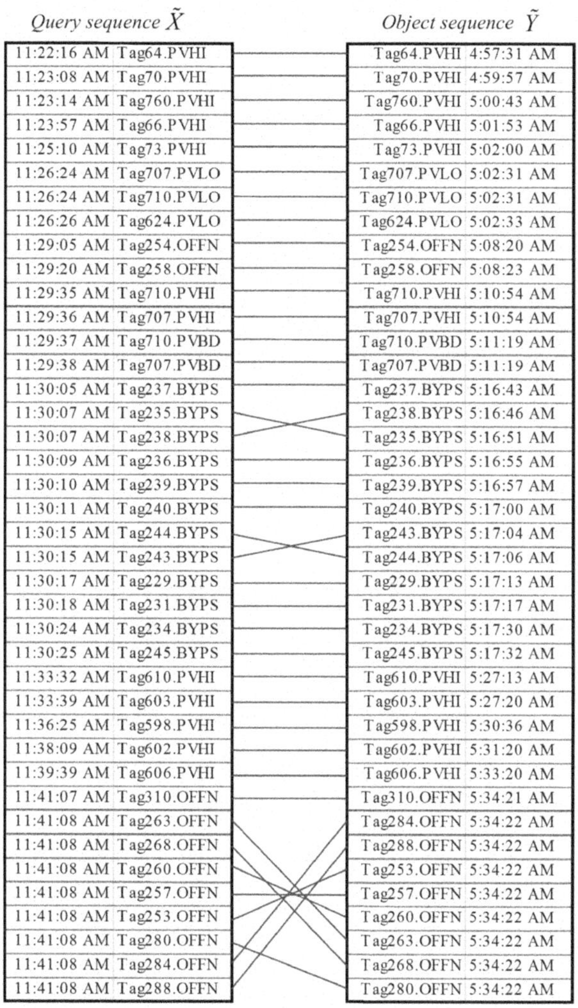

Fig. 5.20 The best sequence alignment between the query alarm flood X and the 117th Y from $\mathbb{B}$, reprinted from Ref. [24], copyright 2016, with permission from Elsevier

5.3.1 Problem Description

The dynamic alarm suppression reduces alarm floods by suppressing irrelevant alarms, so that operators are less overloaded and can focus on more critical alarms [13]. For example, a compressor trip has many associated diagnostic alarms (e.g., 150 in a gas processing facility) [22]. The distraction to operators can be minimized if these non-important alarms are suppressed. To design dynamic alarm suppression strategies, it should be known what alarms to suppress. Usually, unplanned events, such as a compressor trip may happen from time to time, and thus one can explore from historical alarm data to look for a group of associated alarms occurring that repeat-

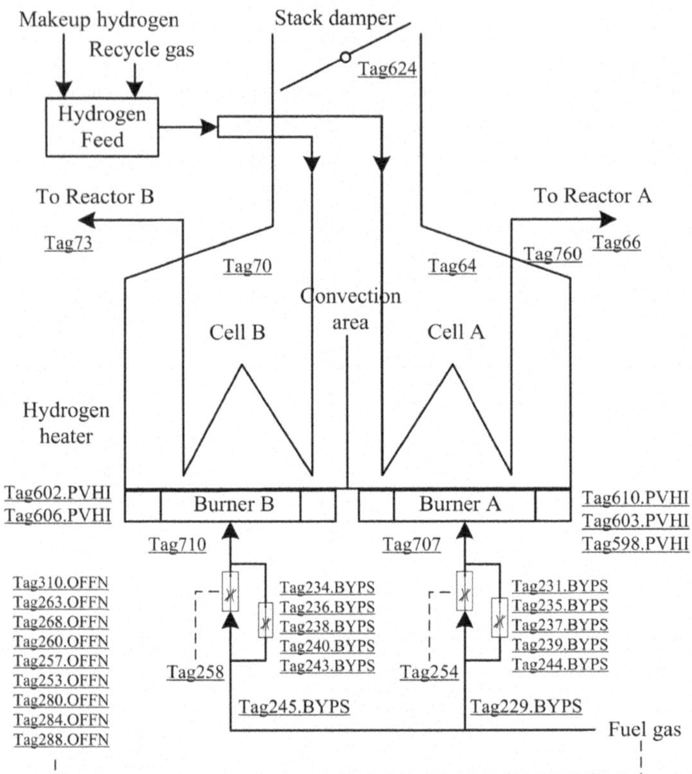

Fig. 5.21 Simplified schematic diagram of the hydrogen heater (the dashed lines indicate data links), reprinted from Ref. [24], copyright 2016, with permission from Elsevier

edly. Then these alarm groups can be used to supervise the configuration of dynamic alarm suppression.

Given an alarm flood (assigned with a unique identifier number k) starting and ending at $t_s^{(k)}$ and $t_e^{(k)}$, respectively, the set of alarms in this alarm flood is denoted as $\mathcal{F}_k = \{a_i \in \mathcal{A}, i = 1, 2, \ldots, |\mathcal{F}_k|\}$, such that the alarm a_i occurred at least once in the time period $(t_s^{(k)} - T, t_e^{(k)}]$, i.e.,

$$\sum_{t=t_s^{(k)}-T+1}^{t_e^{(k)}} x_{a_i}(t) > 0, \tag{5.41}$$

where $i = 1, \ldots, |\mathcal{F}_k|$. It is obvious that $\mathcal{F}_k \subseteq \mathcal{A}$. In the sequel, the alarm set $\mathcal{F}_k$ is called an alarm flood transaction. The identifier number k is called Transaction IDentifier (TID). The collection of historical alarm flood transactions makes up a transaction database $\mathbb{D} = \{\mathcal{F}_k, k = 1, 2, \ldots, |\mathbb{D}|\}$. The complete set of alarm flood TIDs is $\mathcal{K} = \{1, 2, \ldots, |\mathbb{D}|\}$. Given a set of alarms $\mathcal{P} = \{a_i \in \mathcal{A}, i = 1, 2, \ldots, |\mathcal{P}|\}$, its TID set is denoted by $\mathcal{T}(\mathcal{P}) = \{k \in \mathcal{K} : \mathcal{P} \subseteq \mathcal{F}_k, \mathcal{F}_k \in \mathbb{D}\}$. The TID set of $\mathcal{P}$ can be easily found as $\mathcal{T}(\mathcal{P}) = \bigcap_{a_i \in \mathcal{P}} \mathcal{T}(a_i)$, where $\bigcap$ denotes the set intersection.

Definition 5.1 (*Frequent Alarm Pattern*) A set of alarms $\mathcal{P} = \{a_i \in \mathcal{A}, i = 1, 2, \ldots, |\mathcal{P}|\}$ is a frequent alarm pattern, if $\sigma(\mathcal{P}) \geq \sigma_{th}$, where $\sigma(\mathcal{P})$ indicates the count of transactions $\mathcal{F}_k$ containing $\mathcal{P}$.

Notations $\sigma(\mathcal{P})$ and σ_{th} are called support and minimum support, respectively. The support of $\mathcal{P}$ is calculated as $\sigma(\mathcal{P}) = |\mathcal{T}(\mathcal{P})|$. The minimum support σ_{th} is a user specified threshold. Generally, a larger σ_{th} can filter out more non-frequent patterns and reduce the searching space [21]. In view of that some long alarm patterns are rare but more interesting, σ_{th} should be set small. A recommendation of minimum support is $\sigma_{th} = 2$, so that any pattern shared by at least two alarm floods will be discovered. Therefore, there will not be any meaningful long alarm patterns missing.

Given a historical transaction database of alarm floods $\mathbb{D}$ defined on $\mathcal{A}$, the objective is to find a collection of frequent alarm patterns. Differing from the conventional data mining [21], this task has the following special features:

- The mining of frequent alarm patterns should find long patterns rather than short ones. As described in [22] and [13], dynamic alarm suppression in alarm floods usually involves a series of alarms (e.g., 150 alarms in a gas processing facility [22]).
- The redundancy of the detected frequent alarm patterns should be minimized. In alarm floods, the burst of alarm information may cause missing records due to limited capacity of communication and data acquisition. This may cause similar patterns with most alarms being the same in presence of a few distinct.
- The detected results should include TID information. Information about which alarm floods a specific frequent alarm pattern appears in, will assist industrial practitioners to recall the root causes and thus quickly determine the event associated with the pattern.

Defining an IT-pair (pair of itemset and its TID set) as $\mathcal{P}_i \times \mathcal{T}(\mathcal{P}_i)$, the collection of all frequent alarm patterns is $\mathbb{F} = \{\mathcal{P}_i \times \mathcal{T}(\mathcal{P}_i) : \mathcal{P}_i \subseteq \mathcal{A}, \sigma(\mathcal{P}_i) \geq \sigma_{th}, i = 1, 2, \ldots, |\mathbb{F}|\}$. In order to find meaningful alarm patterns and minimize the pattern redundancy, the above conditions must be considered in the detection process. The next section proposes systematic methods to mine such meaningful alarm patterns that fulfill the three conditions.

5.3.2 *Frequent Alarm Pattern Mining*

Frequent pattern mining is an essential data mining task that discovers patterns, such as itemsets, sequences, lattices, and graphs, which appear frequently in a database [21]. The problem of discovering alarm patterns is essentially an itemset mining task, which finds itemset patterns from transactional database. To discover

Algorithm 5.5 CHARM: Mine closed frequent alarm patterns.

Input Argument: $\mathbb{D}$, σ_{th};
Output Argument: $\mathbb{C}$;
$\mathbb{C} = \emptyset$;
$[E] = \{\mathcal{P}_i \times \mathcal{T}(\mathcal{P}_i) : \mathcal{P}_i = \{a_i \in \mathcal{A}\}, \sigma(\mathcal{P}_i) \geq \sigma_{th}\}$;
$\mathbb{C}$ = CHARM-Extend($\mathbb{C}$, $[E]$, σ_{th});

Algorithm 5.6 CHARM-Extend: Search the IT-tree.

Input Argument: $\mathbb{C}$, $[E]$, σ_{th};
Output Argument: $\mathbb{C}$;
for $\mathcal{P}_i \times \mathcal{T}(\mathcal{P}_i) \in [E]$ **do**
  $[E_i] = \emptyset$;
  $\mathbf{P} = \mathcal{P}_i$;
  for $\mathcal{P}_j \times \mathcal{T}(\mathcal{P}_j) \in [E]$ with $\mathcal{P}_j \geq_f \mathcal{P}_i$ **do**
    $\mathbf{P} = \mathbf{P} \cup \mathcal{P}_i$;
    $\mathcal{T}(\mathbf{P}) = \mathcal{T}(\mathcal{P}_i) \cap \mathcal{T}(\mathcal{P}_j)$;
    if $|\mathcal{T}(\mathbf{P})| \geq \sigma_{th}$ **then**
      Check closure properties and update $[E]$, $[E_i]$;
    end if
  end for
  if $[E_i] \neq \emptyset$ **then**
    $\mathbb{C}$ = CHARM-Extend($\mathbb{C}$, $[E_i]$, σ_{th});
  end if
  Delete E_i;
  if then$|\mathcal{T}(\mathbf{P})| > 1$
    $\mathbb{C} = \mathbb{C} \cup (\mathbf{P} \times \mathcal{T}(\mathbf{P}))$;
  end if
end for

itemset patterns, the Apriori, FP-growth, and Equivalence CLAss Transformation (Eclat) algorithms and their variants are the most prevalent techniques [21]. However, down-closure property (i.e., any subset of a frequent itemset must be frequent) of itemset mining, often leads to the explosive number of frequent patterns. For instance, given a frequent pattern containing k items, all its $2^k - 2$ sub-patterns are frequent [38]. To avoid explosion of detected patterns, methods to efficiently mine closed frequent patterns have been proposed [21, 38, 43]. Closed frequent patterns can be discovered through (i) maintaining discovered patterns in a pattern tree, e.g., CLOSET+ [38] and (ii) tracking the TID list of a pattern, e.g., CHARM (Closed Association Rule Mining, 'H' is gratuitous) [43]. In view of the first property of the alarm pattern mining problem in Sect. 5.3.1, closed pattern mining definitely suites such an application of discovering long frequent alarm patterns rather than the complete set of all patterns. The CHARM algorithm is exploited in view of its capability of tracking the TID list of each pattern, which also fulfills the third condition in Sect. 5.3.1.

Analogous to concepts in [21], the definition of a closed frequent alarm pattern is given as follows.

Definition 5.2 (*Closed Frequent Alarm Pattern*) A set of alarms $\mathcal{P} = \{a_i \in \mathcal{A}, i = 1, 2, \ldots, |\mathcal{P}|\}$ is a closed frequent alarm pattern, if $\sigma(\mathcal{P}) \geq \sigma_{th}$, and there is no super frequent alarm pattern $\mathcal{Q} = \{a_i \in \mathcal{A}, i = 1, 2, \ldots, |\mathcal{Q}|\}$ such that $\mathcal{Q} \supset \mathcal{P}$ with $\sigma(\mathcal{Q}) = \sigma(\mathcal{P})$.

The collection of closed frequent alarm patterns is denoted by $\mathbb{C} = \{\mathcal{P}_i \times \mathcal{T}(\mathcal{P}_i) \in \mathbb{F}, i = 1, 2, \ldots, |\mathbb{C}| : \nexists \mathcal{Q} \in \mathbb{F}, \text{s.t. } \mathcal{Q} \supset \mathcal{P}_i, \sigma(\mathcal{Q}) = \sigma(\mathcal{P}_i)\}$. It is obvious that $\mathbb{C} \subseteq \mathbb{F}$. The *closure* $c(\mathcal{P})$ of an alarm set $\mathcal{P}$ is defined as the smallest closed set containing $\mathcal{P}$. An alarm set $\mathcal{P}$ is closed if and only if $\mathcal{P} = c(\mathcal{P})$ [43]. Accordingly, an equality holds, $\sigma(\mathcal{P}) = \sigma(c(\mathcal{P}))$. The CHARM algorithm [43] is exploited to mine closed frequent alarm patterns. CHARM explores both the itemset space $\mathcal{A}$ and transaction space $\mathcal{K}$, which constructs a dual IT-tree (a search tree of itemset and TID set) [43]. Each node of the IT-tree is represented by an *IT-pair* $\mathcal{P} \times \mathcal{T}(\mathcal{P})$ with $\sigma(\mathcal{P}) \geq \sigma_{th}$.

Given a parent node $\mathcal{P} \times \mathcal{T}(\mathcal{P})$, its child node is obtained as $(\mathcal{P} \cup a_i) \times \mathcal{T}(\mathcal{P} \cup a_i)$ s.t. $\sigma(\mathcal{P} \cup a_i) \geq \sigma_{th}$. $\mathcal{P}$ is called the prefix of all its child nodes. A prefix class is defined as $[\mathcal{P}] = \{p_1, p_2, \ldots, p_{|[\mathcal{P}]|}\}$, where each child node $p_i \in [\mathcal{P}]$ represents an IT-pair $(\mathcal{P} \cup a_i) \times \mathcal{T}(\mathcal{P} \cup a_i)$, where $a_i \in \mathcal{A}, a_i \notin \mathcal{P}$. It is noteworthy that the root of an IT-tree corresponds to a prefix class $[\emptyset] = \{a_i \in \mathcal{A} : \sigma(a_i) \geq \sigma_{th}\}$.

The pseudo codes of CHARM are presented in Algorithms 5.5 and 5.6. CHARM searches all closed patterns by firstly initializing the prefix class $[E]$ that contains all frequent single alarms (Line 4 in Algorithm 5.5), and then recursively searching the IT-tree in a depth-first way (Line 5 in Algorithm 5.5) by calling Algorithm 5.6. In Line 6 of Algorithm 5.6, $\leq_f$ indicates the ordering function defined as follows: For any two alarm sets $\mathcal{P}_i$ and $\mathcal{P}_j$, $\mathcal{P}_i \leq_f \mathcal{P}_j$ if and only if $f(\mathcal{P}_i) \leq f(\mathcal{P}_j)$, where $f(\cdot)$ is a bijective mapping function between alarm sets and integers. If the alarm sets are sorted by their supports, then $f(\cdot) = \sigma(\cdot)$. The ordering function $f(\cdot)$ could also denote the lexicographic ordering [43].

In Line 10 of Algorithm 5.6, the closure properties are checked and IT-pairs subsumed by other pairs are deleted. Given $\mathcal{P}_i \times \mathcal{T}(\mathcal{P}_i) \in [\mathcal{P}]$ and $\mathcal{P}_j \times \mathcal{T}(\mathcal{P}_j) \in [\mathcal{P}]$ with $\mathcal{P}_i \leq_f \mathcal{P}_j$, the following closure properties hold:

- $c(\mathcal{P}_i) = c(\mathcal{P}_j) = c(\mathcal{P}_i \cup \mathcal{P}_j)$, if $\mathcal{T}(\mathcal{P}_i) = \mathcal{T}(\mathcal{P}_j)$;
- $c(\mathcal{P}_i) \neq c(\mathcal{P}_j), c(\mathcal{P}_i) = c(\mathcal{P}_i \cup \mathcal{P}_j)$, if $\mathcal{T}(\mathcal{P}_i) \subset \mathcal{T}(\mathcal{P}_j)$;
- $c(\mathcal{P}_i) \neq c(\mathcal{P}_j), c(\mathcal{P}_j) = c(\mathcal{P}_i \cup \mathcal{P}_j)$, if $\mathcal{T}(\mathcal{P}_j) \subset \mathcal{T}(\mathcal{P}_i)$;
- $c(\mathcal{P}_i) \neq c(\mathcal{P}_j) \neq c(\mathcal{P}_i \cup \mathcal{P}_j)$, if $\mathcal{T}(\mathcal{P}_i) \neq \mathcal{T}(\mathcal{P}_j)$.

The prefix classes $[E]$ and $[E_i]$ are updated as follows:

(1) remove $\mathcal{P}_j \times \mathcal{T}(\mathcal{P}_j)$ from $[E]$ and replace all $\mathcal{P}_i \times \mathcal{T}(\mathcal{P}_i)$ with $\mathbf{P} \times \mathcal{T}(\mathbf{P})$, if $\mathcal{T}(\mathcal{P}_i) = \mathcal{T}(\mathcal{P}_j)$;
(2) replace all $\mathcal{P}_i \times \mathcal{T}(\mathcal{P}_i)$ with $\mathbf{P} \times \mathcal{T}(\mathbf{P})$, if $\mathcal{T}(\mathcal{P}_i) \subset \mathcal{T}(\mathcal{P}_j)$;
(3) remove $\mathcal{P}_j \times \mathcal{T}(\mathcal{P}_j)$ from $[E]$ and add $\mathbf{P} \times \mathcal{T}(\mathbf{P})$ to $[E_i]$, if $\mathcal{T}(\mathcal{P}_j) \subset \mathcal{T}(\mathcal{P}_i)$;
(4) add $\mathbf{P} \times \mathcal{T}(\mathbf{P})$ to $[E_i]$, if $\mathcal{T}(\mathcal{P}_i) \neq \mathcal{T}(\mathcal{P}_j)$.

In Lines 17–19, any closed pattern containing only one single alarm variable will be filtered out.

5.3.3 Determination of Representative Alarm Patterns

Although the closed pattern mining is effective in discovering meaningful long patterns, redundancy of detected frequent patterns remains a problem. For instance, given 5 similar frequent patterns as $\mathcal{P}_1 = \{2, 3, 4, 5\}$, $\mathcal{P}_2 = \{1, 3, 4, 5\}$, $\mathcal{P}_3 = \{1, 2, 4, 5\}$, $\mathcal{P}_4 = \{1, 2, 3, 5\}$, $\mathcal{P}_5 = \{1, 2, 3, 4\}$, it can be seen that these patterns resemble but do not contain each other. In alarm floods, it is common that some alarms are missing due to limited capacity of communication and data acquisition. As a result, there could be too many similar alarm patterns. Thus, a natural question is: How to minimize the redundancy of detected frequent alarm patterns and present the user with more abstracted information? A related work can be found in [35], which searches abstract patterns based on strategies that requires additional information, such as the notification type and device name. However, this method is designed to decrease the misclassification, so that its objective is different. Moreover, as described in [35], this method results in a huge set of subsequences. Thus, it is not effective in reducing the number of patterns and cannot be used to achieve pattern compression here.

This section addresses this problem by formulating a pattern compression problem as: Given the collection of closed frequent alarm patterns $\mathbb{C} = \{\mathcal{P}_i \times \mathcal{T}(\mathcal{P}_i) \in \mathbb{F}, i = 1, 2, \dots, |\mathbb{C}| : \nexists \mathcal{Q} \in \mathbb{F}, \text{s.t. } \mathcal{Q} \supset \mathcal{P}_i, \sigma(\mathcal{Q}) = \sigma(\mathcal{P}_i)\}$, the pattern compression in this section is to discover a collection of representative alarm patterns $\mathbb{R} = \{\mathcal{R}_i \times \mathcal{T}(\mathcal{R}_i), i = 1, 2, \dots, |\mathbb{R}|\}$, such that for any $\mathcal{P}_i \times \mathcal{T}(\mathcal{P}_i) \in \mathbb{C}$, there exists $\mathcal{R}_j \times \mathcal{T}(\mathcal{R}_j) \in \mathbb{R}$ subsuming it, namely, $\mathcal{R}_j \supseteq \mathcal{P}_i$. The definition of a representative alarm pattern is given as follows:

Definition 5.3 (*Representative Alarm Pattern*) A set of alarms $\mathcal{R} = \{a_i \in \mathcal{A}, i = 1, 2, \dots, |\mathcal{R}|\}$ is a representative alarm pattern of n alarm patterns $\mathcal{P}_i$, $i = 1, 2, \dots, n$, if the following two conditions are fulfilled:

$$\mathcal{R} = \bigcup_{i=1}^{n} \mathcal{P}_i, \tag{5.42}$$

and

$$S(\mathcal{P}_i, \mathcal{P}_j) \geq \delta, \forall i \neq j, \tag{5.43}$$

where $\bigcup_{i=1}^{n}$ indicates the union of n sets, $S(\cdot, \cdot)$ denotes the pattern similarity between two alarm patterns, and δ is a user specified threshold. The alarm pattern $\mathcal{P}_i$ is called a descendent alarm pattern of $\mathcal{R}$.

The pattern similarity in (5.43) can be calculated based on the Jaccard similarity index, i.e., given two alarm patterns $\mathcal{P}_i$ and $\mathcal{P}_j$, their pattern similarity is

$$S(\mathcal{P}_i, \mathcal{P}_j) = \frac{|\mathcal{P}_i \cap \mathcal{P}_j|}{|\mathcal{P}_i \cup \mathcal{P}_j|}. \tag{5.44}$$

The shortest closed alarm patterns in $\mathbb{C}$ contain only two alarms each and thus have three possible similarities: $S(\mathcal{P}_i, \mathcal{P}_j) = \frac{0}{4} = 0$, $S(\mathcal{P}_i, \mathcal{P}_j) = \frac{1}{3}$, and $S(\mathcal{P}_i, \mathcal{P}_j) = \frac{2}{2} = 1$. $S(\mathcal{P}_i, \mathcal{P}_j) = \frac{1}{3}$ indicates that the two shortest patterns share one common alarm. Thus, if the pattern similarity threshold δ is set greater than $\frac{1}{3}$, there could be a huge number of representative patterns containing only two alarms. In view of this, a threshold $\delta = \frac{1}{3}$ is recommended as a rule of thumb to reduce such short representative patterns and achieve significant pattern compression.

Equations (5.42) and (5.43) indicate that a representative alarm pattern $\mathcal{R}$ is a super set of a group of similar alarm patterns. To find out such groups of similar alarm patterns, a modified δ-cluster algorithm (the pseudo-code is presented in Algorithm 5.7) is proposed based on the work in [40]. This algorithm consists of two main loops. The first main loop in Lines 6–13 calculates the similarities between closed alarm patterns in pairwise and cluster them into groups. As a result, a family $\mathfrak{H}$ of sets $\mathbb{H}_i \subseteq \mathbb{C}, i = 1, 2, \ldots, |\mathfrak{H}|$ is obtained. However, the clustering may lead to one alarm pattern belonging to multiple groups. To reduce redundancy, the second loop in Lines 17–21 is adopted. Given the collection $\mathbb{C}$ of closed alarm patterns and the family $\mathfrak{H}$ of sets containing similar alarm patterns, the objective is to find a minimum size subset $\mathfrak{L}$ of $\mathfrak{H}$,

$$\min_{\mathfrak{L} \subseteq \mathfrak{H}} |\mathfrak{L}|, \text{ s.t. } \bigcup_{\mathbb{L} \in \mathfrak{L}} \mathbb{L} = \mathbb{C}. \tag{5.45}$$

This is a typical set-covering problem and can be solved using a greed set covering method [15]. In Lines 17–21, the algorithm selects an $\mathbb{H}_i \in \mathfrak{H}$ that covers as many as uncovered elements, removes $\mathbb{H}_i$ from $\mathbb{G}$, and adds $\mathbb{H}_i$ to $\mathfrak{L}$ in each loop, until $\mathbb{G}$ becomes empty. Eventually, the family $\mathfrak{L} = \{\mathbb{L}_i, i = 1, 2, \ldots, |\mathfrak{L}|\}$ covering all closed alarm patterns in $\mathbb{C}$ is obtained.

Eventually, the collection of all representative alarm patterns is obtained as $\mathbb{R} = \{\mathcal{R}_i \times \mathcal{T}(\mathcal{R}_i), i = 1, 2, \ldots, |\mathbb{R}|\}$ such that for each $\mathbb{L}_i \in \mathfrak{L}$,

$$\mathcal{R}_i = \bigcup_{k=1}^{|\mathbb{L}_i|} \mathcal{P}_k, \mathcal{T}(\mathcal{R}_i) = \bigcup_{k=1}^{|\mathbb{L}_i|} \mathcal{T}(\mathcal{P}_k), \tag{5.46}$$

where $\mathcal{P}_k \times \mathcal{T}(\mathcal{P}_k) \in \mathbb{L}_i$. Here $\mathcal{R}_i$ is a superset of similar closed alarm patterns contained in each $\mathbb{L}_i \in \mathfrak{L}$. The TID set $\mathcal{T}(\mathcal{R}_i)$ contains the TIDs of alarm floods where all closed alarm patterns in $\mathbb{L}_i$ are extracted from.

The objective aims at detecting interesting patterns from historical data through offline analysis and improving the online alarm monitoring by configuring dynamic alarm suppression modules based on the detected results. Figure 5.22 describes the role of the proposed method in improving alarm systems. The alarm system produces alarm signals and presents them to plant operators. The alarm signals are historized and exploited by the proposed method to detect frequent alarm patterns. As shown in the two colored boxes, the proposed method consists of two critical steps: (i) the identification and extraction of alarm floods generate a transaction database of alarm

Algorithm 5.7 Modified δ-cluster: Cluster closed alarm patterns.

Input Argument: $\mathbb{C}$, δ;
Output Argument: $\mathfrak{L}$;
for $\mathcal{P}_j \times \mathcal{T}(\mathcal{P}_j) \in \mathbb{C}$ **do**
  $\mathbb{H}_j = \{\mathcal{P}_j \times \mathcal{T}(\mathcal{P}_j)\}$;
end for
for $\mathcal{P}_i \times \mathcal{T}(\mathcal{P}_i) \in \mathbb{C}$ **do**
  for $\mathcal{P}_j \times \mathcal{T}(\mathcal{P}_j) \in \mathbb{C}, i \neq j$ **do**
    Calculate $S(\mathcal{P}_i, \mathcal{P}_j)$ using Eq. (5.44);
    if $S(\mathcal{P}_i, \mathcal{P}_j) \geq \delta$ **then**
      $\mathbb{H}_j = \mathbb{H}_j \cup \{\mathcal{P}_i \times \mathcal{T}(\mathcal{P}_i)\}$;
    end if
  end for
end for
$\mathfrak{H} = \{\mathbb{H}_i, i = 1, 2, \cdots, |\mathfrak{H}|\}$;
$\mathbb{G} = \mathbb{C}$;
$\mathfrak{L} = \emptyset$;
while $\mathbb{G} \neq \emptyset$ **do**
  Select an $\mathbb{H}_i \in \mathfrak{H}$ that maximizes $|\mathbb{H}_i \cap \mathbb{G}|$;
  $\mathbb{G} = \mathbb{G} \backslash \mathbb{H}_i$;
  $\mathfrak{L} = \mathfrak{L} \cup \{\mathbb{H}_i\}$;
end while

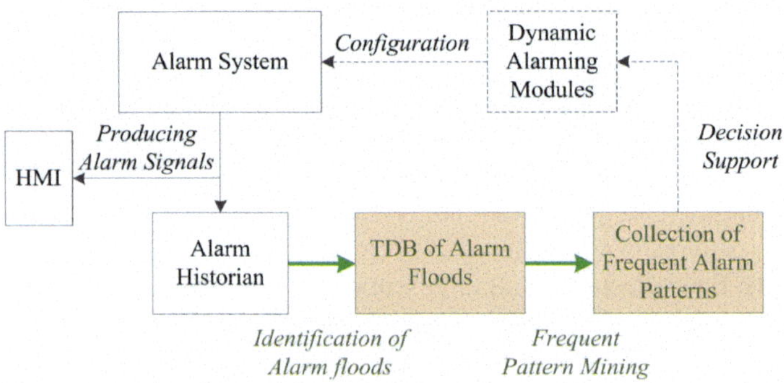

Fig. 5.22 Schematic of the role of the proposed method in improving alarm systems (HMI and TDB are abbreviated for Human-Machine Interface and Transaction DataBase, respectively), reprinted from Ref. [23], copyright 2018, with permission from IEEE

floods, and (ii) the frequent pattern mining detects a collection of meaningful alarm patterns. The detected frequent alarm patterns are validated by industrial practitioners and utilized to supervise configuration of dynamic alarm suppression, which reduces alarm floods by suppressing irrelevant alarms. The implementation of the proposed methods should follow procedures below:

1. Historical alarm data is prepared and preprocessed. Chattering alarms are reduced significantly.
2. Alarm floods are identified and transactions of alarm floods are extracted based on the criteria in (5.41).

3. The set of closed frequent alarm patterns $\mathbb{C} = \{\mathcal{P}_i \times \mathcal{T}(\mathcal{P}_i) \in \mathbb{F}, i = 1, 2, \ldots, |\mathbb{C}| : \nexists \mathcal{Q} \in \mathbb{F}, \text{s.t. } \mathcal{Q} \supset \mathcal{P}_i, \sigma(\mathcal{Q}) = \sigma(\mathcal{P}_i)\}$ is detected using the method in Sect. 5.3.2;
4. The set of representative alarm patterns $\mathbb{R} = \{\mathcal{R}_i \times \mathcal{T}(\mathcal{R}_i), i = 1, 2, \ldots, |\mathbb{R}|\}$ is computed using the method in Sect. 5.3.3.
5. Results including the alarm patterns $\mathcal{R}_i$ and their associated TIDs $\mathcal{T}(\mathcal{R}_i)$, are produced and visualized.

5.3.4 Industrial Examples

This section presents industrial examples to demonstrate the effectiveness of the proposed methods using real industrial alarm data, which was collected in April from a large-scale industrial process. The alarm system in this facility had not been well rationalized, making alarm floods quite common. Hence, this is an exact application that the proposed method can be used to improve the alarm system. The algorithms in Sect. 5.3.2 involve algebra of sets, hash tables, and statistics. The results and discussions for this example are produced as follows:

First, the alarm floods are identified and extracted. The number of unique alarms was 1779. The set of unique alarms is then denoted by $\mathcal{A} = \{a_i, i = 1, 2, \ldots, 1779\}$. The alarm burst plot in Fig. 5.23a visualizes the sequence of alarm rate $\zeta(t)$ (chattering alarms have been removed), based on which 180 alarm floods are identified. The 180 alarm floods are indexed with TIDs from 1 to 180 in a chronological order based on their start time stamps. Thus, the complete set of TIDs is $\mathcal{K} = \{1, 2, \ldots, 180\}$. Table 5.13 presents a portion of the transaction database of alarm floods is $\mathbb{D} = \{\mathcal{F}_k, k = 1, 2, \ldots, 180\}$.

Figure 5.23b presents a small portion of the alarm burst plot with 5 alarm flood periods highlighted. In summary, some conclusions are made from the alarm burst plot in Fig. 5.23:

- The alarm rate in the whole month was high and close to the benchmark threshold of 10 alarms per 10 min, leading to alarm floods occurring frequently.
- There were a few peaks indicating extreme high alarm rates and serious alarm flood situations.

The alarm flood spiral graph in Fig. 5.24 displays the presence of alarm floods (denoted using yellow areas) over the whole month of April. The spiral starts at 00 : 00 : 00 on April 1st at the innermost circle, and goes clockwise until reaching 23 : 59 : 59 on April 30th at the outermost circle. In summary, Fig. 5.23 reveals the following conclusions:

- Alarm floods were more likely to happen during the following time periods: 0 : 00-3 : 00, 6 : 00-7 : 00, 11 : 00-13 : 00, and 16 : 00-17 : 00.
- The following days were likely to have many alarm floods: April 1st–7th and April 19th–21st.

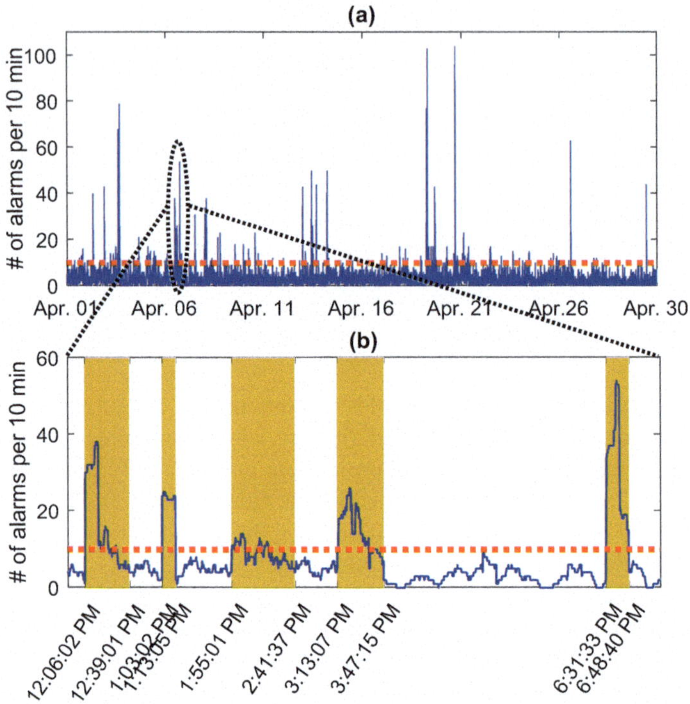

Fig. 5.23 Alarm burst plots: **a** the holistic alarm burst plot over one month (the blue curve describes the sequence of alarm rate calculated based on a 10 min time bin and the red dashed line indicates the benchmark threshold, namely, 10 alarms per 10 min); **b** a portion of the alarm burst plot with 5 alarm flood periods highlighted by yellow areas, reprinted from Ref. [23], copyright 2018, with permission from IEEE

Table 5.13 The transaction database of alarm floods, reprinted from Ref. [23], copyright 2018, with permission from IEEE

TID(k)	Alarm indices in alarm flood $\mathcal{F}_k$
1	25, 71, 253, 254, 527, 612, 639, 738, 739, 770, 897, 903, 1139, 1522, 1523, 1603, 1651, 1741
2	381, 662, 670, 723, 729, 772, 876, 899, 943, 1004, 1005, 1007, 1025, 1026, 1113, 1418, 1469, 1643
...	...
180	134, 135, 138, 140, 141, 142, 143, 169, 170, 171, 172, 181, 183, 186, 192, 220, 502, 512, 820, 1194, 1241, 1394, 1523

Next, the methods in Sect. 5.3.2 are applied to discover frequent alarm patterns from $\mathbb{D} = \{\mathcal{F}_k, k = 1, 2, \ldots, 180\}$. Firstly, the set $\mathbb{C}$ of 921 closed frequent alarm patterns is detected. The shortest and longest patterns contain 2 and 65 alarms,

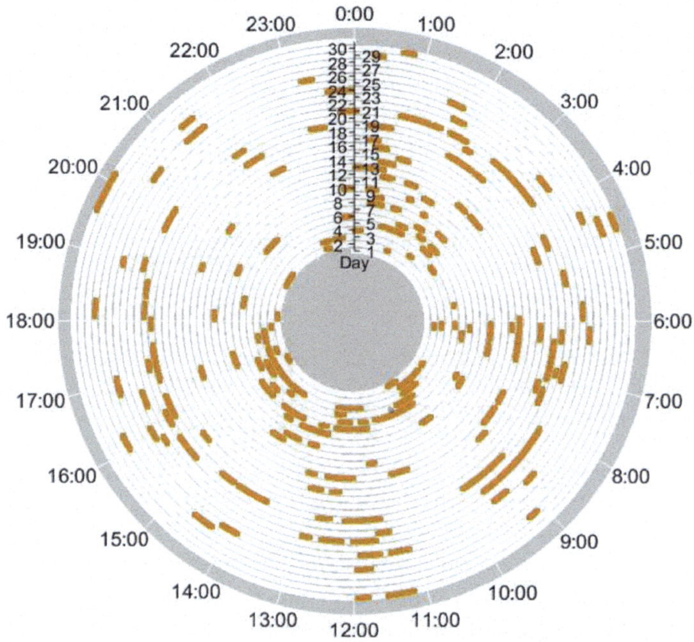

Fig. 5.24 Alarm flood spiral graph: there are 30 circles corresponding to 30 days; each circle represents 24 h; the yellow areas indicate the presence of alarm floods; the spiral goes clockwise from the innermost circle to the outermost circle, reprinted from Ref. [23], copyright 2018, with permission from IEEE

respectively. The least and most frequent pattern exist in 2 and 21 alarm floods, respectively. Table 5.14 presents an example of three closed alarm patterns and their alarm flood TIDs. It can be found that the three closed alarm patterns resemble each other, but there isn't any super alarm pattern $\mathcal{Q}$ containing any of the three patterns with the same support. For instance, the first pattern is different from the second one by one alarm, namely, alarm 1596. The first pattern is a subset of the second one, but has different set of alarm flood TIDs. In view of that the three closed alarm patterns in Table 5.14 are quite similar, it is necessary to determine a representative alarm pattern, rather than to present the three patterns to the user.

Using the method in Sect. 5.3.3, the set $\mathbb{R}$ of 207 representative alarm patterns is discovered. The representative alarm pattern for the three closed patterns in Table 5.14 is $\mathcal{R}_1 = \{1220, 1224, 1226, 1228, 1229, 1230, 1232, 1233, 1234, 1270, 1271, 1272, 1279, 1322, 1323, 1324, 1325, 1334, 1352, 1382, 1383, 1385, 1386, 1417, 1477, 1507, 1508, 1509, 1510, 1511, 1513, 1514, 1539, 1596, 1647, 1686, 1701, 1702, 1724, 1725, 1728, 1729, 1738, 1739\}$ and the corresponding TID set is $\mathcal{T}(\mathcal{R}_i) = \{12, 93, 178\}$. The alarm pattern color map for $\mathcal{R}_1 \times \mathcal{T}(\mathcal{R}_i)$ is shown in Fig. 5.25a. It can be noticed that alarms 1279, 1325, 1728, and 1738 were absent in alarm flood 12 but present in alarm floods 93 and 178. Alarm 1596 was absent

Table 5.14 An example of closed alarm patterns and their TID set, reprinted from Ref. [23], copyright 2018, with permission from IEEE

Closed alarm pattern	Alarm flood TIDs
1220, 1224, 1226, 1228, 1229, 1230, 1232, 1233, 1234, 1270, 1271, 1272, 1322, 1323, 1324, 1334, 1352, 1382, 1383, 1385, 1386, 1417, 1477, 1507, 1508, 1509, 1510, 1511, 1513, 1514, 1539, 1647, 1686, 1701, 1702, 1724, 1725, 1729, 1739	12, 93, 178
1220, 1224, 1226, 1228, 1229, 1230, 1232, 1233, 1234, 1270, 1271, 1272, 1322, 1323, 1324, 1334, 1352, 1382, 1383, 1385, 1386, 1417, 1477, 1507, 1508, 1509, 1510, 1511, 1513, 1514, 1539, 1596, 1647, 1686, 1701, 1702, 1724, 1725, 1729, 1739	12, 93
1220, 1224, 1226, 1228, 1229, 1230, 1232, 1233, 1234, 1270, 1271, 1272, 1279, 1322, 1323, 1324, 1325, 1334, 1352, 1382, 1383, 1385, 1386, 1417, 1477, 1507, 1508, 1509, 1510, 1511, 1513, 1514, 1539, 1647, 1686, 1701, 1702, 1724, 1725, 1728, 1729, 1738, 1739	93, 178

in alarm flood 178 but present in alarm floods 12 and 93. In contrast to $\mathcal{R}_1$, the representative pattern $\mathcal{R}_2$ shown as the alarm pattern color map in Fig. 5.25b has no missing alarms. $\mathcal{R}_2$ consists of 65 unique alarms. The two alarm floods containing $\mathcal{R}_2$ happened at 14:13:19 and 16:28:21 on April 3rd, respectively. They might be triggered by a repeated root cause. Further analysis requires process knowledge for validation, which is beyond the scope in this context, due to confidentiality issues as it would require significant process disclosures. Figure 5.25c presents the alarm pattern color map for the third representative pattern $\mathcal{R}_3$, which is determined from 12 similar closed alarm patterns. The compression of patterns is quite significant in this case. Thus, the user only needs to observe 1 alarm pattern rather than 12 similar patterns.

5.4 Prediction of Alarm Floods

This section is on prediction of upcoming alarms for an occurring alarm flood. In practice, it is desirable to predict upcoming alarms when several alarms occur as a precursor of alarm floods. Alarm prediction implies that proactive actions may be taken to prevent a further deterioration or propagation of occurring abnormalities. A maximum-entropy-based method is proposed here for upcoming alarm prediction in alarm floods [41]. The method has two important features: all currently-occurred alarms are exploited for prediction, and upcoming alarms are given with quantitative probabilities. By contrast, existing alarm prediction methods either use most-recent (not all) occurred alarms in prediction, or cannot predict specific alarms with quantitative probabilistic values. The method takes all historical alarm flood sequences into account to establish relationships between currently-occurred alarms and upcoming

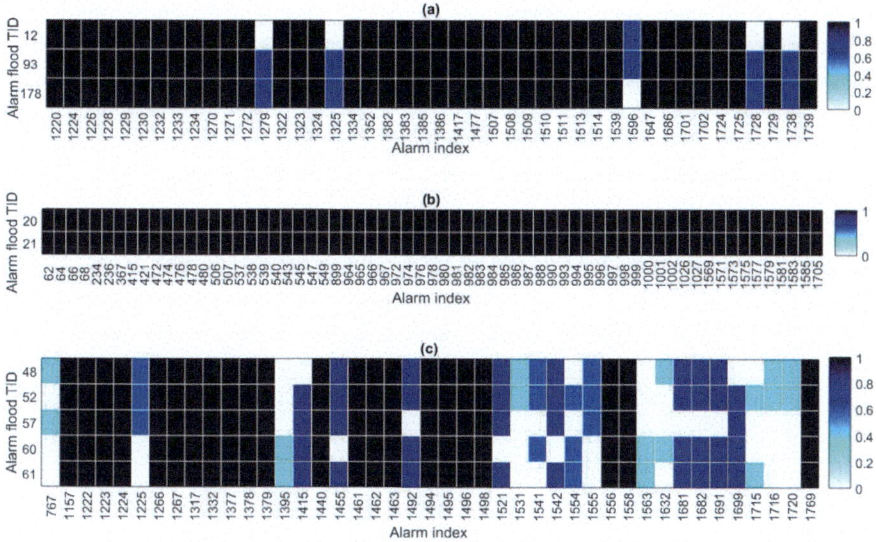

Fig. 5.25 Alarm pattern color maps for three representative alarm patterns. The x-axis and y-axis represent the alarm index and alarm flood TID, respectively. Each colored block at the i-th row and j-th column indicates the presence of alarm j in flood i whereas a white bock indicates the absence, reprinted from Ref. [23], copyright 2018, with permission from IEEE

alarms, and formulates an optimization problem to maximize conditional entropies of upcoming alarms.

5.4.1 Problem Description

An alarm event $\dot{x}_i$ is defined as the event that an alarm variable goes into the alarm state when some faults occur. A tuple with two attributes is used to represent $\dot{x}$, that is,

$$\dot{x} = (x, t) . \tag{5.47}$$

Here x is the alarm tag and t is the corresponding time stamp. Since the alarm events during an alarm flood are caused by some abnormalities. It means that these alarm events will not be removed until the corresponding abnormalities are resolved. Therefore, the time that an alarm variable first goes into the alarm state is taken as the time stamp of this alarm event. The set of historical alarm flood sequences is assumed to be given as $\{X_h(n)\}_{n=1}^{N}$, and the n-th historical alarm flood sequence $X_h(n) := \left[\dot{x}_{h,n,1}, \dot{x}_{h,n,2}, \ldots, \dot{x}_{h,n,l}\right]$ is composed of chronologically sorted alarm events, where $\dot{x}_{h,n,l}$ denotes the l-th alarm event in $X_h(n)$.

The set of all alarm tags occurred in $\{X_h(n)\}_{n=1}^{N}$ is assumed to be given as X_a. Each alarm tag in X_a is assumed to be unique. In order to satisfy this assumption, the alarm data should be preprocessed by γ-sample alarm delay timers [19]. The negative effects of chattering alarms can be reduced greatly by the alarm delay timer. For the chattering alarms that are not completely removed, the time instance that the corresponding alarm variable first goes into the alarm state is taken as the time stamp of the alarm events. The current alarm sequence is given as $X_{cs} = [\dot{x}_{c,1}, \dot{x}_{c,2}, \ldots, \dot{x}_{c,i-1}]$, where $\dot{x}_{c,1}, \dot{x}_{c,2}, \ldots, \dot{x}_{c,i-1}$ are in chronological order. All alarm tags in X_{cs} are referred to all currently-occurred alarms. The last m alarm tags in X_{cs}, that is, $\{x_{c,i-m}, x_{c,i-m+1}, \ldots, x_{c,i-1}\}$, are referred to the last m currently-occurred alarms. The set of all currently-occurred alarms is denoted as X_c. The set of alarm tags contained in X_a but not included in X_c is given as X_r. The objective is to predict the alarm x_r to be appeared next in X_{cs} by using all currently-occurred alarm events.

5.4.2 Main Idea Based on Maximum Entropy

An example is given to illustrate the main idea of the proposed method. The current alarm sequence is assumed to be $X_{cs} = [x_1, x_2]$. Twenty historical alarm sequences used for prediction are shown in Table 5.15. The schematic diagram of historical alarm sequences is shown in Fig. 5.26. The upcoming alarm could be x_3, x_4 or x_5. The aim of the proposed method is to calculate the probability distribution $P(x_j|X)$ where $j=3, 4, 5$ and $X=\{x_1, x_2, \{x_1, x_2\}\}$.

The principle of the proposed method is to obtain the probability distribution $P(x_j|X)$ that maximizes the conditional entropy H by considering all known information. The conditional entropy is defined as the weight sum of $H(x_j|X = x_q)$'s, where x_q is one of elements in X and $P(x_q)$ as the probability of x_q is used as the weight [17],

$$\begin{aligned} H &= -\sum_{x_q} P(x_q) \sum_{j=3}^{5} P(x_j|x_q) \log P(x_j|x_q) \\ &\approx -\sum_{x_q} \tilde{P}(x_q) \sum_{j=3}^{5} P(x_j|x_q) \log P(x_j|x_q). \end{aligned} \quad (5.48)$$

Table 5.15 Historical alarm sequences, reprinted from Ref. [41], copyright 2021, with permission from Elsevier

#	Alarm sequences	Number
1	$\{x_1, x_3\}$	3
2	$\{x_2, x_1, x_4\}$	16
3	$\{x_2, x_5\}$	1

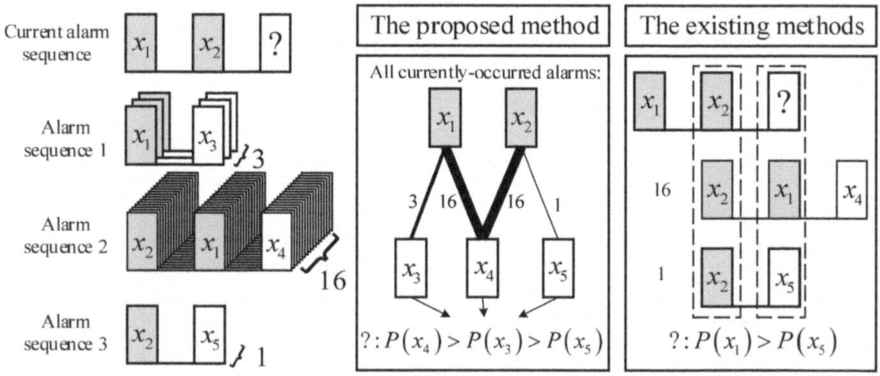

Fig. 5.26 Schematic diagram of historical alarm sequences, reprinted from Ref. [41], copyright 2021, with permission from Elsevier

The empirical probability $\widetilde{P}(x_q)$ is used to approximate $P(x_q)$. The empirical probabilities obtained from alarm sequences #1, #2 and #3 in Table 5.15 are $\widetilde{P}(x_1) = \frac{3}{20}$, $\widetilde{P}(\{x_1, x_2\}) = \frac{4}{5}$ and $\widetilde{P}(x_2) = \frac{1}{20}$. The empirical conditional probability $\widetilde{P}\left(x_j|x_q\right)$ obtained from alarm sequences #1, #2 and #3, respectively, are considered to be the known information,

$$\begin{aligned} \widetilde{P}\left(x_3|x_1\right) &= \frac{3}{20}, \\ \widetilde{P}\left(x_4|\{x_1, x_2\}\right) &= \frac{4}{5}, \\ \widetilde{P}\left(x_5|x_2\right) &= \frac{1}{20}. \end{aligned} \tag{5.49}$$

Here the empirical conditional probability $\widetilde{P}\left(x_j|x_q\right)$ is used to approximate the real conditional probability $P\left(x_j|x_q\right)$ in (5.48). In addition, the equation

$$\sum_{j=3}^{5} P\left(x_j|X = x_q\right) = 1 \tag{5.50}$$

is also known information. The above-known information in (5.49), (5.50) and the maximum entropy expression in (5.48) are combined to form an unconstrained optimization function through the Lagrange function,

$$
\begin{aligned}
L(P, w) &= \sum_{x_q}\sum_{j=3}^{5} \widetilde{P}(x_q) P(x_j|x_q) \log P(x_j|x_q) \\
&+ w_0\left(1-\sum_{j=3}^{5} P(x_j|x_q)\right) + w_1\left(P(x_3|x_1) - \frac{3}{20}\right) \\
&+ w_2\left(P(x_4|\{x_1, x_2\}) - \frac{4}{5}\right) + w_3\left(P(x_5|x_2) - \frac{1}{20}\right).
\end{aligned}
\tag{5.51}
$$

Here $w = [w_0, w_1, w_2, w_3]$ is referred to as Lagrangian multiplier [8]. By setting the partial of $L(P, w)$ with respect to $P(x_j|x_q)$ equals to 0, the expression of $P(x_j|x_q)$ with respect to w can be obtained as

$$
\begin{aligned}
&P(x_3|x_1) = e^{\frac{20}{3}(w_0-w_1)-1}, \\
&P(x_4|x_1) = P(x_5|x_1) = e^{\frac{20}{3}w_0-1}, \\
&P(x_4|\{x_1, x_2\}) = e^{\frac{5}{4}(w_0-w_2)-1}, \\
&P(x_3|\{x_1, x_2\}) = P(x_5|\{x_1, x_2\}) = e^{\frac{5}{4}w_0-1}, \\
&P(x_5|x_2) = e^{20(w_0-w_1)-1}, \\
&P(x_3|x_2) = P(x_4|x_2) = e^{20w_0-1}.
\end{aligned}
\tag{5.52}
$$

Inserting (5.52) in (5.50) gives $P(x_j|x_q)$ without w_0 as

$$
\begin{aligned}
&P(x_3|x_1) = \frac{e^{w_1}}{e^{w_1}+2}, \\
&P(x_4|x_1) = P(x_5|x_1) = \frac{1}{e^{w_1}+2}, \\
&P(x_4|\{x_1, x_2\}) = \frac{e^{w_2}}{e^{w_2}+2}, \\
&P(x_3|\{x_1, x_2\}) = P(x_5|\{x_1, x_2\}) = \frac{1}{e^{w_2}+2}, \\
&P(x_5|x_2) = \frac{e^{w_3}}{e^{w_3}+2}, \\
&P(x_3|x_2) = P(x_4|x_2) = \frac{1}{e^{w_3}+2}.
\end{aligned}
\tag{5.53}
$$

Substituting (5.53) in (5.51) and setting the partial of $L(P, w)$ with respect to w equal to 0 yield value of w as

$$
w_1 = -1.0414,\ w_2 = 2.0794,\ w_3 = -2.2513. \tag{5.54}
$$

Finally the probability distribution $P\left(x_j|X\right)$ can be calculated by substituting (5.54) in (5.53), and the optimal alarm prediction is x_4 with the largest value in $P\left(x_j|X\right)$, i.e.,

$$\max\{P\left(x_j|X\right)\} = P\left(x_4|\{x_1, x_2\}\right) = \frac{e^{w_1}}{e^{w_1}+2} = 0.7999.$$

The proposed method gives the optimal predicted alarm x_4 with the probability of 0.7999. This is a reasonable result since x_4 occurs 16 times after x_1 and x_2 in historical alarm sequences, while x_3 and x_5 only occur 3 times after x_1 and once after x_2, respectively. Therefore, the most possible upcoming alarm in the current alarm sequence should be x_4. In the existing methods [10, 42, 45], the prediction of alarm is limited by exploiting identical alarm subsequences. As shown in Fig. 5.26, the identical alarm subsequence only contains one alarm x_2. If the existing methods are used, only x_2 is used for prediction. As a result, x_1 is incorrectly predicted as the next upcoming alarm. In fact, x_4 is more likely to occur.

5.4.3 Steps of the Proposed Method

If some historical alarm flood sequences in $\{X_h(n)\}_{n=1}^{N}$ are similar to the current alarm sequence X_{cs}, possibly from the same fault, but no subsequences of alarm events are exactly the same, these sequences should not be removed from predicting the upcoming alarm event in X_{cs}; otherwise, the prediction accuracy may be compromised. Thus, the proposed method needs to calculate the historical influence of all currently-occurred alarms for the upcoming alarms, and predict the upcoming alarms by exploiting historical alarm floods. The prediction is developed in the following steps.

First, the set of currently-occurred alarms X_c and the set of upcoming alarms $X_r := \{x_{r,1}, x_{r,2}, \ldots, x_{r,K}\}$ are obtained. The conditional entropy is

$$\begin{aligned} H\left(X_r|X_c\right) &= -\sum_{j=1}^{i-1} P\left(x_{c,j}\right) H\left(X_r|X_c = x_{c,j}\right) \\ &= -\sum_{j=1}^{i-1} P\left(x_{c,j}\right) \sum_{k=1}^{K} P\left(x_{r,k}|x_{c,j}\right) \log P\left(x_{r,k}|x_{c,j}\right). \end{aligned} \quad (5.55)$$

For the notation simplicity, $P\left(x_{r,k}|x_{c,j}\right)$ is denoted by $P := P\left(x_{r,k}|x_{c,j}\right)$. The aim is to find the conditional probabilities $P^*\left(x_{r,k}|x_{c,j}\right)$'s that maximize $H\left(X_r|X_c\right)$ according to the maximum entropy principle,

$$P^*\left(x_{r,k}|x_{c,j}\right) = \underset{P(x_{r,k}|x_{c,j})}{\arg\max} H\left(X_r|X_c\right). \quad (5.56)$$

Second, the quantitative relationships between $x_{c,j}$ and $x_{r,k}$ are extracted from the set of historical alarm flood sequences. If $x_{c,j}$ has occurred in some historical alarm floods, it may be connected with subsequent alarms in those alarm floods. Hence, a characteristic function f_1 is used to represent the quantitative relationship that $x_{r,k}$ occurs after $x_{c,j}$ in historical alarm flood sequences,

$$f_1\left(x_{c,j}, x_{r,k}\right) = \sum_{n=1}^{N} f_1^{(n)}\left(x_{c,j}, x_{r,k}\right), \tag{5.57}$$

where

$$f_1^{(n)}\left(x_{c,j}, x_{r,k}\right) = \begin{cases} 1, & \text{if } \dot{x}_{c,j}, \dot{x}_{r,k} \in X_h(n) \text{ and } t_{c,j} < t_{r,k}, \\ 0, & \text{otherwise.} \end{cases}$$

The characteristic function f_1 counts the number of times that $x_{r,k}$ occurs after $x_{c,j}$ in historical alarm floods. However, it is possible that some closely related alarms arise almost simultaneously, but the order of them in alarm sequences varies from time to time [14]. As a result, the alarm events occurred after an alarm event should also be considered for prediction. In order to distinguish from the normal chronological order situation, a new characteristic function with a penalty is introduced in calculating this quantitative relationship. Two concepts 'time distance' and 'time weight' defined in [14] are used to explain this penalty.

The time distance between $\dot{x}_{c,j}$ and $\dot{x}_{r,k}$ is defined as

$$d_{c,j}^{r,k} = \left|t_{c,j} - t_{r,k}\right|, \tag{5.58}$$

and the corresponding time weight is

$$w_{c,j}^{r,k} = e^{-\frac{\left(d_{c,j}^{r,k}\right)^2}{2\sigma^2}}. \tag{5.59}$$

Here the scaled Gaussian function is used as the time weighting function. The choice of σ in (5.59) is a user-selected parameter. The value of σ shows how much the time difference can be tolerated. A larger σ leads to a less important of the orders of two alarm events [14]. A so-called penalty characteristic function f_2 is introduced

$$f_2\left(x_{c,j}, x_{r,k}\right) = \sum_{n=1}^{N} f_2^{(n)}\left(x_{c,j}, x_{r,k}\right), \tag{5.60}$$

where

$$f_2^{(n)}\left(x_{c,j}, x_{r,k}\right) = \begin{cases} w_{r,k}^{c,j}, & \text{if } \dot{x}_{c,j}, \dot{x}_{r,k} \in X_h(n) \text{ and } t_{c,j} > t_{r,k}, \\ 0, & \text{otherwise.} \end{cases}$$

The characteristic function f_2 reflects that in the prediction of $x_{r,k}$, the influence of the alarm events occurred after $x_{r,k}$ in historical alarm flood sequences is also taken into account.

For the notation simplicity, $f_v\left(x_{c,j}, x_{r,k}\right)$ is denoted by $f_v := f_v\left(x_{c,j}, x_{r,k}\right)$ for $v = 1, 2$. The expectation of f_v corresponding to the joint distribution $P\left(x_{c,j}, x_{r,k}\right)$ is

$$E_P\left(f_v\right) = \sum_{j=1}^{i-1}\sum_{k=1}^{K} P\left(x_{c,j}, x_{r,k}\right) f_v, \tag{5.61}$$

and the expectation of f_v corresponding to the empirical distribution $\widetilde{P}(x_{c,j}, x_{r,k})$ is

$$E_{\widetilde{P}}\left(f_v\right) = \sum_{j=1}^{i-1}\sum_{k=1}^{K} \widetilde{P}\left(x_{c,j}, x_{r,k}\right) f_v. \tag{5.62}$$

$E_P\left(f_v\right)$ in (5.61) is the proportion of the sequences that satisfy characteristic function f_v in real but unknown situations. $E_{\widetilde{P}}\left(f_v\right)$ in (5.62) is the proportion of the sequences that satisfy characteristic functions f_v's in historical alarm flood sequences. The values of two expectations should be approximately equal [7],

$$\left|E_P\left(f_v\right) - E_{\widetilde{P}}\left(f_v\right)\right| < \varepsilon, \tag{5.63}$$

where a very small value ε is the error between $E_P\left(f_v\right)$ and $E_{\widetilde{P}}\left(f_v\right)$. A maximum optimization problem with constraints is formulated as

$$\begin{aligned} P^*\left(x_{r,k}|x_{c,j}\right) &= \underset{P(x_{r,k}|x_{c,j})}{\arg\max}\, H\left(X_r|X_c\right), \\ s.t.\quad & E_P\left(f_v\right) \approx E_{\widetilde{P}}\left(f_v\right), v = 1, 2, \\ & \sum_{k=1}^{K} P\left(x_{r,k}|x_{c,j}\right) = 1. \end{aligned} \tag{5.64}$$

This optimization problem is solved as follows. Each element $x_{h,n,l}$ in $X_h(n)$ is combined with $x_{r,k}$ into pairs

$$\left\{X_h\left(n; x_{h,n,l}, x_{r,k}\right)\right\} := \{\left[x_{h,n,1}, x_{r,1}\right], \left[x_{h,n,1}, x_{r,2}\right], \ldots, \left[x_{h,n,l}, x_{r,K}\right]\}. \tag{5.65}$$

The empirical distribution of $\widetilde{P}\left(x_{c,j}, x_{r,k}\right)$ is

$$\widetilde{P}\left(x_{c,j}, x_{r,k}\right) = \frac{\sum_{n=1}^{N}\left|\left[x_{c,j}, x_{r,k}\right] \equiv \left\{X_h\left(n; x_{h,n,i}, x_{r,k}\right)\right\}\right|}{N}, \tag{5.66}$$

where the symbol "≡" is the true value '1' if the compared pairs are the same, and the operation $|\cdot|$ counts the number of true values. The empirical distribution $\widetilde{P}\left(x_{c,j}\right)$ is

$$\widetilde{P}\left(x_{c,j}\right)=\frac{\sum_{n=1}^{N}\left|x_{c,j}\equiv\left\{X_h\left(n;x_{h,n,l}\right)\right\}\right|}{N}, \tag{5.67}$$

where

$$\left\{X_h\left(n;x_{h,n,l}\right)\right\}:=\left\{x_{h,n,1},x_{h,n,2},\ldots,x_{h,n,l}\right\}. \tag{5.68}$$

Since the joint distribution $P\left(x_{c,j},x_{r,k}\right)$ in (5.61) is unknown, the empirical distribution $\widetilde{P}\left(x_{c,j}\right)$ is used for approximation,

$$P\left(x_{c,j},x_{r,k}\right)\approx\widetilde{P}\left(x_{c,j}\right)\cdot P\left(x_{r,k}|x_{c,j}\right). \tag{5.69}$$

A negative operation is given to $H\left(X_r|X_c\right)$ in order to make $-H\left(X_r|X_c\right)$ become a convex function. A loss function $L_{r,c}$ is defined as

$$L_{r,c}=\sum_{j=1}^{i-1}\sum_{k=1}^{K}\widetilde{P}\left(x_{c,j}\right)P\left(x_{r,k}|x_{c,j}\right)\log P\left(x_{r,k}|x_{c,j}\right).$$

Since $L_{r,c}=-H\left(X_r|X_c\right)$ is a convex function, the optimization problem can be converted to an unconstrained optimization function through Lagrange function. The Lagrange function $L_{r,c}\left(w\right)$ is

$$\begin{aligned}L_{r,c}\left(w\right)=&-H\left(X_r|X_c\right)+w_0\left(1-\sum_{k=1}^{K}P\left(x_{r,k}|x_{c,j}\right)\right)\\&+\sum_{v=1}^{2}w_v\left(E_{\widetilde{P}}\left(f_v\right)-E_P\left(f_v\right)\right).\end{aligned} \tag{5.70}$$

Here $w=\left[w_0,w_1,w_2\right]$ is the Lagrangian multiplier. All constraints in (5.64) should be satisfied, so that the optimization problem becomes

$$\begin{aligned}\min_{P\in C}\max_{w}L_{r,c}\left(w\right)&=\min_{P\in C}L_{r,c}+\begin{cases}0, & P\in C\\ \infty, & \text{otherwise}\end{cases}\\&=\min_{P\in C}L_{r,c},\ \text{if } P\in C,\end{aligned}$$

where $C=\{P|E_P(f_v)\approx E_{\widetilde{P}}(f_v),\ v=1,2\}$ is the set of constraints. Then the original optimization problem is equivalent to

$$\min_{P\in C}\max_{w}L_{r,c}(w). \tag{5.71}$$

Since $L_{r,c}(w)$ is a convex function, the problem in (5.71) can be converted to the solution of the Lagrange duality problem [8],

$$\min_{P\in C}\max_{w} L_{r,c}(w) = \max_{w}\min_{P\in C} L_{r,c}(w). \tag{5.72}$$

The first step in solving the problem in (5.72) is to find $\min_{P\in C} L_{r,c}(w)$. For the notation simplicity, $\min_{P\in C} L_{r,c}(w)$ is denoted by $\varphi(w) := \min_{P\in C} L_{r,c}(w)$, so that the solution of $\varphi(w)$ is

$$P_w := P_w\left(x_{r,k}|x_{c,j}\right) = \arg\min_{P\in C} L_{r,c}(w). \tag{5.73}$$

The partial derivative of $L_{r,c}(w)$ in (5.70) with respect to $P\left(x_{r,k}|x_{c,j}\right)$ is

$$\begin{aligned}\frac{\partial L_{r,c}(w)}{\partial P\left(x_{r,k}|x_{c,j}\right)} &= \sum_{j=1}^{i-1} \widetilde{P}\left(x_{c,j}\right)\left(\log P\left(x_{r,k}|x_{c,j}\right)+1\right) \\ &\quad - \sum_{j=1}^{i-1} \widetilde{P}\left(x_{c,j}\right)\left(w_0 + \sum_{v=1}^{2} w_v f_v\right).\end{aligned}$$

By setting the partial derivative equal to 0, the expression of $P\left(x_{r,k}|x_{c,j}\right)$ with respect to w is

$$\begin{aligned}P_w\left(x_{r,k}|x_{c,j}\right) &= \exp\left(\sum_{v=1}^{2} w_v f_v + w_0 - 1\right) \\ &= \frac{\exp\left(\sum_{v=1}^{2} w_v f_v\right)}{\exp(1-w_0)}.\end{aligned} \tag{5.74}$$

Since $\sum_{k=1}^{K} P_w\left(x_{r,k}|x_{c,j}\right) = 1$, the denominator in (5.74) is

$$Z_w := \exp(1-w_0) = \sum_{k=1}^{K} \exp\left(\sum_{v=1}^{2} w_v f_v\right).$$

The solution of $\varphi(w)$ in (5.73) is

$$P_w := P_w\left(x_{r,k}|x_{c,j}\right) = \frac{1}{Z_w}\exp\left(\sum_{v=1}^{2} w_v f_v\right). \tag{5.75}$$

Thus, $\varphi(w)$ can be calculated by substituting (5.75) into (5.70),

$$\varphi(w) = \sum_{j=1}^{i-1}\sum_{k=1}^{K} \widetilde{P}\left(x_{c,j}\right) P_w \log P_w + \sum_{v=1}^{2} w_v\left(E_{\widetilde{P}}(f_v) - E_P(f_v)\right).$$

As an equality holds,

$$\log P_w = \sum_{v=1}^{2} w_v f_v - \log Z_w, \tag{5.76}$$

$\varphi(w)$ is simplified by substituting (5.55), (5.61), (5.62) and (5.69) into (5.76),

$$\varphi(w) = \sum_{v=1}^{2} w_v f_v \sum_{j=1}^{i-1} \sum_{k=1}^{K} \widetilde{P}\left(x_{r,k}, x_{c,j}\right) - \sum_{j=1}^{i-1} \widetilde{P}\left(x_{c,j}\right) \log Z_w.$$

Finally, the Lagrangian multiplier that needs to be calculated in (5.72) is

$$w^* = \left[w_0^*, w_1^*, w_2^*\right] = \arg\max_w \varphi(w). \tag{5.77}$$

The optimal solution of the ME-based method in (5.56) is obtained as

$$P^*\left(x_{r,k}|x_{c,j}\right) = \frac{1}{Z_{w^*}} \exp\left(\sum_{v=1}^{2} w_v^* f_v\right), \tag{5.78}$$

where

$$Z_{w^*} = \sum_{k=1}^{K} \exp\left(\sum_{v=1}^{2} w_v^* f_v\right).$$

There is no analytic solution to the estimation of $P^*\left(x_{r,k}|x_{c,j}\right)$ since (5.78) is a log-linear model. However, $\varphi(w)$ is a convex function. Thus the solution can be obtained via standard optimization algorithms [18]. Here an optimization algorithm called Generalized Iterative Scaling (GIS) algorithm is used to obtain w^*. The GIS algorithm is composed of the following steps [31]:

1. Initialize all w_v to 0, that is,

$$w_v^{(\delta=0)} = 0, v = 1, 2,$$

   where the superscript (δ) means the δ-th iteration.
2. Update the following equation

$$w_v^{(\delta+1)} = w_v^{(\delta)} + \frac{1}{\lambda} \log \frac{E_{\widetilde{P}}(f_v)}{E_p(f_v)}. \tag{5.79}$$

   Here λ is the number of alarm events in the longest historical alarm flood sequence.
3. Repeat the above step #2 until w_v converges

$$w_v^{(\delta+1)} - w_v^{(\delta)} < \varepsilon_w, \tag{5.80}$$

where ε_w means an admissible error and the value of ε_w is a user-selected parameter. The optimal result $w^* = [w_0^*, w_1^*, w_2^*]$ in (5.77) is the one that all w_v's are convergent.

Thus, $P^*(x_{r,k}|x_{c,j})$ in (5.78) can be calculated. The optimal alarm prediction is x_r with the largest value in $P^*(x_{r,k}|x_{c,j})$, i.e.,

$$P(x_r|x_{c,j}) = \max P^*(x_{r,k}|x_{c,j}). \tag{5.81}$$

The time difference between the occurrence of x_r and the current time instant is taken as the mean of all time differences between x_r and $x_{c,i-1}$ in the historical alarm flood sequences,

$$T_r = \frac{1}{N_{r,c}} \sum_{n=1}^{N} \left(t_r^{(n)} - t_{c,i-1}^{(n)}\right), \tag{5.82}$$

where

$$\begin{aligned} N_{r,c} &= \sum_{n=1}^{N} \left|\left[x_r, x_{c,i-1}\right] \equiv \left\{X_h\left(n; x_{h,n,i}, x_{r,k}\right)\right\}\right|, \\ t_r^{(n)} - t_{c,i-1}^{(n)} &= \begin{cases} t_r - t_{c,i-1}, & \text{if } \dot{x}_r, \dot{x}_{c,i-1} \in X_h(n), \\ 0, & \text{otherwise.} \end{cases} \end{aligned} \tag{5.83}$$

Note that $\{X_h(n; x_{h,n,i}, x_{r,k})\}$ is defined in (5.65) and $N_{r,c}$ means the total number of $[x_r, x_{c,i-1}]$ occurred in the historical alarm flood sequences.

Given the current alarm sequence X_{cs} and $\{X_h(n)\}_{n=1}^{N}$ as the set of historical alarm flood sequences. Here the start of an alarm flood is indicated by the first event that at least 10 alarm variables run into the alarm state per 10 min, and the end of an alarm flood is indicated by the first event that less than 5 alarm variables remain in the alarm state per 10 min. The alarm sequence within a 10-min sliding window is chosen as the current alarm sequence X_{cs}. The step of the sliding window is set as the sampling period. The proposed method is composed of the following steps:

1. Obtain the currently-occurred alarm tags set X_c and the upcoming alarm tags set X_r.
2. Calculate the characteristic functions set $\{f_v\}_{v=1}^{2}$ to establish the constraints of the proposed method in (5.64).
3. Determine the expression of $P_w(x_{r,k}|x_{c,j})$ in (5.76) by minimizing the Lagrange function $L_{r,c}(w)$ in (5.70).
4. Obtain the convergent Lagrangian multiplier w^* via GIS algorithm and substitute w^* into (5.78).
5. Select the optimal alarm prediction x_r as the one with the largest prediction probability $P(x_r|x_{c,j})$.

The computational cost of the proposed method is analyzed here. The computation complexity in calculating the constraint functions f_v's is equal to $O\left((i-1)\times K\right)$, where i is the number of alarm tags in X_c and K is the number of alarm tags in X_r. The computation complexity in calculating the empirical distribution of $\widetilde{P}\left(x_{c,j}, x_{r,k}\right)$ and $\widetilde{P}\left(x_{c,j}\right)$ is equal to $O\left(N\right)$. Here N is the total number of historical alarm flood sequences used for prediction. The computation complexity in calculating w^* is $O\left(\Delta\right)$. Here Δ is the total iteration times until the Lagrangian multipliers are convergent. The computation complexity of the rest calculations is minor as $O\left(1\right)$. Because K, N, T and i usually take small values, the computational cost is quite minor, being acceptable for practical applications. For instance, the proposed method in the second example at Sect. 5.4.4.2 costs 0.192 s in processing historical alarm floods at a personal computer with Intel i5 2.4-GHz CPU and 16.0-GB memory.

5.4.4 Numerical Examples

This section presents two numerical examples to illustrate the proposed method for prediction of upcoming alarms.

5.4.4.1 Example 1

This example illustrates the necessity of introducing the characteristic functions f_1 and f_2 in the proposed method. Let the current alarm sequence be $X_{cs} = [(x_1, 0s),$ $(x_2, 2s), (x_3, 3s), (x_4, 312s), (x_5, 313s)]$. The historical alarm flood sequence is obtained as $X_h(1)$ in Table 5.16.

The first step of the proposed method obtains the set of currently-occurred alarm tags $X_c = \{x_1, x_2, x_3, x_4, x_5\}$ and the set of all alarm tags $X_t = \{x_1, x_2,$ $x_3, x_4, x_5, x_6, x_7\}$. Hence, the set of remaining alarm tags is $X_r = \{x_6, x_7\}$. The second step of the proposed method calculates the constraint functions f_v's for $v = 1, 2$. The value of σ is taken as 30. Since x_1 occurs before x_6 and x_7 in $X_h(1)$, the known information given by $X_h(1)$ is that the occurrence of x_1 may lead to x_6 or x_7. Thus, the values of $f_1(x_1, x_6)$ and $f_1(x_1, x_7)$ are taken as 1, and the values of $f_2(x_1, x_6)$ and $f_2(x_1, x_7)$ are taken as 0. The effects of x_2 and x_3 on the prediction are the same as x_1. The values of constraint functions are shown in the second and third rows of

Table 5.16 The set of historical alarm flood sequences $\{X_h\left(n\right)\}_{n=1}^{N=2}$, reprinted from Ref. [41], copyright 2021, with permission from Elsevier

n	Alarm flood sequences
1	$(x_3, 0s), (x_2, 4s), (x_1, 5s), (x_7, 54s), (x_6, 332s),$ $(x_5, 334s), (x_4, 336s)$

Table 5.17 The constraint functions extracted from $X_h(1)$, reprinted from Ref. [41], copyright 2021, with permission from Elsevier

$x_{c,j}$	$f_1(x_{c,j}, x_7)$	$f_2(x_{c,j}, x_7)$	$f_1(x_{c,j}, x_6)$	$f_2(x_{c,j}, x_6)$
x_1	1	0	1	0
x_2	1	0	1	0
x_3	1	0	1	0
x_4	0	1.21×10^{-18}	0	0.997
x_5	0	6.5×10^{-19}	0	0.991

Table 5.17. Although x_4 occurs after x_6 and x_7 in $X_h(1)$, the time distance d_4^7 is 282s while d_4^6 is only 4s. It means that x_4 may be highly related to x_6 but irrelevant with x_7. Hence, the value of $f_2(x_4, x_6)$ is close to 1 while the value of $f_2(x_4, x_7)$ is close to 0 as shown in Table 5.17. The effect of x_5 on the prediction is similar to x_4. The values of characteristic functions are shown in the last row of Table 5.17. The rest steps in the proposed method yield the prediction probability of upcoming alarms as

$$P^*\left(x_6|x_{c,j}\right) = 0.9712,\ \ P^*\left(x_7|x_{c,j}\right) = 0.0288.$$

The proposed method gives the optimal predicted alarm x_6 with the probability as 0.9712. This clearly is a reasonable result. A larger value of σ leads to a less important order between two alarms with a large time difference. The necessity of using the time weighting function is transparent by looking at the results for $\sigma = \infty$. If $\sigma = \infty$, the upcoming alarm predicted by the proposed method is x_6 or x_7, both with the probability as 0.5. However, such a result is not reasonable, because x_7 occurs much earlier than x_4 and x_5 in $X_h(1)$, which means x_4 and x_5 should not have a major impact on the occurrence of x_7.

5.4.4.2 Example 2

The effectiveness of the proposed method for alarm prediction has been tested on simulated data from the Tennessee Eastman Process (TEP), which is a nonlinear open-loop unstable process that has been used in many studies as a benchmark for process control and fault diagnosis. The diagram of the TEP model is shown in Fig. 4.26. The TEP consists of five major units: reactor, condenser, compressor, separator and stripper, as well as some auxiliary devices such as pumps, energy exchangers, valves and sensors. There are mainly four irreversible exothermic chemical reactions,

$$\begin{cases} A + C + D \rightarrow G,\ A + C + E \rightarrow H, \\ A + E \rightarrow F,\ 3D \rightarrow 2F. \end{cases}$$

Here A, C, D and E are four feed gases, F is a liquid byproduct, and G and H are two gas products.

There are 6 operation conditions and 21 preprogrammed faults in the TEP [32], and the faults are classified into the step, sticking and random fault. The step faults make significant changes to several process variables. By contrast, the random faults make only minor changes to several process variables. When a sticking fault occurs, a valve is stuck, some process variables are affected with significant changes. Due to the nonstationary characteristics of the production process, different alarm floods occur under different operation conditions [44]. Therefore, in this context, the set of historical alarm flood sequences and the current alarm sequence are simulated by step faults and sticking faults under different operation conditions, while the alarms induced by random faults are treated as the 'noise' to the alarm floods simulated by step faults and sticking faults.

In particular, seven faults are considered under three operation conditions here [32], namely, the fault 3 in the D feed temperature, the step fault 4 in the reactor cooling water inlet temperature, the step fault 7 in the C header pressure loss-reduced availability, the random fault 8 in the D feed temperature, the random fault 9 in the C feed temperature, the random fault 12 in the condenser cooling water inlet temperature, and the step fault 14 in the reactor cooling water valve. Twenty-six Monte Carlo simulations are implemented by activating different step faults after different random faults under operation conditions 1, 2, and 3, respectively, and the step fault 7 is activated to simulate nuisance alarms x_{11}, x_{31} and x_{32}. The alarms and their associated process variables, alarm thresholds, tags, and units are shown in Table 5.18. The alarm thresholds are set according to the control limits being designed in [33]. Table 5.19 shows the set of historical alarm flood sequences. The sample period is set as 1s. The time stamp of the alarm flood sequence has been recorded since the first alarm occurs.

The proposed method is ready to predict the upcoming alarm for a current alarm event sequence X_{cs}. Let X_{cs}=[$(x_1, 0)$,$(x_2, 18s)$,$(x_3, 32s)$,$(x_4, 197s)$, $(x_5, 227s)$, $(x_6, 235s)$] be the current alarm sequence. The time sequences of the corresponding process variables and the alarm thresholds are presented in Fig. 5.28. Here the step faults 3 and 14 are activated under operation condition 1 to simulate an occurring alarm flood. The time sequences of alarm floods 1, 2, 3, and the current alarm sequence X_{cs} are presented in Fig. 5.27. The first step of the proposed method is to calculate the set of currently-occurred alarms and the set of upcoming alarms by dividing X_{cs} and $\{X_h(n)\}_{n=1}^{N=26}$. The set of currently-occurred alarms is $\{x_1, x_2, x_3, x_4, x_5, x_6\}$, and the set of upcoming alarms is $\{x_7, x_9, x_{10}, x_{11}, x_{14}, x_{15}, x_{16}, x_{17}, x_{18}, x_{19}, x_{20}, x_{21}, x_{22}, x_{31}, x_{32}\}$. Since the alarms in the #4 set of alarm floods are totally different from the alarms in X_{cs} so that the alarms in the alarm flood #4 are not considered in calculating the characteristic functions. The characteristic functions f_v's for $v = 1, 2$ are given in Tables 5.20 and 5.21. The value of σ is taken as 60. Since the alarms in X_{cs} always occur before x_7 in 16 historical alarm floods. The values of characteristic functions $f_1(x_{c,j}, x_7)$ are counted as 16. The alarms in X_{cs} only occur together with x_9 and x_{10} in the alarm flood #1, so that the values of characteristic functions $f_1(x_{c,j}, x_9)$ and $f_1(x_{c,j}, x_{10})$ are counted as 1. It is easy to observe from $\{X_h(n)\}_{n=1}^{N=26}$

Table 5.18 Alarm descriptions, reprinted from Ref. [41], copyright 2021, with permission from Elsevier

Process variable	Alarm threshold (Type)	Tag	Unit
Stripper temp.	65 (low)	x_1	°C
Product separator temp.	88.7 (low)	x_2	°C
Separator coolant temp.	89.8 (low)	x_3	°C
Reactor feed rate	49 (high)	x_4	kscm/h
Product separator level	60 (high)	x_5	%
Compressor power	280 (high)	x_6	kw
Stripper level	47 (low)	x_7	%
Product separator pressure	2680 (low)	x_8	kPa
Recycle flow rate	29.5 (low)	x_9	kscm/h
Product separator level	42 (low)	x_{10}	%
Reactor pressure	2462 (low)	x_{11}	kPa
Stripper pressure	3360 (high)	x_{12}	kPa
Stripper underflow rate	24 (high)	x_{13}	m3/h
Reactor coolant temp.	102 (low)	x_{14}	°C
Product separator temp.	95 (high)	x_{15}	°C
Separator coolant temp.	96 (high)	x_{16}	°C
Recycle flow rate	34 (high)	x_{17}	kscm/h
Compressor power	250 (low)	x_{18}	kw
Stripper level	60 (high)	x_{19}	%
Purge rate	0.7 (high)	x_{20}	kscm/h
Stripper temp.	70 (high)	x_{21}	°C
Reactor feed rate	45 (low)	x_{22}	kscm/h
Reactor level	60 (low)	x_{23}	%
Purge rate	0.4 (low)	x_{24}	kscm/h
Recycle flow rate	34.5 (high)	x_{25}	kscm/h
Reactor pressure	2820 (high)	x_{26}	kPa
Stripper pressure	3020 (low)	x_{27}	kPa
Separator underflow rate	22.3 (low)	x_{28}	m3/h
Stripper underflow rate	20.5 (low)	x_{29}	m3/h
Reactor coolant temp.	130.5 (high)	x_{30}	°C
Separator underflow rate	27 (high)	x_{31}	m3/h
Product separator pressure	2995 (high)	x_{32}	kPa
Reactor temp.	120 (low)	x_{33}	°C

that $\{x_4, x_5, x_6\}$ occurs much later than x_{11}, while $\{x_1, x_2, x_3\}$ always occurs before x_{11}. It means that $\{x_4, x_5, x_6\}$ has a much smaller impact on the prediction of x_{11} than $\{x_1, x_2, x_3\}$. Hence the values of $f_2(x_4, x_{11})$, $f_2(x_5, x_{11})$ and $f_2(x_6, x_{11})$ are

Table 5.19 The set of historical alarm flood sequences $\{X_h(n)\}_{n=1}^{N=26}$, reprinted from Ref. [41], copyright 2021, with permission from Elsevier

Operation conditions	Faults	Alarm flood sequences	Numbers
1	3,7,8,14	$(x_1, 0s)$, $(x_2, 20s)$, $(x_3, 36s)$, $(x_{11}, 110s)$, $(x_4, 202s)$, $(x_5, 226s)$, $(x_6, 250s)$, $(x_9, 376s)$, $(x_{10}, 402s)$, $(x_7, 500s)$	1
2	3,7,9,14	$(x_1, 0s)$, $(x_3, 22s)$, $(x_2, 30s)$, $(x_{11}, 110s)$, $(x_{31}, 112s)$, $(x_{32}, 115s)$, $(x_5, 224s)$, $(x_6, 242s)$, $(x_4, 274s)$, $(x_7, 460s)$	5
3	3,7,12,14	$(x_1, 0s)$, $(x_2, 24s)$, $(x_3, 34s)$, $(x_{11}, 110s)$, $(x_{31}, 112s)$, $(x_{32}, 115s)$, $(x_6, 218s)$, $(x_4, 256s)$, $(x_5, 282s)$, $(x_7, 476s)$	10
1	4	$(x_{14}, 0s)$, $(x_{15}, 40s)$, $(x_{16}, 74s)$, $(x_{17}, 152s)$, $(x_{18}, 176s)$, $(x_{19}, 268s)$, $(x_{20}, 272s)$, $(x_{21}, 404s)$, $(x_{22}, 446s)$	10

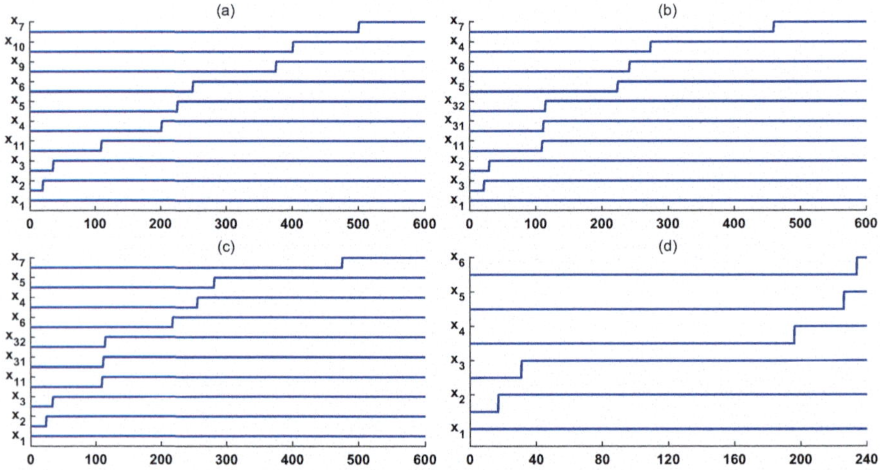

Fig. 5.27 **a** The time sequence of the alarm flood #1, **b** the time sequence of the alarm flood #2, **c** the time sequence of the alarm flood #3, **d** the time sequence of the current alarm sequence X_{cs}, reprinted from Ref. [41], copyright 2021, with permission from Elsevier

much smaller than 16 as shown in the last three rows of Table 5.21. The analysis of $f_v(x_{c,j}, x_{31})$'s and $f_v(x_{c,j}, x_{32})$'s is similar to $f_v(x_{c,j}, x_{11})$'s. The rest steps in the proposed method yield the prediction results as listed in Table 5.22.

The predicted probability of x_7 is significantly larger than the other three alarms. This is a reasonable result, because only the alarm flood #1 contains alarms x_9 and x_{10}. By contrast, x_7, x_{11}, x_{31} and x_{32} occur in not less than 15 alarm floods. However, the values of characteristic functions for x_{11}, x_{31} and x_{32} are confined by the time weight function. In fact, the alarms x_{11}, x_{31} and x_{32} induced by the step fault 7 are treated as the nuisance alarms in the alarm floods caused by the step faults 3 and 14. Hence, the

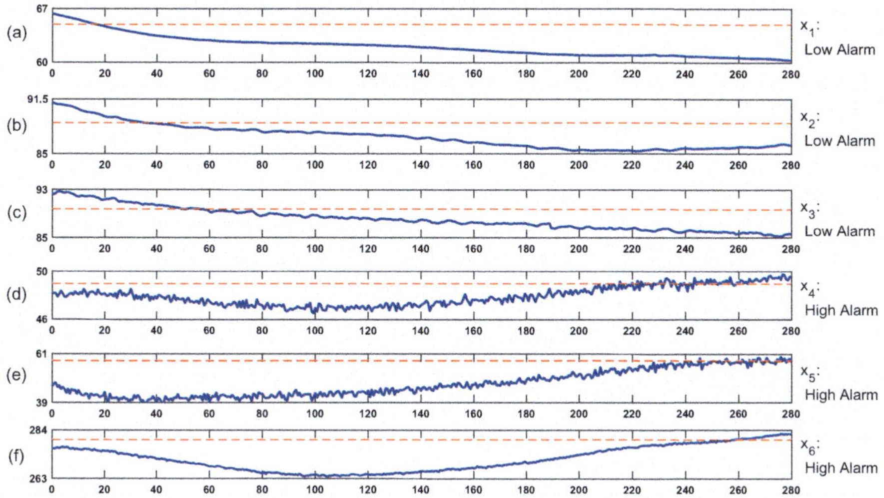

Fig. 5.28 Time sequences of process variables for current occurred alarms in X_{cs} (solid) and the alarm thresholds (dashed): **a** stripper temperature, **b** product separator temperature, **c** separator coolant temperature, **d** reactor feed rate, **e** product separator level, **f** compressor power, reprinted from Ref. [41], copyright 2021, with permission from Elsevier

Table 5.20 The constraint function f_1, reprinted from Ref. [41], copyright 2021, with permission from Elsevier

$x_{c,j}$	$f_1(x_7)$	$f_1(x_9)$	$f_1(x_{10})$	$f_1(x_{11})$	$f_1(x_{31})$	$f_1(x_{32})$
x_1	16	1	1	16	15	15
x_2	16	1	1	16	15	15
x_3	16	1	1	16	15	15
x_4	16	1	1	0	0	0
x_5	16	1	1	0	0	0
x_6	16	1	1	0	0	0

Table 5.21 The constraint function f_2, reprinted from Ref. [41], copyright 2021, with permission from Elsevier

$x_{c,j}$	$f_2(x_7)$	$f_2(x_9)$	$f_2(x_{10})$	$f_2(x_{11})$	$f_2(x_{31})$	$f_2(x_{32})$
x_1	0	0	0	0	0	0
x_2	0	0	0	0	0	0
x_3	0	0	0	0	0	0
x_4	0	0	0	0.9462	0.6919	0.7814
x_5	0	0	0	1.141	1.1613	1.168
x_6	0	0	0	2.4786	2.5784	2.8235

Table 5.22 Prediction results for X_{cs} while $\sigma = 60$, reprinted from Ref. [41], copyright 2021, with permission from Elsevier

Alarm prediction	Probability
x_7	0.8084
x_{11}	0.0578
x_{32}	0.053
x_{31}	0.0472
x_9	0.0168
x_{10}	0.0168

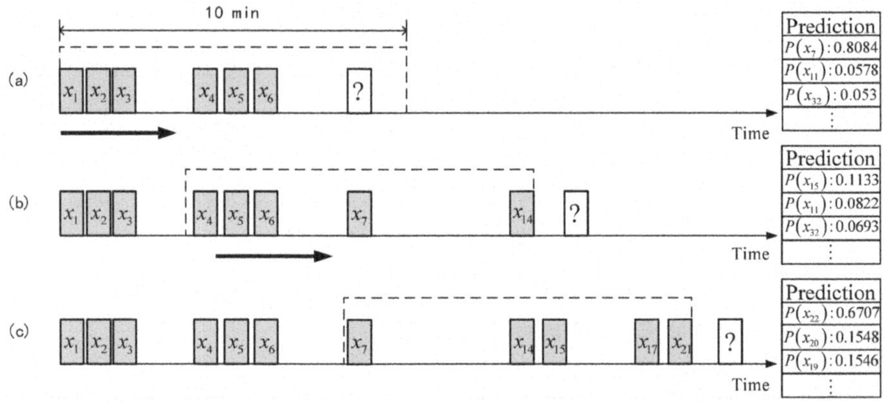

Fig. 5.29 Prediction results for two alarm floods, reprinted from Ref. [41], copyright 2021, with permission from Elsevier

proposed method gives the optimal predicted alarm x_7 with the probability as 0.8084. The time difference between x_7 and x_6 in the current alarm sequence is predicted as 245s. Actually, the time difference between x_6 and x_7 is 250s. The prediction is shown in Fig. 5.29a. After 5 min, the step fault 4 after the random fault 8 is activated to simulate a new abnormal condition, and a new alarm x_{14} is arisen from this abnormal condition. The proposed method gives x_{15} as the optimal predicted alarm. However, the probability $P\{x_{15}|x_{c,j}\}$ is only 0.1133, because the 10-min sliding window in Fig. 5.29b contains x_4, x_5, x_6, x_7 and x_{14}. The alarms $\{x_4, x_5, x_6, x_7\}$ are actually from the step faults 3 and 14, instead of the step fault 4 causing x_{14}. As a result, the corresponding probability is quite low because of the influence of $\{x_4, x_5, x_6, x_7\}$. In other words, the prediction result is not reliable. After 71s, the new occurring alarm events increase as $[(x_{14}, 785s), (x_{15}, 796s), (x_{17}, 841s), (x_{21}, 856s)]$. The 10-min sliding window contains x_7, x_{14}, x_{15}, x_{17} and x_{21} as shown in Fig. 5.29c. Compared with Fig. 5.29b, x_{14}, x_{15}, x_{17} and x_{21} are from the new fault while only x_7 is from another fault. Hence the proposed method gives a confident prediction result as x_{22} with a probability of 0.6707.

As comparisons, four existing methods [10, 28, 42, 45] are used here to predict the upcoming alarm for X_{cs}=[$(x_1, 0)$,$(x_2, 18s)$,$(x_3, 32s)$,$(x_4, 197s)$, $(x_5, 227s)$,$(x_6,$

$235s)$]. The method in [45] takes the n-gram model to yield the result. Since the various order of alarm events is not taken into account therein, the predicted result is x_9 with $n = 4$, solely based on the alarm flood #1. The method in [28] extracts templates from historical alarm flood sequences and predicts alarms by comparing with the extracted patterns. The extracted pattern similar to X_{cs} is $\{x_1, x_2, x_3, x_{11}, x_{31}, x_{32}, x_4, x_5, x_6, x_7\}$. Thus, the upcoming alarm may be one of the alarms in $\{x_{11}, x_{31}, x_{32}, x_7\}$. However, no specific results are predicted and no quantitative values are provided. The method in [10] uses the trained LSTM network to yield the result. The predicted result is x_9 since only one alarm sequence contains $\{x_4, x_5, x_6, x_9\}$ in the set of historical alarm flood sequences. This method is similar to the method in [10], which means that the method cannot handle the various order of alarm events. The method in [42] takes the Bayesian estimator to yield the result as $x_9 = 0.6678$ with confidence interval [0.3189, 1]. However, the alarm x_9 is arisen from the random fault 8, which is not the one being activated to simulate the current alarm sequence. The two methods [42, 45] cannot deal with the various order of alarm events and use all previous alarms for prediction, which lead to a false prediction. By contrast, the proposed method says that the predicted result is x_7 with the probability as 0.8558, which is consistent with the fact that alarm floods #1, #2, #3 and the current alarm sequence are arisen from the same faults. Therefore, the predicted result in the proposed method is more reasonable than the ones in the literature [10, 28, 42, 45].

Finally, the sample size has an important influence on prediction results. Since the characteristic functions f_v's for predicting the optimal alarm x_r are actually the quantitative relationships between $x_{c,j}$'s and x_r in historical alarm flood sequences. Hence, the sample size N_s is referred to as the number of all currently-occurred alarms $x_{c,j}$'s that occurred in $\{x_h(n)\}_{n=1}^{N}$,

$$N_s = \sum_{j=1}^{i-1} \sum_{n=1}^{N} \left| x_{c,j} \equiv \left\{ x_h\left(n; x_{h,n,l}\right) \right\} \right|. \tag{5.84}$$

Note that $\left\{ x_h\left(n; x_{h,n,l}\right) \right\}$ is defined in (5.68). The symbol "$\equiv$" is the true value 1 if the compared alarms are the same and the operation $|\cdot|$ counts the number of true values. In order to analyze the influence of sample size on prediction results, 10-fold cross-validation [25] is used to evaluate the prediction result. Each historical alarm sequence $X_h(n)$ with the last element $\dot{x}_{h,n,l}$ removed is denoted as $X_h'(n) := \left[\dot{x}_{h,n,1}, \dot{x}_{h,n,2}, \ldots, \dot{x}_{h,n,l-1}\right]$. In the 10-fold cross-validation, $X_h'(n)$ is treated as the current alarm sequence, and $x_{h,n,l}$ is used to verify the prediction result x_r. The remaining alarms are collectively denoted as $x_{\bar{r}}$,

$$x_{\bar{r}} := X_r - x_r.$$

In order to analyze the influence of sample size on prediction results, three indicators are introduced:

$$N_T = \sum_{n=1}^{N} \left| x_r \equiv x_{h,n,l} \right|, \; N_F = \sum_{n=1}^{N} \left| x_r \neq x_{h,n,l} \right|, \; N_L = \sum_{n=1}^{N} \left| x_{\bar{r}} \equiv x_{h,n,l} \right|,$$

where the symbol "$\equiv$" is the true value 1 if the compared alarms are the same, the symbol "$\neq$" is the true value 1 if the compared alarms are not the same, and the operation $|\cdot|$ counts the number of true values. N_T is the times that the predicted alarm did occur. N_F refers to the times that the predicted alarm did not occur. N_L stands for the times that the unpredicted alarms did occur. The experimental indicators to verify the proposed method is denoted as η, that is,

$$\eta = \frac{2\alpha \times \beta}{(\alpha + \beta)}. \tag{5.85}$$

where

$$\alpha = \frac{N_T}{N_T + N_F},$$

$$\beta = \frac{N_T}{N_T + N_L}.$$

η is the harmonic mean value of α and β. A large value of η indicates that the proposed method is effective [20].

100 Monte Carlo simulations are implemented by activating the same faults in this example. For different sample sizes, Fig. 5.30 presents the corresponding η. When the sample size increases, the corresponding η continues to rise. Generally, the prediction result is regarded to be effective if η is not less than 0.8 [30]. Figure 5.31 shows the empirical conditional probability and the real conditional probability of the optimal prediction result x_7. It is obvious that when the sample size N_s is no less than 100,

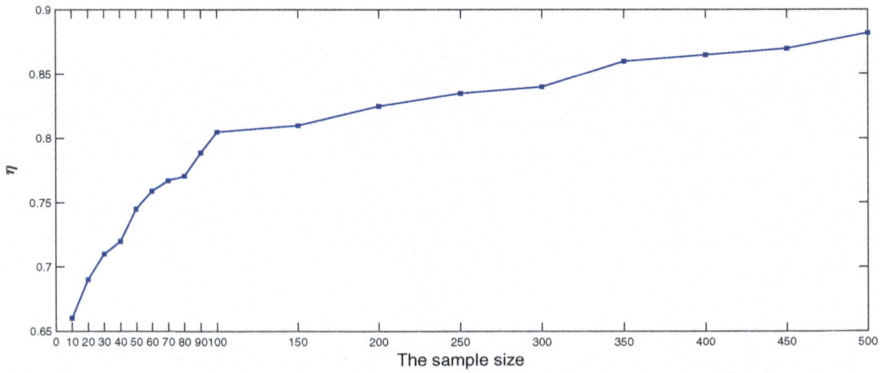

Fig. 5.30 The value of η for different sample sizes N_s's, reprinted from Ref. [41], copyright 2021, with permission from Elsevier

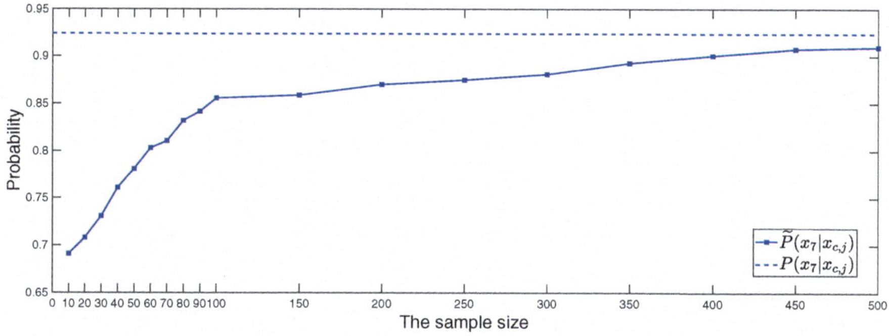

Fig. 5.31 The empirical conditional probability (solid line) for different sample sizes N_s's and the real conditional probability (dotted line) of x_7, reprinted from Ref. [41], copyright 2021, with permission from Elsevier

$\widetilde{P}(x_7|x_{c,j})$ approximates $P(x_7|x_{c,j})$. Therefore, the proposed method can obtain an effective result while more than 100 historical alarms are used for prediction.

References

1. Ahmed, K., Izadi, I., Chen, T., Joe, D., & Burton, T. (2013). Similarity analysis of industrial alarm flood data. *IEEE Transactions on Automation Science and Engineering, 10*(2), 452–457.
2. Al-Dabbagh, A. W., Hu, W., Lai, S., Shah, S., & Hollender, M. (2018). Toward the advancement of decision support tools for industrial facilities: Addressing operation metrics, visualization plots, and alarm floods. *IEEE Transactions on Automation Science and Engineering, 15*, 1883–1896.
3. Altschul, S., Gish, W., Miller, W., Myers, E., & Lipman, D. (1990). Basic local alignment search tool. *Journal of Molecular Biology, 215*, 403–410.
4. Altschul, S., Madden, T., Schäffer, A., Zhang, J., Zhang, Z., Miller, W., & Lipman, D. (1997). Gapped BLAST and PSI-BLAST: A new generation of protein database search programs. *Nucleic Acids Research, 25*, 3389–3402.
5. ANSI, ISA-18.2. (2016). *Management of Alarm Systems for the Process Industries*. Triangle Park, NC, USA: International Society of Automation.
6. Babich, I. V., & Moulijn, J. A. (2003). Science and technology of novel processes for deep desulfurization of oil refinery streams: A review. *Fuel, 82*, 607–631.
7. Berger, A., Della Pietra, S. A., & Della Pietra, V. J. (2018). A maximum entropy approach to natural language processing. *IEEE Transactions on Automation Science and Engineering, 15*, 1883–1896.
8. Bertsekas, D. P. (2009). *Convex Optimization Theory*. Athena Scientific.
9. Bransby, M. L., & Jenkinson, J. (1998). *The Management of Alarm Systems*. Suffolk, UK: HSE Books.
10. Cai, S., Palazoglu, A., Zhang, L., & Hu, J. (2019). Process alarm prediction using deep learning and word embedding methods. *ISA Transactions, 85*, 274–283.
11. Charbonnier, S., Bouchair, N., & Gayet, P. (2015). A weighted dissimilarity index to isolate faults during alarm floods. *Control Engineering Practice, 45*, 110–122.
12. Charbonnier, S., Bouchair, N., & Gayet, P. (2016). Fault template extraction to assist operators during industrial alarm floods. *Engineering Applications of Artificial Intelligence, 50*, 32–44.

13. Charlie, F. (2012). Using alarm suppression: Effective alarm systems improve operations. *InTech Magazine*.
14. Cheng, Y., Izadi, I., & Chen, T. (2013). Pattern matching of alarm flood sequences by a modified Smith-Waterman algorithm. *Chemical Engineering Research and Design, 91*, 1085–1094.
15. Cormen, T. H., Leiserson, C. E., Rivest, R. L., & Stein, C. *Introduction to Algorithms* (3rd ed.). Cambridge: MIT Press.
16. Couper, J. R., Penney, W. R., & Fair, J. R. (2009). *Chemical Process Equipment: Selection and Sesign*. Gulf Professional Publishing.
17. Cover, T. M., & Thomas, J. A. (2009). *Elements of Information Theory*. Hoboken, N. J.: Wiley-Interscience.
18. Darroch, J. N., & Ratcliff, D. (1972). Generalized iterative scaling for log-linear models. *The Annals of Mathematical Statistics, 43*, 1470–1480.
19. EEMUA (Engineering Equipment and Materials Users' Association). (2013). *Alarm Systems: A Guide to Design, Management and Procurement*. London: EEMUA Publication 191.
20. Fawcett, T. (2006). An introduction to ROC analysis. *Pattern Recognition Letters, 27*, 861–874.
21. Han, J., Cheng, H., Xin, D., & Yan, X. (2007). Frequent pattern mining: Current status and future directions. *Data Mining and Knowledge Discovery, 15*(1), 55–86.
22. Hollifield, B., & Habibi, E. (2010). *The Alarm Management Handbook* (2nd ed.). Houston, USA: PAS.
23. Hu, W., Chen, T., & Shah, S. L. (2018). Detection of frequent alarm patterns in industrial alarm floods using itemset mining methods. *IEEE Transactions on Industrial Electronics, 65*(9), 7290–7300.
24. Hu, W., Wang, J., & Chen, T. (2016). A local alignment approach to similarity analysis of industrial alarm flood sequences. *Control Engineering Practice, 55*, 13–25.
25. James, G., Witten, D., Hastie, T., & Tibshirani, R. (2013). *An Introduction to Statistical Learning*. London, UK: Springer.
26. Kent, W. J. (2002). BLAT - The BLAST-like alignment tool. *Genome Research, 4*, 656–664.
27. Lai, S., & Chen, T. (2017). A method for pattern mining in multiple alarm flood sequences. *Chemical Engineering Research and Design, 117*, 831–839.
28. Lai, S., Yang, F., & Chen, T. (2017). Online pattern matching and prediction of incoming alarm floods. *Journal of Process Control, 56*, 69–78.
29. Nimmo, I. (2005). Rescue your plant from alarm overload. *Chemical Processing*, 28–33.
30. Power, D. M. W. (2013). Evaluation from precision, recall and F-measure to ROC, informedness, markedness and correlation. *Chemical Engineering Research & Design, 91*, 1085–1094.
31. Ratnaparkhi, A. (1997). A simple introduction to maximum entropy models for natural language processing. *IRCS Technical Reports Series*.
32. Ricker, N. L. (1996). Decentralized control of the Tennessee Eastman challenge process. *Journal of Process Control, 6*, 205–221.
33. Scallan, P. (2003). *Process Planning*. Butterworth-Heinemann.
34. Smith, T., & Waterman, M. (1981). Identification of common molecular subsequences. *Journal of Molecular Biology, 147*, 195–197.
35. Vogel-Heuser, B., Schutz, D., & Folmer, J. (2015). Criteria-based alarm flood pattern recognition using historical data from automated production systems (aPS). *Mechatronics, 31*, 89–100.
36. Wang, J., & Chen, T. (2014). An online method to remove chattering and repeating alarms based on alarm durations and intervals. *Computers and Chemical Engineering, 67*, 43–52.
37. Wang, J., & Chen, T. (2016). Main causes of long-standing alarms and their removal by dynamic state-based alarm systems. *Journal of Loss Prevention in the Process Industries, 43*, 106–119.
38. Wang, J., Han, J., & Pei, J. (2003). Closet+: Searching for the best strategies for mining frequent closed itemsets. In *Proceedings of the Ninth ACM-SIGKDD International Conference on Knowledge Discovery and Data Mining* (pp. 236–245).
39. Wang, J., Zhao, Y., & Bi, Z. (2018). Criteria and algorithms for online and offline detections of industrial alarm floods. *IEEE Transactions on Control Systems and Technology, 26*, 1722–1731.
40. Xin, D., Han, J., Yan, X., & Cheng, H. (2005). Mining compressed frequent-pattern sets. In *Proceedings of the 31st International Conference on Very large Data Bases* (pp. 709–720).

41. Xu, Y., & Wang, J. (2021). A maximum-entropy-based method for alarm flood prediction. *Journal of Process Control, 107*, 58–69.
42. Xu, Y., Wang, J., & Yu, Y. (2019). Alarm event prediction from historical alarm flood sequences based on Bayesian estimators. *IEEE Transactions on Automation Science and Engineering, 17*, 1–6.
43. Zaki, M. J., & Hsiao, C. J. (2002). CHARM: An efficient algorithm for closed itemset mining. In *Proceedings of the 2002 SIAM International Conference on Data Mining* (pp. 457–473).
44. Zhao, C., Chen, J., & Jing, H. (2020). Condition-driven data analytics and monitoring for wide-range nonstationary and transient continuous processes. *IEEE Transactions on Automation Science and Engineering, 18*(4), 1563–1574.
45. Zhu, J., Wang, C., Li, C., Gao, X., & Zhao, J. (2016). Dynamic alarm prediction for critical alarms using a probabilistic model. *Chinese Journal of Chemical Engineering, 24*, 881–885.

Chapter 6
Alarm Visual Analytics and Applications

Abstract This chapter is on the techniques and applications of alarm visual analytics. The availability of massive data in complex industrial facilities has prompted the development of alarm management techniques and also resulted in a great demand to transform such data and derived results into effective visual forms. Some well established visualization techniques are presented in three categories, including visual analytics for alarm system performance, visual analytics for alarms and related events, as well as visual analytics for alarm flood sequences. Examples are provided to demonstrate the effectiveness and utility of these techniques.

6.1 Overview of Alarm Data and Analytics

This section introduces a smart alarm analytics framework that integrates various advanced methods for alarm system design and alarm data analysis. Then, the basics about alarm and event data, such as the data structure and properties, are elaborated with examples. Last, the design problems and principles for alarm visual analytics are discussed.

(1) Framework for Advanced Alarm Analytics

The operation of modern industrial facilities has become highly automated with the deployment of computerized control systems, such as the distributed control system (DCS) and the supervisory control and data acquisition (SCADA) system. Massive amount of data in the DCS and SCADA systems contain rich information about process operations, making it indispensable assets for decision making processes [39]. However, without effective analytics tools, such data would only be compressed and archived for record keeping rather than being turned into valuable resources to facilitate decision-making. The discovery and learning from process and alarm data refer to a set of tools and techniques for modeling and understanding of complex data sets. Such data sets generally include numerical (or non-categorical) data from historical databases, but should also take into account categorical (or non-numerical or qualitative) data from Alarm and Event (A&E) logs combined with process connectivity or topology information [25]. The latter refers to the capture of

J. Wang et al., *Intelligent Industrial Alarm Systems*,
https://doi.org/10.1007/978-981-97-6516-4_6

material flow streams in industrial process units as well as information flow-paths in the processes due to control loops.

The emphasis in this chapter is on tools that help in the process of understanding data and discovering information that would lead to improvement of alarm system performance and enhancement of operational safety.

Figure 6.1 presents the architecture of an industrial alarm system and the role of advanced techniques for improvement of alarm management. An alarm system is an integration of hardware and software that connect processes and operators through several components, including the data Inputs/Outputs (I/O) server, control and safety systems, alarm logs, and human-machine interface (HMI) systems. The control and safety systems are typically associated with two major parts, namely, the basic process control system (BPCS) and the safety instrumented system (SIS), which produce alarms based on process measurements or logic conditions. The HMI system (namely, through a computer screen or an annunciator panel) presents alarms to operators. The alarm logs store alarms and related events in certain structures. To improve the performance of alarm system and the safety of process operation, the advanced alarm analytics techniques serve to analyze the operational data and extract useful knowledge, which can be then utilized to supervise alarm rationalization and alarm system design, or to directly assist operators with decision-making during critical situations or alarm floods.

The preceding chapters in this book have introduced a multitude of sophisticated alarm analytics methods, encompassing univariate alarm system design, multivariate alarm system design, alarm root cause analysis, and alarm flood analysis. In practical applications, it is feasible to construct a smart alarm analytics platform, that amalgamates these methodologies for deployment in various real industrial sectors.

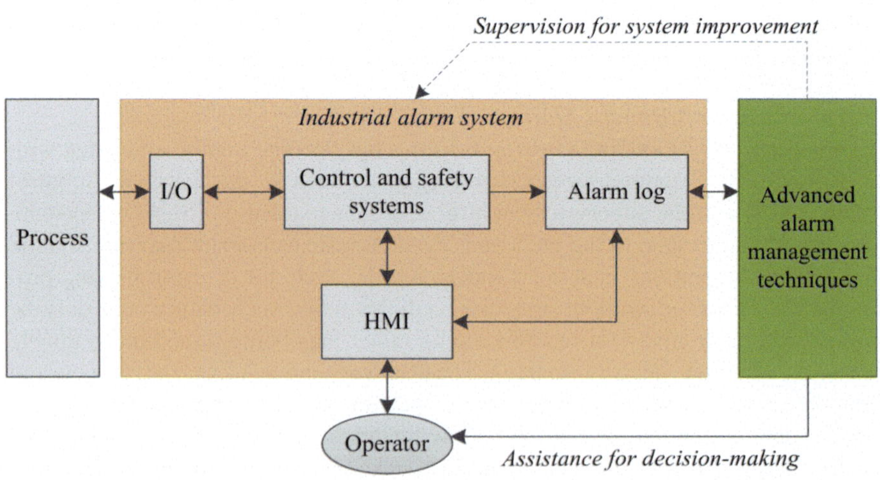

Fig. 6.1 The architecture of an industrial alarm system and the role of advanced techniques for improvement of alarm management, reprinted from Ref. [21], copyright 2018, with permission from Elsevier

This platform would harness diverse data sources for analysis, as discussed in [25]. Fundamental components of this analytics platform could comprise functions for ingesting and pre-processing process or alarm data, functions for computing statistical indicators and discerning patterns from alarm data, functions for the design and evaluation of alarm systems, as well as functions for pinpointing the underlying causes of abnormalities.

Figure 6.2 presents an example for the framework of such a smart alarm analytics platform, which mainly includes the following four typical functional parts [25]:

(i) "Data Importing" imports the alarm & event data, continuous-valued process data, and preprocessed data in a certain format. In addition, it can merge process data with associated alarm data for a certain alarm tag. It can also export data and clear data whenever needed.
(ii) "Alarm Data Analysis" provides analytics and reporting functions to analyze and visualize alarm data, such as performance evaluation of an industrial alarm system, detection of nuisance alarms, alarm flood analysis, and operator action analysis.
(iii) "Alarm System Design" provides functions, such as the design of univariate alarm system such as alarm delay timers and alarm deadbands for a specific process variable.
(iv) "Connectivity and Causality Analysis" uncovers correlations and causal relations between process variables based on diverse data resources, including process signals and process connectivity graphs.

(2) Basics of Alarm & Event Data

Alarm signals are produced as binary time series in alarm systems to indicate the associated processes deviating from their normal specifications or violation of some safety conditions. When an alarm occurs, it is presented to operators in an audible or visual form, and is displayed on a computer screen as a textual message in a scrolling

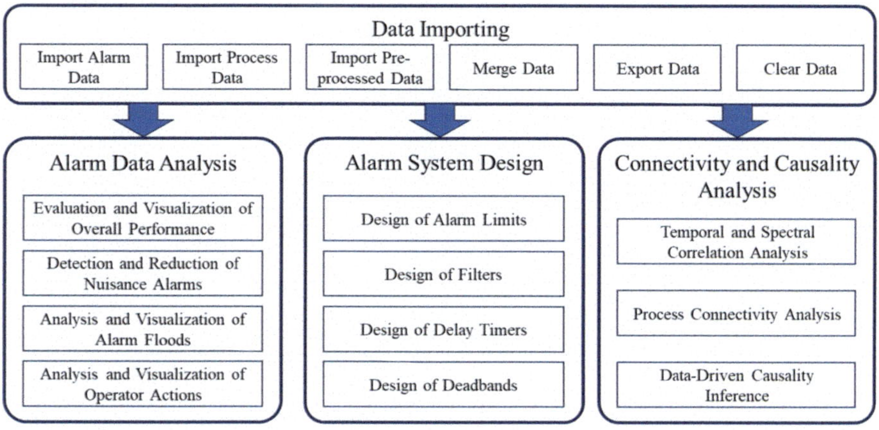

Fig. 6.2 Framework for a smart alarm analytics platform

list [6, 16, 19, 29]. Then, the operator responds to this alarm and takes corrective actions to bring the system back to normal. There are two types of operator responses: (i) A response involving state transitions of alarms, for example, acknowledging, shelving/unshelving, suppressing/unsuppressing, and disabling/enabling alarms, as shown in the alarm state transition diagram in [6]; (ii) a response involving operator actions dealing with the process or equipment, for example, manipulating a valve, adjusting control parameters, and changing over to a standby pump [16, 19]. All these messages involving alarms and operator responses can be historized and stored in an alarm & event (A&E) database. In addition to alarm messages and operator responses, events about automatic process changes, for example, the variation of operating rates and transition of operating modes [24], are also available in many A&E databases.

Figure 6.3 offers a detailed schematic of the event categorization within an A&E database along with some real-world examples for each category. It is also noteworthy that each event is associated with many attributes to uniquely identify the process, describe the functionality of the alarm, and denote real-time messages.

Figure 6.4 displays some commonly used data attributes in industrial alarm systems. These attributes are important bricks constituting the A&E database that describe the alarms. For example, a tag name uniquely identifies a basic element, such as a process variable, a device, and a control loop, which may have various built-in alarm functions distinguished by alarm types, such as low (LO), Low-Low (LL), High (HI), High-High (HH), Low Deviation (LD), and High Deviation (HD) alarms [7]. Priorities, such as critical, high, medium, and low, are essential attributes indicating the importance of alarms, and are usually assigned based on safety, financial, and environmental factors [48]. In addition to these configuration attributes, which do not usually change over time, the other data attributes, such as alarm messages, time stamps, and event IDs, are real-time messages indicating if an alarm occurs (ALM)

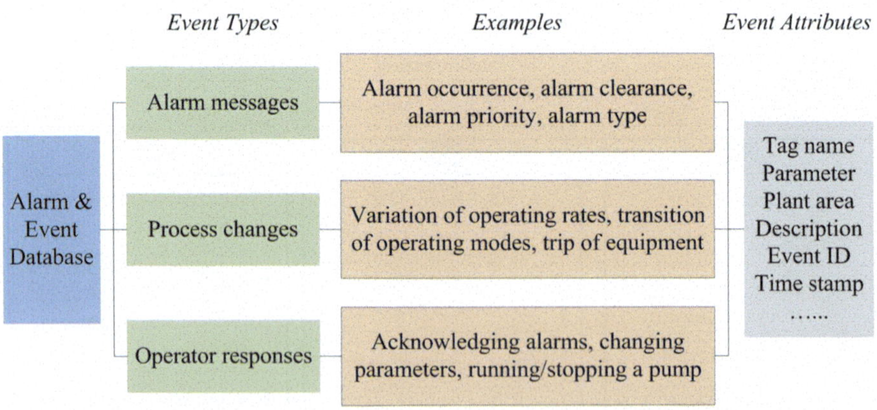

Fig. 6.3 An overview of messages stored in an A&E database, reprinted from Ref. [21], copyright 2018, with permission from Elsevier

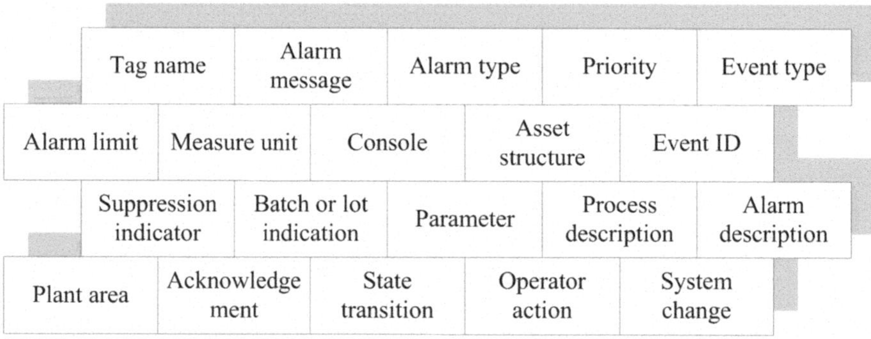

Fig. 6.4 An overview of data attributes associated with alarms, reprinted from Ref. [21], copyright 2018, with permission from Elsevier

or returns-to-normal (RTN), along with the time as well as the numerical identifier information related to the event [19].

Table 6.1 presents an illustrative example of a structured A&E database, where each column header describes an attribute and each row corresponds to an event displayed through the textual message. Here, all the events occurred on the same date, and thus, the date information is omitted. More specifically, the column "Event type" identifies if the event is an alarm or an action; the column "Message" presents the changes in alarm states or operator actions; the column "Description" describes the functionality of each alarm or action. For example, for the process tag "LV2" measuring the fluid level located in "Area B", its high limit "HI" alarm configured with a "Critical" priority was triggered at 14:21:19. Then, the operator responded by opening a drainage valve "VAV2" at 14:26:47. This action brought the process back to normal and the alarm cleared at 14:27:33. It can be observed that the availability of such detailed information stored in an A&E database, such as that of Table 6.1, allows for further analysis to gain a deeper understanding on the operation of complex industrial facilities. To facilitate such efforts, data visualization must be considered an instrumental step to give quick and straightforward observations of the insights in massive data.

(3) Discussions on Design of Alarm Visual Analytics Methods

Visualization is a powerful tool that is overlooked but critical for industrial applications. Thus, this chapter introduces the visual analytics methods for alarms, related events, and alarm flood sequences.

Specifically, given an A&E data set, it can visualize the raw information, which does not involve any statistical calculation or advanced data analytics; its main objective is related to two aspects, namely, providing an overview picture of the performance of industrial alarm systems for offline analysis, and tracking the annunciation of alarms for real-time monitoring. Further, it can also display the statistical results, which are not directly available from the A&E database and usually involve certain calculations. The basic statistical analysis derives performance metrics, such as the

Table 6.1 An example of a structured A&E database, reprinted from Ref. [21], copyright 2018, with permission from Elsevier

Time stamp	Event type	Tag name	Alarm type	Message	Priority	Area	Description
14:21:19	Alarm	LV2	HI	ALM	Critical	B	High limit alarm for fluid level
14:26:47	Action	VAV2		OPEN		B	Open a drainage valve
14:27:33	Alarm	LV2	HI	RTN	Critical	B	High limit alarm for fluid level
17:55:27	Action	PMP1		STOP		C	Stop a lube oil pump
17:56:31	Alarm	PI1	LO	ALM	Low	C	Low limit alarm for pressure
21:17:23	Action	PMP1		START		C	Start a lube oil pump
21:18:02	Alarm	PI1	LO	RTN	Low	C	Low limit alarm for pressure

alarm count, average alarm rate, and peak alarm rate, to judge whether an industrial alarm system experienced alarm overloading or bad alarm flood situations. Next, it can also show alarm patterns that refer to the results, such as correlations, similarities, association rules, and sequential patterns. These alarm patterns can help industrial practitioners with decision-making under different scenarios, such as configuring state-based alarming, coping with alarm floods, and reducing redundant alarm configurations.

To help industrial practitioners to have a better understanding the behavior of alarm systems and take actions to enhance industrial alarm management, the development of visualization techniques of alarm & event data should incorporate the following requirements [21]:

(i) The capability of being easily understandable to industrial practitioners, such as alarm analysts, process engineers and plant operators.
(ii) The capability of presenting the overall performance of an alarm system from multiple perspectives and revealing hidden problems in the historical data.
(iii) The capability of making comparisons between groups, variables, time periods, as well as metrics and the corresponding benchmarks.
(iv) A better design methodology to deliver effective display of the data and catch users' attentions immediately.

Due to the availability of large volumes of historical data in modern industrial facilities, the demand of more and better visualization techniques is huge, so as to augment human capabilities to discover problems and patterns from the alarm & event data.

In this chapter, the visual analytics tools are categorized into three groups, namely, visual analytics for alarm system performance, visual analytics for alarms and related events, as well as visual analytics for alarm flood sequences. Details about these visual analytics tools are discussed in the following sections.

6.2 Visual Analytics for Alarm System Performance

Alarm visual analytics functions target at evaluating the overall performance of an industrial alarm system. The number of alarm variables is usually quite large, e.g., more than 2000 in a larger-scale coal-fired power generation unit [52]. Thus, it is infeasible for users to look at each of these alarm variables in a manual manner. Performance metrics are calculated and visualized to allow users to gain abstract information from alarm data, quickly and easily assess the overall performance, and make comparisons to benchmarks. Here, several useful visual analytics methods are presented, including the performance bubble chart, hierarchical alarm treemap, alarm analytics graph, top bad actor ranking chart, layered alarm count radar chart, multi-layered performance comparison radar chart, high density alarm plot, and dynamic 3D bar chart [4, 21, 53].

(1) Performance Bubble Chart

This visualization plot is a combination of the trendy scatter/bubble chart and the segmented area plot. Compared to conventional scatter/bubble charts, the designed bubble chart is capable of comparing metrics with industrial benchmarks, making the performance evaluation more intuitive. Compared to the segmented area plot [16, 17], it provides additional information about the scale of each console or alarm system, and enables comparisons between different consoles or alarm systems in three dimensions. Targets can be set for future alarm management based on the zone location of each alarm system in the plot and the size of its scale. It can also be used in realtime to track the performance change of an alarm system. An unhealthy status, e.g., "Reactive" and "Overloaded", will lead to interventions to improve the alarm system performance.

A bubble chart is a variant of the scatter plot, but displays data in three dimensions. It is adopted here and combined with the segmented area plot [16, 17] to evaluate the overall performance of an industrial alarm system. Hereby, this bubble chart consists of two essential parts:

(i) A segmented area plot divides the performance of industrial alarm systems into 5 zones based on the KPI (key performance indicator) recommendations in the EEMUA guideline [16].
(ii) Bubbles are displayed in three dimensions, including the average alarm rate γ_a as the vertical coordinate, the peak alarm rate γ_p as the horizontal coordinate, and the number of unique alarms γ_n as the area of the bubble.

Five colors are used to distinguish the five performance zones. Given the average alarm rate γ_a and peak alarm rate γ_p (the number of alarms per 10 min for each operator), the coordinates (x, y) (the center of each bubble) are calculated as

$$(x, y) = \left(\min(f(\gamma_a), 4), \min(f(\gamma_p), 5)\right). \tag{6.1}$$

Here $f(\cdot)$ is given by

$$f(\gamma) = \begin{cases} 1 + \log_{10}\gamma, & \text{if } \gamma > 1, \\ \gamma, & \text{if } 0 \leq \gamma \leq 1, \end{cases} \tag{6.2}$$

where γ is either γ_a or γ_p. The minimization in (6.1) guarantees that the center of a bubble never exceeds the boundary of the area plot, even if γ_a and γ_p take large values. The area A of a bubble is proportional to the number of unique alarms γ_n, i.e., $A = K\gamma_n$, where K is a user defined parameter to adjust the size of the bubble. It can be set to a large value, so as to avoid overspreading of large bubbles.

Figure 6.5 gives an example of the bubble chart, which displays the metrics calculated based on the data over one full month. Each bubble represents one console. The five colors correspond to five performances, namely, predictive, robust, stable, reactive, and overloaded statuses, from the best to the worst. The center of each bubble corresponds to the average alarm rate and the peak alarm rate of each console. The area of the bubble is proportional to the number of unique alarms in each console. A bubble could be as small as a point, representing only one unique alarm. It can be observed from the plot that the alarm system in Console 1 has a predictive performance, meaning that the average alarm rate is acceptable to the operator and there does not exist any alarm flood problem. By contrast, the alarm systems in Consoles 4 and 5 have extremely high alarm rates, making the alarm systems lose their effectiveness and thus overloading operators. The alarm rates in Consoles 2 and

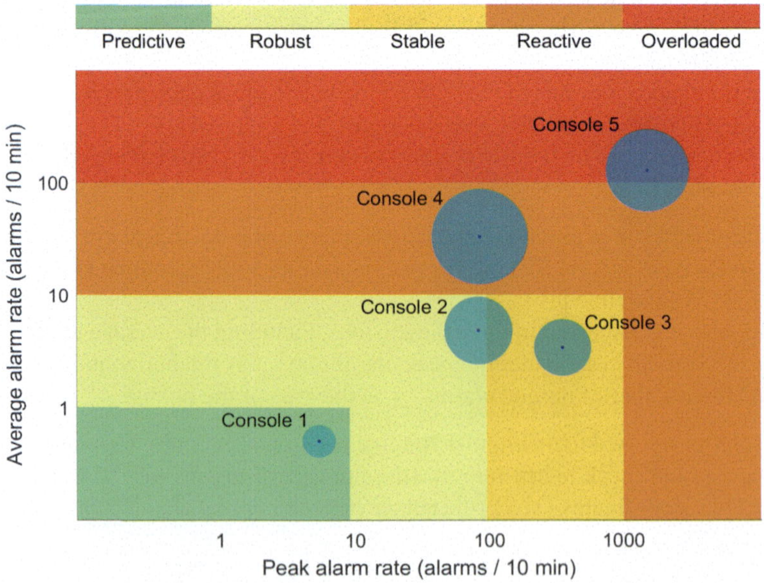

Fig. 6.5 An example of the bubble chart to evaluate and compare the overall performance of industrial alarm systems, reprinted from Ref. [21], copyright 2018, with permission from Elsevier

3 are high but tolerable. Further efforts can be taken to bring the alarm rates down targeting at a predictive performance.

(2) Hierarchical Alarm Treemap

This visualization plot is designed to observe statistical results from different perspectives and quickly locate problems at rectangular areas. It can be concluded quickly from the treemap which groups (e.g., units, priorities, and types) tend to receive more alarms and which alarms account for a high alarm rate. As a result, the problems with these groups and alarms will be given priorities for investigation and solution.

A treemap is a simple visualization plot displaying data of multiple categories in a hierarchical structure [42]. Each category is represented by a rectangular area with its size determined by the associated quantity value. Rectangle areas representing subcategories are nested inside of each rectangle in the parent level. The sum of the sizes of all child rectangles in each parent category is usually equal to the size of their parent rectangle. Here, the treemap is adapted to visualize statistical information, such as alarm counts. It consists of the following essential parts:

(i) The hierarchical structure is used to show categories (e.g., plant units, priorities, and alarm types) and items (e.g., alarm tags) in the first and second levels of the treemap, respectively.
(ii) The interactive design is incorporated to allow observation from different perspectives, to enable zooming-in on the second level of the treemap, and to show the associated basic information as data tips (e.g. the tag name, priority, unit, alarm count, and chattering index).
(iii) Colors are effectively used to highlight certain groups or items, and texts are adjusted based on sizes of rectangles.

Compared to bar charts and pie charts, this treemap presents more information involving groups and individual alarms through the hierarchical structure, and makes more efficient use of space. Compared to conventional tables, it enables direct comparisons based on the rectangular areas, and allows flexible grouping and highlighting.

Assume the alarms in a complex industrial facility are grouped into M different categories and denote the set of alarm variables in the jth category as $\mathcal{C}^j = \{a_i^j, i = 1, 2, \ldots, |\mathcal{C}^j|\}$, where a_i^j represents the ith unique alarm variable in the jth category, $|\mathcal{C}^j|$ denotes the number of unique alarm variables in $\mathcal{C}^j$, and $j = 1, 2, \ldots, M$. Given a time period $[\mathcal{T}_s, \mathcal{T}_e]$, the alarm count of each unique alarm variable in this period is denoted by $\psi_{a_i^j}$. Then, the percentage area for alarm a_i^j in the jth category in the second level of a treemap is calculated as

$$A_i^j = \frac{\psi_{a_i^j}}{\sum_{i=1}^{|\mathcal{C}^j|} \psi_{a_i^j}}. \tag{6.3}$$

The percentage area for the jth category in the first level of a treemap is calculated as

$$A^j = \frac{\sum_{i=1}^{|C^j|} \psi_{a_i^j}}{\sum_{j=1}^{M} \sum_{i=1}^{|C^j|} \psi_{a_i^j}}. \tag{6.4}$$

An example of the treemap displaying alarm counts categorized by industrial plant units is shown in Fig. 6.6. There are 48 categories corresponding to 48 areas in a thermal power plant. The size of each rectangle (or child rectangle) corresponds to the total number of alarm occurrences in a plant unit (or of an alarm tag). The name of each plant unit and the alarm count in proportion to the whole alarm count are shown at the header of each rectangle. Top 10 bad actors are displayed with tag names and highlighted using an orange color. It can be observed from the plot where the most alarms were from as well as who the top bad actors were. In addition, an interactive design is incorporated: two dropdown menus are included to facilitate observations from different perspectives rather than just to display top 10 bad actors. The first dropdown menu enables grouping alarms based on different categories, such as alarm priorities and alarm types. The second dropdown menu allows highlighting alarms based on different criteria, e.g., users can choose to highlight all chattering alarms.

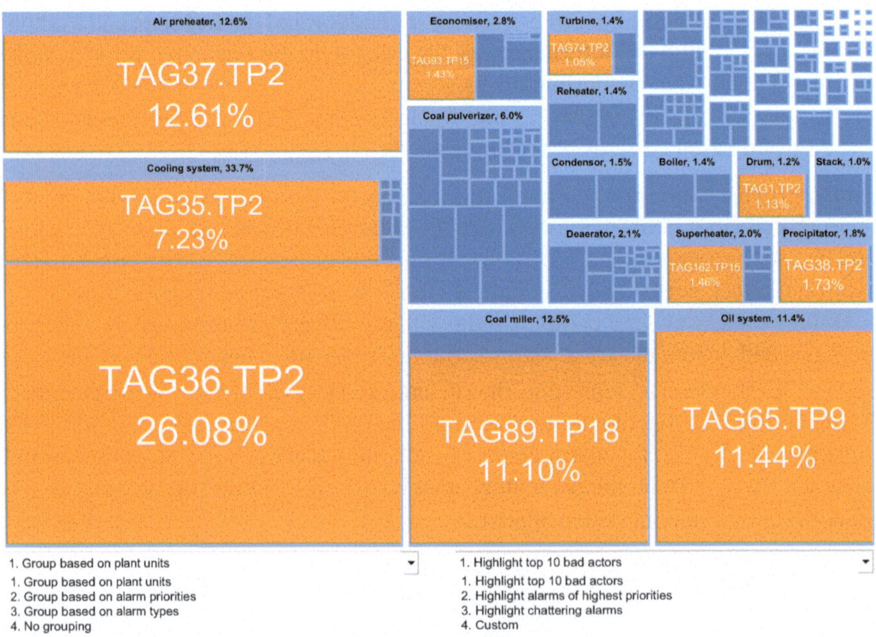

Fig. 6.6 An example of the treemap of alarm counts categorized by industrial plant units, reprinted from Ref. [21], copyright 2018, with permission from Elsevier

The second interactive design is incorporated: The treemap in the second level is generated by clicking on the appropriate rectangle in the first level. For example, Fig. 6.7 presents the second level of the treemap in Fig. 6.6 by selecting the plant unit "Coal pulverizer". Moreover, in order to observe detailed information, the third interactive design is incorporated: a plot shows up to present detailed information of one alarm tag when the mouse pointer moves over its associated rectangle. For example, moving the mouse pointer over the rectangle of "TAG165.TP18", this rectangular area is highlighted by a golden frame; meanwhile a data tip window shows up to display the basic information of this alarm tag, including the tag name, alarm priority, plant unit, alarm count, chattering index, and the line graph of time series, shown in Fig. 6.8. The orange color is used for a "High" priority. Moving off this rectangle, this data tip window for "TAG165.TP18" disappears automatically. Such an interactive design is helpful for users to quickly observe the detailed information hidden in the treemap. It can also be incorporated in the first level of the treemap.

(3) Alarm Analytics Graph

This graph is used for visualization of some standard and new metrics that are organized in two dimensions, namely, the variable dimension and the time dimension. The variable dimension is about whether a single alarm variable is investigated or multiple alarm variables are considered. For the time dimension, a large time period is separated into several time windows with the equal length Δt. The objective is to describe the operating status of alarm variables in the given time period. By looking at the analytical graphs, industrial plant operators are aware of the status of alarm

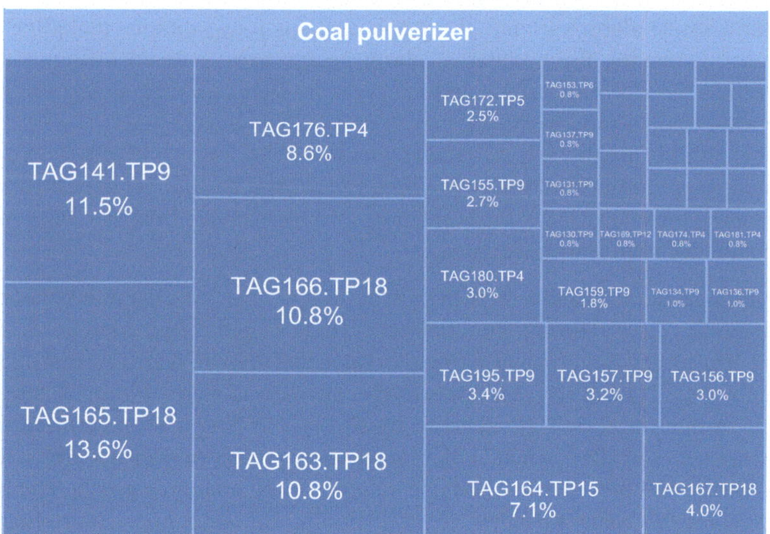

Fig. 6.7 An example of the treemap in the second layer, reprinted from Ref. [21], copyright 2018, with permission from Elsevier

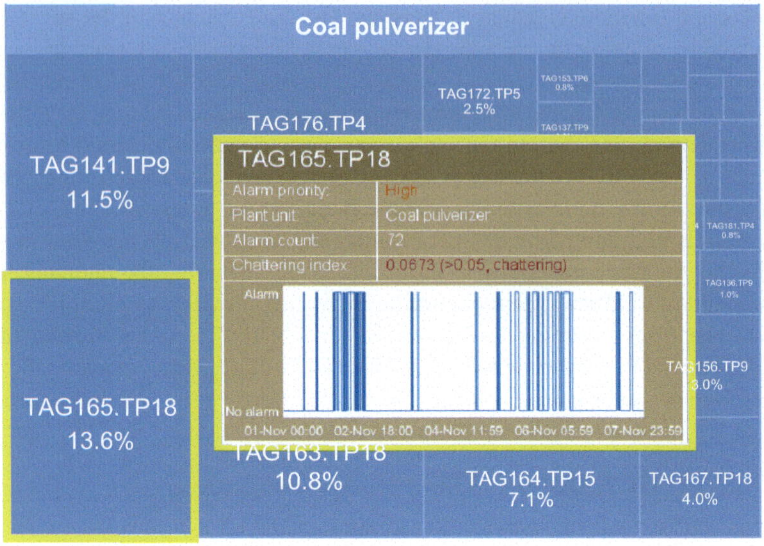

Fig. 6.8 An example of the interactive design that displays the basic information of an alarm in a data tip window, reprinted from Ref. [21], copyright 2018, with permission from Elsevier

variables in a concise and clear manner, instead of via a high-demanding approach by visualizing massive operation data [53]. The graph consists of the following essential parts:

(i) For a single alarm variable, the graph contains four subplots, including a bar chart displaying the number of alarm occurrences, colored histograms presenting the alarm durations and intervals, as well as a time series plot of the corresponding process signal.
(ii) For multiple alarm variables, the graph consists of four bar charts showing the number of alarm variables having occurred alarms, the number of new variables running into alarm state, the number of variables staying in alarm state, and the number of total alarm occurrences.

For a single alarm variable x_a in the jth period $[t_j, t_j + \Delta t]$, the number of alarm occurrences is $N_o(j)$. The number of alarm clearances in the jth period $[t_j, t_j + \Delta t]$ is $N_c(j)$. The alarm duration and alarm interval in the jth period are two real numbers denoted by $D_{a,j}$ (1) and $I_{a,j}$ (1), respectively. Two sets of alarm durations and alarm intervals in the jth period are denoted by $S_D(j)$ and $S_I(j)$. The alarm durations and alarm intervals reflect whether alarms occur frequently or occasionally in a time period. Figure 6.9 presents a schematic sample of the analytics graph. The top of Fig. 6.9 gives the basic attributes of x_a, including the tag name, description, and threshold. Figure 6.9a presents the number of alarm occurrences, $\{N_o(j)\}_{j=1}^{J}$, for J consecutive time periods $[t_1, t_1 + \Delta t]$, ..., $[t_j, t_j + \Delta t]$, ..., $[t_J, t_J + \Delta t]$. The red, green, and blue colors in Fig. 6.9a are respectively used for $N_o(j) > N_o(j-1)$,

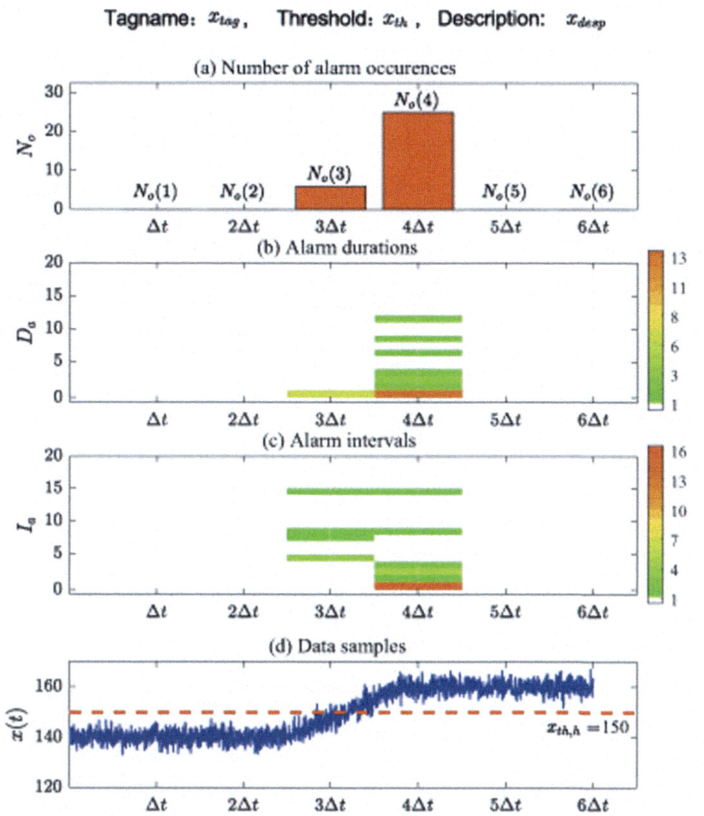

Fig. 6.9 A schematic sample of the analytics graph for an alarm variable (the red color means the increasing trend in (**a**)), reprinted from Ref. [53], copyright 2022, with permission from Elsevier

$N_o(j) < N_o(j-1)$ and $N_o(j) = N_o(j-1)$. Since both $N_o(j)$ and $N_c(j)$ reflect the number of alarms, and the difference is no more than 1, Fig. 6.9a presents $N_o(j)$ only. Figure 6.9b and c provide the colored histogram of alarm durations $S_D(j)$ and alarm intervals $S_I(j)$ for the periods $[t_j, t_j + \Delta t]$ ($j \in [1, J]$) with N_o is not zero. The color bar denotes the frequency of alarm durations (or the counterpart of alarm intervals). Figure 6.9d shows data samples of x and the alarm threshold for x_a.

For an alarm set X_a being composed of multiple alarm variables in the jth period $[t_j, t_j + \Delta t]$, the number $N_{var}(j)$ of alarm variables having occurred alarms is the total number of alarm variables in X_a, having experienced the alarm state in the jth period. The number $N_{new}(j)$ of new alarm variables running into alarm state is the total number of alarm variables in X_a, being in the non-alarm state in the (j-1)th period and running into the alarm state in the jth time period. Here, $I_k(j)$ indicates whether the kth alarm variable runs into the alarm state. The number $N_{sta}(j)$ of alarm variables staying in alarm state all the time is the total number of alarm variables in X_a, being in the alarm state all the time during the jth period. The number $N_{occ}(j)$

of total alarm occurrences is the total number of alarm occurrences for all alarm variables belonging to X_a in the jth period.

Figure 6.10 gives a schematic sample of the analytics graph. The top of Fig. 6.10 lists the basic attributes of X_a in terms of the description X_{desp} for the alarm system and the number K of alarm variables in X_a. Figure 6.10a–d show the four status metrics $\{N_{var}(j), N_{new}(j), N_{sta}(j), N_{occ}(j)\}_{j=1}^{J}$ for J consecutive time periods. The symbols $m_{ax,i}$, μ_i, $m_{in,i}$, and σ_i in Fig. 6.10 are the maximum, mean, minimum, and standard deviation of these metrics. Similar to Fig. 6.9, the status metrics $N_{var}(j)$, $N_{new}(j)$, $N_{sta}(j)$, and $N_{occ}(j)$ have different changing trends shown in the multiple consecutive time intervals. The operating status can be summarized from these changing trends.

(4) Top Bad Actor Ranking Chart

This visualization plot is designed to track top bad actors, which is an important task of periodic alarm audit. Compared to conventional bar charts and pie charts, this ranking chart allows for comparisons between different time periods, which involve two aspects, including the total alarm count of top bad actors and that of individual alarms. Potential users will take actions to deal with top bad actors based on the decreasing or increasing of alarm counts, e.g, evaluating whether some implemented

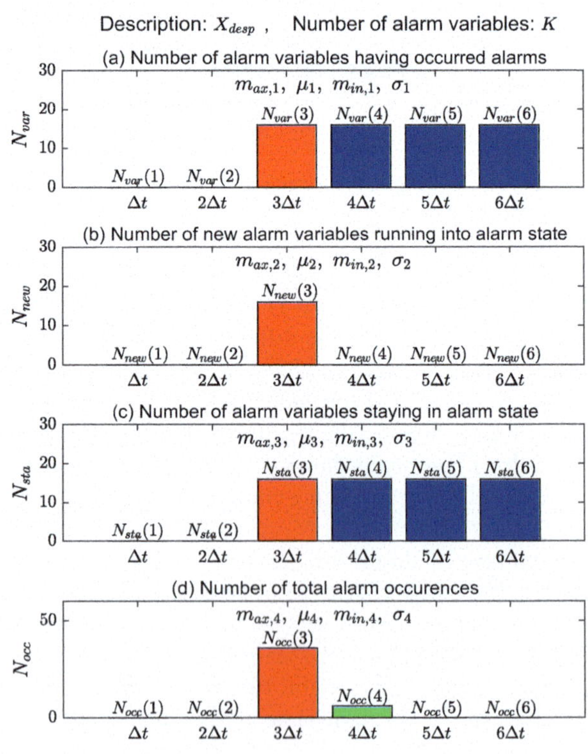

Fig. 6.10 A schematic sample of an analytics graph for multiple alarm variables (red, green and blue colors represent the increasing, decreasing and stable trends in consecutive time intervals, respectively), reprinted from Ref. [53], copyright 2022, with permission from Elsevier

solutions are effective in reducing nuisance alarms, and determining what alarms require further efforts to bring the alarm counts down.

As an essential stage of the alarm management life cycle described in [6], an audit provides periodic reviews of the performance of an industrial alarm system and may reveal gaps that are not apparent from routine monitoring. A typical task of alarm audit is to track top bad actors, which are known as alarms with highest alarm counts. The top bad actors are often the most contributing factors of high alarm rate. Tracking and solving top bad actors help to relieve alarm overloading. To track the changes of top bad actors, a ranking chart is proposed. It consists of four essential components:

(i) Two zones display the top bad actors over two time periods, and the bad actors in each zone are ranked in a decreasing order of the alarm count.
(ii) Each zone consists of a group of stacked blocks corresponding to different bad actors.
(iii) Different colors are used to indicate disappearing, appearing, increasing rank, and decreasing rank of the bad actors.
(iv) Alarm counts are displayed beside the colored blocks.

The alarm count of each unique alarm variable is denoted by ψ_{a_i} in a time period of $[\mathcal{T}_s, \mathcal{T}_e]$. Given alarm data over two time periods, two zones of top bad actors are drawn based on the alarm counts. In each zone (denoted by $j = 1$ or 2), the top N bad actors over the corresponding time period are found as $\hat{a}_i^j$, $i = 1, 2, \ldots, N$ with their alarm counts denoted by $\psi_{\hat{a}_i^j}$. Then, the total alarm count of the top N bad actors in each zone is

$$\Psi^j = \sum_{i=1}^{N} \psi_{\hat{a}_i^j}. \tag{6.5}$$

The height H^j of each zone of the ranking chart is calculated as

$$H^j = \frac{\Psi^j}{\max(\Psi^1, \Psi^2)}. \tag{6.6}$$

Accordingly, the block height h_i^j of each bad actor $\hat{a}_i^j$ is obtained as

$$h_i^j = H^j * \frac{\psi_{\hat{a}_i^j}}{\Psi^j}. \tag{6.7}$$

An example of the ranking chart to track the changes of the top 10 bad actors over two months is presented in Fig. 6.11. The height of each block is proportional to the alarm count of the bad actor. Two colors are used: orange represents the appearing or increasing rank; green indicates the disappearing or decreasing rank. The numbers on the side are alarm counts for the top 10 bad actors over the week. It can be observed from the plot that the total alarm count of the top 10 bad actors decreases in the current month compared to that in the previous month. Four bad actors disappeared

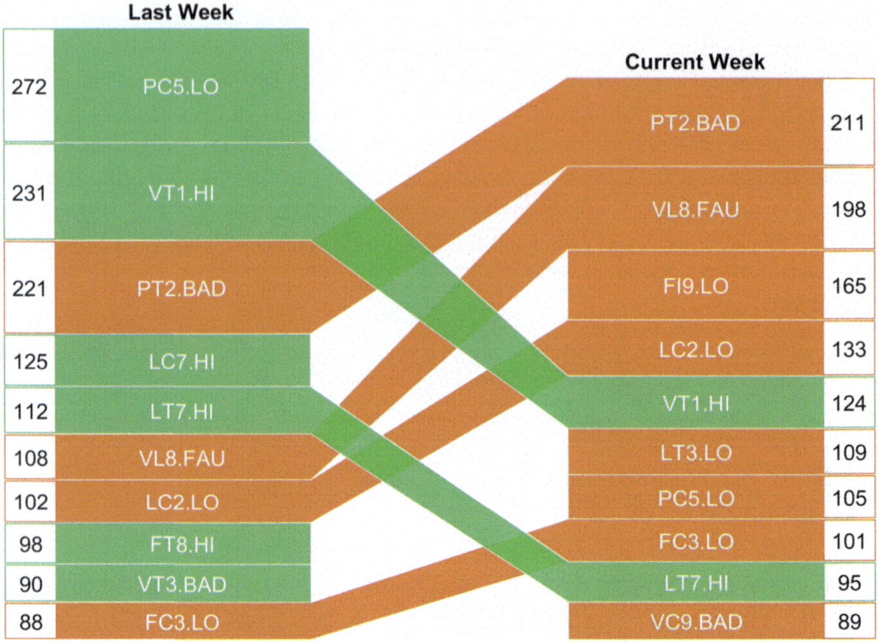

Fig. 6.11 An example of the ranking chart to track top bad actors, reprinted from Ref. [21], copyright 2018, with permission from Elsevier

whereas four new bad actors appear. In addition, four bad actors have increasing ranks whereas the ranks of two other bad actors decrease.

(5) Layered Alarm Count Radar Chart

This layered radar chart is specifically designed to track the changes of alarm counts for a large number of individual alarms [4]. It is powerful in showing the alarm counts of a large number of unique alarms using bars arranged in a circular manner. It consists of three essential components:

(i) The inner layer presents the alarm counts for all individual alarms in the current time period;
(ii) The outer layer shows the alarm count increments in comparison to the previous time period;
(iii) The alarm counts and their increments are represented using bars.

In order to have a better view of the increase and decrease of alarm counts, a further improvement has been incorporated. Two different colors are used to mark the bars: The red and green colors are used to indicate the increase and decrease of alarm counts, respectively. In addition, the maximum and minimum alarm counts are shown in the center, so that users can approximate the alarm count represented by each bar.

In a given time period of $[T_s, T_e]$, the alarms are ranked based on the count of their alarm occurrences. Denote the alarm count for the ith alarm variable a_i in the ranking list as ψ_{a_i} in the current time period. Meanwhile, the alarm count for a_i in the previous time period is denoted by $\tilde{\psi}_{a_i}$. Then, the bar length in the inner layer for a_i is

$$B_i^I = r_1 * \psi_{a_i} / \psi_{a_1}, \tag{6.8}$$

where r_1 denotes the user-defined radius of the inner circle. The increment of alarm count for a_i is $\psi_{a_i} - \tilde{\psi}_{a_i}$. Denote the maximum increment as $\psi_{\max} = \max_{i=1,2,\ldots,N} |\psi_{a_i} - \tilde{\psi}_{a_i}|$, where N indicates the number of selected alarm variables for display in the chart. Then, the bar length in the outer layer for a_i is

$$B_i^O = (r_2 - r_1) \frac{\psi_{a_i} - \tilde{\psi}_{a_i}}{\psi_{\max}}, \tag{6.9}$$

where r_2 stands for the user-defined radius of the outer circle.

Figure 6.12 presents an example of the layered radar chart to track alarm counts of 100 unique alarms. The names of the alarms are shown outside the chart in a circle. The center of the plot indicates that the minimum and maximum alarm counts are 1 and 443, respectively. The red and green colors indicate the increase and decrease of the alarm counts in the current time period in comparison to the previous period. The alarms are ranked in a counter clockwise fashion based on their alarm counts, with the topmost bad actor shown at 3 o'clock. It can be observed that many alarms, especially the top bad actors, have increasing alarm counts.

(6) Multilayered Performance Comparison Radar Chart

This visualization plot compares the overall behavior of the metrics for a single or multiple variables of interest (e.g., alarm, group of alarms, and average of alarms) and in different periods of time (e.g., hours, shift, day, week, and month) [4]. The variables of interest as well as the periods of time can be selected by the user, and the computed values of the metrics can be displayed and compared. It consists of following essential components:

(i) In the inner layer, the pie segments of the metrics represent the value of the variables of interest for a selected period of time, and the + signs represent the values of the variables to be compared against for a period of time (namely, may be different than the period of time of the variable of interest).
(ii) In the middle and outer layers, the name of the metric and the category are specified, respectively. In addition, a layer may be added after the inner layer with segments for each metric to represent the change in the values of the variables with respect to the period of time.
(iii) The plot contains a color bar consisting of colors ranging from green to yellow and finally to red to represent the transition of values from 0 to 100%. For each metric, a green colored pie segment indicates a good value, and a red colored pie segment indicates a bad value.

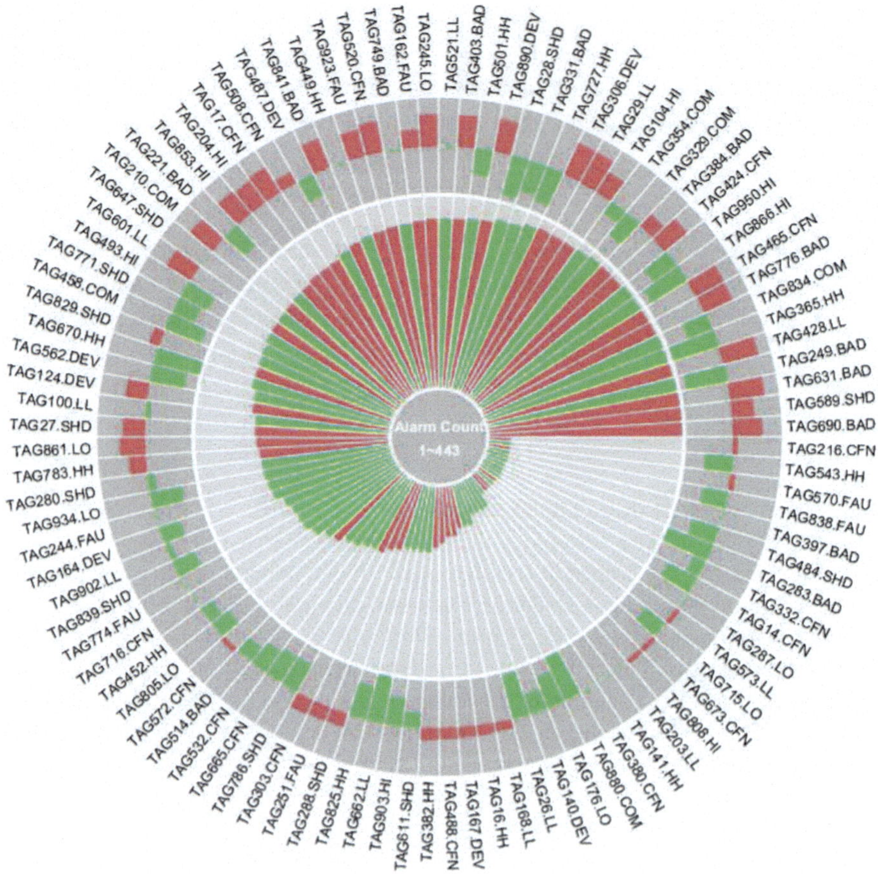

Fig. 6.12 An example of the layered radar chart to track alarm counts

The values of the metrics are logarithmically scaled to accommodate large variations in the data. They are normalized according to the following expression with V_n denoting the normalized value, V denoting the actual value to be normalized, and $V_{\max}$ denoting the maximum value of V, i.e.,

$$V_n = \frac{V + 1}{V_{\max} + 1}. \tag{6.10}$$

The metrics are grouped into categories to distinguish their relations. There can be metric groups for annunciated alarms, alarm floods, and operator actions. The number of variables and metrics to be compared can be specified by users and is limited by the visibility of information on the plot.

Figure 6.13 presents an example of the multilayered performance comparison radar chart, with the description of the variables given in Table 6.2. Meanwhile, the

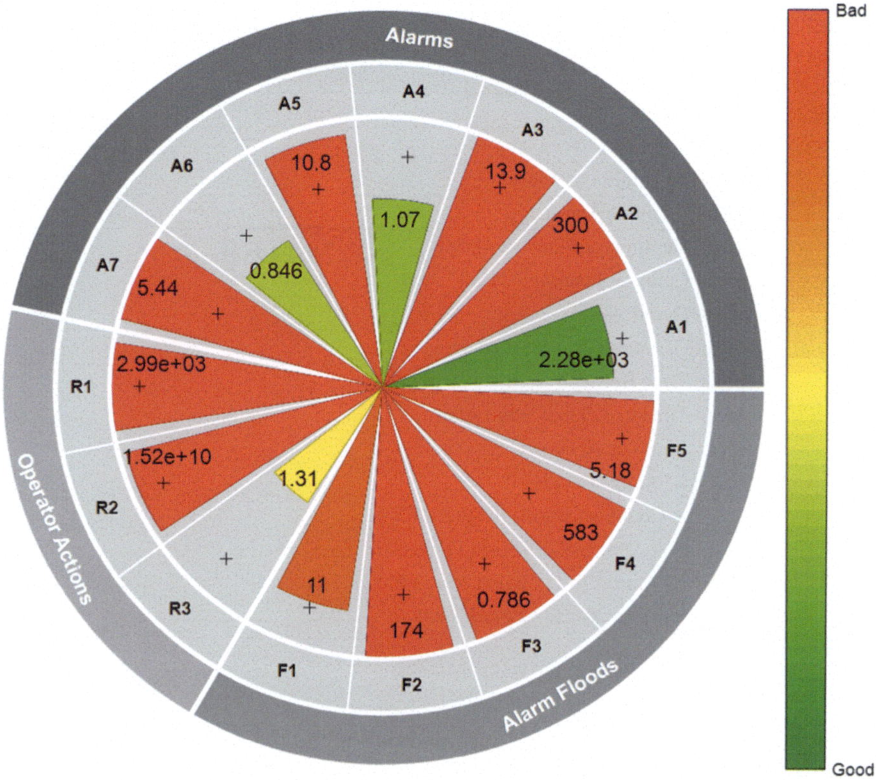

Fig. 6.13 An example of the multilayered performance comparison radar chart. A single (colored pie segments) against the average of the entire week before (+ signs), reprinted from Ref. [4], copyright 2018, with permission from IEEE

color of the pie segments indicates the relative performance from the week before (e.g., green indicates an improving performance, while red indicates a deteriorating performance). As can be observed from Fig. 6.13 for a single day, alarms did not occur frequently (A1), the durations of both fleeting and chattering alarms were not too short (A4 and A6), and the operators were responsive (R3). For example, the mean durations of fleeting and chattering alarms were only 1.07 and 0.846 s (A4 and A6), respectively. However, there were severe problems as observed from other metrics. Many alarms were active, and for long periods (A2 and A3), there were many fleeting and chattering alarms (A5 and A7), operators were not quick to respond (R1 and R2), and there were significant alarm floods (F1–F5). Those metrics are colored in red, which can easily draw attention. For example, the metrics related to alarm floods (F1–F5) were quite high. The alarm system experienced a count of 11 alarm floods (F1), a mean alarm count of 174 in alarm floods (F2), an average time duration of 0.786 h in alarm floods (F3), and a serious worst alarm flood with 583 alarms (F4) and 5.18 h (F5). Apparent from the plot, the alarm flood problem was much more

Table 6.2 Description of variables in Fig. 6.13. The unit of A1, A2, A4, A6, R1, and R2 is seconds, A3 is alarms per 10 min, and F5 is hours, reprinted from Ref. [4], copyright 2018, with permission from IEEE

Variable	Metric description
A1	Median Time Between Alarms
A2	Median Duration of Alarm
A3	Mean Count of Alarms
A4	Mean Duration of Fleeting Alarms
A5	Mean Count of Fleeting Alarms
A6	Mean Duration of Chattering Alarm
A7	Mean Count of Chattering Alarms
R1	Mean Time Between Acknowledgments
R2	Mean Time to Acknowledge Alarm
R3	Mean Count of Acknowledgments
F1	Count of Alarm Floods
F2	Average Count of Alarms in Alarm Floods
F3	Percentage Duration of Alarm Floods over the Operating Period
F4	Count of Alarms in the Worst Alarm Flood
F5	Duration of the Worst Alarm Flood

serious than the average performance, by comparing the red bars and the + signs for the metrics F2–F5 in Fig. 6.13.

(7) High Density Alarm Plot

This visualization plot provides an overview picture of an industrial alarm system by tracking the occurrences of individual alarms in a given time period [33]. A high density alarm plot (HDAP) displays alarm counts for top bad actors using a color map. By observing this plot, users can identify if the alarm system has a good performance, what the top bad actors are, as well as whether there are any chattering alarms, repeating alarms, standing alarms, correlated alarms, and alarm floods. Based on such observations, actions will be taken to reduce nuisance alarms and improve alarm management. It consists of two essential components:

(i) A main plot displays the alarm counts in each 10-min period for top bad actors over a certain time period.
(ii) A color bar maps the alarm counts into colors, such that a large number corresponds to dark red and a small number tends to be green.

Given alarm data over a given period $[\mathcal{T}_s, \mathcal{T}_e]$, the alarms are firstly ranked based on their alarm occurrence counts. Denote a_i as the ith bad actor in the ranking list. The whole time period $[\mathcal{T}_s, \mathcal{T}_e]$ is partitioned into short time segments; each segment lasts for $\Delta t = 10$ min. Denote the width and height of the main plot as L_w and L_h, respectively. Then, each cell in the main plot takes the size $\Delta L_w \times \Delta L_h$,

$$\Delta L_w = \frac{L_w}{\lceil (\mathcal{T}_e - \mathcal{T}_s)/10 \rceil},$$
$$\Delta L_h = \frac{L_h}{N},$$

where $\lceil \cdot \rceil$ returns the smallest integral value that is not less than the operand; N denotes the number of displayed top bad actors. The time period $\mathcal{T}_e - \mathcal{T}_s$ should be transformed to take minute as the time unit. Then, each cell is filled with a certain color corresponding to the number of alarm occurrences for a_i in the current time segment of 10 min.

Figure 6.14 shows a high density alarm plot of the top 20 bad actors over the selected time period of 10 days. Totally, 173 alarm tags are found to occur with an average of 8.5 alarms over a 10 min period, which is much higher than the benchmark threshold of an efficient alarm system, namely, 1 alarm over 10 min. This figure contains 864,000 bits of alarm data for each of the 20 tags. The color bar denotes the number of alarms in each 10 min time bin. The red and orange colors indicate high alarm rates, implying chattering or repeating alarms. It is obvious that "Tag102.CFN" and "Tag101.CFN" are likely chattering over some short time periods, and "Tag192.FAILED" keeps repeating for the whole time period. Moreover, it can be seen that two alarms, "Tag64.COMM" and "Tag60.IOF", are annunciated almost simultaneously in the first two days of the selected time period. Around 11pm on Aug. 23rd, there is a high chance that most top bad actors occur, implying a plant upset around this time instant.

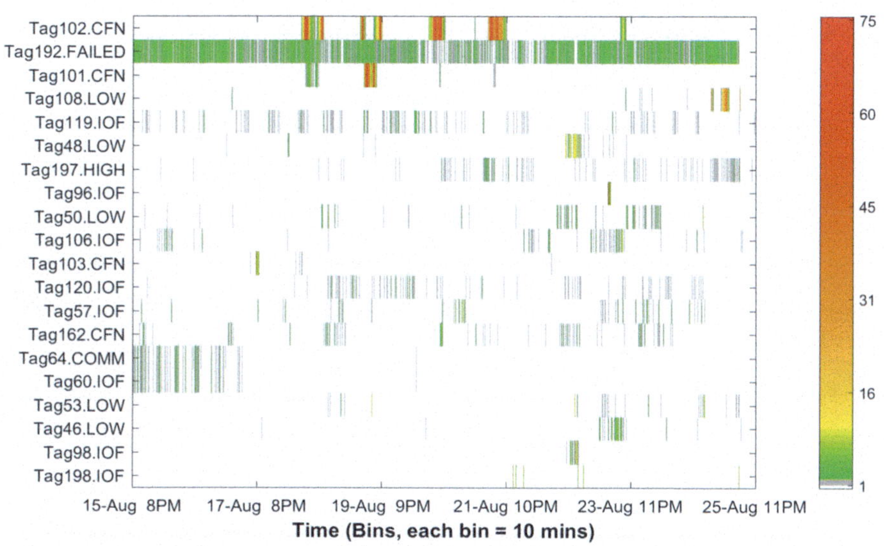

Fig. 6.14 An example of the high density alarm plot, reprinted from Ref. [25], copyright 2018, with permission from Elsevier

This static plot can also be turned into a dynamic plot for realtime applications. Comparing to the static one in [33], the new dynamic HDAP has the following special features:

(i) The plot presents the alarm counts for different alarm variables in a short recent period (e.g., hours or days).
(ii) The time window, color bins, and alarm orders update with time (alarm tags are sorted by their alarm counts in a descending order).
(iii) An alarm burst plot is shown to indicate alarm flood situations, and benchmark thresholds are included and shown as horizontal lines.

Figure 6.15 presents an example of the dynamic HDAP. The upper subplot shows the burst alarm rate, namely, the number of alarm occurrences in each 10 min time bin, and compares it to the red line, which indicates the benchmark threshold of alarm floods (namely, 10 alarms per 10 min) in [6]. Depending on the application, the threshold of the acceptable alarm rate (namely, 1 alarm per 10 min) can also be included in the alarm burst plot. The lower subplot displays alarm counts for the top 20 bad actors in a time window of 24 h. The alarm count for one specific alarm

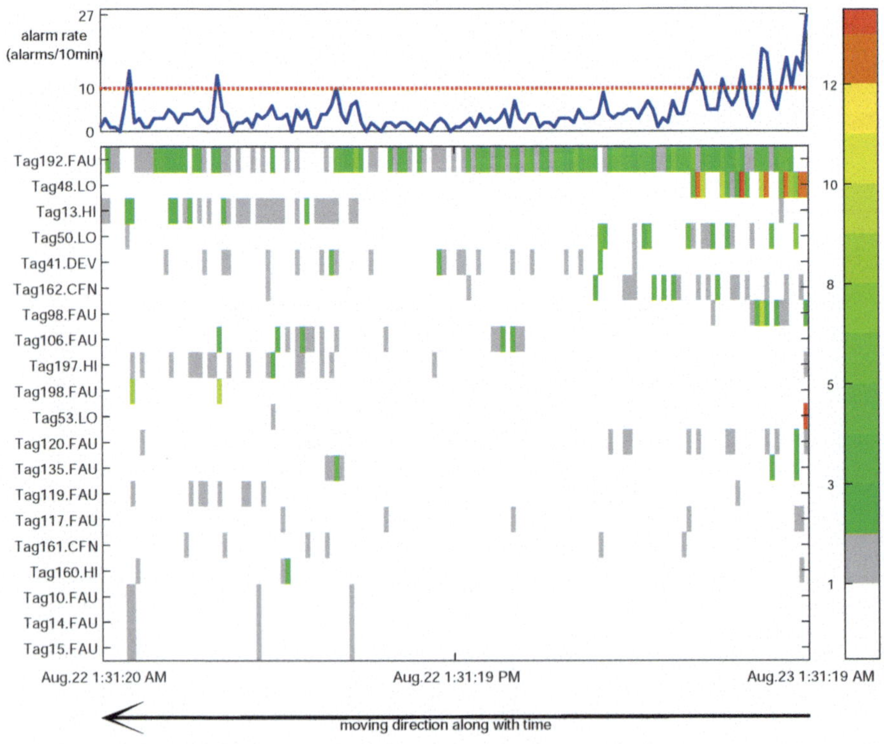

Fig. 6.15 An example of the dynamic high density alarm plot, reprinted from Ref. [21], copyright 2018, with permission from Elsevier

variable in a time bin of 10 min is color coded and shown as a color bin. The upper and lower subplots update with time in a direction as indicated by the black arrow at the bottom of Fig. 6.15. It can be observed from this plot that the alarm system has a good performance in the past day according to the low burst alarm rate in the upper plot. However, the increase of the burst alarm rate in the most recent hour indicates the emerging of new alarm floods. This can be also observed from the lower plot when more bins of red and orange colors appear. Moreover, it can be also observed that there is a group of correlated alarms, namely, "Tag10.FAU", "Tag14.FAU", and "Tag15.FAU", since they appear almost simultaneously.

(8) Dynamic 3D Bar Chart

This visualization plot is designed to present an overview picture of an industrial alarm system in a real-time manner. Compared to the previous high density alarm plot, this plot has similar functionality, but presents data in a different visual form. Analogous to the dynamic HDAP, the 3D bar chart has the following features:

(i) The chart presents the alarm information in three dimensions, including the time on the x-axis, the names of alarms on the y-axis, and the alarm counts on the z-axis.
(ii) The time window, bar lengths, and alarm orders update with time (alarm tags are sorted by their alarm counts in a descending order).
(iii) An alarm burst plot is integrated to denote alarm flood situations, and benchmark thresholds are included and shown as horizontal lines.

Denote the current time as t, the time period for each bar as Δt, and the number of bars as M. Then, for each alarm variable a_i, its alarm counts for the M bars corresponding to M consecutive time segments are denoted by a sequence as

$$[\psi_{a_i}(t-(M-1)\Delta t), \dots, \psi_{a_i}(t-(M-j)\Delta t), \dots, \psi_i(t)], \tag{6.11}$$

where $\psi_{a_i}(t-(j-1)\Delta t)$ counts the alarm occurrences of a_i in the jth time period from $t-(M-j-1)\Delta t$ to $t-(M-j)\Delta t$. Assume that the bar chart is updating with a period of ΔW. Then, the alarm counts for a_i in all the M bars are updated as

$$[\psi_{a_i}(t+\Delta W-(M-1)\Delta t), \dots, \psi_{a_i}(t+\Delta W-(M-j)\Delta t), \dots, \psi_i(t+\Delta W)]. \tag{6.12}$$

Figure 6.16 presents an example of the dynamic 3D bar chart. Each bar indicates the alarm count for one alarm variable over a 10 min time period. The black curve indicates the burst alarm rate. The green and red lines represent benchmark thresholds of 1 and 10 alarms over a 10 min time period, respectively. From the plot, the alarms that have high alarm rates in the most recent time period can be observed from the vertical bars, and the changes of the burst alarm rate can be observed from the black curve. The time window moves in a direction as indicated by the black arrow at the bottom of Fig. 6.16. It can be found from the black line that the burst alarm rate is fluctuating around the benchmark threshold of alarm floods, indicating a bad performance of the alarm system. It can also be found that the alarm "Tag48.LO" is

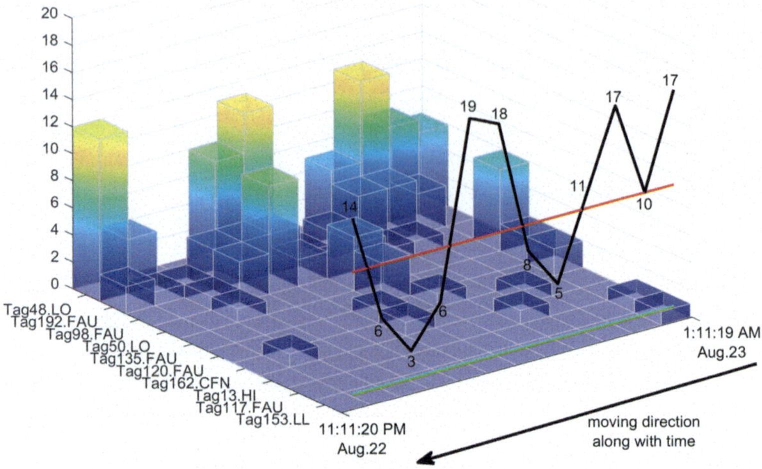

Fig. 6.16 An example of the dynamic 3D bar chart, reprinted from Ref. [21], copyright 2018, with permission from Elsevier

chattering or repeating, which accounts for the peaks of the black line. Overall, the industrial alarm system has alarm floods and alarm chattering problems in the recent 2 h time period, and thus corrective actions are required to solve the two problems.

6.3 Visual Analytics for Related Alarms and Events

These visual analytics functions target at mining and visualizing relations for alarms and related events, such as temporal and spectral correlations, causal relations, and alarm response workflows. Three useful visual analytics methods are presented, including correlation color map, alarm response workflow diagram, as well as alarm and response event flow chart.

(1) Correlation Color Map

This visualization plot provides a straightforward view of correlations for a number of process or alarm variables. Correlation analysis is a task commonly used in process monitoring and alarm management. There exist various definitions and metrics for correlations for either continuous-valued process signals or binary-valued alarm signals. For instance, the correlation between two process signals is measured as the absolute maximum value of correlation coefficients [55, 58]. As another example, spectral correlation is used as a specific application of correlation coefficient to the power spectra of process signals [44]; it indicates the similarities in the spectral "shapes" of all signals and can reveal all process variables that are oscillating at the same frequency. In addition, alarm correlation analysis detects whether two alarms are correlated or not and measures the correlation level with a certain metric [26, 33, 38, 58, 60]. Correlated alarms are referred to as alarms occurring almost simultaneously in high frequency. They could be either redundant or overlapping in indicating the same abnormality. Based on the alarm correlations, redundant alarms can be removed and related alarms can be grouped. Given the calculated correlation values for all different pairs of process or alarm variables, a correlation color map can be drawn. Such a plot consists of the following two parts:

(i) A main plot presents the correlations for all variable pairs using a heating map, where the horizontal axis contains the same tick labels (variable names or indices) as the vertical axis.
(ii) A color bar is provided to indicate the correlation values corresponding to the colors shown in the main plot.

Specifically, given the time series of variables $x_1, x_2, \ldots, x_M$, the correlation C_{ij} between a pair of variables x_i and x_j is calculated as

$$C_{ij} = h(x_i, x_j),\ i, j \in [1, M], \tag{6.13}$$

where M denotes the number of variables, and $h(\cdot)$ denotes a correlation measure function, such as the Pearson's correlation [26], Jaccard coefficient [60], and spectral correlation [44]. Then, a correlation matrix is formed as

$$\mathbf{C} = \begin{bmatrix} C_{11} & C_{12} & \cdots & C_{1M} \\ C_{21} & C_{22} & \cdots & C_{2M} \\ \vdots & \vdots & \ddots & \vdots \\ C_{M1} & C_{M2} & \cdots & C_{MM} \end{bmatrix}, \tag{6.14}$$

where the diagonal element is a constant indicting the highest correlation, as it indicates the correlation between a variable and itself, i.e., $C_{11} = C_{22} = \cdots = C_{MM}$. For instance, $C_{ii} = 1$ if the Pearson's correlation is applied. In addition, the correlation matrix $\mathbf{C}$ is symmetric, i.e., $C_{ij} = C_{ji}$.

Figure 6.17 presents an example of spectral correlation color map, which is used to analyze plant oscillations leading to the presence of oscillating alarms [44, 50]. The process data is collected from a distillation chemical plant [46]. The collected data has 48 process variables and 5040 observations sampled at the 20 s interval. The power spectra of there process variables are calculated, and then variables with similar power spectra are clustered. The color bar on the right side of the color map indicates the strength of correlation. The red and orange colors indicate strong correlations and the green color indicates a weak correlation. The diagonal of the color map represents the correlation between one variable and itself. Based on Fig. 6.17, it is easy to identify all process variables that share the same oscillating feature.

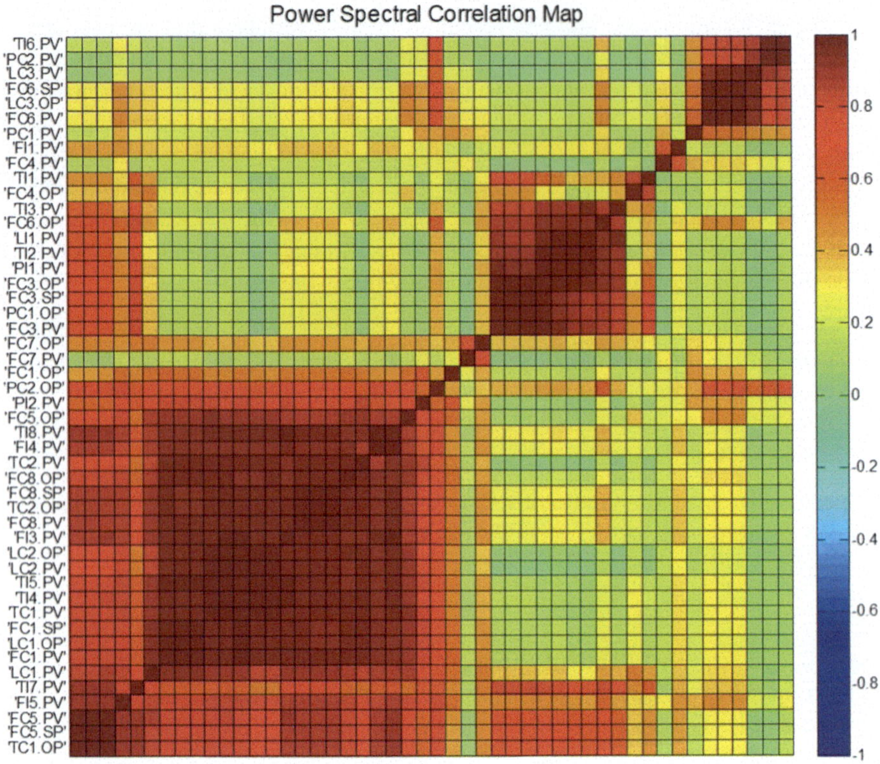

Fig. 6.17 Spectral correlation color map, reprinted from Ref. [25], copyright 2018, with permission from Elsevier

Figure 6.18 shows an example of alarm correlation color map for 40 alarm tags, which are clustered based on their correlations. The darker color of a block indicates a higher correlation between alarms. The diagonal of the map consists of 1's (black squares), indicating the highest correlations of alarm tags with themselves. In this example, it can identify several groups of correlated alarms shown by blue blocks in Fig. 6.18. Specifically, the alarms "FI402.PVHI" and "FI311.PVHI" hold a considerable correlation; in addition, the alarms "TC202.SVHI" and "TC202.SVHI" are detected to hold a strong correlation, indicating they always occur simultaneously.

(2) Alarm Response Workflow Diagram

This diagram is designed to visualize how operators take actions to respond to alarms, which are summarized in the form of a workflow diagram, representing the relations between alarms and actions. Historical records about such relations between alarm systems and operators make it possible to find out how operators respond to alarms and how operational procedures are conducted routinely. The workflow mining, or also known as process mining [49], provides solutions to gain such insights

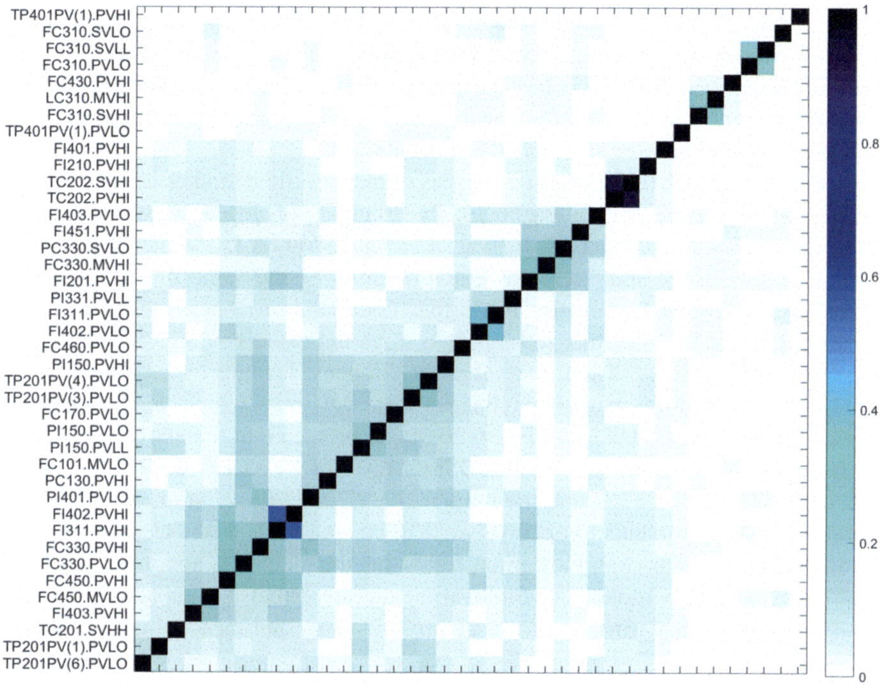

Fig. 6.18 Alarm correlation color map

from historical data. By extracting all variations of procedures and turning them into understandable models, the real execution can be discovered [17, 20, 22, 23]. The detected relations of operator responses to alarms can be represented by workflow models, which are graphical representations in different forms [13]. The graphical representation of workflow models of operator responses to alarms is presented [22]. It consists of two main parts:

(i) Nodes: Each node indicates a unique event, such as the occurrence of an unique alarm, its return-to-normal, or an associated alarm response. In this framework, nodes related to alarm occurrences or return-to-normal events, are representing by triangle symbols. Nodes related to operator responses or other related events are represented by ellipse symbols.
(ii) Edges: Each edge indicates the dependency between two events. Three types of edges are defined in this graphical representation: a green solid arrow indicating the forward dependency, a red solid arrow indicating the reverse or feedback dependency, and a dotted arrow indicating a portion of the process model which is connected to the other portion of the process model. In most cases, a well structured or ideal process model should not contain self loops and reverse dependencies. Dashed lines are useful to display large and complex process models.

The list in Table 6.3 presents the graphical notations for the representation of workflow models of operator responses to alarms. To distinguish between different alarms or responses, different colors are used for the above symbols. Since the occurrence and return-to-normal events usually appear in pair for the same alarm tag, the same color should be used for such alarm events.

Figures 6.19, 6.20, 6.21 present examples of the workflow models extracted from a real A&E data set [22]. In Fig. 6.19, the triangles of different colors indicate three alarms, namely, "Var1.LO", "Var1.LL" and "Var1.VL", which correspond to "low", "low low", and "very low" alarm limits of the same process variable "Var1". As discovered from the A&E log, commands to open the valve "V2" are usually made by operators through two ways indicated by "V2.OPENCMD1" and "V2.OPENCMD2". Then, the valve "V2" opens, leading to the clearance of "Var1.LO". Meanwhile, an operator changes the alarm limits, which clears the alarms "Var1.LL" and "Var1.VL". From Fig. 6.20, it can be seen that the alarm "Var2.LO" is triggered by the command "V3.CLOSECMD1" or "V3.CLOSECMD1" that is used to close valve "V3". Then, the pumps "PM1" and "PM2" stop. After that, an operator makes commands to open valve "V3", which leads to the clearance of the alarm "Var2.LO". This example demonstrates that the method is not only capable of discovering workflow models for multivariate alarms, but also is capable of finding those for univariate alarms. Figure 6.21 presents a special case, where no opera-

Table 6.3 Graphical notations for the representation of workflow models of operator responses to alarms

Graph symbol	Notation
▷	A hollow triangle connecting to the start of an edge represents the occurrence of an alarm
◁	A hollow triangle connecting to the end of an edge represents the return-to-normal event of an alarm
▶	A solid triangle connecting to the start of an edge indicates the repeated occurrence of an alarm
◀	A solid triangle connecting to the end of an edge indicates repeated return-to-normal event of an alarm
→	A green solid arrow represents an edge connecting two events in a forward direction
←	A red solid arrow represents an edge connecting two events in a feedback direction
...	A dotted line (either a green or red arrow) connects two partial process models
⬭	An hollow ellipse represents a unique operator response, and a solid ellipse indicates a repeated operator response

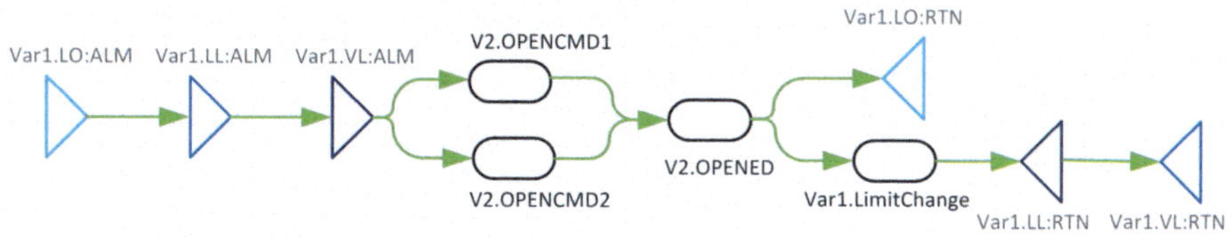

Fig. 6.19 Example 1 of constructed workflow model

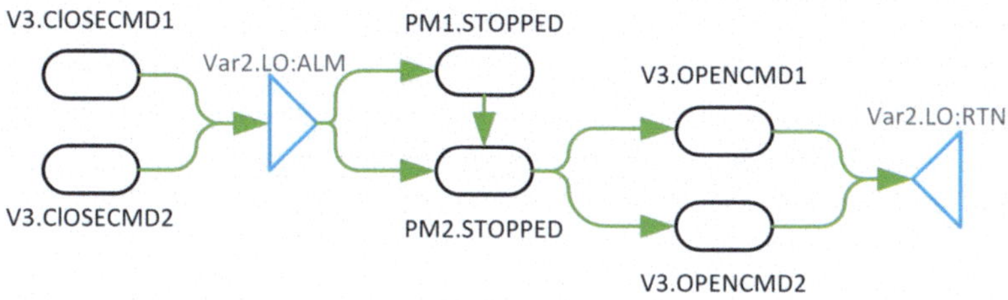

Fig. 6.20 Example 2 of constructed workflow model

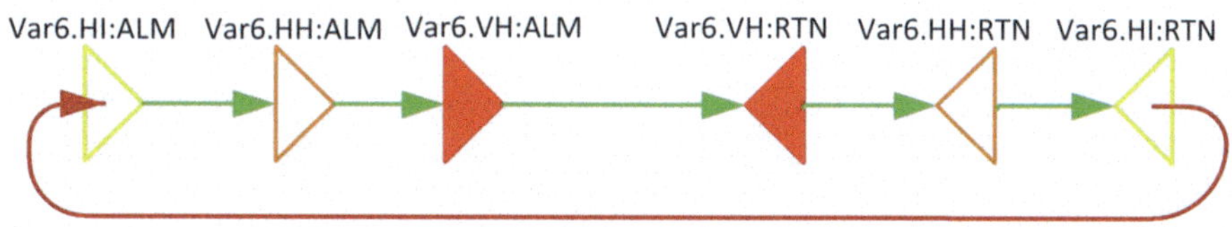

Fig. 6.21 Example 3 of constructed workflow model

tor actions or system changes appear in the extracted workflow model. The alarms "Var6.HI", "Var6.HH", and "Var6.VH" occur in a sequential order and clear in a reverse order. After that "Var6.HI" alarm appears again, indicating the repeating of these alarms without the intervention of operators. As indicated by the solid symbols, "Var6.VH:ALM" and "Var6.VH:RTN" are likely to repeat by themselves.

(3) Alarm and Response Event Flow Chart

This visualization plot is used to observe the relations between alarms and their associated responses by presenting information, such as orders, frequencies, and time intervals, in a good visual form. An event flow chart transforms an entire data set of temporal event records into an aggregated display, allowing analysis of patterns and trends in a population-level [35]. It is exploited to present the interactions between alarms and responses. From such an event flow chart, it can tell what actions are usually taken to respond to certain alarms and whether or not they are effective. An event flow chart consists of three essential components:

(i) A sequence of alarms and related response events is shown as an event flow in one row of the chart.
(ii) Alarms and related response events are presented in a chronological order, and different alarms and events are denoted by different symbols and colors.

(iii) An interactive design is incorporated to enable the alignment of events in different perspectives and the simplification of the event flow.

Given an alarm variable x_a, its pairwise time-stamped alarm occurrence and clearance events are extracted from the historical A&E data and denoted as $[(t_1^a, x_a), (t_1^r, x_a)]$, $[(t_2^a, x_a), (t_2^r, x_a)]$, ..., $[(t_N^a, x_a), (t_N^r, x_a)]$, where N indicates how many times the alarm x_a occurs in the given time period; t_i^a and t_i^r represent the time stamps for the ith occurrence and clearance events of x_a, respectively. Then, in each period from t_i^a to t_i^r, all the time-stamped response events are found and denoted by $[(t_i^{e1}, x_{e1}), (t_i^{e2}, x_{e2}), \ldots, (t_i^{eM}, x_{eM})]$, where t_i^{ej} and x_{ej} respectively indicate the time stamp and response name of the jth response event in the duration from t_i^a to t_i^r. In the event flow chart, each period from the occurrence to the clearance of x_a is displayed as a horizontal bar with the bar length representing the time period. The related responses are shown as vertical short lines in the bar with the positions indicating the relative time compared to the occurrence time of x_a.

Figure 6.22 presents an example of the event flow chart to visualize the alarm state transitions and the related operator responses [21]. The vertical axis corresponds to the sequence index number that indicates a pair of alarm occurrence and clearance over different periods. The horizontal axis corresponds to the time that indicates the duration of an alarm state. Two distinct color bars within the alarm states represent two different operator responses (green for feed pump 1 and blue for feed pump 2). It can be observed from the plot that two types of operator responses are effective in clearing the alarms.

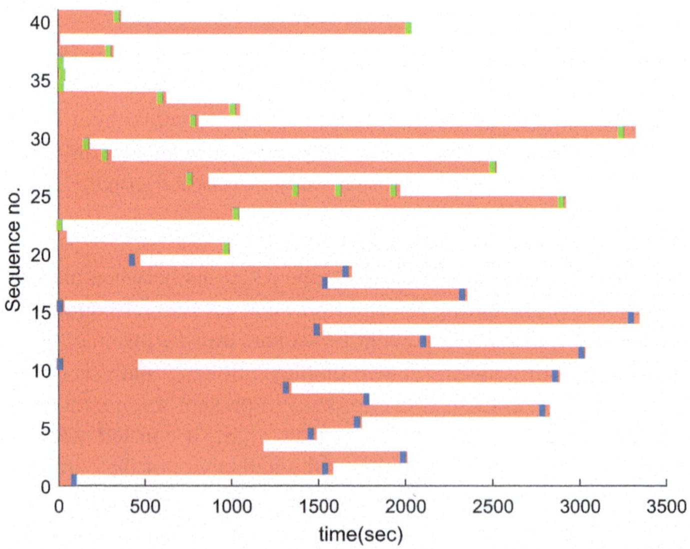

Fig. 6.22 An example of the event flow chart of alarm states and operator responses for the low alarm of a tank level, reprinted from Ref. [21], copyright 2018, with permission from Elsevier

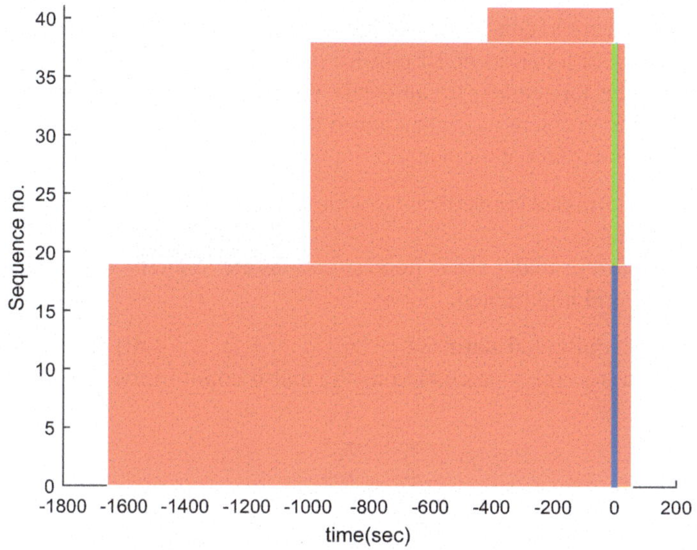

Fig. 6.23 An example of the simplified event flow chart presented in Fig. 6.22, reprinted from Ref. [21], copyright 2018, with permission from Elsevier

In order to have abstracted information and compare different cases, the event flow chart is simplified [35]. Figure 6.23 presents such a simplified graph. The height of each area indicates the frequency of the associated operator response marked by a blue or green color. The topmost area indicates that there are no responses for three alarm occurrences. The width of the red area before the blue or green bar indicates the time to respond to the alarm occurrences. The width of that after the blue or green bar denotes the time for the alarm to return to normal after a response. It can be observed from the plot that the response marked by the green color is more timely and effective than the response marked by the blue color.

6.4 Visual Analytics for Alarm Flood Sequences

These visual analytics functions target at analyzing alarm flood sequences and visualizing the patterns. Three useful visual analytics methods are presented, including alarm burst plot, alarm flood similarity color map, as well as alarm flood spiral graph.

(1) Alarm Burst Plot

This plot visualizes the trend of the burst alarm rate (namely, the number of alarms within the duration of 10 min) along with time, and can be used to identify periods of alarm flood situations. A common way for alarm flood detection is to compare the burst alarm rate with the thresholds defined in industrial standards [6, 16, 29]:

an alarm flood starts and ends when the burst alarm rate reaches 10 alarms and drops below 5 alarms over a period of 10 min for each operator. By visualising the burst alarm rate and the thresholds, it can easily observe whether an alarm system has alarm overloading problem and when alarm floods are present. An alarm burst plot consists of two essential components:

(i) The burst alarm rate is calculated in a moving time window of 10 min and drawn as a trend curve.
(ii) The time periods under alarm flood situations are found by comparing with the thresholds and highlighted.

Given a set of configured alarms $\mathcal{A} = \{x_{a,i}, i = 1, 2, \ldots, |\mathcal{A}|\}$ in an alarm system, the burst alarm rate $\nu(t)$ is calculated as the alarm count in the past time bin $[t - T + 1, t]$, i.e.,

$$\nu(t) = \sum_{i=1}^{|\mathcal{A}|} \sum_{k=t-T+1}^{t} x_{a,i}(k), \tag{6.15}$$

where $x_{a,i}(k) = 1$ indicates the occurrence of alarm $x_{a,i}$ at time instance k; otherwise, $x_{a,i}(k) = 0$. The window size T takes the value of 600 s (namely, 10 min) based on benchmarks in the ANSI/ISA-18.2 standard [6]. Based on the alarm rate $\nu(t)$, it can be identified when an alarm flood begins and ends by comparing $\nu(t)$ with predefined thresholds. An indexing variable ψ indicating the presence of alarm floods is defined as

$$\psi(t) = \begin{cases} 1, & \text{if } \zeta(t) \geq \Gamma_s \ \& \ \psi(t-1) = 0, \\ 0, & \text{if } \zeta(t) < \Gamma_e \ \& \ \psi(t-1) = 1, \\ \psi(t-1), & \text{otherwise,} \end{cases} \tag{6.16}$$

where 1 and 0 stand for the presence and absence of alarm floods, respectively. The initial sample $\psi(0)$ is set to be 0. According to the ANSI/ISA-18.2 standard [6], the benchmark thresholds to identify the start and end of an alarm flood are 10 and 5 alarms over a 10 min period for each operator, respectively. Hereby, the two thresholds are $\Gamma_s = 10$ and $\Gamma_e = 5$ based on a time bin of size $T = 600$ s. To avoid false identification of alarm floods, chattering alarms should be reduced. Commonly used chattering removal techniques include alarm delay timers and alarm deadbands, among which alarm delay timers can be applied to alarm variables directly [1, 6, 51].

An example is shown in Fig. 6.24 to illustrate the identification of alarm floods based on the benchmark thresholds in the industrial standards [6, 16, 29]. The black line indicates the threshold of identifying the occurrence of an alarm flood, namely 10 alarms over 10 min. The alarm floods are highlighted by blue blocks. It can be seen that this industrial plant is in the alarm flood situation for a considerable time period. Figure 6.25 shows the alarm burst plot for the alarm data with chattering alarms reduced by applying alarm delay timers with the factor 40 s to all alarm variables. It can be seen that significant reduction is achieved by applying the alarm delay timers. Originally, there are 66 alarm floods in Fig. 6.24 and they occur for almost 25.4% of the entire time period of 10 days. After applying the alarm delay timers, there are only 38 alarm floods left and they occur during 6.4% of this time period.

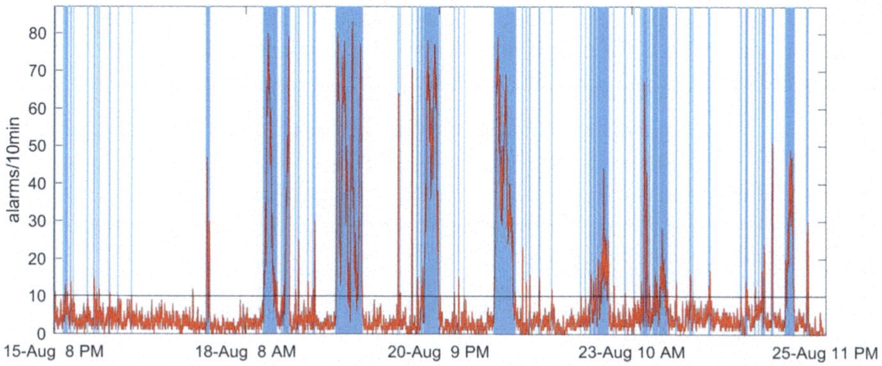

Fig. 6.24 Alarm burst plot for raw alarm data, reprinted from Ref. [25], copyright 2018, with permission from Elsevier

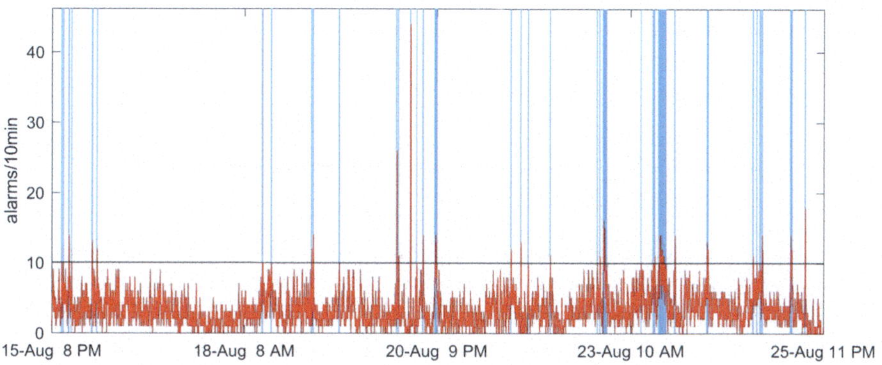

Fig. 6.25 Alarm burst plot for alarm data with chattering alarms reduced, reprinted from Ref. [25], copyright 2018, with permission from Elsevier

Having removed the chattering alarms, these alarm floods can be considered as true alarm floods, requiring some further analysis, so as to prevent the re-occurrence of the same underlying issue.

(2) Alarm Flood Similarity Color Map

This plot is specifically designed to visualize the similarities for a group of alarm flood sequences. Alarm floods caused by the same root cause should assemble each other and contain common subsequences of alarms. Similarity analysis of alarm floods aims at calculating the similarities scores between each pair or alarm floods and identifying the clusters of analogous sequences. The results are helpful for further root cause analysis and early prediction of alarm floods. Generally, alarm floods are compared in pairwise using sequence alignment algorithms, such as dynamic time warping (DTW) [3, 28], modified Smith-Waterman (SW) algorithm [11, 43], and BLAST-like algorithm [5, 27]. Sequence patterns of alarm floods can be found from a cluster of alarm floods using a multiple sequence alignment algorithm [34]. An alarm flood similarity color map is used to observe which alarm flood sequences are analogous to each other. Such a plot consists of the following two parts:

(i) A main plot displays the similarity matrix for clustered alarm flood sequences; the horizontal and vertical axis have the same tick labels, indicating the numerical indices of alarm flood sequences after clustering.
(ii) A similarity color bar is shown to indicate the similarity scores corresponding to the colors in the main plot.

Given a set of extracted alarm flood sequences $F_1, F_2, \ldots, F_M$ (M denotes the number of sequences), the similarity score is calculated for each pair of sequences. Specifically, for two different sequences F_i and F_j, sequence alignment methods, such as those in [3, 11, 27, 28] can be applied to calculate a similarity score $s(F_i, F_j)$ and normalize it into [0, 1], such that 1 indicates F_i and F_j containing the same alarm sequence and 0 implies no common alarms shared by the two sequences. Accordingly, a similarity matrix **S** is

$$\mathbf{S} = \begin{bmatrix} 1 & s(F_1, F_2) & \cdots & s(F_1, F_M) \\ s(F_2, F_1) & 1 & \cdots & s(F_2, F_M) \\ \vdots & \vdots & \ddots & \vdots \\ s(F_M, F_1) & s(F_M, F_2) & \cdots & 1 \end{bmatrix}, \tag{6.17}$$

where all the diagonal elements are 1's, as the similarity between an alarm flood and itself should be the highest. Meanwhile, the matrix is symmetric, i.e., $s(F_i, F_j) = s(F_j, F_)$ for $i \neq j$. In the similarity color map, a cell with the similarity score closer to 1 is filled with a darker color. By contrast, if it is closer to 0, a lighter color is applied to the cell.

Figure 6.26 presents an example of similarity indices for 38 different alarm floods. The modified Smith-Waterman algorithm in [11] is exploited for similarity analysis of alarm flood sequences. In the plot, the vertical and horizontal axes display the numeric indices of the 38 alarm floods. A smaller event number index refers to an earlier occurrence in time. The color bar at the right side of the cluster map indicates the strength of similarity indices. The diagonal of the color map represents the similarity between each alarm flood event and itself.

(3) Alarm Flood Spiral Graph

This visualization tool is used to help users to observe alarm floods and discover hidden problems from historical data. A spiral graph [54] can visualize data over a very long time period without breaking its continuity with respect to time. The spiral graph has several advantages: It makes an effective use of screen space and can visualize data over a very long time period; it supports effective comparisons in two dimensions, namely, comparisons in a neighborhood and comparisons of circles; it allows for an easy observation of periodic behaviors and trends. Using this graph, users not only can identify when an alarm flood starts and ends, but also can determine which day and what time tends to receive more alarm floods. The spiral graph consists of the following two parts:

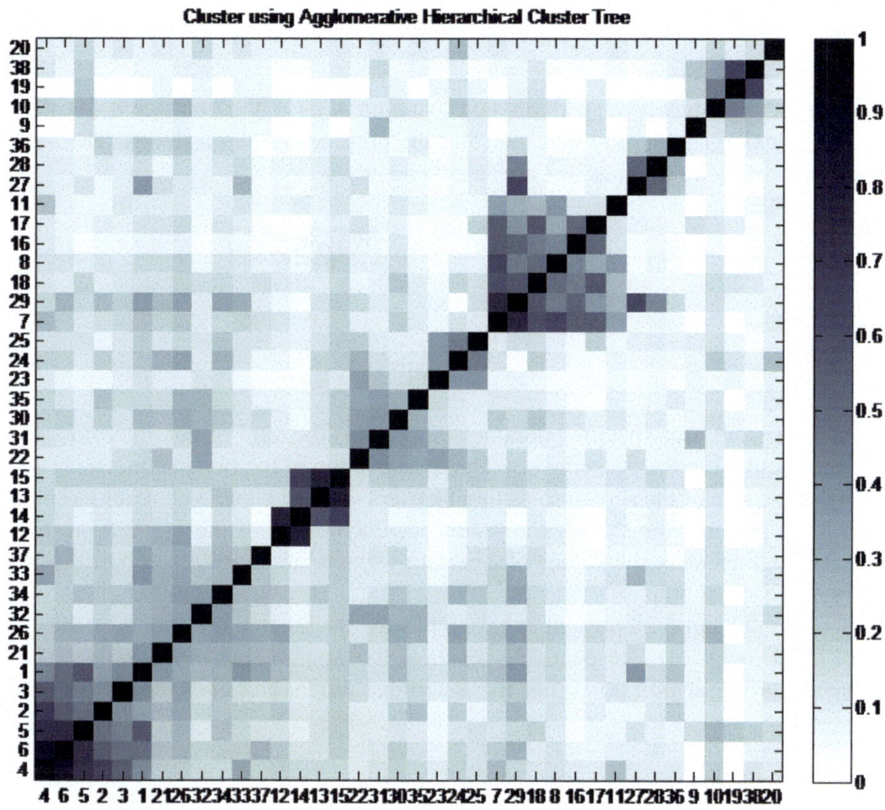

Fig. 6.26 Similarity color map of clustered alarm floods, reprinted from Ref. [25], copyright 2018, with permission from Elsevier

(i) A spiral grows clockwise and continuously in circles with each circle representing a certain time period, e.g., one day; each point on one circle represents a time instant of a smaller time resolution depending on the sampling rate, e.g., one second.

(ii) Colors in the spiral graph represent values of the time series related to alarm flood situations. A color bar is used to denote the values.

To observe alarm floods, two metrics are useful, including the burst alarm rate $\nu(t)$ that counts the number of alarms within a time window of 10 min, and the alarm flood index ψ that indicates whether it is under an alarm flooding period or not. The index variable $\psi(t) = 1$ and $\psi(t) = 0$ represent the presence and absence of an alarm flood at time instant t, respectively. The spiral graph to visualize alarm flood situations is implemented based on the following discussion. Given an alarm data spanning N days, the time instant t is scaled to the range $[1, 86 \times N, 400]$ (1 day is equal to 86,400 s). Then the angle in the spiral graph for each time instant t is

$$\theta(t) = \frac{2\pi t}{\alpha} + 2\beta\pi. \tag{6.18}$$

More specifically, given an alarm data spanning N days, the first step is to calculate the indexing signal $\psi(t)$ that indicates the presence of alarm floods. Then, the coordinates at each time instant t are calculated as

$$\begin{aligned} x(t) &= \gamma\theta(t)\cos\left(\theta(t) + \tfrac{\pi}{2}\right), \\ y(t) &= \gamma\theta(t)\sin\left(\theta(t) + \tfrac{\pi}{2}\right), \end{aligned} \tag{6.19}$$

where

$$\theta(t) = \frac{2\pi t}{86400} + 2\beta\pi. \tag{6.20}$$

Here β and γ are user-specified parameters to adjust the radii of the innermost and outermost spiral circles, respectively. The specification of the two parameters β and γ is up to the size of the screen so as to give users the best view. As a result, a spiral graph with N continuous circles is produced. Each circle of the graph represents 24 h. As the final step, the values of a certain variable are represented by different colors in the spiral graph.

Figures 6.27 and 6.28 present spiral graphs for the burst alarm rate $\nu(t)$ and the alarm flood index $\psi(t)$, respectively. Each of the graph displays a time series of 30 days, starting at 00 : 00 : 00 at the innermost circle, and goes clockwise until

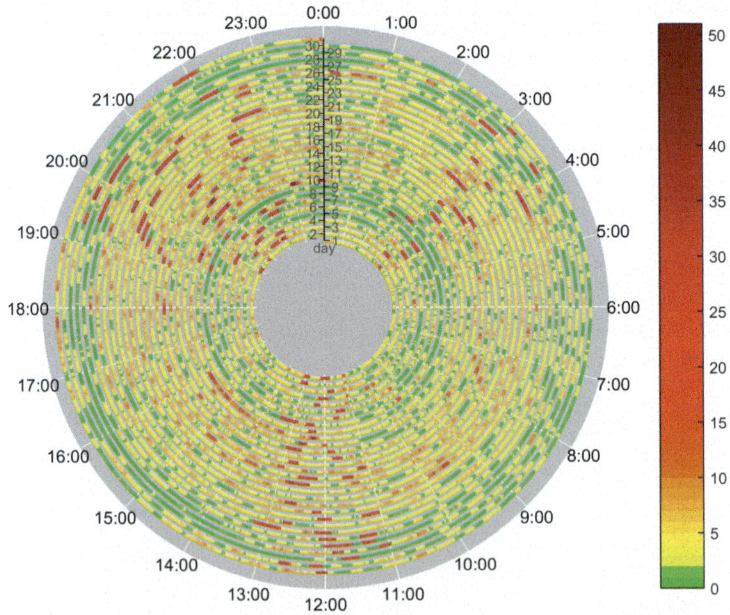

Fig. 6.27 An example of the spiral graph to show burst alarm rates, reprinted from Ref. [21], copyright 2018, with permission from Elsevier

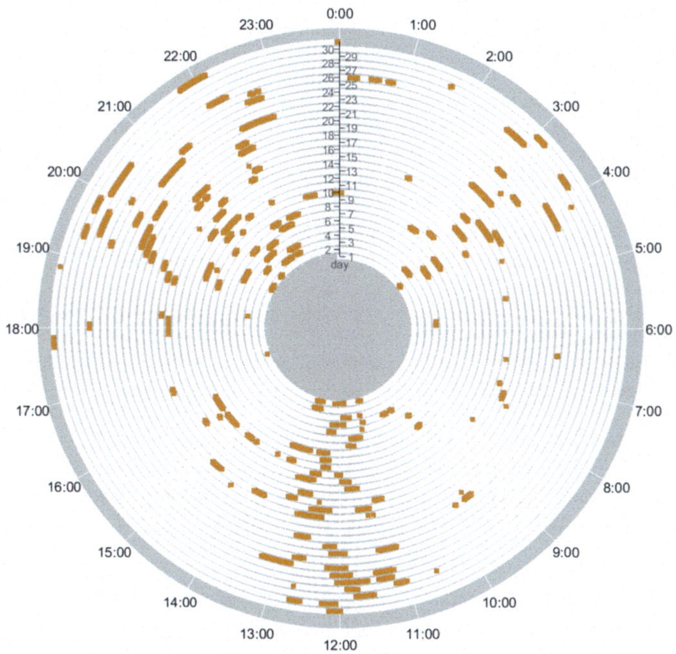

Fig. 6.28 An example of the spiral graph to show alarm flood periods, reprinted from Ref. [21], copyright 2018, with permission from Elsevier

reaching 23 : 59 : 59 at the outermost circle. In Fig. 6.27, green colors denote lower burst alarm rates, whereas yellow and orange colors represent higher burst alarm rates. Red colors indicate alarm flood situations. From Fig. 6.27, the time periods (marked by green colors) which are likely to receive less alarms and time periods which tend to have high alarm rates (marked by red colors) can be observed. It can also be found that most points on the spiral are green and yellow, implying the alarm system was healthy in most of the time period. But there are still quite a few spots with red colors, indicating high alarm rates and alarm flood situations. According to Fig. 6.28, alarm floods are quite common and they are more likely to appear during three periods: 3:00–4:00 hrs, 11:00–13:00 hrs, and 19:00–23:00 hrs.

References

1. Adnan, N., Cheng, Y., Izadi, I., & Chen, T. (2013). Study of generalized delay-timers in alarm configuration. *Journal of Process Control, 23*, 382–395.
2. Adnan, N., Izadi, I., & Chen, T. (2011). On expected detection delays for alarm systems with deadbands and delay-timers. *Journal of Process Control, 21*, 1318–1331.
3. Ahmed, K., Izadi, I., Chen, T., Joe, D., & Burton, T. (2013). Similarity analysis of industrial alarm flood data. *IEEE Transactions on Automation Science and Engineering, 10*, 452–457.
4. Al-Dabbagh, A., Hu, W., Lai, S., Chen, T., & Shah, S. (2017). Towards the advancement of decision support tools for industrial automation: Operation metrics, visualization plots, and

ranking in alarm floods. *IEEE Transactions on Automation Science and Engineering, 15*(4), 1883–1896.
5. Altschul, S., Gish, W., Miller, W., Myers, E., & Lipman, D. (1990). Basic local alignment search tool. *Journal of Molecular Biology, 215*, 403–410.
6. ANSI/ISA-18.2. (2016). *ANSI/ISA-18.2: Management of Alarm Systems for the Process Industries*. ISA (International Society of Automation).
7. Arnold, M., & Darius, I. (1989). Alarm management in batch process control. *ISA Transactions, 28*, 33–40.
8. Bauer, M., Cox, J., Caveness, M., Downs, J., & Thornhill, N. (2006). Finding the direction of disturbance propagation in a chemical process using transfer entropy. *IEEE Transactions on Control Systems Technology, 15*, 12–21.
9. Cheng, Y. (2013). Data-driven techniques on alarm system analysis and improvement. University of Alberta (Canada).
10. Cheng, Y., Izadi, I., & Chen, T. (2013). Optimal alarm signal processing: Filter design and performance analysis. *IEEE Transactions on Automation Science and Engineering, 10*, 446–451.
11. Cheng, Y., Izadi, I., & Chen, T. (2013). Pattern matching of alarm flood sequences by a modified Smith-Waterman algorithm. *Chemical Engineering Research and Design, 91*, 1085–1094.
12. Chiang, L., & Braatz, R. (2003). Process monitoring using causal map and multivariate statistics: Fault detection and identification. *Chemometrics and Intelligent Laboratory Systems, 65*, 159–178.
13. de Medeiros, A. K. A., van der Aalst, W. M., & Weijters, A. J. M. M. (2003). Workflow mining: Current status and future directions. In *OTM Confederated International Conference on the Move to Meaningful Internet Systems 2003*, Catania, Sicily, Italy, Nov 3–7, 2003.
14. Duan, P., Yang, F., Chen, T., & Shah, S. (2013). Direct causality detection via the transfer entropy approach. *IEEE Transactions on Control Systems Technology, 21*, 2052–2066.
15. Duan, P., Yang, F., Shah, S., & Chen, T. (2014). Transfer zero-entropy and its application for capturing cause and effect relationship between variables. *IEEE Transactions on Control Systems Technology, 23*, 855–867.
16. EEMUA (Engineering Equipment and Materials Users' Association). (2013). *Alarm Systems: A Guide to Design, Management and Procurement*. London: EEMUA Publication 191.
17. Gao, H., Xu, Y., Gu, X., Lin, X., & Zhu, Q. (2015). Systematic rationalization approach for multivariate correlated alarms based on interpretive structural modeling and Likert scale. *Chinese Journal of Chemical Engineering, 23*, 1987–1996.
18. Granger, C. (1969). Investigating causal relations by econometric models and cross-spectral methods. *Econometrica*, 424–438.
19. Hollifield, B., & Habibi, E. (2010). *The Alarm Management Handbook* (2nd ed.). Houston, USA: PAS.
20. Hu, W., Al-Dabbagh, A. W., Chen, T., & Shah, S. L. (2016). Process discovery of operator actions in response to univariate alarms. *IFAC-PapersOnLine, 49*, 1026–1031.
21. Hu, W., Al-Dabbagh, A. W., Chen, T., & Shah, S. L. (2018). Design of visualization plots of industrial alarm and event data for enhanced alarm management. *Control Engineering Practice, 79*, 50–64.
22. Hu, W., Al-Dabbagh, A. W., Li, D., & Chen, T. (2018). Extraction and graphical representation of operator responses to multivariate alarms in industrial facilities. *IFAC-PapersOnLine, 51*(18), 25–30.
23. Hu, W., Chen, T., & Meyer, G. (2019). Constructing workflow models of alarm responses via trace labeling and dependency analysis. In *58th Annual Conference of the Society of Instrument and Control Engineers of Japan (SICE)* (pp. 1615–1620).
24. Hu, W., Chen, T., & Shah, S. (2018). Discovering association rules of mode-dependent alarms from alarm and event logs. *IEEE Transactions on Control Systems Technology, 26*(3), 971–983.
25. Hu, W., Shah, S. L., & Chen, T. (2018). Framework for a smart data analytics platform towards process monitoring and alarm management. *Computers & Chemical Engineering, 114*, 225–244.

26. Hu, W., Wang, J., & Chen, T. (2015). A new method to detect and quantify correlated alarms with occurrence delays. *Computers & Chemical Engineering, 80*, 189–198.
27. Hu, W., Wang, J., & Chen, T. (2016). A local alignment approach to similarity analysis of industrial alarm flood sequences. *Control Engineering Practice, 55*, 13–25.
28. Hu, W., Zhang, X., Wang, J., Yang, G., & Cai, Y. (2023). Pattern matching of industrial alarm floods using word embedding and dynamic time warping. *IEEE/CAA Journal of Automatica Sinica, 10*(4), 1096–1098.
29. IEC-62682. (2014). *Management of Alarm Systems for the Process Industries*. IEC (International Electrotechnical Commission).
30. Izadi, I., Shah, S., Shook, D., Kondaveeti, S., & Chen, T. (2009). A framework for optimal design of alarm systems. *IFAC Proceedings Volumes, 42*, 651–656.
31. Jiang, H., Patwardhan, R., & Shah, S. (2009). Root cause diagnosis of plant-wide oscillations using the concept of adjacency matrix. *Journal of Process Control, 19*, 1347–1354.
32. Kaiser, A., & Schreiber, T. (2002). Information transfer in continuous processes. *Physica D: Nonlinear Phenomena, 166*, 43–62.
33. Kondaveeti, S., Izadi, I., Shah, S., Black, T., & Chen, T. (2012). Graphical tools for routine assessment of industrial alarm systems. *Computers & Chemical Engineering, 46*, 39–47.
34. Lai, S., & Chen, T. (2017). A method for pattern mining in multiple alarm flood sequences. *Chemical Engineering Research and Design, 117*, 831–839.
35. Monroe, M., Lan, R., Lee, H., Plaisant, C., & Shneiderman, B. (2013). Temporal event sequence simplification. *IEEE Transactions on Visualization and Computer Graphics, 19*, 2227–2236.
36. Naghoosi, E., Izadi, I., & Chen, T. (2011). Estimation of alarm chattering. *Journal of Process Control, 21*, 1243–1249.
37. Nimmo, I. (2005). Alarm overload. *Chemical Processing, 68*, 28–33.
38. Noda, M., Higuchi, F., Takai, T., & Nishitani, H. (2011). Event correlation analysis for alarm system rationalization. *Asia-Pacific Journal of Chemical Engineering, 6*, 497–502.
39. Qin, S. (2014). Process data analytics in the era of big data. *AIChE Journal, 60*, 3092–3100.
40. Schleburg, M., Christiansen, L., Thornhill, N., & Fay, A. (2013). A combined analysis of plant connectivity and alarm logs to reduce the number of alerts in an automation system. *Journal of Process Control, 23*, 839–851.
41. Schreiber, T. (2000). Measuring information transfer. *Physical Review Letters, 85*, 461.
42. Shneiderman, B. (1992). Tree visualization with tree-maps: 2-d space-filling approach. *ACM Transactions on Graphics (TOG), 11*, 92–99.
43. Smith, T., & Waterman, M. (1981). Identification of common molecular subsequences. *Journal of Molecular Biology, 147*, 195–197.
44. Tangirala, A., Shah, S., & Thornhill, N. (2005). PSCMAP: A new tool for plant-wide oscillation detection. *Journal of Process Control, 15*, 931–941.
45. Thambirajah, J., Benabbas, L., Bauer, M., & Thornhill, N. (2009). Cause-and-effect analysis in chemical processes utilizing XML, plant connectivity and quantitative process history. *Computers & Chemical Engineering, 33*, 503–512.
46. Thornhill, N., Cox, J., & Paulonis, M. (2003). Diagnosis of plant-wide oscillation through data-driven analysis and process understanding. *Control Engineering Practice, 11*, 1481–1490.
47. Thornhill, N., & Hagglund, T. (1997). Detection and diagnosis of oscillations in control loops. *Control Engineering Practice, 5*, 1343–1354.
48. Timms, C. (2009). Hazards equal trips or alarms or both. *Process Safety and Environmental Protection, 87*, 3–13.
49. Van der Aalst, W., Weijters, T., & Maruster, L. (2004). Workflow mining: Discovering process models from event log. *IEEE Transactions on Knowledge and Data Engineering, 16*(9), 1128–1142.
50. Wang, J., & Chen, T. (2013). An online method for detection and reduction of chattering alarms due to oscillation. *Computers & Chemical Engineering, 54*, 140–150.
51. Wang, J., & Chen, T. (2014). An online method to remove chattering and repeating alarms based on alarm durations and intervals. *Computers & Chemical Engineering, 67*, 43–52.

52. Wang, J., Yang, Z., Chen, K., & Zhou, D. (2017). Practices of detecting and removing nuisance alarms for alarm overloading in thermal power plants. *Control Engineering Practice, 67*, 21–30.
53. Wang, Z., & Wang, J. (2022). Analytical graphs to describe operating status of industrial alarm variables. *Control Engineering Practice, 118*, 104961.
54. Weber, M., Alexa, M., & Muller, W. (2011). Visualizing time-series on spirals. *Proceedings of IEEE Symposium on Information Visualization INFOVIS, 2001*, 7–13.
55. Welch, L. (1974). Lower bounds on the maximum cross correlation of signals. *IEEE Transactions on Information Theory, 20*, 397–399.
56. Xu, J., Wang, J., Izadi, I., & Chen, T. (2012). Performance assessment and design for univariate alarm systems based on FAR, MAR, and AAD. *IEEE Transactions on Automation Science and Engineering, 9*, 296–307.
57. Yang, F., Duan, P., Shah, S., & Chen, T. (2014). *Capturing Connectivity and Causality in Complex Industrial Processes*. Springer.
58. Yang, F., Shah, S., Xiao, D., & Chen, T. (2012). Improved correlation analysis and visualization of industrial alarm data. *ISA Transactions, 51*, 499–506.
59. Yang, F., Xiao, D., & Shah, S. (2010). Qualitative fault detection and hazard analysis based on signed directed graphs for large-scale complex systems. *Fault Detection*, 15–50.
60. Yang, Z., Wang, J., & Chen, T. (2013). Detection of correlated alarms based on similarity coefficients of binary data. *IEEE Transactions on Automation Science and Engineering, 10*, 1014–1025.

Index

J. Wang et al., *Intelligent Industrial Alarm Systems*,
https://doi.org/10.1007/978-981-97-6516-4

GPSR Compliance

The European Union's (EU) General Product Safety Regulation (GPSR) is a set of rules that requires consumer products to be safe and our obligations to ensure this.

If you have any concerns about our products, you can contact us on ProductSafety@springernature.com

In case Publisher is established outside the EU, the EU authorized representative is:

Springer Nature Customer Service Center GmbH
Europaplatz 3
69115 Heidelberg, Germany

The manufacturer's authorised representative in the EU is Springer Nature Customer Service Centre GmbH, Europaplatz 3, 69115 Heidelberg, Germany. If you have any concerns regarding our products, please contact ProductSafety@springernature.com

Printed and bound by CPI Group (UK) Ltd, Croydon, CR0 4YY
30/09/2026
02225304-0016